Modern Engineering Statistics

BICENTENNIAL
1807
WILEY
2007
BICENTENNIAL

The Wiley Bicentennial—Knowledge for Generations

Each generation has its unique needs and aspirations. When Charles Wiley first opened his small printing shop in lower Manhattan in 1807, it was a generation of boundless potential searching for an identity. And we were there, helping to define a new American literary tradition. Over half a century later, in the midst of the Second Industrial Revolution, it was a generation focused on building the future. Once again, we were there, supplying the critical scientific, technical, and engineering knowledge that helped frame the world. Throughout the 20th Century, and into the new millennium, nations began to reach out beyond their own borders and a new international community was born. Wiley was there, expanding its operations around the world to enable a global exchange of ideas, opinions, and know-how.

For 200 years, Wiley has been an integral part of each generation's journey, enabling the flow of information and understanding necessary to meet their needs and fulfill their aspirations. Today, bold new technologies are changing the way we live and learn. Wiley will be there, providing you the must-have knowledge you need to imagine new worlds, new possibilities, and new opportunities.

Generations come and go, but you can always count on Wiley to provide you the knowledge you need, when and where you need it!

William J. Pesce

WILLIAM J. PESCE
PRESIDENT AND CHIEF EXECUTIVE OFFICER

Peter Booth Wiley

PETER BOOTH WILEY
CHAIRMAN OF THE BOARD

Modern Engineering Statistics

THOMAS P. RYAN

Acworth, Georgia

BICENTENNIAL
1807
WILEY
2007
BICENTENNIAL

WILEY-
INTERSCIENCE

A JOHN WILEY & SONS, INC., PUBLICATION

Published by John Wiley & Sons, Inc., Hoboken, New Jersey.
Published simultaneously in Canada.

For general information on our other products and services or for technical support, please contact our Customer
Care Department within the United States at 800-762-2974, outside the United States at
317-572-3993 or fax 317-572-4002.

Wiley also publishes its books in a variety of electronic formats. Some content that appears in print may not be
available in electronic formats. For more information about Wiley products, visit our web site at www.wiley.com

Wiley Bicentennial Logo: Richard J. Pacifico

Library of Congress Cataloging-in-Publication Data:

Ryan, Thomas P., 1945–
    Modern engineering statistics / Thomas P. Ryan.
            p.   cm.
    Includes bibliographical references and index.
    ISBN 978-0-470-08187-7
    1.   Engineering–Statistical methods.   I. Title.
    TA340.R93   2007
    620.0072–dc22

                                                    20060521558

Printed in the United States of America

10 9 8 7 6 5 4 3 2 1

Contents

Preface

Statistical methods are an important part of the education of any engineering student. This was formally recognized by the Accreditation Board for Engineering and Technology (ABET) when, several years ago, education in probability and statistics became an ABET requirement for all undergraduate engineering majors. Specific topics within the broad field of probability and statistics were not specified, however, so colleges and universities have considerable latitude regarding the manner in which they meet the requirement. Similarly, ABET's *Criteria for Accrediting Engineering Programs*, which were to apply to evaluations during 2001–2002, were not specific regarding the probability and statistics skills that engineering graduates should possess.

Engineering statistics courses are offered by math and statistics departments, as well as being taught within engineering departments and schools. An example of the latter is The School of Industrial and Systems Engineering at Georgia Tech, whose list of course offerings in applied statistics rivals that of many statistics departments.

Unfortunately, many engineering statistics courses have not differed greatly from mathematical statistics courses, and this is due in large measure to the manner in which many engineering statistics textbooks have been written. This textbook makes no pretense of being a "math stat book." Instead, my objective has been to motivate an appreciation of statistical techniques, and to do this as much as possible within the context of engineering, as many of the datasets that are used in the chapters and chapter exercises are from engineering sources. I have taught countless engineering statistics courses over a period of two decades and I have formulated some specific ideas of what I believe should be the content of an engineering statistics course. The contents of this textbook and the style of writing follow accordingly.

Statistics books have been moving in a new direction for the past fifteen years, although books that have beaten a new path have often been overshadowed by the sheer number of books that are traditional rather than groundbreaking.

The optimum balance between statistical thinking and statistical methodology can certainly be debated. Hoerl and Snee's book, *Statistical Thinking*, which is basically a book on business statistics, stands at one extreme as a statistics book that emphasizes the "big picture" and the use of statistical tools in a broad way rather than encumbering the student with an endless stream of seemingly unrelated methods and formulas.

This book might be viewed as somewhat of an engineering statistics counterpart to the Hoerl and Snee book, as statistical thinking is emphasized throughout, but there is also a solid dose of contemporary statistical methodology.

This book has many novel features, including the connection that is frequently made (but hardly ever illustrated) between hypothesis tests and confidence intervals. This connection is illustrated in many places, as I believe that the point cannot be overemphasized.

I have also written the book under the assumption that statistical software will be used (extensively). A somewhat unusual feature of the book is that computing equations are kept to a minimum, although some have been put in chapter appendixes for readers interested in seeing them. MINITAB is the most frequently used statistical software for college and university courses. Minitab, Inc. has been a major software component of the Six Sigma movement and has made additions to the MINITAB software to provide the necessary capabilities for Six Sigma work. Such work has much in common with the field of engineering statistics and with the way that many engineers use statistics. Therefore, MINITAB is heavily relied on in this book for illustrating various statistical analyses, although JMP from SAS Institute, Inc. is also used.

This is not intended, however, to be a book on how to use MINITAB or JMP, since books have been written for that purpose. Nevertheless, some MINITAB code is given in certain chapters and especially at the textbook Website to benefit users who prefer to use MINITAB in command mode. Various books, including the *MINITAB User's Guide*, have explained how to use MINITAB in menu mode, but not in command mode. The use of menu mode is of course appropriate for beginners and infrequent users of MINITAB, but command mode is much faster for people who are familiar with MINITAB and there are many users who still use command mode. Another advantage of command mode is that when the online help facility is used to display a command, all of the subcommands are also listed, so the reader sees all of the options, whereas this view is not available when menu mode is used. Rather, the user has to navigate through the various screens and mentally paste everything together in order to see the total capability relative to a particular command.

There are, however, some MINITAB routines for which menu mode is preferable, due in part to the many subcommands that will generally be needed just to do a standard analysis. Thus, menu mode does have its uses.

Depending on how fast the material is covered, the book could be used for a two-semester course as well as for a one-semester course. If used for the latter, the core material would likely be all or parts of Chapters 1–6, 8, 11, 12, and 17. Some material from Chapters 7 and 14 might also be incorporated, depending on time constraints and instructor tastes.

For the second semester of a two-semester course, Chapters 7, 9, 10, 13, 14, and 15 and/or 16 might be covered, perhaps with additional material from Chapters 11 and 12 that could not be covered in the first semester. The material in Chapter 12 on Analysis of Means deserves its place in the sun, especially since it was developed for the express purpose of fostering communication with engineers on the subject of designed experiments. Although Chapter 10 on mechanistic models and Chapter 7 on tolerance intervals and prediction intervals might be viewed as special topics material, it would be more appropriate to elevate these chapters to "core material chapters," as this is material that is very important for engineering students. At least some of the material in Chapters 15 and 16 might be covered, as time permits. Chapter 16 is especially important as it can help engineering students and others realize that nonparametric (distribution-free) methods will often be viable alternatives to the better-known parametric methods.

There are reasons for the selected ordering of the chapters. Standard material is covered in the first six chapters and the sequence of those chapters is the logical one. Decisions had to be made starting with Chapter 7, however. Although instructors might view this as a special topics chapter as stated, there are many subject matter experts who believe that tolerance intervals and prediction intervals should be taught in engineering statistics courses. Having a chapter on tolerance intervals and prediction intervals follow a chapter on confidence intervals is reasonable because of the relationships between the intervals and the need for this to be understood. Chapter 9 is an extension of Chapter 8 into multiple linear regression and it is reasonable to have these chapters followed by Chapter 10 since nonlinear regression is used in this chapter. In some ways it would be better if the chapter followed Chapter 14 since reliability models are used, but the need to have it follow Chapters 8 and 9 seems more important. The regression chapters should logically precede the chapter on design of experiments, Chapter 12, since regression methods should be used in analyzing data from designed experiments. Processes should ideally be in a state of statistical control when designed experiments are performed, so the chapter on control chart methods, Chapter 11, should precede Chapter 12. Chapters 13 and 14 contain subject matter that is important for engineering and Chapters 15 and 16 consider topics that are generally covered in a wide variety of introductory type statistics texts. It is useful for students to be able to demonstrate that they have mastered the tools they have learned in any statistics course by knowing which tool(s) to use in a particular application after all of the material has been presented. The exercises in Chapter 17 provide students with the opportunity to demonstrate that they have acquired such skill.

The book might also be used for self-study, aided by the Answers to Selected Exercises, which is sizable and detailed. A separate *Solutions Manual* with solutions to all of the chapter exercises is also available. The data in the exercises, including data in MINITAB files (i.e., the files with the .MTW extension), can be found at the website for the text: ftp:// ftp.wiley.com/public/ sci_med/engineering_statistics.

I wish to gratefully acknowledge the support and assistance of my editor, Steve Quigley, associate editor Susanne Steitz, and production editor Rosalyn Farkas, plus various others, including the very large number of anonymous reviewers who reviewed all or parts of the manuscript at various stages and made helpful comments.

THOMAS P. RYAN

Acworth, Georgia
May 2007

CHAPTER 1

Methods of Collecting and Presenting Data

People make decisions every day, with decision-making logically based on some form of data. A person who accepts a job and moves to a new city needs to know how long it will take him/her to drive to work. The person could guess the time by knowing the distance and considering the traffic likely to be encountered along the route that will be traveled, or the new employee could drive the route at the anticipated regular time of departure for a few days before the first day of work.

With the second option, an experiment is performed, which if the test run were performed under normal road and weather conditions, would lead to a better estimate of the typical driving time than by merely knowing the distance and the route to be traveled.

Similarly, engineers conduct statistically designed experiments to obtain valuable information that will enable processes and products to be improved, and much space is devoted to statistically designed experiments in Chapter 12.

Of course, engineering data are also available without having performed a designed experiment, but this generally requires a more careful analysis than the analysis of data from designed experiments. In his provocative paper, "Launching the Space-Shuttle *Challenger*—Disciplinary Deficiencies in the Analysis of Engineering Data," F. F. Lighthall (1991) contended that "analysis of field data and reasoning were flawed" and that "staff engineers and engineering managers . . . were unable to frame basis questions of covariation among field variables, and thus unable to see the relevance of routinely gathered field data to the issues they debated before the *Challenger* launch." Lighthall then states "Simple analyses of field data available to both Morton Thiokol and NASA at launch time and months before the *Challenger* launch are presented to show that the arguments against launching at cold temperatures could have been quantified. . . ." The author's contention is that there was a "gap in the education of engineers." (Whether or not the *Columbia* disaster will be similarly viewed by at least some authors as being a deficiency in data analysis remains to be seen.)

Perhaps many would disagree with Lighthall, but the bottom line is that failure to properly analyze available engineering data or failure to collect necessary data can endanger lives—on a space shuttle, on a bridge that spans a river, on an elevator in a skyscraper, and in many other scenarios.

Intelligent analysis of data requires much thought, however, and there are no shortcuts. This is because analyzing data and solving associated problems in engineering and other areas is more of an art than a science. Consequently, it would be impractical to attempt to give a specific step-by-step guide to the use of the statistical methods presented in succeeding chapters, although general guidelines can still be provided and are provided in subsequent chapters. It is desirable to try to acquire a broad knowledge of the subject matter and position oneself to be able to solve problems with powers of reasoning coupled with subject matter knowledge.

The importance of avoiding the memorization of rules or steps for solving problems is perhaps best stated by Professor Emeritus Herman Chernoff of the Harvard Statistics Department in his online algebra text, *Algebra 1 for Students Comfortable with Arithmetic* (`http://www.stat.harvard.edu/People/Faculty/Herman_Chernoff/ Herman_Chernoff_Algebra_1.pdf`).

> Memorizing rules for solving problems is usually a way of avoiding understanding. Without understanding, great feats of memory are required to handle a limited class of problems, and there is no ability to handle new types of problems.

My approach to this issue has always been to draw a rectangle on a blackboard and then make about 15–20 dots within the rectangle. The dots represent specific types of problems; the rectangle represents the body of knowledge that is needed to solve not only the types of problems represented by the dots, but also any type of problem that would fall within the rectangle. This is essentially the same as what Professor Chernoff is saying.

This is an important distinction that undoubtedly applies to any quantitative subject and should be understood by students and instructors, in general.

Semiconductor manufacturing is one area in which statistics is used extensively. International SEMATECH (SEmiconductor MAnufacturing TECHnology), located in Austin, Texas, is a nonprofit research and development consortium of the following 13 semiconductor manufacturers: Advanced Micro Devices, Conexant, Hewlett-Packard, Hyundai, Infineon Technologies, IBM, Intel, Lucent Technologies, Motorola, Philips, STMicroelectronics, TSMC, and Texas Instruments. Intel, in particular, uses statistics extensively.

The importance of statistics in these and other companies is exemplified by the *NIST/SEMATECH e-Handbook of Statistical Methods* (Croarkin and Tobias, 2002), a joint effort of International SEMATECH and NIST (National Institute of Standards and Technology), with the assistance of various other professionals. The stated goal of the handbook, which is the equivalent of approximately 3,000 printed pages, is to provide a Web-based guide for engineers, scientists, businesses, researchers, and teachers who use statistical techniques in their work. Because of its sheer size, the handbook is naturally much more inclusive than this textbook, although there is some overlap of material. Of course, the former is not intended for use as a textbook and, for example, does not contain any exercises or problems, although it does contain case studies. It is a very useful resource, however, especially since it is almost an encyclopedia of statistical methods. It can be accessed at `www.itl.nist.gov/div898/handbook` and will henceforth often be referred to as the *e-Handbook of Statistical Methods* or simply as the *e-Handbook*.

There are also numerous other statistics references and data sets that are available on the Web, including some general purpose Internet statistics textbooks. Much information, including many links, can be found at the following websites: `http://www.utexas.edu/cc/stat/world/softwaresites.html` and `http://my.execpc.com/~helberg/statistics.html`. The *Journal of Statistics Education* is a free, online statistics publication devoted to statistics education. It can be found at `http://www.amstat.org/publications/jse`.

Statistical education is a two-way street, however, and much has been written about how engineers view statistics relative to their work. At one extreme, Brady and Allen (2002) stated: "There is also abundant evidence—for example, Czitrom (1999)—that most practicing engineers fail to consistently apply the formal data collection and analysis techniques that they have learned and in general see their statistical education as largely irrelevant to their professional life." (It is worth noting that the first author is an engineering manager in industry.) The Accreditation Board for Engineering and Technology (ABET) disagrees with this sentiment and several years ago decreed that *all* engineering majors must have training in probability and statistics. Undoubtedly, many engineers would disagree with Brady and Allen (2002), although historically this has been a common view.

One relevant question concerns the form in which engineers and engineering students believe that statistical exposition should be presented to them. Lenth (2002), in reviewing a book on experimental design that was written for engineers and engineering managers and emphasizes hand computation, touches on two extremes by first stating that "... engineers just will not believe something if they do not know how to calculate it ...," and then stating "After more thought, I realized that engineers are quite comfortable these days—in fact, far too comfortable—with results from the blackest of black boxes: neural nets, genetic algorithms, data mining, and the like."

So have engineers progressed past the point of needing to see how to perform all calculations that produce statistical results? (Of course, a world of black boxes is undesirable.) This book was written with the knowledge that users of statistical methods simply do not perform hand computation anymore to any extent, but many computing formulas are nevertheless given for interested readers, with some formulas given in chapter appendices.

1.1 OBSERVATIONAL DATA AND DATA FROM DESIGNED EXPERIMENTS

Sports statistics are readily available from many sources and are frequently used in teaching statistical concepts. Assume that a particular college basketball player has a very poor free throw shooting percentage, and his performance is charted over a period of several games to see if there is any trend. This would constitute observational data—we have simply observed the numbers. Now assume that since the player's performance is so poor, some action is taken to improve his performance. This action may consist of extra practice, visualization, and/or instruction from a professional specialist. If different combinations of these tasks were employed, this could be in the form of a designed experiment. In general, if improvement is to occur, there should be experimentation. Otherwise, any improvement that seems to occur might be only accidental and not be representative of any real change.

Similarly, W. Edwards Deming (1900–1993) coined the terms *analytic studies* and *enumerative studies* and often stated that "statistics is prediction." He meant that statistical methods should be used to improve future products, processes, and so on, rather than simply "enumerating" the current state of affairs as is exemplified, for example, by the typical use of sports statistics. If a baseball player's batting average is .274, does that number tell us anything about what the player should do to improve his performance? Of course not, but when players go into a slump they try different things; that is, they *experiment*. Thus, experimentation is essential for improvement.

This is not to imply, however, that observational data (i.e., enumerative studies) have no value. Obviously, if one is to travel/progress to "point B," it is necessary to know the starting point, and in the case of the baseball player who is batting .274, to determine if the starting point is one that has some obvious flaws.

When we use designed experiments, we must have a way of determining if there has been a "significant" change. For example, let's say that an industrial engineer wants to determine if a new manufacturing process is having a significant effect on throughput. He/she obtains data from the new process and compares this against data that are available for the old process. So now there are two sets of data and information must be extracted from those two sets and a decision reached. That is, the engineer must compute *statistics* (such as averages) from each set of data that would be used in reaching a decision. This is an example of *inferential statistics*, a subject that is covered extensively in Chapters 4–15.

DEFINITION

A *statistic* is a summary measure computed from a set of data.

One point that cannot be overemphasized (so the reader will see it discussed in later chapters) is that experimentation should generally not be a one-time effort, but rather should be repetitive and sequential. Specifically, as is illustrated in Figure 1.1, exprimentation should in many applications be a never-ending learning process. Mark Price has the highest free throw percentage in the history of the National Basketball Association (NBA) at .904, whereas in his four-year career at Georgia Tech his best year was .877 and he does not even hold the career Georgia Tech field goal percentage record (which is held by Roger Kaiser at .858). How could his professional percentage be considerably higher than his college percentage, despite the rigors of NBA seasons that are much longer than college seasons? Obviously, he had to experiment to determine what worked best for him.

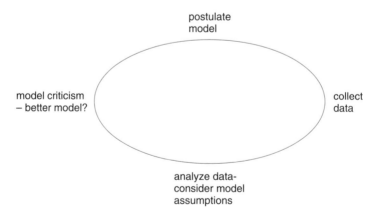

Figure 1.1 Repetitive nature of experimentation.

1.2 POPULATIONS AND SAMPLES

Whether data have been obtained as observational data or from a designed experiment, we have obtained a *sample* from a *population*.

DEFINITION

A *sample* is a subset of observations obtained from a larger set, termed a *population*.

To the layperson, a population consists of people, but a statistical population can consist of virtually anything. For example, the collection of desks on a particular college campus could be defined as a population. Here we have a *finite* population and one could, for example, compute the average age of desks on campus. What is the population if we toss a coin ten times and record the number of heads? Here the population is conceptually infinite as it would consist of all of the tosses of the coin that could be made. Similarly, for a manufacturing scenario the population could be all of the items of a particular type produced by the current manufacturing process—past, present, and future.

If our sample is comprised of observational data, the question arises as to how the sample should be obtained. In particular, should we require that our sample be *random*, or will we be satisfied if our sample is simply *representative*?

DEFINITION

A *random sample* of a specified size is one for which every possible sample of that size has the same chance of being selected from the population.

A simple example will be given to illustrate this concept. Suppose a population is defined to consist of the numbers 1, 2, 3, 4, 5, and 6, and you wish to obtain a random sample of size two from this population. How might this be accomplished? What about listing all of the possible samples of size two and then randomly selecting one? There are 15 such samples and they are given below.

$$
\begin{array}{ccccc}
12 & 15 & 24 & 34 & 45 \\
13 & 16 & 25 & 35 & 46 \\
14 & 23 & 26 & 36 & 56
\end{array}
$$

Following the definition just given, a random sample of size two from this population is such that each of the possible samples has the same probability of being selected.

There are various ways to obtain a random sample, once a *frame*, a list of all of the elements of a population, is available. Obviously, one approach would be to use a software program that generates random numbers. Another approach would be to use a random number table such as Table A at the end of the book. That table could be used as follows. In general, the elements in the population would have to be numbered in some way. In this example the elements are numbers, and since the numbers are single-digit numbers, only one column of Table A need be used. If we arbitrarily select the first column in the first set of four columns, we could proceed down that column; the first number observed is 1 and the second is 5. Thus, our sample of size two would consist of those two numbers.

Now how would we proceed if our population is defined to consist of all transistors of a certain type manufactured in a given day at a particular facility? Could a random sample be obtained?

In general, to obtain a random sample we do need a frame, which as has been stated is a list of all of the elements in the population. It would certainly be impractical to "number" all of the transistors so that a random sample could be taken. Consequently, a *convenience sample* is frequently used instead of a random sample. The important point is that the sample should be representative, and more or less emulate a random sample since common statistical theory is based on the assumption of random sampling.

For example, we might obtain samples of five units from an assembly line every 30 minutes. With such a sampling scheme, as is typical when control charts (see Chapter 11) are constructed, every item produced will not have the same probability of being included in any one of the samples with this *systematic sampling* approach, as it is called.

Such a sampling approach could produce disastrous results if, unbeknown to the person performing the sampling, there was some cyclicality in the data. This was clearly shown in McCoun (1949, 1974) in regard to a tooling problem. If you imagine data that would graph approximately as a sine curve, and if the sampling coincided with the periodicity of the curve, the variability of the data could be greatly underestimated and the trend that would be clearly visible for the entire set of data would be hidden.

As a second example, assume that every twenty-first unit of a particular product is nonconforming. If samples of size three happen to be selected in such a way (perhaps every 15 minutes) that one nonconforming unit is included in each sample, the logical conclusion would be that one out of every three units produced is nonconforming, instead of one out of twenty-one.

Because of these possible deleterious effects, how can we tell whether or not convenience samples are likely to give us a true picture of a particular population? We cannot, unless we have some idea as to whether there are any patterns or trends in regard to the units that are produced, and we may not know this unless we have a time sequence plot of historical data.

Another point to keep in mind is that populations generally change over time, and the change might be considerable relative to what we are trying to estimate. Hence, a sample that is representative today may not be representative six months later. For example, the racial composition of a particular high school could change considerably in a relatively short period of time, as could the sentiments of voters in a particular district regarding who they favor for political office.

Consequently, it is highly desirable to acquire a good understanding of the processes with which you will be working before using any "routine" sampling procedure.

1.3   VARIABLES

When we obtain our sample, we obtain data values on one or more *variables*. For example, many (if not most) universities use regression modeling (regression is covered in Chapters 8 and 9) as an aid in predicting what a student's GPA would be after four years if the student were admitted, and use that predicted value as an aid in deciding whether or not to admit the student. The sample of historical data that is used in developing the model would logically have the student's high school grade point average, aptitude test scores, and perhaps several other variables.

If we obtained a random sample of students, we would expect a list of the aptitude test scores, for example, to vary at random; that is, the variable would be a *random variable*. A mathematically formal definition of a random variable that is usually found in introductory statistics books will be avoided here in favor of a simpler definition. There are two general types of random variables and it is important to be able to distinguish between them.

DEFINITIONS

A *random variable* is a statistic or an individual observation that varies in a random manner. A *discrete* random variable is a random variable that can assume only a finite or countably infinite number of possible values (usually integers), whereas a *continuous* random variable is one that can theoretically assume any value in a specified interval (i.e., continuum), although measuring instruments limit the number of decimal places in measurements.

The following simple example should make clear the idea of a random variable. Assume that an experiment is defined to consist of tossing a single coin twice and recording the number of heads observed. The experiment is to be performed 16 times. The random variable is thus the number of heads, and it is customary to have a random variable represented by an alphabetical (capital) letter. Thus, we could define

$$X = \text{the number of heads observed in each experiment}$$

Assume that the 16 experiments produce the following values of X:

$$0 \quad 2 \quad 1 \quad 1 \quad 2 \quad 0 \quad 0 \quad 1 \quad 2 \quad 1 \quad 1 \quad 0 \quad 1 \quad 1 \quad 2 \quad 0$$

There is no apparent pattern in the sequence of numbers so based on this sequence we would be inclined to state that X (apparently) *varies* in a *random* manner and is thus a random variable.

Since this is an introductory statistics text, the emphasis is on *univariate* data; that is, data on a single random variable. It should be kept in mind, however, that the world is essentially *multivariate*, so any student who wants to become knowledgeable in statistics must master both univariate and multivariate methods. In statistics, "multivariate" refers to more than one response or dependent variable, not just more than one variable of interest; researchers in other fields often use the term differently, in reference to the independent variables. The graphs in Section 1.4.4 are for two variables.

1.4 METHODS OF DISPLAYING SMALL DATA SETS

We can enumerate and summarize data in various ways. One very important way is to graph data, and to do this in as many ways as is practical. Much care must be exercised in the use of graphical procedures, however; otherwise, the impressions that are conveyed could be very misleading. It is also important to address at the outset what one wants a graph to show as well as the intended audience for the graph.

We consider some important graphical techniques in the next several sections. There are methods that are appropriate for displaying essential information in large data sets and there are methods for displaying small data sets. We will consider the latter first.

■ EXAMPLE 1.1

The data in Table 1.1, compiled by GaSearch Pricing Data for November 2001, is a sample of natural gas purchasers in the state of Texas with over 1,500,000 Mcf throughput.

Data

TABLE 1.1   **Gas Pricing Data for November 2001**

Purchaser Name	Average Cost per Mcf
Amoco Production Corp.	$2.78
Conoco Inc.	2.76
Duke Energy Trading and Marketing	2.73
Exxon Corporation	2.71
Houston Pipe Line Co.	3.07
Mitchell Gas Services LP	2.95
Phillips Petroleum Co.	2.65
Average Top State of Texas Production	2.79

Discussion

With data sets as small as this one, we really don't need to rely on summary measures such as the average because we can clearly see how the numbers vary; we can quickly see the largest and smallest values, and we could virtually guess the average just by looking at the numbers. Thus, there is no need to "summarize" the data, although a graphical display or two could help identify any outliers (i.e., extreme observations) and one such graphical display of these data is given in Section 1.4.6.

There is, however, a need to recognize that sampling variability exists whenever one takes a sample from a population, as was done in this case. That is, if a different sample of gas purchasers had been selected, the largest and smallest values would have been different, and so would any other computed statistic. Sampling variability is introduced in Chapter 4 and plays an important role in the material presented in subsequent chapters.   ■

1.4.1   Stem-and-Leaf Display

Few data sets are as small as this one, however, and for data sets that are roughly two to three times this size, we need ways of summariing the data, as well as displaying the data. Many college students have had the experience of seeing the final exam scores for their class posted on their professor's office door in a format such as the following.

$$5|\ 0113568$$
$$6|\ 012446778$$
$$7|\ 001223345667899$$
$$8|\ 12233455679$$
$$9|\ 014567$$

A student who had not been to class for awhile might suffer the double indignation of (1) not being sure how to read the display, and (2) after being told how to do so, discovering that he/she made the lowest grade in the class, which is 50. This display is an example of a *stem-and-leaf display*, which is a popular and established way of displaying a small-to-moderate-sized data set. There are many different ways to create a stem-and-leaf display, depending on the type of data that are to be displayed and what the user wishes to show. In this example, the "stem" is the tens digit and the "leaf" is the units digit. We may thus observe that, for example, two students in the class made a 70 on the final exam.

Velleman and Hoaglin (1981, p. 14) discussed an example in which the pulse rates of 39 Peruvian Indians were displayed in a histogram and in a stem-and-leaf display. The latter revealed that all of the values except one were divisible by four, thus leading to the conjecture that 38 of the values were obtained by taking 15-second readings and multiplying the results by 4, with the other value obtained by doubling a 30-second reading resulting (perhaps) from missing the 15-second mark. Thus, the stem-and-leaf display provided some insight into how the data were obtained, whereas this information was not provided by the histogram.

Variations of the basic method of constructing stem-and-leaf displays are given in Velleman and Hoaglin (1981).

Although stem-and-leaf displays were originally intended to be a pencil-and-paper tool in an arsenal of *exploratory data analysis* (EDA) techniques (Tukey, 1977), if we had, say, 80 observations, we would certainly not want to have to manually sort the numbers into ascending or descending order and then construct the display, or alternatively to construct the display without first sorting the numbers. Computer use in engineering statistics courses has been common for the last few decades and, as was emphasized in the Preface, will be emphasized in this book.

1.4.2 Time Sequence Plot and Control Chart

A *time sequence plot* is a plot of data against time. Ideally, the time values, which are plotted on the horizontal axis, should be equally spaced; otherwise the plot is harder to interpret.

A time sequence plot will often reveal peculiarities in a data set. It is an important graphical tool that should routinely be used whenever data have been collected over time, and the time order has been preserved. A convincing argument of the importance of this type of plot can be found in Ott (1975, pp. 34–36). Specifically, a student completed a course assignment by recording the amount of time for sand to run through a 3-minute egg timer. A time sequence plot of the times exhibited a perfect sawtooth pattern, with hardly any point being close to the median time. This should suggest that the two halves of the egg timer differed noticeably, a difference that might not be detected when the egg timer was used in the intended manner. Since the two halves must differ more than slightly, this means that at least one of the two halves is not truly a 3-minute egg timer—a discovery that could be of interest when the timer is applied to eggs instead of sand!

In engineering applications, a time sequence plot is often an invaluable aid in detecting a change in a process. This is illustrated in the following example. (*Note*: Datafiles for all examples and exercises in this text can be found at the text website, ftp://ftp. wiley.com/public/sci_tech_med/engineering_statistics.)

■ EXAMPLE 1.2

Data

The following are coded data, to save space. Letting W represent one of the coded values, the original data were given by $Y = 2 + W/10000$. The data are transmittance figures collected during the 1970s from an automatic data acquisition system for a filter transmittance experiment. The coded values are, in time order: 18, 17, 18, 19, 18, 17, 15, 14, 15, 15, 17, 18, 18, 19, 19, 21, 20, 16, 14, 13, 13, 15, 15, 16, 15, 14, 13, 14, 15, 14, 15, 16, 15, 16, 19, 20, 20, 21, 21, 22, 23, 24, 25, 27, 26, 26, 27, 26, 25, and 24.

Time Sequence Plot

Figure 1.2 is the time sequence plot of the coded data, which was produced using MINITAB, as were the other graphs in this chapter.

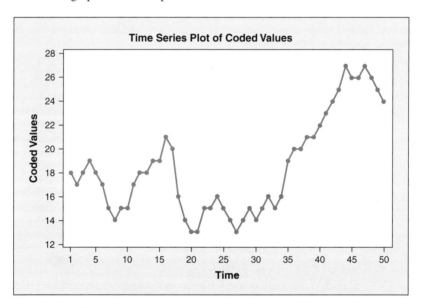

Figure 1.2 Time sequence plot of transmittance data.

Graph Interpretation

The graph shows that there was an upward trend in the coded transmittance values that becomes apparent in the vicinity of observation number 45. ■

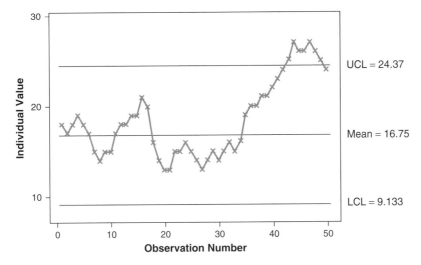

Figure 1.3   Control chart of transmittance data.

In this instance, the change is clear because of the large shift. In other cases, a change may be much smaller and thus not be so obvious. Consequently, it would be desirable to use some decision-making criterion. For example, assume that the data represented in Figure 1.2 were obtained in real time and it was decided to use the first 40 observations to construct a baseline. That is, the average of those observations was computed and decision lines greater than and less than the average were constructed such that any point outside one of the lines would signal a probable shift in the average.

When decision lines are displayed on a time sequence chart and the chart is used for "control" (such as controlling a manufacturing process), the chart is referred to as a *control chart*. (The computation of the decision lines, referred to as *control limits*, is explained in Chapter 11.)

Figure 1.3 is the control chart for these data, with the control limits computed using the first 40 observations.

Here the shift is made even more apparent by the points that fall above the upper control limit (UCL). As previously stated, for this example control limits are not needed as an aid to detect the shift, but in manufacturing and in other applications shifts can be small and in such instances control charts can be an invaluable aid in detecting shifts.

1.4.3 Lag Plot

Either a time sequence plot or a control chart of individual observations can be used to detect nonrandomness in data. A plot that can be used to identify correlation between consecutive observations is a *lag plot*. This is a plot in which observations are lagged by one or more observations. The original observations are then plotted against the lagged observations. For example, if we lag the observations by one, we are plotting the second observation against the first, the third against the second, and so on. If consecutive observations are correlated, the resulting plot will show a strong linear relationship. Such a relationship is evident in Figure 1.4, which is a plot of observations lagged by one with the correlation

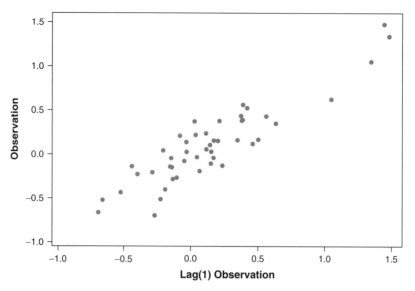

Figure 1.4   Lag plot of autocorrelated observations.

between consecutive observations being 0.8. (This is called autocorrelation and is discussed in more detail in Section 8.2.6.1.) That is, the ith observation is plotted on the vertical axis and the $(i-1)$st observation is plotted on the horizontal axis. [An autocorrelation of 0.8 is a high positive correlation, meaning that the relationship between consecutive observations is not far from being a functional relationship. A high negative correlation (say, -0.8) would result if consecutive observations were far apart, fluctuating up and down.]

   If consecutive observations were uncorrelated, the plot would exhibit a random configuration. Observations might be correlated at something other than one unit apart, however, so a plot such as Figure 1.4 is not sufficient for detecting autocorrelation. A graphical method for detecting such relationships is given in Section 8.2.6.1. Figure 1.4 is also a type of scatter plot, which is covered in the next section.

1.4.4   Scatter Plot

Both a time sequence plot and a control chart are forms of a *scatter plot*. The latter is a frequently used graph that is a two-dimensional plot with a vertical axis label and a horizontal axis label. With a time sequence plot, time is always the horizontal axis label and the measurement of interest is the vertical axis label. Of course, these are the same labels for a control chart. In general, the labels on a scatter plot can be anything.

   A rule of thumb that should be followed regarding graphical displays is that the amount of information per square inch of the display should be maximized. By "maximize" we don't mean putting everything imaginable in the display, as there would be essentially no information if the display were so cluttered that it was practically unreadable. *U.S. News and World Report* has an annual issue in which they provide statistics on colleges and universities, including the schools that it rates as its top 50 national universities. We would expect to see a configuration of points with a negative slope if we plot the 75th percentile SAT score for each school on the vertical axis against the school's acceptance rate on the

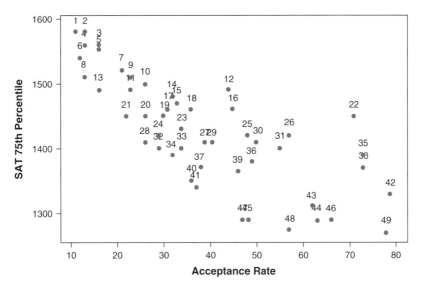

Figure 1.5　Data on top 50 national universities for 2002.

horizontal axis. We know that should happen, so constructing a scatter plot that has such a configuration of points does not tell us anything we don't already know. Labeling the plotted points would provide additional information, especially for schools that lie apart from the other schools. Of course, there would not be sufficient space on the graph to list the name of each school, but a number would require much less space, although that isn't a perfect solution.

Figure 1.5 shows the graph, with the numbers denoting the rank of the school in terms of 75th percentile SAT score for the 49 schools in the top 50 (actually 52 schools because 5 schools tied for 48th place), for which the SAT test is the primary aptitude test. These data are from the 2002 rankings. The graph was produced with MINITAB code that is listed at the textbook website: `ftp://ftp.wiley.com/public/ sci_tech_med/engineering_statistics`.

The numbers denote the ranks of the schools in terms of the 75th percentile SAT score. Although some of the numbers are difficult to read, it is not hard to determine what the number is, and of course the datafile could be used to resolve any ambiguities. (The schools are not listed either here or in the file, `75-25PERCENTILES2002.MTW`, at the Wiley website so as to avoid discussions about individual universities, although a few schools are mentioned later in regard to unusual data points. *U.S. News and World Report* no longer provides complete online data on schools free of charge. Some information, including the ranking of the top 50 national universities, is still available without charge, however, and the interested reader can obtain the data for the current year online at the magazine website.)

We would expect to see the smallest numbers in the upper-left portion of the graph and the largest numbers in the lower-right portion. We do see that, but we see one point (#22) that is noticeably apart from the other points, although the vertical spread of the points is obviously much greater at high acceptance rates than at low acceptance rates—as we would expect. If we extend a horizontal line from point #22, we see that the line will go through points 19–21, and that these points are well removed from point #22. That is, all four points have the same 75th percentile SAT score but the acceptance rate for the school corresponding to point #22 differs greatly from the acceptance rates for schools 19–21. A scatter plot can be used to identify outliers and outliers are sometimes bad data points. One would have to ask if this point is in fact a bad data point or simply part of the pattern of increased spread at high acceptance rates.

There are various other ways to construct scatter plots; good references on the construction of graphical displays that maximize the informational content, as well as graphical displays in general, are Tufte (1990, 1997, 2001). A compact way to show multiple scatter plots is through the use of a scatterplot matrix, which is illustrated in Section 9.4.1.

1.4.5 Digidot Plot

The usefulness of a stem-and-leaf display should be apparent, but such a display does not show the time order of the data. A runner who is preparing for a race would not find a stem-and-leaf display that showed his times for the past three months to be nearly as meaningful as a display that showed when he registered each of the times. Obviously he would want to see progress leading up to the race, so having the running times in sequence would be of paramount importance. Similarly, a major league manager would have little use for a stem-and-leaf display of a player's weekly batting average over the course of the season without knowing how performance varied over the days and weeks of the season.

For these and many other scenarios, it is very important that the time sequence of events be indicated, and in fact a time sequence plot will be of much greater value than a stem-and-leaf display or any other type of display that does not show the time sequence.

The two types of displays can be used together, however, and Hunter (1988) developed a plot, which he termed a *digidot plot*, that is a combination of a time sequence plot and a stem-and-leaf display. Unfortunately, the plot is not widely available in statistical software.

1.4.6 Dotplot

Another way to display one-dimensional data is through the use of dot diagrams. A dotplot is simply a display in which a dot is used to represent each point. Figure 1.6 is a dotplot of the data that were given in Example 1.1.

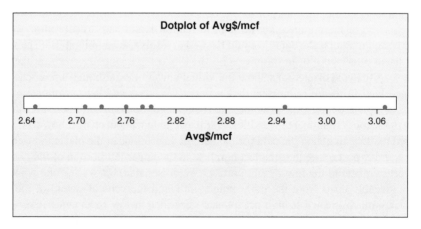

Figure 1.6 Dotplot of Example 1.1 data.

As stated in Example 1.1, when there is only a small number of observations (8 in this case), we can "see" the data reasonably well without any visual aids. However, Figure 1.6 shows that the two largest numbers are clearly separated from the rest of the data (and from themselves)—facts that may or may not have been obvious just by looking at the list of numbers.

Multiple dotplots are also useful, as the following example illustrates.

■ EXAMPLE 1.3

Study Description

Mitchell, Hegeman, and Liu (1997) described a gauge reproducibility and repeatability study, using data that were collected by a process engineer at Texas Instruments.

Purpose

Operator effect is one of the focal points in such studies (see Chapter 13) and this study involved three operators. The pull strength in grams of a single wire was measured, using 10 wires for each operator. A multiple dotplot can provide a picture of how the operators compared and the multiple dotplot is shown in Figure 1.7.

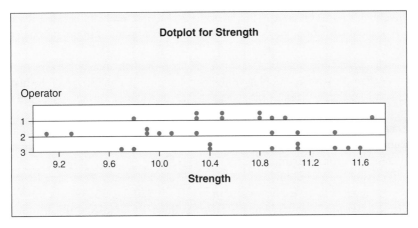

Figure 1.7 Dotplot for pull strength.

Graphical Analysis and Conclusion

The plot in Figure 1.7 shows that the operators are quite comparable.　　■

A dotplot does have some limitations, however. In particular, we would not want to use it if we have a large number of observations [Box, Hunter, and Hunter ((1978), p. 25) suggest at most 20 observations], as the dots would tend to run together and give rather poor resolution. We also cannot accurately determine the individual values from the diagram, which simply shows us the relationship between the numbers for a small set of numbers.

1.5 METHODS OF DISPLAYING LARGE DATA SETS

The graphical techniques described in Section 1.4 break down when there are much more than 100 data values. Specifically, the dotplot will become blurred, the stem-and-leaf display will have too many leaves, and there will be resolution problems with the digidot plot since a large number of leaves will compress the plot horizontally and thus compress the horizontal axis of the time sequence plot portion of the display.

So other methods are needed and are presented in the following sections.

1.5.1 Histogram

A *histogram* is probably the most commonly used way of displaying data. Simply stated, a histogram is a bar chart with the height of the "bars" representing the frequency of each class after the data have been grouped into classes. (For example, the classes might be 10–19, 20–29, 30–39, and 40–49, and the classes along with the accompanying frequencies would constitute the *frequency distribution*.)

Selection of the classes is nontrivial and should be given some thought, as one objective in using a histogram is to try to gain insight into the shape of the distribution for the population of data from which the sample has been obtained. As shown by Scott (1979), however, we need to know the shape of that distribution in order to construct the classes in such a way that the histogram will be likely to reflect the distribution shape. (This should be intuitively apparent.) Of course, this presents a Catch-22 problem because if we knew the shape of the distribution, we wouldn't be using a histogram to try to gain insight into that shape!

If no such *a priori* information is available, this is a problem about which we should not be overly concerned, but we should nevertheless recognize that the shape of the histogram might be a misleading indicator of the shape of the distribution of the population values. We should also recognize that a histogram is "random" if it is constructed from data in a random sample. That is, if we take another sample we will have a second histogram with a shape that is at least slightly different from the first histogram, even if the same number of classes is used for each histogram.

If at all possible, we would like to use "natural" class intervals. For example, if data in the sample ranged from 10 to 99, we might use 10–19, 20–29, 30–39, ..., 90–99 as the class intervals, *provided* that the sample size was sufficiently large. Specifically, this would make sense if the sample size was approximately 100 or more, but would be illogical if the sample size were only 50. Why? If we have too many classes, there may be some empty classes. Sizable gaps between numbers would almost certainly not exist in a population, so we would not want a histogram constructed from a sample to exhibit gaps.

A reasonable approach would be to use a rule of thumb for determining the number of classes, and then perhaps alter that number slightly, if necessary, to permit natural intervals. One rule that works reasonably well is called the "power-of-2 rule": for n observations we would use a classes with $2^{a-1} < n \leq 2^a$. Thus, for $n = 100$ we have $2^6 < 100 < 2^7$, so that seven classes would be used. Another rule of thumb that has been advanced is to let the number of classes equal $\sqrt{n}$, but this will produce a large number of classes when n is well in excess of 100. At the other extreme, Terrell and Scott (1985) suggested that the number of classes be $(2n)^{1/3}$. The number of classes that will result from the use of this rule will generally be one less than the number that results from using the power-of-2 rule for

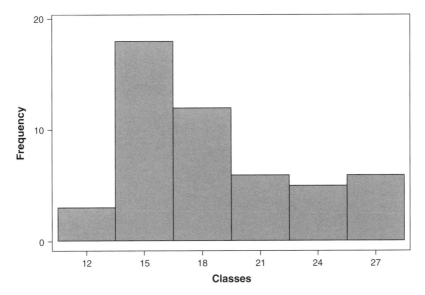

Figure 1.8 Histogram of Example 1.2 data with six classes.

moderate values of n. Each of these two rules seems preferable to having the number of classes equal $\sqrt{n}$.

Since we will not likely be constructing a histogram by hand, the natural question to ask is: Why not just let the software that we are using determine the number of classes? Unfortunately, this is not necessarily a good idea. We will use the data given in Example 1.2 to illustrate why this is the case. (Admittedly, this is not a large data set, but it is useful later for making an important point.)

Use of the power-of-2 rule leads to the selection of six classes, and the histogram with this number of classes is given in Figure 1.8, with the class midpoints shown on the horizontal axis and the class frequencies on the vertical axis. It is also possible to display the relative frequency (percentage) on the vertical axis. This will be desirable in many applications, but if such an approach were used, it would be desirable to show the total number of observations somewhere on or near the display. The reason for this is if the histogram appears to have too many classes (as evidenced by at least one class with a zero frequency) or too few classes, the reader will want to compare the number of observations with the number of classes and see if the latter conforms with one of the suggested rules of thumb.

Note that the rectangles are "joined," even though there is a gap between the end of intervals such as 10–19, 20–29, and so on. (For example, 19.5 is not accounted for with these intervals.) Technically, histograms are generally constructed using the *class boundaries*, which are defined as the average of adjacent class limits (e.g., 29.5 is the average of 29 and 30). Class boundaries are not possible values, so there is no question about what to do if a value is on a boundary since that cannot happen. More specifically, if the data are in integers, the boundaries will be in tenths; if the data are in tenths, the boundaries will be in hundredths, and so on.

The histogram shows the data to be "skewed." That is, the class with the largest frequency is not in the center. Instead, the data are "right skewed," meaning that the data "tail off" on

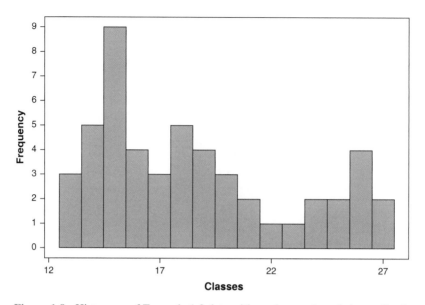

Figure 1.9   Histogram of Example 1.2 data without the number of classes fixed.

the right. (If the tail were on the left, the data would be left skewed. The latter occurs much less frequently than does right-skewed data and is rather uncommon.)

Compare the histogram in Figure 1.8 with the histogram in Figure 1.9, with the latter constructed using MINITAB, but without specifying the number of classes.

This histogram has a jagged appearance because there are too many classes. The histogram also has three peaks, but the population from which the data were obtained almost certainly does not have a distribution with three peaks. Therefore, the histogram is misleading in regard to the shape of the population distribution.

In general, histograms can also be very misleading regarding the symmetry of a distribution, especially for only a moderate number of observations. Minitab, Inc. gives nine examples of histograms of data from a particular symmetric distribution (a normal distribution, covered in Section 3.4.3) in the online *MINITAB User's Guide*, with eight of the histograms exhibiting asymmetry. (A symmetric distribution is such that the distribution of values above the mean is a mirror image of the distribution of values below the mean.)

Histograms can also be constructed with unequal class intervals and with open-ended classes (e.g., "less than 10"). Indeed, classes with unequal width will often be necessary. Consider two scenarios: (1) we want to use a histogram of the starting salaries of civil engineering graduates in 2006, and (2) we want to use a histogram of the salaries of employees at General Motors in 2006.

For the second scenario, what is going to happen if we try to use equal class intervals? If we take the number of employees and divide by the number of classes we decide on so as to give us the approximate class width, we will have some empty classes and other classes with one or two observations because of the high salaries of the top executives. In general, empty classes mean that too many classes were used, as indicated previously. Furthermore, in this case the classes will also be too wide in the center of the histogram because the highest salary is used in determining the class width.

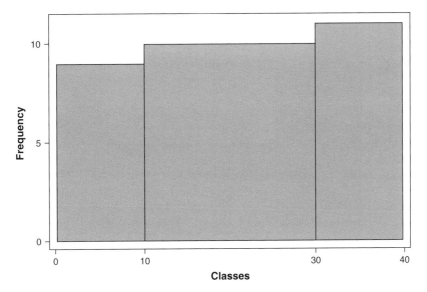

Figure 1.10   Histogram with unequal class widths and unequal areas.

Clearly, this is not the way that the histogram should be constructed. We know that there are some very extreme salaries at GM, just as there are at any large company, so we don't need to have a histogram prove this to us. Instead, we should be interested in the distribution of the rest of the salaries, and we don't want the extreme salaries to alter our view of those salaries. Therefore, a class such as "over $200,000" would be desirable and perhaps also classes of unequal widths.

When classes of unequal widths are used, a decision must be made as to how the heights of the rectangles will be determined. That is, should the heights represent frequencies? Consider the histogram in Figure 1.10. The class frequencies are 9, 10, and 11, respectively, but the middle class appears to be much larger than the other two classes because the area of the rectangle is almost twice the area of each of the other two rectangles. Compare the histogram in Figure 1.10 with the one in Figure 1.11.

This histogram is constructed to make the areas of the rectangles have their proper relationships by taking into consideration the widths of the classes relative to the frequencies. More specifically, this is a "density histogram" such that the total area is 1.

There is nothing wrong with the first of the two histograms as long as the user understands what it represents, but certainly the second histogram will generally be preferable. See Nelson (1988) for additional comments on constructing histograms with unequal class widths.

The construction of a histogram in the popular statistical software JMP for something other than the default number of classes requires manually altering the number by sliding a cursor, as the number of classes cannot be specified directly.

Histograms can also be constructed using frequencies of individual values. Velleman and Hoaglin (1981) provided a histogram of the chest measurements of 5738 Scottish militiamen; the measurements were recorded to the nearest inch and ranged from 33 to 48 inches. With only 16 different values (33–48) there is certainly no need to group them into classes, and, in fact, the "power-of-2" rule would specify 13 classes anyway. We would

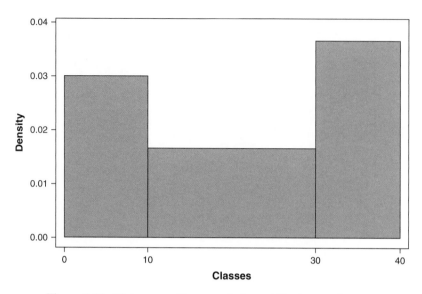

Figure 1.11 Histogram with unequal class widths but equal areas.

expect such anthropometric measurements to be roughly normally distributed, and the histogram did have that general shape.

Histograms have many engineering uses. For example, when used in process capability studies, specification limits (i.e., tolerance limits) can be displayed on a histogram to show the portion of the data that exceeds the specifications. Ishikawa (1976) displayed these as dotted vertical lines.

1.5.2 Boxplot

Another way to provide a picture of a set of data is through use of a *boxplot*. The plot derives its name from the fact that the middle half of a set of data is depicted by the region between the top and bottom of a box (rectangle). Thus, the top of the box (if the software displays the box vertically, some software products display the box horizontally) is the 75th percentile (equivalently the third quartile, Q_3) and the bottom of the box is the 25th percentile (the first quartile, Q_1). A horizontal line is drawn at the 50th percentile (equivalently the median and second quartile, Q_2). Vertical lines are then drawn from the box to the largest and smallest observations. These lines might be viewed as "whiskers"; hence, the other name for the plot is the box-and-whiskers plot. There are several ways to construct a boxplot and the method just described has been termed a *skeletal boxplot* by Velleman and Hoaglin (1981, p. 66).

We will use the data given in Example 1.2, as this will illustrate an important point. Figure 1.6 showed the data to be skewed, yet the boxplot of the data given in Figure 1.12 suggests a symmetric distribution.

How do we rectify the conflicting signals from the histogram and boxplot? The data are indeed skewed, as the reader is asked to show in Exercise 1.4 by constructing a dotplot. What is not apparent from the boxplot is the number of observations that are above the box

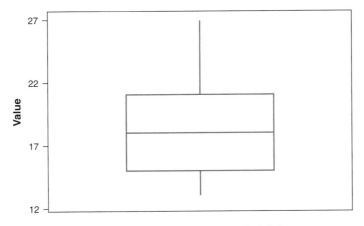

Figure 1.12   Boxplot of the Example 1.2 data.

(i.e., greater than Q_3). There are 11 such observations, whereas there are only 8 observations below Q_1.

So how do we interpret the boxplot? Does Figure 1.12 imply that the middle half of the data are symmetric? Actually, the middle half is highly asymmetric as $Q_1 = 14$, $Q_2 = 15$, and $Q_3 = 19$, whereas for the full data set the numbers are 15, 18, and 21.25, respectively. Thus, a boxplot constructed using the middle half (observations 13–37) would suggest that the middle half is highly skewed since the midline would be much closer to Q_1 than to Q_3. The fact that there are more observations above Q_3 than below Q_1 for the full data set tends to bring the midline up (relative to the boxplot for half the data) and create the false impression that the full data set is symmetric.

Thus, a boxplot cannot be relied on to suggest symmetry or asymmetry because we don't know how many points are outside the box. The data might be symmetric except for one large observation, or there might be several more observations above Q_3 than below Q_1.

There are other variations of boxplots that are more sophisticated and provide more information than skeletal boxplots. Two uses of these other types of boxplots deserve special mention: (1) the determination of outliers (i.e., extreme observations) and (2) the comparison of groups. The latter will be illustrated in the chapter on design of experiments (Chapter 12); the former can be illustrated as follows.

Assume that the largest observation in the Example 1.2 data had been 47 instead of 27. As such, the point would be almost twice the next largest value and would thus be suspect. Clearly, a decision criterion is needed for classifying an observation as an extreme observation or not. The difference $Q_3 - Q_1$, termed the *interquartile range* in Chapter 2, is a measure of the variability in the data. If a large observation is far above Q_3 relative to the variation, then it should be classified as an extreme observation. Therefore, the decision rule is of the form $Q_3 + k(Q_3 - Q_1)$, with the value of k to be selected. Two values of k are typically used: 1.5 and 3.0. An observation is considered to be mildly outlying if it exceeds $Q_3 + 1.5(Q_3 - Q_1)$ or is less than $Q_1 - 1.5(Q_3 - Q_1)$, and is considered to be an extreme outlying observation if it is outside the computed values when 1.5 is replaced by 3.0.

For the present example, $Q_3 - Q_1 = 21.25 - 15 = 6.25$, and 47 is greater than $Q_3 + k(Q_3 - Q_1) = 21.25 + 6.25k$ regardless of whether k is 1.5 or 3.0. Thus, 47 is an extreme outlying observation.

When outlier detection methods are presented, there is generally the tacit assumption that the distribution of possible values is symmetric and at least approximately bell-shaped. Often, however, the distribution will be strongly skewed. When this is the case, extreme observations will not be uncommon and cannot be considered outliers. Statistical distributions are covered in Chapter 3.

1.6 OUTLIERS

An extreme (i.e., outlying) observation is generally called an *outlier*. It is not easy to define an outlier in mathematical terms, even for one-dimensional data, and it is very difficult to define outliers for multidimensional data. As stated by Rousseeuw and Van Zomeren (1990), "Outliers are an empirical reality, but their exact definition is as elusive as the exact definition of a cluster (or, for that matter, an exact definition of data analysis itself)." We want to detect outliers, but it is not possible to provide a general rule for how outliers should be handled once they have been detected. One thing is certain: we do not want to automatically discard them. A good example of the harm this has caused in regard to the South Pole ozone depletion problem is described in Section 1.3.3.14.8 of the *e-Handbook* (Croarkin and Tobias, 2002), as it is stated therein that ozone depletion over the South Pole would have been detected years earlier if outliers had not been discarded automatically by the data collection and recording system.

1.7 OTHER METHODS

Of the methods presented so far in this chapter, which one probably is used the most frequently in a publication like *USA Today*? The answer is undoubtedly a histogram (or bar chart), but there is another method, not yet discussed in this chapter, that is probably used more often than a histogram. A *pie chart* can be used for either a small or large amount of data since what is displayed are percentages, such as the percentage of all crimes by category, with "other" being one of the categories. When used with "other" as one of the categories, a pie chart is thus similar to a *Pareto chart*. The latter can be viewed as an "ordered histogram," with the labels for the bars being qualitative rather than quantitative. For example, the datafile Exh_qc.MTW comes with MINITAB and contains data on defects and causes of them, in addition to other information. Figure 1.13 is a graph that shows the major causes of defects for that dataset.

Note the line in the top part of the display, which shows the cumulative frequencies. This of course is not used with a histogram and is often not used with a Pareto chart. A graph of cumulative frequencies has been called an *ogive*.

Another chart that has considerable potential for engineering applications is a *muli-vari chart*. This is a chart that depicts the variability in whatever is being charted relative to certain selected factors and is thus a way of identifying the factors that affect variation.

A multi-vari chart can be used to considerable advantage in the analysis of data from designed experiments as it can show the factors that are contributing the most to the variability of the response variable. The chart is discussed in certain books on quality improvement (e.g., see Ryan, 2000, Section 11.7).

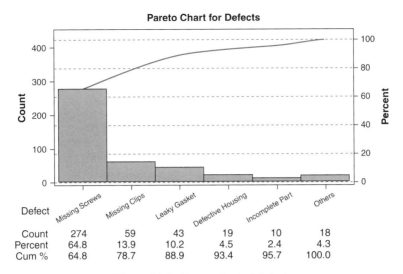

Defect	Missing Screws	Missing Clips	Leaky Gasket	Defective Housing	Incomplete Part	Others
Count	274	59	43	19	10	18
Percent	64.8	13.9	10.2	4.5	2.4	4.3
Cum %	64.8	78.7	88.9	93.4	95.7	100.0

Figure 1.13   Pareto chart of defects.

1.8   EXTREMELY LARGE DATA SETS: DATA MINING

Very large datasets can be compiled with our present data collection capabilities. Specifi-cally, datasets with over one million data values are routinely maintained. How can such huge datasets be portrayed graphically, if at all? Many users of graphical techniques make the mistake of using common graphical methods for large data sets. What would a time sequence plot look like if applied to 500 observations when there is considerable vari-ability? It would essentially be a blur and thus be of virtually no value. Obviously, any type of plotting display will not work with very large data sets. A histogram would work, however, provided that a user was content with the power-of-2 rule. For example, that rule would specify 20 classes for a data set with a million observations, whereas the $\sqrt{n}$ rule would suggest 1,000 classes! (This illustrates the statement made in Section 1.5.1 that the latter rule specifies too many classes when n is large.) Even with a smaller data set of 500 observations, there would be a considerable discrepancy as the $\sqrt{n}$ rule would specify 23 classes, whereas the power-of-2 rule would specify 9 classes.

Because of the difficulty in working with huge data sets, such data sets must generally be sampled. Huge data sets with observations on many characteristics do present sampling challenges, however. This general problem is essentially unsolved. Data mining, as it is called, has become an immensely popular field, with many jobs available for "data miners" both in industry and in academia, software that has been developed for data mining, and even data mining competitions. Readers who wish additional information on data mining should visit, in particular, the website www.kdnuggets.com.

1.9   GRAPHICAL METHODS: RECOMMENDATIONS

The emphasis in this chapter has been on graphical methods. The appropriate type(s) of graphs to use in particular situations depends primarily on the number of observations that are to be plotted and whether or not data have been collected over time. If the latter is true, then data should be plotted over time in some manner. Possible candidates would then be a time sequence plot and a digidot plot. If there is only a small number of univariate observations, a dotplot can be very useful if the data are not collected over time, and the individual values might be shown in a stem-and-leaf display. A control chart, which is a type of time sequence plot, is used for data collected over time and the objective is to determine if processes are in control.

For large datasets, a histogram is a common choice, although it is judicious not to use a large number of classes as this could result in some classes having few, if any, observations. If so, a distorted picture of the shape of the population from which the data came would likely result. Similarly, too few classes should also be avoided as this could also obfuscate the shape: for example, a histogram with only one class (impractical of course) has no shape. Considerable space was devoted to illustrating possible problems along these lines in Section 1.5.1 because histograms are extensively used and it isn't always wise to rely on software, without intervention, to guide histogram construction. Use of the power-of-2 rule generally gives good results.

For bivariate data, the scatter plot is the most frequently used graph and various enhancements, such as was illustrated in Figure 1.5, add to its utility. When there are more than two variables of interest, a Pareto chart can be useful for spotlighting the "vital few and trivial many." This chart can be used very effectively in analyzing data from designed experiments.

1.10 SUMMARY

In this chapter we introduced some terminology that is needed for subsequent chapters, and we focused attention on graphical methods for displaying one-dimensional (i.e., univariate) and two-dimensional data. The graphical methods that were presented are the ones that are most commonly used for low-dimensional data. There is one other popular graphical technique, probability plotting, that is introduced in Chapter 3. Other specialized methods for displaying and analyzing data will be introduced in later chapters. The general rule in constructing graphical displays is to plot the data in as many different ways as seems meaningful, recognizing that as more meaningful graphical displays are constructed, more discoveries are apt to be made about the data. The reader should also bear in mind that this chapter is just an introduction to the most frequently used graphical methods. There is also the need to display data in higher dimensions and continuing advances in computing capabilities make this possible.

REFERENCES

Box, G. E. P., W. G. Hunter, and J. S. Hunter (1978). *Statistics for Experimenters*. New York: Wiley. (The current edition is the second edition, 2005.)

Brady, J. E. and T. T. Allen (2002). Case study based instruction on DOE and SPC. *The American Statistician*, **56**(4), 312–315.

Croarkin, C. and P. Tobias, eds. (2002). *NIST/SEMATECH e-Handbook of Statistical Methods* (http://www.itl.nist.gov/div898/handbook).

Czitrom, V. (1999). One factor at a time versus designed experiments. *The American Statistician*, **53**, 126–131.

Hunter, J. S. (1988). The digidot plot. *The American Statistician*, **42** (1), 54.

Ishikawa, K. (1976). *Guide to Quality Control*. White Plains, NY: UNIPUB.

Lenth, R. V. (2002). Review of *Experimental Quality: A Strategic Approach to Achieve and Improve Quality* by J. Anthony and M. Kaye. *Journal of the American Statistical Association*, **97**, 924.

Lighthall, F. F. (1991). Launching the space-shuttle *Challenger*—disciplinary deficiencies in the analysis of engineering data. *IEEE Transactions on Engineering Management*, **38**(1), 63–74.

McCoun, V. E. (1974). The case of the perjured control chart. *Quality Progress*, **7**(10), 17–19. (Originally published in *Industrial Quality Control*, May 1949.)

Mitchell, T, V. Hegemann, and K. C. Liu (1997). GRR methodology for destructive testing and quantitative assessment of gauge capability for one-side specifications. In *Statistical Case Studies for Industrial Process Improvement* (V. Czitrom and P. D. Spagon, eds.). Society for Industrial and Applied Mathematics and American Statistical Association.

Nelson, L. S. (1988). Notes on the histogram II: Unequal class intervals. *Journal of Quality Technology*, **20**, 273–275.

Ott, E. R. (1975). *Process Quality Control*. New York: McGraw-Hill.

Rousseeuw, P. J. and B. C. Van Zomeren (1990). Unmasking multivariate outliers and leverage points. *Journal of the American Statistical Association*, **85**, 648–651.

Ryan, T. P. (2000). *Statistical Methods for Quality Improvement*, 2nd ed. New York: Wiley.

Scott, D. W. (1979). On optimal and data-based histograms. *Biometrika*, **66**, 605–610.

Tufte, E. R. (1990). *Envisioning Information*. Cheshire, CT: Graphics Press.

Tufte, E. R. (1997). *Visual Explanations: Images and Quantities, Evidence and Narrative*. Cheshire, CT: Graphics Press.

Tufte, E. R. (2001). *The Visual Display of Quantitative Information*, 2nd ed. Cheshire, CT: Graphics Press.

Terrell, G. R. and D. W. Scott (1985). Oversmooth nonparametric density estimates. *Journal of the American Statistical Association*, **80**, 209–214.

Tukey, J. W. (1977). *Exploratory Data Analysis*. Reading, MA: Addison-Wesley.

Velleman, P. V. and D. C. Hoaglin (1981). *ABC of EDA*. Boston: Duxbury.

EXERCISES

Note: The data in the following exercises, including data in MINITAB files (i.e., the files with the .MTW extension), can be found at the website for the text: `ftp://ftp.wiley.com/public/sci_tech_med/engineering_statistics`. This also applies to the other chapters in this book.

1.1. Given below are the earned run averages (ERAs) for the American League for 1901–2003 (in ERAMC.MTW), with the years 1916 and 1994 corrected from the source, *Total Baseball*, 8th edition, by John Thorn, Phil Birnbaum, and Bill Deane, since those two years were obviously in error. (The league started in 1901.)

Construct a time sequence plot, either by hand or using software such as MINITAB, or equivalently a scatter plot with ERA plotted against Year. Does the plot reveal a random pattern about the overall average for these 103 years, or does the plot indicate nonrandomness and/or a change in the average?

Year	1901	1902	1903	1904	1905	1906	1907	1908	1909	1910
ERA	3.66	3.57	2.96	2.6	2.65	2.69	2.54	2.39	2.47	2.51

Year	1911	1912	1913	1914	1915	1916	1917	1918	1919	1920
ERA	3.34	3.34	2.93	2.73	2.93	2.82	2.66	2.77	3.22	3.79

Year	1921	1922	1923	1924	1925	1926	1927	1928	1929	1930
ERA	4.28	4.03	3.98	4.23	4.39	4.02	4.14	4.04	4.24	4.64

Year	1931	1932	1933	1934	1935	1936	1937	1938	1939	1940
ERA	4.38	4.48	4.28	4.5	4.45	5.04	4.62	4.79	4.62	4.38

Year	1941	1942	1943	1944	1945	1946	1947	1948	1949	1950
ERA	4.15	3.66	3.29	3.43	3.36	3.5	3.71	4.29	4.2	4.58

Year	1951	1952	1953	1954	1955	1956	1957	1958	1959	1960
ERA	4.12	3.67	3.99	3.72	3.96	4.16	3.79	3.77	3.86	3.87

Year	1961	1962	1963	1964	1965	1966	1967	1968	1969	1970
ERA	4.02	3.97	3.63	3.62	3.46	3.43	3.23	2.98	3.62	3.71

Year	1971	1972	1973	1974	1975	1976	1977	1978	1979	1980
ERA	3.46	3.06	3.82	3.62	3.78	3.52	4.06	3.76	4.21	4.03

Year	1981	1982	1983	1984	1985	1986	1987	1988	1989	1990
ERA	3.66	4.07	4.06	3.99	4.15	4.17	4.46	3.96	3.88	3.9

Year	1991	1992	1993	1994	1995	1996	1997	1998	1999	2000
ERA	4.09	3.94	4.32	4.80	4.71	5.00	4.57	4.65	4.86	4.91

Year	2001	2002	2003
ERA	4.47	4.46	4.52

1.2. Consider the data given in Example 1.2. As stated in that example, the values that were used to produce the time sequence plot in Figure 1.2 were coded values, not the actual values. Would the plot have looked any different if the actual values had been used? Explain. Use appropriate software to produce the time sequence plot using the actual data and compare the results.

1.3. Construct a dotplot for the data in Example 1.2.

1.4. The sample datafile EXAM.MTW in MINITAB contains the grades on each of two exams for a small graduate statistics class of 8 students. The data are given below.

Row	Exam 1	Exam 2
1	89	83
2	56	77
3	78	91
4	88	87
5	94	99
6	87	80
7	96	85
8	72	75

Construct a dotplot of the grades for each exam. Does either dotplot reveal anything unusual? Is a dotplot a good graphical procedure for identifying unusual observations? Explain.

1.5. Statistical literacy is important not only in engineering but also as a means of expression. There are many statistical guffaws that appear in lay publications. Some of these are given in the "Forsooth" section of *RSS News* (Royal Statistical Society News) each month. Others can be found online at *Chance News*, whose website is at the following URL: `http://www.dartmouth.edu/~chance/chance_news/news.html`. The following two statements can be found at the latter website. Explain what is wrong with each statement.

(a) Migraines affect approximately 14% of women and 7% of men, that's one-fifth the population (*Herbal Health Newsletter Issue 1*).

(b) Researchers at Cambridge University have found that supplementing with vitamin C may help reduce the risk of death by as much as 50% (*Higher Nature Health News* No. HN601, 2001).

(*Comment*: Although the errors in these two statements should be obvious, misstatements involving statistical techniques are often made, even in statistics books, that are not obvious unless one has a solid grasp of statistics.)

1.6. Consider the following salary survey data as compiled by Human Resources Program Development and Improvement (HRPDI), which was current as of October 1, 2002.

Job Title	Number of Companies Responding	Number of Employees in Position	Mean Average Salary
Engineering lab assistant	153	822	$20,887
Electronics tech	119	2,266	25,476
Hardware engineer	116	806	39,889
Civil engineering	218	3,998	51,211
Civil engineering supervisor	110	720	68,993
Mechanical engineer	201	2,104	50,998
Mechanical engineering supervisor	182	785	67,322
Chemical engineer	165	1,765	61,839
Chemical engineering supervisor	141	307	83,292
Electrical engineer	100	905	51,899
Electrical engineering supervisor	87	174	71,884
Vice president, engineering	217	270	149,002

(a) What type of graph would you use to display job title and mean average salary? Be specific.

(b) Note that the average of the average salaries was given: that is, the average salary was determined for each company and then averaged over the companies. Assume that you were asked to estimate the total payroll for a company that employs 8 civil engineers. What would your estimate be if you were not told how many civil engineers work for a particular company? Would it be practical to provide such an estimate?

1.7. Consider Figure 1.5. The data are as follows (in 75-25PERCENTILES2002.MTW):

Row	SAT 75th percentile	acceptance rate	SAT 25th percentile
1	1540	12	1350
2	1580	11	1410
3	1550	16	1380
4	1580	13	1450
5	1560	16	1410
6	1560	13	1360
7	1490	23	1310
8	1500	26	1300
9	1510	13	1310
10	1520	21	1330
11	1490	44	1280
12	1470	33	1290
13	1510	23	1310
14	1450	30	1290
15	1490	16	1290
16	1480	32	1300
17	1460	45	1300
18	1460	31	1270
19	1430	34	1270
20	1450	26	1200
21	1410	39	1200
22	1400	55	1220
23	1460	36	1280
24	1400	29	1170
25	1380	49	1220
26	1410	26	1240
27	1340	37	1130
28	1410	41	1230
29	1370	38	1160
30	1450	22	1280
31	1420	29	1250
32	1420	48	1220
33	1400	34	1210
34	1410	50	1240
35	1390	32	1220
36	1450	71	1240
37	1365	46	1183
38	1420	57	1250
39	1290	63	1060
40	1275	57	1070
41	1330	79	1130
42	1270	78	1050
43	1290	48	1080
44	1350	36	1150
45	1370	73	1180
46	1290	66	1070
47	1290	47	1090
48	1310	62	1090
49	1390	73	1210

(a) Construct the graph of the acceptance rate against the 75th percentile SAT score with the latter on the horizontal axis. Is the slope exactly the same as the slope of Figure 1.5? Explain why the slope should or should not be the same.

(b) Construct the graph of the 25th percentile SAT score against the acceptance rate with the former on the vertical axis. (The data on the 25th percentile are in the third column in the file.) Does the point that corresponds to point #22 in Figure 1.5 also stand out in this graph?

(c) Compute the difference between the 75th percentile and 25th percentile for each school and plot those differences against the acceptance rate. Note that there are two extreme points on the plot, with differences of 250 and 130, respectively. One of these schools is for a prominent public university and the other is a private university, both in the same state. Which would you guess to be the public university?

1.8. A boxplot is constructed for a set of data. The top of the box is at 68.3 and the bottom of the box is at 34.8. What is the numerical value of the interquartile range? Interpret it.

1.9. Consider different amounts of one-dimensional data. What graphical display would you recommend for each of the following numbers of observations: (a) 10, (b) 100, and (c) 1000?

1.10. Should points be connected in a time sequence plot? Why or why not?

1.11. The following numbers are the first 50 of 102 chemical data measurements of color from a leading chemical company that were given in Ryan (2000): 0.67, 0.63, 0.76, 0.66, 0.69, 0.71, 0.72, 0.71, 0.72, 0.72, 0.83, 0.87, 0.76, 0.79, 0.74, 0.81, 0.76, 0.77, 0.68, 0.68, 0.74, 0.68, 0.68, 0.74, 0.68, 0.69, 0.75, 0.80, 0.81, 0.86, 0.86, 0.79, 0.78, 0.77, 0.77, 0.80, 0.76, 0.67, 0.73, 0.69, 0.73, 0.73, 0.74, 0.71, 0.65, 0.67, 0.68, 0.71, 0.69, and 0.73.

(a) What graphical display would you suggest if it was suspected that there may be some relationship between consecutive measurements (which would violate one of the assumptions of the statistical methods presented in later chapters)?

(b) Construct the display that you suggested in part (a). Do consecutive observations appear to be related?

1.12. Lewis, Montgomery, and Myers (*Journal of Quality Technology*, July 2001, Vol. 33, pp. 265–298) gave some data on advance rate from a drill experiment, with the data having been previously analyzed by other authors using different approaches. The data are as follows: 1.68, 1.98, 3.28, 3.44, 4.98, 5.70, 9.97, 9.07, 2.07, 2.44, 4.09, 4.53, 7.77, 9.43, 11.75, and 16.30. These data were collected under different experimental conditions (i.e, they are not all from the same population).

(a) Construct a dotplot for the data. Without knowing anything about the experimental conditions under which the data were obtained, can any preliminary conclusions be drawn? (More sophisticated methods of analyzing such data are given in Chapter 12.)

(b) Construct a boxplot. Does the plot indicate the presence of an outlier? How well-determined are the top and bottom of the box relative to what the top and bottom would be for a population of data values? (that is, how well-determined are the 25th and 75th percentiles from 16 data values?)

1.13. This exercise illustrates how the choice of the number of intervals greatly influences the shape of a histogram. Construct a histogram of the first 100 positive integers for the following numbers of classes: 3, 4, 6, 7, 10, and 12. (The number of classes can be specified in MINITAB, for example, by using the `NINT` subcommand with the `HIST` command, and the sequence `MTB>SET C1, DATA>1:100, DATA>END` will place the first 100 integers in the first column of the worksheet.) We know that the distribution of numbers is uniform over the integers 1–100 because we have one of each. We also have the same number of observations in the intervals 1–9, 10–19, 20–29, and so on. Therefore, the histograms should theoretically be perfectly flat. Are any of the histograms flat? In particular, what is the shape when only three classes are used? Explain why this shape results. What does this exercise tell you about relying on a histogram to draw inferences about the shape of the population of values from which the sample was obtained?

1.14. State the population(s) that would correspond to each of the following samples, if in fact a population exists. If a population does not exist, explain why.
(a) The batting average of a particular player for one season.
(b) The number of votes cast for Al Gore in the 2000 presidential election.
(c) The number of votes cast for Al Gore in Vermont in the 2000 presidential election.
(d) A student's randomly selected test score from among the student's test scores in a statistics course.
(e) The number of nonconforming transistors of a certain type produced by a specific company on a given day.

1.15. Explain why consecutive observations that are correlated will be apparent from a digidot plot but not from a dotplot, histogram, stem-and-leaf display, scatter plot, or boxplot. Is there another plot that you would recommend for detecting this type of correlation? Explain.

1.16. Chart your driving time to school or work for a month. Do any of the plotted points seem to deviate significantly from the other points? If so, is there an explanation for the aberrant points? Can a manufacturing process "explain" why it is out of control? Then what should be added to a time sequence plot to detect problems when inanimate objects are involved?

1.17. Construct a boxplot of your driving times from the previous exercise. Do any of your times show as an outlier? If the box doesn't exhibit approximate symmetry, try to provide an explanation for the asymmetry.

1.18. If we had a sample of $n = 1000$ observations, which of the graphical displays presented in the chapter would be potentially useful and which ones should be avoided?

1.19. Given in file `NBA2003.MTW` are the scoring averages for the top 25 scorers in the National Basketball Association (NBA) in 2002. The data are given below.

	Name		Scoring Average
1	Tracy	McGrady	32.1
2	Kobe	Bryant	30.0
3	Allen	Iverson	27.6
4	Shaquille	O'Neal	27.5
5	Paul	Pierce	25.9
6	Dirk	Nowitzki	25.1
7	Tim	Duncan	23.3
8	Chris	Webber	23.0
9	Kevin	Garnett	23.0
10	Ray	Allen	22.5
11	Allan	Houston	22.5
12	Stephon	Marbury	22.3
13	Antawn	Jamison	22.2
14	Jalen	Rose	22.1
15	Jamal	Mashburn	21.6
16	Jerry	Stackhouse	21.5
17	Shawn	Marion	21.2
18	Steve	Francis	21.0
19	Glenn	Robinson	20.8
20	Jermaine	O'Neal	20.8
21	Ricky	Davis	20.6
22	Karl	Malone	20.6
23	Gary	Payton	20.4
24	Antoine	Walker	20.1
25	Michael	Jordan	20.0

What type of graphical display would you recommend for displaying the data? Construct the display, but before doing so, would you expect the averages to exhibit asymmetry? Why or why not?

1.20. Compute and list in juxtaposition the number of classes for a histogram using the power-of-2 rule and the square root rule for $n = 50, 100, 200, 300, 400, 500,$ and 1000. What pattern or relationship do you observe? (Construct a scatter plot if the pattern is not apparent.) Name one (bad) thing that could happen if the square root rule were used with $n = 500$.

1.21. With a conventional scatter plot, two variables are displayed—one on the vertical axis and one on the horizontal axis. How many variables were displayed in the scatter plot in Figure 1.5? Can you think of how additional variables might be displayed?

1.22. Use statistical software to first generate 100 observations from a particular symmetric distribution (a normal distribution covered in Chapter 3), and then construct the histogram. The following sequence of MINITAB commands, for example, will accomplish this: `MTB>RAND 100 C1 MTB> HIST C1`. Do this ten times. How many of your histograms are symmetric? What does this suggest about relying on a histogram to determine whether or not a population has a symmetric distribution?

1.23. A data set contains 25 observations. The median is equal to 26.8, the range is 62, $Q_1 = 16.7$, and $Q_3 = 39.8$. What is the numerical value of the interquartile range?

1.24. When would the interquartile range have greater value as a measure of variability than either the range or the standard deviation?

1.25. Would a histogram of the data given in Exercise 1.1 be a meaningful display? Why or why not?

1.26. Assume that you work for a company that makes ball bearings of various sizes and you want to sample periodically from the total number of each type that are manufactured in a day. Would it be practical to take a random sample? Explain. If not, what type of sample would you expect to take?

1.27. What graphical display discussed in this chapter would be best suited for showing the breakdown of the number of Nobel Prize winners by country for a specified time period?

1.28. A Pareto chart can be used for many purposes. Use the baseball salary data in file EX1-28ENGSTAT.MTW to construct a Pareto chart using players' positions as the categories. Are the results what you would expect?

1.29. Toss a coin twice and record the number of tails; then do this nine more times. Does the string of numbers appear to be random?

1.30. In their paper, "Criminal Violence of NFL Players Compared to the General Population" (*Chance*, **12**(3), pp. 12–15, Summer 1999), Alfred Blumstein and Jeff Benedict presented data that demonstrated that the violent crime rate among professional football players is actually less than that among other males of the same age and race. Was this an observational study or could it have been an experiment? Explain. (*Source*: http://www.amstat.org/publications/chance/123.nflviol. pdf.)

1.31. In an article in the Winter 2001 issue of *Chance* magazine, Derek Briggs found that SAT and ACT preparation courses had a limited impact on students' test results, contrary to what companies that offer these courses have claimed. Read this article and write a report explaining how an experiment would have to be conducted before any claim of usefulness of these courses could be made. (*Source*: http:// www.amstat.org/publications/chance/141.briggs.pdf.)

1.32. Given in file BASKET03.MTW are the team 3-point field goal percentages for all NCAA Division I teams at the end of the 2002–2003 season. Construct a scatter plot of the field goal percentage against the average number of shot attempts per game. Does there appear to be a relationship between these two variables? Would you expect a relationship to exist? Could the individual schools be designated in some manner on the plot? If so, how would you construct the plot?

1.33. The following data are frequency distributions of weights of cars and trucks sold in the United States in 1975 and 1990. (*Source*: U.S. Environmental

Protection Agency, *Automotive Technology and Full Economic Trends through 1991*, EPA/AA/CTAB/91–02, 1991.)

WT	WT (L)	WT (U)	CA75	TR75	CA90	TR90
1750	1625	1875	0	0	1	0
2000	1875	2125	105	0	109	0
2250	2125	2375	375	0	107	0
2500	2375	2625	406	0	1183	34
2750	2625	2875	281	204	999	45
3000	2875	3250	828	60	3071	428
3500	3250	3750	1029	55	2877	784
4000	3750	4250	1089	1021	1217	1260
4500	4250	4750	1791	386	71	797
5000	4750	5250	1505	201	0	457
5500	5250	5750	828	59	1	46
6000	5750	6250	0	1	0	32

Variable Names:

WT: Weight in pounds, class midpoint

WT(L): Weight in pounds, class lower limit

WT(U): Weight in pounds, class upper limit

CA75: Cars sold, 1975 (thousands)

TR75: Trucks sold, 1975 (thousands)

CA90: Cars sold, 1990 (thousands)

TR90: Trucks sold, 1990 (thousands)

(a) Compare the distributions of CA75 and CA90 by constructing a histogram of each. Comment on the comparison. In particular, does there appear to have been a significant change in the distribution from 1975 to 1990? If so, what is the change? (In MINITAB, the histograms can be constructed using the CHART command with the C1*C2 option; that is, CHART C1 C2 with C1 containing the data and C2 being a category variable, and these two column numbers being arbitrary designations.)

(b) Construct the histograms for TR75 and TR90 and answer the same questions as in part (a).

(c) Having constructed these four histograms, is there any problem posed by the fact that the intervals are not of equal width? In particular, does it create a problem relative to the 1975 and 1990 comparisons? If so, how would you correct for the unequal widths? If necessary, make the appropriate correction. Does this affect the comparison?

(d) In view of the small number of observations, would it be better to use another type of graphical display for the comparison? If so, use that display and repeat the comparisons.

1.34. Construct a dotplot for the data in Table 1.1. Do any of the values appear to be outliers? Explain.

1.35. Use appropriate software, such as MINITAB, or Table A in the back of the book to generate three samples of size 20 from the first 50 positive integers. Compare the three samples. Is there much variability between the three samples? If so, would you have anticipated this amount of variability?

1.36. Assume that you want to take a sample of 200 Scholastic Aptitude Test (SAT) total scores at your university and use the distribution of the sample observations to represent the distribution of all of the SAT scores. Would you simply take a random sample from the list of all students, or would you use some other approach?

1.37. Consider the Lighthall (1991) article that was discussed at the beginning of the chapter. If you are presently taking engineering courses, can you think of data that should be collected and analyzed on some aspect in an engineering discipline, but that are usually not collected and analyzed? Explain.

1.38. The datafile DELTALATE.DAT contains the number of minutes that each of the 566 Delta flights departing from Hartsfield International Airport in Atlanta (now Hartsfield–Jackson International Airport) was late on June 30, 2002, with a negative value indicating the number of minutes that the flight departed early. (source: Department of Transportation.)

   (a) Construct a histogram of the data, using the power-of-2 rule to determine the number of classes. Comment on the shape of the histogram. In particular, is the shape unexpected? Explain.

   (b) Construct a boxplot. Note the large number of observations that are spotlighted as outliers. Would you recommend that all of these points be investigated, or should a different outlier-classification rule be used in light of the shape of the histogram in part (a)? Explain.

   (c) Construct a dotplot. What is one major advantage of the dotplot relative to the histogram for this type of data, regarding what we would like to glean from the display?

1.39. Given the following stem-and-leaf display,

$$
\begin{array}{l|lllllll}
3 & 1 & 2 & 2 & 4 & 5 & 7 \\
4 & 1 & 3 & 5 & 7 & 7 & 9 \\
5 & 2 & 4 & 5 & 6 & 8 & 9 & 9 \\
6 & 1 & 3 & 3 & 4 & 7 & 8 \\
\end{array}
$$

   determine the median.

1.40. The data in CLINTON.MTW contain a conservative score/rating that had been assigned to each U.S. senator at the time of the Senate votes on perjury and obstruction of justice alleged against former president Bill Clinton. (See impeach.txt at http://www.amstat.org/publications/jse/jse_data_archive. html for variable descriptions.)

   (a) What graphical display would you use to show the overlap of scores between Democrats and Republicans? Would any other displays suffice?

   (b) Construct the display.

1.41. Scatter plots of binary data have limited value. To see this, use the CLINTON.MTW datafile and construct a scatter plot of each senator's conservatism score against whether or not the person is a first-time senator. Does the plot show any relationship between the two plotting variables? If not, how would the plot have appeared if there were a strong relationship between the variables?

1.42. Relative to the preceding exercise, the graphing of binary variables becomes even more problematic when both variables are binary. Using the CLINTON.MTW data set, construct a scatter plot using the vote on each of the two questions, perjury and obstruction of justice, as the plotting variables.

   (a) Is the plot of any value when counts of each point are not shown? Explain.
   (b) Construct a table of the counts (if using MINITAB in command mode, use the TABLE command with two columns specified). Is this a preferred substitute to constructing a scatter plot with two binary variables? Explain.

1.43. The file BASKETBALL.MTW contains the NCAA Division 1 highest yearly men's team field goal percentages from 1972 through 2002, which are given below (the years go across the rows):

```
52.8   52.7   53.0   54.7   53.7   54.5   54.6   55.5   57.2   56.4   56.1
55.6   55.2   54.8   56.1   54.1   54.6   56.6   53.3   53.5   53.6   52.2
50.6   51.7   52.8   52.0   51.8   52.3   50.0   51.1   50.1
```

   (a) Construct either a time series plot of the data or a scatter plot using the year (72, 73, etc.) on the horizontal axis. What type of pattern, if any, would you expect the graph to exhibit? Does the graph appear different from what you expected?
   (b) Determine (from an Internet search if necessary) the underlying cause, if any, for the configuration of plotted points. If a cause was discovered, how would you recommend that the graph be reconstructed to show a change?

1.44. Figure 1.4 was a lag plot for observations that were positively correlated. What should be the general appearance of the plot (of lag one) when observations (a) have a high negative correlation and (b) are uncorrelated?

1.45. What chart would you use (and in fact is used by brokerage companies) to show an investor the breakdown of the total value of his/her stock portfolio into the dollar value of the component parts (plus cash, if applicable)?

   The material in Exercises 46–50 appeared in the "Forsooth" section of *RSS News*, a publication of the Royal Statistical Society, being chosen selectively by the editorial staff from various published sources. The statements therein illustrate mistakes made in the presentation and interpretation of data, and some examples are given in the following problems.

1.46. The proper presentation of data is obviously important. The following example made the "Forsooth!" section of the June 2001 issue of *RSS* (Royal Statistical Society) *News*, as an example of something that went wrong.

Top 50 highest vehicle theft rates by city in 2000

CITY	NO. OF THEFTS	THEFT RATE
Phoenix	29, 506	976.1
Miami	20, 812	956.6
Detroit	40, 685	909.2
Jersey City	4, 502	814.4
Tacoma	5, 565	807.9
....	....	....

(The vehicle theft rate is the number of stolen vehicles divided by the city's population, then divided again by 100,000.) (*Original source*: http://www.insure.com/auto/thefts/cities401.html.)

(a) First, what is wrong with the numbers?

(b) If the intent was to show the number of thefts in each city for each 100,000 people, what should have been the theft rate for Phoenix?

(c) How would you have presented the data so as to facilitate an easy comparison of the cities?

1.47. The following statement appeared in the June 2001 issue of *RSS News*. "The number of official investigations into accidents on building sites is expected to have risen by more than 200 per cent in the last five years while routine inspections have fallen by in excess of 100 per cent over the same period, a Commons written reply showed.". (*Original source*: *The Times*, April 2001.) What is wrong with the statement?

1.48. The following statement appeared in the February 2002 issue of *RSS News*: "A rail route takes up four times less land than a motorway." (*Original source*: "Railtrack advert at Euston Station," December 2001.) What is the error in this statement?

1.49. The following statement appeared in the January 2002 issue of *RSS News*: "A skilled teacher in a state school can tell if a child is troubled because of difficulties in the family; an experienced teacher will say if more than half the classroom have a problem, from divorce to drug abuse, then it materially affects the education of the other half." (*Source*: *Spectator*, June 30, 2001.) What is wrong with this statement?

1.50. According to folklore, there will be a long winter if a groundhog sees its shadow on Groundhog Day. The "representative" in the Atlanta area for this event is a groundhog named Gen. Beauregard Lee. The caption for a picture that appeared in the February 3, 2003 edition of the *Atlanta Journal and Constitution* stated, in part: "Lee, who didn't see his shadow on Sunday morning, has a 99% forecast accuracy rate". The average lifespan for a groundhog in captivity is 10 years. Is there anything wrong with the picture caption?

1.51. Why are "percent unfavorable" and "percent favorable" not both needed as columns in a table?

1.52. The following eight resistivity measurements for wires were given by Gunst (*Quality Progress*, October 2002): 0.141, 0.138, 0.144, 0.142, 0.139, 0.146, 0.143, 0.142. If you were greatly concerned about the possibility of outliers in data sets such as this one, what graphical display would you use to try to detect and spotlight an outlier?

1.53. The following numbers are grades for a particular class on a test that I gave over 20 years ago: 98, 96, 94, 93, 93, 91, 91, 90, 88, 88, 85, 84, 84, 83, 82, 81, 79, 79, 78, 78, 78, 78, 76, 75, 75, 74, 74, 73, 72, 67, 67, 65, 63, 63, 62, 62, 58, 54, 49, and 44. Indicate what graphical display you would use for each of the following objectives, and then construct each display.

    (a) You want to see if there are any large gaps between certain scores. (Of course, here the scores are ordered, but numbers won't always be ordered when received.)

    (b) You want to obtain an idea of the shape of the distribution of the scores.

1.54. There are many ways to construct a stem-and-leaf display. Explain how you would construct such a display, in particular, how the stems would be constructed, if you had a sample of 50 numbers that ranged from 1.12 to 1.98.

1.55. Assume that someone has constructed a time sequence plot for a particular manufacturing process, but *without* a constant time increment given on the horizontal axis. If a time sequence plot for a similar manufacturing process (same product) was also constructed, could the two plots be meaningfully compared? Explain.

1.56. Although not discussed in this chapter, it is possible to construct a scatter plot so as to easily show the largest and smallest values for each of the two plotted variables. Give one way that this could be done.

1.57. Do finite populations occur and are they of interest in your field of engineering or science? If so, give three examples; if not, explain why they do not occur.

1.58. You are shown a histogram that has two empty classes. What would you recommend to the person who constructed the histogram?

1.59. Consider the 75th percentile SAT scores for the data given in Exercise 1.7. What would you expect for the shape of a histogram of these scores? Construct a histogram of the scores using 8 classes. Is the shape what you expected?

1.60. Give an example of what would be an enumerative study in your field, and give an example with the same subject matter that would be an analytic study.

1.61. To illustrate a stem-and-leaf display that is more sophisticated than the one given in Section 1.4.1, use MINITAB (or other software) to generate 100 random integers between 1 and 10, with the integers all being equally likely to be selected. The MINITAB command for a stem-and-leaf display is STEM. Use that (or other software) to produce a stem-and-leaf display of these 100 random integers. Is the appearance of the display unexpected? Explain. If MINITAB was used, interpret the numbers to the left of the display.

1.62. Consider the 25th percentile SAT scores for the data given in Exercise 1.7. If you wanted to construct a graphical display that would spotlight any outliers with small values, what display would you use? Construct that display for these data. Are there any outliers? Explain.

1.63. The distribution of grade-point averages (GPAs) for all 1,999 sorority members during the second semester of the 2001–2002 academic year at Purdue University was shown by a histogram to be highly left-skewed.

    (a) Five classes were used for the histogram. Would you suggest that more or fewer classes be used in an effort to present a clear picture of the shape of the distribution? If you believe that a different number should be used, how many class intervals would you suggest?

    (b) With the highest GPA being 3.222, the lowest 2.538, Q_1 and Q_3 being 3.148 and 2.839, respectively, would you expect the distribution to still be left-skewed if a different number of classes had been used? Explain.

    (c) It was stated in Section 1.5.1 that data are generally not left-skewed, and indeed the distribution of GPAs for members of a fraternity that semester was close to being symmetric. In particular, the highest GPA was 3.130 and the value of Q_3 was 2.7535, with the difference between these two numbers being much greater than for sorority members. What graphical device would you use (which was not used at the Purdue website) to show the difference in the distributions for fraternity and sorority members, especially the difference in skewness?

1.64. The U.S. Consumer Product Safety Commission reported in an October 23, 2002 memorandum (`http://www.cpsc.gov/library/toydth01.pdf`) that for 2001, 79% (202,500) of the estimated number of toy-related injuries were sustained by children under the age of 15; 30% (77,100) were under the age of 5.

    (a) What were the number of toy-related injuries for children at least 15 years old and the total number of injuries?

    (b) What graphical display would you suggest be used to show breakdown of toy-related injuries among these age groups?

1.65. Consider the statistics for teams in the National Basketball Association (NBA) that are available at `http://www.nba.com/index.html` and at `http://sports.espn.go.com/nba/statistics?stat=teamstatoff&season=2003&seasontype=2`.

    (a) Construct a scatter plot of overall field goal percentage versus 3-point field goal percentage for the 29 teams. Would we expect the scatter plot to show a relationship? Why or why not?

    (b) Construct a scatter plot of 3-point field goal percentage versus 2-point percentage, after performing the appropriate calculations to obtain the 2-point percentages. Does the scatter plot show a relationship, and if so, is the relationship expected?

    (c) Construct a boxplot of the average number of 3-point field goal attempts per game for the 29 teams. Does the boxplot show any outlier teams? Explain.

1.66. Critique the following statement made by a plant manager: "I know there are only five possible causes of nonconforming tesla coil capacitors in my plant, so I've told my people not to bother with constructing Pareto charts for that nonconforming data, as I can certainly obtain the information that I need from the numbers; a Pareto chart really wouldn't add anything."

1.67. What can a time sequence plot of individual observations reveal that could be at least partially hidden in a time sequence plot of averages of 5 observations (as in quality control work)?

1.68. Consider a histogram with short rectangle heights not close to the extremes of the data. What does that signify?

1.69. Name three engineering continuous random variables that you would expect to result in a moderately skewed distribution if 100 values were obtained on each variable. Name three discrete random variables that relate to a company's clerical operations that you would also expect to be skewed. Should any of these variables be left-skewed (tail to the left) or should they all be right-skewed?

1.70. Explain why a class frequency should not be the sole determinant of the height of a rectangle in a histogram when the class intervals are of unequal widths.

1.71. The Browser Summary report for hits to the U.S. Geological Survey website for November 2002 showed Internet Explorer accounted for 69.59% of the hits, 7.30% for Netscape, 4.14% for Inktomi Search, and 4.13% for Googlebot, in addition to the percentages for 29 other browsers. What graphical display given in this chapter would seem to be best suited for displaying this information?

1.72. A graph very similar to the one given below appeared in the *Atlanta Journal-Constitution* on January 28, 2003. The graph showed the number of applicants (the top line) and the number of applicants accepted (the bottom line) at a local university over a period of several years. The objective of the graph was to show that admission to the school is becoming more difficult over time. Does the increasing distance between the lines over time actually show this, however? Given the objective, would you have constructed the same type of graph differently, or would you have used a different type of graph? Explain.

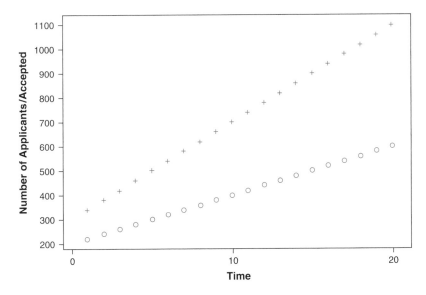

1.73. Consider the data given in Exercise 1.7. Plot the 75th percentile scores against the 25th percentile scores and then plot the 25th percentile scores against the 75th percentile scores.

(a) Compare the configuration of points and comment. Would you expect the plots to differ by very much? Explain.

(b) In general, under what conditions would you expect a plot of Y versus X to differ noticeably from a plot of X versus Y?

1.74. A student in your engineering statistics class intends to study for the final exam solely by memorizing rules for working each of the many different types of problems that were presented in the course. What would be your advice for him or her?

1.75. Would you recommend that a pie chart be constructed with 50 "slices"? Why or why not? In particular, assume that there are 50 different causes of defects of a particular product. If a pie chart would not be suitable for showing this breakdown, what type of graph would you recommend instead?

1.76. Assume that you work in the Office of the Dean of Student Affairs at a large southern university. The dean would like to look at the difference between the average GPA for female students who are in sororities and female students who are not in a sorority, relative to the difference between the GPA for male students who are in fraternities and those who are not in fraternities, and to look at the numbers over time. What type of graph would you recommend and how should the graph be constructed? Be specific.

1.77. The Boise Project furnishes a full irrigation water supply to approximately 400,000 acres of irrigable land in southwestern Idaho and eastern Oregon (see `http://www.usbr.gov/dataweb/html/boise.html`). The following data are from that website.

Year	Actual Area Irrigated (Acres)	Crop Value (Dollars)
1983	324,950	168,647,200
1984	327,039	149,081,226
1985	325,846	135,313,538
1986	320,843	151,833,166
1987	309,723	153,335,659
1988	308,016	157,513,694
1989	326,057	185,990,361
1990	323,241	177,638,311
1991	323,241	177,145,462
1992	325,514	182,739,499

What graphical display (if any) would you use to determine if there seems to be a relationship between area irrigated and crop value over time? Does there appear to be a relationship between the two?

1.78. The following intervals are used by the National Water and Climate Center of the U.S. Department of Agriculture in giving its weather map streamflow forecasts as a percentage of the average: >150, 130–150, 110–129, 90–109, 70–89, 50–69, <50.

Note that there are two open-ended intervals and the others do not have a constant width. Would you suggest any changes in the format, or do you believe this is the best choice? Explain.

1.79. One of the sample datafiles that comes with MINITAB, MASSCOLL.MTW in the STUDNT12 subdirectory of datafiles, contains various statistics for the 56 four-year colleges and universities in Massachusetts in 1995. Consider the variable Acceptance Rate, with the following data:

```
%Accept
  76.9718   23.0590   80.8571   79.6253   56.7514   67.2245   76.7595
  47.1248   64.4166   74.4921   65.5280   67.9552   71.3197   56.4796
  76.4706   79.2636   84.8980   66.4143   85.8000   73.1537   55.9285
  83.8279   74.3666   15.6149   87.3786   81.1475   33.3801   77.7890
  73.0000   85.0737   90.0000   64.2994   71.3553   91.2429   90.1639
  79.7382   77.9661   57.0000   54.6325   71.3453   63.0828   78.5886
  47.3470   85.9814   66.8306   77.5919   75.9157   43.1433   84.9324
  89.1515   69.3548   78.9989   83.9599   29.7420   83.5983   68.9577
```

If you wanted to show the distribution of this variable over the 56 schools and spotlight extreme observations, what graphical technique would you employ? Produce the graph and comment. Over the years there have been suspicions (not necessarily for Massachusetts) that self-reporting of certain statistics such as average SAT score has resulted in exaggerated numbers being reported.

(a) Recognizing that there are many different variables in this datafile, how would you check to determine if any of the numbers looked out of line?

(b) Which pair of variables would you expect to have the strongest linear relationship? Check your answer by constructing all possible scatter plots. (This can be done in MINITAB by using the command MATR followed by the columns for which the set of scatter plots is to be constructed.

1.80. One of the sample datafiles that comes with MINITAB is UTILITY.MTW, which is in the STUDENT1 subdirectory. The data in the fourth column (given below) show a family's electricity bill for approximately four years.

```
KWH $
   82.02   133.98   114.47    97.25    75.44    51.33    73.71   101.27
   90.93    65.10    58.22    82.32   165.55   173.02   116.76   144.09
  100.00    45.64    79.46    71.61   104.83    49.26    57.72   106.04
  125.97   194.22   146.50    93.35    90.33    32.17    56.63    78.61
   71.37    71.97    99.88    93.84   151.96   184.62   110.06   113.21
   79.44    61.88    77.05    74.61
```

Construct the appropriate plot to show the cost of electricity over time for this family. What do you conclude from the graph?

1.81. One of the sample datafiles that comes with MINITAB is GAGELIN.MTW. The dataset is reprinted with permission from the *Measurement Systems Analysis*

Reference Manual (Chrysler, Ford, General Motors Supplier Quality Requirements Task Force). Five parts were selected by a plant foreman that covered the expected range of measurements and each part was randomly measured 12 times by an operator. The data are given below.

Row	Part	Master	Response
1	1	2	2.7
2	1	2	2.5
3	1	2	2.4
4	1	2	2.5
5	1	2	2.7
6	1	2	2.3
7	1	2	2.5
8	1	2	2.5
9	1	2	2.4
10	1	2	2.4
11	1	2	2.6
12	1	2	2.4
13	2	4	5.1
14	2	4	3.9
15	2	4	4.2
16	2	4	5.0
17	2	4	3.8
18	2	4	3.9
19	2	4	3.9
20	2	4	3.9
21	2	4	3.9
22	2	4	4.0
23	2	4	4.1
24	2	4	3.8
25	3	6	5.8
26	3	6	5.7
27	3	6	5.9
28	3	6	5.9
29	3	6	6.0
30	3	6	6.1
31	3	6	6.0
32	3	6	6.1
33	3	6	6.4
34	3	6	6.3
35	3	6	6.0
36	3	6	6.1
37	4	8	7.6
38	4	8	7.7
39	4	8	7.8
40	4	8	7.7
41	4	8	7.8
42	4	8	7.8
43	4	8	7.8
44	4	8	7.7
45	4	8	7.8

Row	Part	Master	Response
46	4	8	7.5
47	4	8	7.6
48	4	8	7.7
49	5	10	9.1
50	5	10	9.3
51	5	10	9.5
52	5	10	9.3
53	5	10	9.4
54	5	10	9.5
55	5	10	9.5
56	5	10	9.5
57	5	10	9.6
58	5	10	9.2
59	5	10	9.3
60	5	10	9.4

What graphical device would you use to show the variability in measurements by the operator relative to the reference value for each part that is given in the adjacent column? Be specific. Would a histogram of the measurements show anything of value? Explain.

1.82. One of the sample datafiles that comes with MINITAB is EMPLOY.MTW, which gives the number of employees in three Wisconsin industries, Wholesale and Retail Trade, Food and Kindred Products, and fabricated Metals, measured each month over five years.

Row	Trade	Food	Metals
1	322	53.5	44.2
2	317	53.0	44.3
3	319	53.2	44.4
4	323	52.5	43.4
5	327	53.4	42.8
6	328	56.5	44.3
7	325	65.3	44.4
8	326	70.7	44.8
9	330	66.9	44.4
10	334	58.2	43.1
11	337	55.3	42.6
12	341	53.4	42.4
13	322	52.1	42.2
14	318	51.5	41.8
15	320	51.5	40.1
16	326	52.4	42.0
17	332	53.3	42.4
18	334	55.5	43.1
19	335	64.2	42.4
20	336	69.6	43.1
21	335	69.3	43.2

Row	Trade	Food	Metals
22	338	58.5	42.8
23	342	55.3	43.0
24	348	53.6	42.8
25	330	52.3	42.5
26	326	51.5	42.6
27	329	51.7	42.3
28	337	51.5	42.9
29	345	52.2	43.6
30	350	57.1	44.7
31	351	63.6	44.5
32	354	68.8	45.0
33	355	68.9	44.8
34	357	60.1	44.9
35	362	55.6	45.2
36	368	53.9	45.2
37	348	53.3	45.0
38	345	53.1	45.5
39	349	53.5	46.2
40	355	53.5	46.8
41	362	53.9	47.5
42	367	57.1	48.3
43	366	64.7	48.3
44	370	69.4	49.1
45	371	70.3	48.9
46	375	62.6	49.4
47	380	57.9	50.0
48	385	55.8	50.0
49	361	54.8	49.6
50	354	54.2	49.9
51	357	54.6	49.6
52	367	54.3	50.7
53	376	54.8	50.7
54	381	58.1	50.9
55	381	68.1	50.5
56	383	73.3	51.2
57	384	75.5	50.7
58	387	66.4	50.3
59	392	60.5	49.2
60	396	57.7	48.1

Construct a time sequence plot for Food. Can the pattern of points be easily explained?

1.83. It was stated in Exercise 1.1 that there were two errors in the source from which the data were obtained. For that type of data, what method(s) would you use to try to identify bad data points?

CHAPTER 2

Measures of Location and Dispersion

To most people the word "location" refers to a position of some sort, especially in terms of geography, such as a particular geographical location. The expressions "this would be a good location for our business" and "his fastball had good location" (such as spoken by a baseball announcer) are common expressions in which the word is used.

The word is used similarly in statistics as the numerical value of a *location parameter* determines where the population distribution is positioned along the number line. The population *mean*, denoted by μ, is one location parameter. If we had a finite population that was small enough so that it was computationally practical to compute the mean, it would be computed by summing all of the elements in the population and dividing by the number of such elements. It is rare when a population is small enough so as to make this practical, however.

We might also consider the population *median* as a measure of location. This is the number such that approximately 50% of the values in the population are above it and approximately 50% are below it. The *mode*, the most frequently occurring value, is sometimes mentioned as a location parameter, but some distributions are bimodal. In general, the mode is not a good measure of location.

Although we may speak conceptually of the mean and mode of a population, we generally won't know their numerical values unless the population is narrowly defined.

The terms *dispersion* and *variability* are used interchangeably. They each refer to the spread of values from the center (location) of the data. The *population standard deviation* is denoted by σ, and of course it is not practical to calculate it except in a rare case of a small finite population. We may still define it, however, as

$$\sigma = \sqrt{\frac{\sum_{i=1}^{N}(X_i - \mu)^2}{N}}$$

for a population of size N. The square of this quantity, denoted by σ^2, is the *population variance*.

Although σ and σ^2 are commonly used measures of variability, the magnitude of σ generally depends on the magnitude of the numbers in the population. For example, let the

Modern Engineering Statistics   By Thomas P. Ryan
Copyright © 2007 John Wiley & Sons, Inc.

population be defined as the number of 10-year-old male children in a small city. Assume that a member of this population who is 5 feet tall has his height recorded as 5.00 feet and also as 60 inches, and similarly for the other members of the population. What will be the relationship between the value of the standard deviation when height is recorded in inches (call it σ_1) versus the standard deviation when height is recorded in feet (call it σ_2)? Obviously, $\sigma_1 > \sigma_2$ because the heights in inches will have greater spread than the heights in feet. It is easy to show by substituting in the expression for σ (as the reader is asked to do in Exercise 2.5) that $\sigma_1 = 12\sigma_2$.

This suggests that if we wanted to compare, say, the variability of 10-year-old males in this particular city with the heights of 15-year-old males, some adjustment would be necessary, even when both measurements are recorded in inches. The adjustment consists of dividing the standard deviation by the mean (i.e., σ/μ) and this is called the population *coefficient of variation*. Such an adjustment permits the comparison of (relative) variability for different populations.

It is worth noting, however, that we cannot make definitive statements about the relationship between two standard deviations based solely on the magnitude of the numbers in the respective populations. For example, if we added the number 100 to the heights (in inches) of every 10-year-old, all of the resultant numbers would be greater than the heights of the 15-year-olds. Yet, the value of σ for the resultant numbers is the same as the value of σ for the original numbers. (The reader is asked to show the general relationship in Exercise 2.6.) In general, however, the larger the numbers, the larger the standard deviation.

Another measure of variability is the *range*. It might seem logical to define the population range as the largest number in the population minus the smallest number. This typically isn't done, however, as "largest minus smallest" is used in defining the sample range, not the population range. Actually, it would be impossible to define a population range in many instances anyway as there is no explicit upper bound in certain applications. For example, if nonconformities on printed circuit boards (PCBs) are modeled with a Poisson distribution (see Section 3.3.3), the random variable can assume any positive integer. Of course, in some instances, such as SAT scores, the population range is known or at least can be closely approximated.

2.1 ESTIMATING LOCATION PARAMETERS

Since location parameters can only rarely be computed, they must be estimated. The term "estimators of location" is short for estimating a location parameter. We would logically estimate a population mean with a sample mean (average), a population median with a sample median, and so on. (Estimation is discussed further in Chapter 4.) Thus, we obtain a sample and compute various sample *statistics*, each of which is used to estimate the corresponding population parameter. (Recall the definition of a statistic in Section 1.1.) Often they are "mixed," such as when a population standard deviation is estimated using sample ranges, as is done in quality improvement work.

We find averages all around us: a baseball player has a batting average, a punter in football has a punting average, there is the Dow Jones Industrial Average, an engineer is interested in average cycle times, and so on.

DEFINITION

The sample mean (average) of n observations $x_1, x_2, \ldots, x_n$ is the sum of the observations divided by n: that is,

$$\bar{x} = \frac{x_1 + x_2 + \cdots + x_n}{n} = \frac{\sum_{i=1}^{n} x_i}{n}$$

The sample average $\bar{x}$ is a statistic that is read "x-bar." The Greek letter $\sum$ is read as (capital) sigma and is used to indicate summation. Specifically, what lies to the right of $\sum$ is to be summed. The letter i in X_i is a subscript, which in this case varies from 1 to n. The number at the bottom of $\sum$ indicates where the summation is to start, and the number (or symbol) at the top indicates where it is to end. (Please note that sometimes a capital letter will be used to represent a statistic and sometimes a lowercase letter will be used, depending partly on personal preference in each context and on the desire to have the symbols stand out. An example of this are the two symbols, capital and lowercase, used on this page to denote the sample average.)

Thus, the average for the sample

$$11 \quad 13 \quad 15 \quad 17 \quad 19$$

could be expressed as

$$\bar{x} = \frac{\sum_{i=1}^{n} X_i}{n}$$

$$= \frac{\sum_{i=1}^{5} X_i}{5}$$

$$= \frac{X_1 + X_2 + X_3 + X_4 + X_5}{5}$$

$$= \frac{11 + 13 + 15 + 17 + 19}{5}$$

$$= 15.0$$

Another type of average that is frequently used in industry to estimate location is a *moving average*, which is also referred to as a *rolling average*. As a simple example, assume that we have the following 15 observations: 5.3, 5.1, 4.9, 4.5, 4.8, 4.1, 3.9, 4.4, 3.7, 5.0, 4.7, 4.6, 4.2, 4.0, and 4.3. A moving average of size 5 would then be the average of observations 1–5, 2–6, 3–7, and so on, so that the moving averages would be 4.92, 4.68, 4.44,.... A moving average is generally used to smooth out the data while portraying "current" estimates of location.

The dispersion counterpart to a moving average is a moving range, which is discussed in Section 2.2.

Another possibility would be to use the median, which is the middle value (after the numbers have been arranged in ascending order) if the sample consists of an odd number of observations, as it does in this case. Thus, the median is 15. If the sample had consisted of these five numbers plus the number 21, there would then not be a single middle number, but rather two middle numbers. The median would then be defined as the average of the two middle values, which in that case would be 16. Thus, when there is an even number of observations, the median will always be a number that is not observed unless the two middle numbers happen to be the same. That might seem strange but it should be remembered that the objective is to *estimate* the middle value of the population from which the sample was drawn.

DEFINITION

The *sample median* is the middle value in an ordered list of an odd number of observations. If the number is even, the median is defined as the average of the two middle values.

For this example the average is also 15, so the average and the median have the same value. This will usually not be the case, however. In fact, the average might not even be close to the center of the sample when the values are ordered from the smallest to largest observation. For the sample

$$28 \quad 39 \quad 40 \quad 50 \quad 97$$

the average (50.8) is between the fourth and fifth numbers, so it is not particularly close to the middle value. This is the result of the fact that one observation, 97, is considerably larger than the others. Thus, although the average is often referred to as a *measure of center* or *measure of central tendency*, it often will not be very close to the middle of the data.

Consequently, during the past twenty years, in particular, there has been considerable interest in the statistical community in developing estimators that are insensitive to extreme observations (numerical values that differ considerably from the others in the sample; i.e., outliers). The median is one such estimator (although not developed within the past twenty years), and for this simple example there is a considerable difference between the median, 40, and the average, 50.8. So which is a better measure of "center"? Stated differently, which number best represents a typical value? Certainly the median will always be in the center of the sample, but whereas the mean and median will be the same when the distribution of population values is symmetric, this will not be true when the distribution is skewed. (Distributions are discussed in detail in Chapter 3.) Specifically, if the distribution is right skewed (i.e., the tail of the asymmetric distribution is on the right), the mean will exceed the median. Conversely, the relationship is reversed when the data are left skewed, but as mentioned in Section 1.5.1, left-skewed data are rather uncommon. Although the sample mean is often a better estimator of the population mean than is the sample median, the result of the comparison in a particular case depends on the shape of the distribution. For this example we might go with the median to represent a typical observation in the absence of any external information.

The choice between median and mean becomes easier for the following scenario. If we wanted a number to represent a typical salary at a small company, which would we prefer to use, the median or the mean? Even at a small company, the salaries of the top executives will far exceed the salary of a typical worker, so the executive salaries will pull the average above the median, with the result that the average may not well-represent a typical salary. Note that the salary example is also an example of a right-skewed distribution. Of course, a left-skewed distribution for this scenario would mean that not very many workers would have a low salary, with most workers having an above average salary. Because of federal minimum wage regulations and high executive salaries in general, a company with a left-skewed salary distribution would be extremely rare.

Professional organizations such as the American Statistical Association and Western Mine Engineering, Inc. give salary distributions that are obtained from sample surveys. Such distributions are invariably skewed, at least slightly, as a salary distribution (or any other type of distribution) that is perfectly symmetric would be extremely rare.

■ **EXAMPLE 2.1**

The Western Mine Engineering, Inc. website gives various statistics on its union and non-union mines, in addition to listing reports that can be purchased.

Data

The following vacation data in days per year relative to years of service were part of the 2004 Annual Mining Cost Service U.S. Mine Labor Survey Wage and Benefit Summary Surface and Underground Coal, Metal, and Industrial Mineral Mines.

	Days of Vacation			
	Union Mines		Non-union Mines	
Years of Service	Interval	Average	Interval	Average
1 yr	5–16	8	0–15	8
3 yr	9–16	12	0–15	9
5 yr	10–16	13	0–17	12
10 yr	14–21	17	0–22	16
15 yr	14–26	19	0–27	17
20 yr	16–29	22	0–27	18
25+ yr	20–30	24	0–30	19

Analysis

As expected, the non-union mines have the widest intervals for the various years of service. If we add the lower and upper end points of each interval, then divide by 2 and compare that number with the average, we can see whether or not the data are skewed within each interval. We can see that there is considerable skewness for the non-union mines, with the degree of skewness increasing with years of service, whereas there is hardly any skewness for the union mines. ■

For highly skewed distributions with extreme observations, a *trimmed average* would be an alternative to the median if the objective is to give a typical value such as a typical salary. If, for example, 10% of the observations in a sample are trimmed from each end (after the observations have been ordered), extreme observations, which could have a considerable effect on the average of all of the observations, would thus be deleted. If there are no extreme observations, such "trimming" should have very little effect on the average. For example, if the smallest and largest values are deleted from the sample

$$11 \quad 13 \quad 15 \quad 17 \quad 19$$

the average remains unchanged, but if the same trimming is done for the sample

$$28 \quad 39 \quad 40 \quad 50 \quad 97$$

the average changes from 50.8 to 43.0.

Extreme observations might very well be values that have been recorded incorrectly. In any event, they are not typical observations unless a distribution is extremely skewed or has "heavy tails" (see Chapter 3). If the trimming is not done haphazardly, but rather some trimming procedure is consistently applied, a better estimate of the center of the corresponding population is apt to result.

2.2   ESTIMATING DISPERSION PARAMETERS

Just as we cannot compute the values of location parameters, we similarly cannot compute the values of dispersion parameters. So we must estimate them.

Variation (dispersion) is unavoidable, as there is "natural" variation in almost everything. Is your driving time to work or school precisely the same every day? Of course not. It will depend on factors such as weather conditions and traffic conditions. Assume that your time varies slightly for a particular week, but at the beginning of the second week an accident on the expressway causes you to be 30 minutes late for work. Your travel time for that day is not due to natural, random variation, but rather to an "assignable cause"—the accident.

With statistical procedures, in general, and control charts (Chapter 11) in particular, a primary objective is to analyze components of variability so that variability due to assignable causes can be detected. If you are the "unit" that is being "measured" for travel time, you know why you were late for work and can thus explain the cause. A ball bearing, however, cannot explain why its diameter is considerably larger than the diameter of the preceding 500 ball bearings that have rolled off the assembly line. Thus, statistical procedures are needed to spotlight the abnormal variation and, we hope, to enable the contributing factor(s) to be pinpointed.

Before we can speak of normal and abnormal variation, however, we must have one or more objective measures of variation. The simplest such measure is the sample *range*, which is defined to be the largest observation in a sample minus the smallest observation. (Note that in the field of statistics the range is a number, not an interval.) For the sample

$$11 \quad 13 \quad 15 \quad 17 \quad 19$$

the range is 8. It should be observed that only two of the five values are used in obtaining this number; the other three are essentially "thrown away." Because of its simplicity and ease of calculation by hand, the range has been used extensively in quality control work, but with modern computational aids there is no need to use a statistic that is computed from only two of the observations in a sample.

Another type of range that has been used extensively in quality control work is the *moving range*. Analogous to a moving average, a moving range is computed from "moving" groups of data, with a moving range of two observations being most frequently used. For the above example of five observations, the moving ranges of size 2 would be 2, 2, 2, and 2 since there is a difference of 2 in the consecutive observations. The absolute value of the difference is what is actually used, so if the 17 and 15 were switched, the moving ranges would be 2, 4, 2, and 4. The use of moving ranges in quality control/improvement work, with the average moving range being used as a measure of variability, is discussed and illustrated in Section 11.5.

Although the range and moving range have the advantage of simplicity, the range is wasteful of information and will be inferior to good measures of variability that use all of the observations. Similarly, the average moving range is also not the best measure of variability, especially when data are correlated.

It is also worth noting that the sample range is not used to estimate the population range. This is because it is very difficult to estimate extreme population percentiles with sample data, and the largest and smallest observations are the most extreme of all! More specifically, assume that a population consists of 10,000 numbers and we obtain a random sample of 30. Do you think the largest of the 30 numbers would be anywhere near the largest of the 10,000 numbers? Historically, the sample range has been used to estimate the population standard deviation, but it should be apparent that any estimator that uses only two observations in a sample is going to be inferior to an estimator that uses all of the observations. This is discussed further in Section 11.5.

I stated in the Preface that this book is written under the assumption that its readers will be using statistical software for data analysis. Therefore, simple methods that are amenable to hand computation will not be recommended over more efficient, but involved, procedures when they can both be handled with approximately equal ease on a computer.

If we were to start from scratch and devise a measure of variability that uses all of the sample observations, it would seem logical that we should construct a measure that shows how the data vary from the average. If we wanted to construct a measure of variability, we might *attempt* to use $\sum_{i=1}^{n} (X_i - \overline{X})$. However, it can be shown that this sum will equal zero for any sample. This is due to the fact that some of the deviations $(X_i - \overline{X})$ will be positive whereas others will be negative, and the positive and negative values add to zero. For the present sample of five numbers,

$$\sum_{i=1}^{5} (X_i - \overline{X})$$
$$= -4 - 2 + 0 + 2 + 4$$
$$= 0$$

One obvious way to eliminate the negative deviations would be to square all of the deviations, and this is what is done.

DEFINITION

The *sample variance*, S^2, is usually defined as

$$S^2 = \frac{\sum_{i=1}^{n} (X_i - \overline{X})^2}{n - 1}$$

with the *sample standard deviation* $S = \sqrt{S^2}$.

A few authors have chosen to divide by n instead of $n - 1$. Arguments can be given in support of each choice, but a discussion of the merits of each is delayed until Section 4.4.2. The form with a divisor of $n - 1$ is what will be used in this book.

If we were to compute S^2 using a calculator, we would prefer a computationally simpler form than the one given above. The problem with the latter is that the deviation of each observation from the mean must be computed. The numerator of S^2 can also be written as $\sum_{i=1}^{n} X_i^2 - \left(\sum_{i=1}^{n} X\right)^2/n$, as the reader is asked to show in Exercise 2.9.

The sample variance is the (natural) estimator of the population variance mentioned at the beginning of the chapter. Furthermore, it is a "good" estimator in the sense that if we had an extremely large population and computed S^2 for each of the $\binom{N}{n}$ samples, the average of the S^2 values would be (essentially) equal to σ^2. (Technically, equality holds only for an infinite population; the *finite population correction factor*, shown in Section 4.2.2, must be used to adjust for a finite population.)

As indicated previously, it could be shown that S^2 can also be calculated as

$$S^2 = \frac{\sum X^2 - \left(\sum X\right)^2/n}{n - 1}$$

For the same sample of five numbers,

$$S^2 = \frac{\sum_{i=1}^{n} (X_i - \bar{x})^2}{n - 1}$$

$$= \frac{\sum_{i=1}^{5} (X_i - 15)^2}{4}$$

$$= \frac{(-4)^2 + (-2)^2 + (0)^2 + (2)^2 + (4)^2}{4}$$

$$= \frac{40}{4}$$

$$= 10$$

The sample variance is not as intuitive as the sample range, nor is it in the same unit of measurement. If the unit of measurement is inches, the range will be in inches but the variance will be in inches squared. This is not sufficient cause for discarding the variance in favor of the range, however.

Another measure of variability that is in terms of the original units is the *sample standard deviation*, which is simply the square root of the sample variance and which estimates σ. It does not estimate σ in quite the same way that S^2 estimates σ^2, however. That is, the average of $\binom{N}{n}$ values of S would *not* be essentially equal to σ. In quality control work, S is divided by a constant so as to produce a better estimator of σ, whereas the constant is ignored in other methodological applications. We will follow convention in both cases. Therefore,

$$S = \sqrt{\frac{\sum\limits_{i=1}^{n}(X_i - \overline{X})^2}{n-1}}$$

is the sample standard deviation. For this example, $S = \sqrt{10} = 3.16$. The standard deviation is also not as intuitive as the range but will generally be of the same order of magnitude as the average deviation between the numbers, although there is no direct relationship between the two measures. Thus, with the (ordered) numbers

$$11 \quad 13 \quad 15 \quad 17 \quad 19$$

the deviation between each pair of adjacent numbers is 2, so the average deviation is also 2. Therefore, a standard deviation of 3.16 is well within reason; a value of, say, 31.6 or 0.316 should lead us to check our calculations. This can be helpful as a rough check, regardless of the calculating device that is used to obtain the answer, as numbers are often entered incorrectly into a computer.

Since σ is estimated by s and μ is estimated by $\overline{x}$, it might seem logical that the population coefficient of variation, σ/μ, would be estimated by $s/\overline{x}$. In fact, this is what is done and the latter quantity is termed the *sample coefficient of variation*. (We note in passing that substituting the usual individual estimators into quotients or products isn't always a good idea as the estimator formed by the quotient or product might not have the best properties.)

It is important to note that the variance, standard deviation, and range are greatly influenced by outliers. To see this, consider a sample of five numbers consisting of 10, 20, 30, 40, and 50. The variance can be seen to be 250, so the standard deviation is $\sqrt{250} = 15.8$. If the 50 is misrecorded as 150, the variance is 3,250 and the standard deviation is 57. Only one number is different, yet the variance is much larger and the standard deviation is 3.6 times what it is without the error. Similarly, the range would be misreported as 140, which is much larger than the correct value of 40. The *interquartile range* (IQR), the difference between the third and first quartiles of the ordered data ($Q_3 - Q_1$ in the usual notation) would not be affected, and in general will not be affected by outliers.

Similarly, the IQR could be used to estimate σ, dividing by the appropriate constant that makes the estimator of σ unbiased. (Unbiasedness is a property of estimators that is discussed in Section 4.2.1.) There is disagreement in the literature on how the interquartile range is defined, however (e.g., to interpolate or not interpolate), and the unbiasing constant of course depends on the definition that is used. For example, Rocke (1989) does not interpolate but acknowledges that his definition of the IQR for control chart applications differs from most definitions, which do involve interpolation.

Because of the large effect that a single erroneous number can have on the computations, bad data points should be identified and removed before a variance or standard deviation is computed. Alternatively, a *trimmed variance* and/or *trimmed standard deviation* might be computed. These would be the counterparts to a *trimmed mean* and would be computed in the usual way after trimming a prespecified percentage of observations off each end. These statistics are part of a body of statistics known as *robust statistics*. Some statistical software and computing environments provide trimmed variances and trimmed standard deviations, such as S-Plus.

We may note that the magnitude of the (conventional) variance will not automatically be influenced by the magnitude of the numbers, although this will generally be the case, as was stated previously. Specifically, the variance is unaffected if we add 1000 to each of the numbers 10, 20, 30, 40, and 50, as the reader is asked to show in Exercise 2.2. This is not true if a number is multiplied *times* each of the original numbers, however. In general, let the original numbers be denoted by X and the transformed numbers be given as $Y = a + bX$. We obtain the average, variance, and standard deviation of Y as $\overline{Y} = a + b\overline{X}$, $S_y^2 = b^2 S_x^2$, and $S_y = |b|S_x$, respectively. To solve for the average, variance, and standard deviation of X, we would just solve these equations. That is, $\overline{X} = (1/b)(\overline{Y} - a)$, $S_x^2 = (1/b^2)S_y^2$, and $S_x = (1/|b|)S_y$. The following two examples illustrate the mechanics involved.

■ **EXAMPLE 2.2**

These results for $Y = a + bX$ are quite useful. To illustrate, the data given in Exercise 2.66 contain decimal fractions. If someone decided to compute the standard deviation of those grade-point averages using software, it would be easier to enter the numbers without the decimal points, have the software compute the standard deviation of those numbers, and then make the appropriate adjustment. Specifically, the standard deviation would be (1/100) times the standard deviation of the numbers that were entered without the decimal points.   ■

■ **EXAMPLE 2.3**

Data

This method of coding numbers can be used very efficiently when there is very little variation in the numbers. For example, S. Liu and R. G. Batson in their paper "A Nested Experimental Design for Gauge Gain in Steel Tube Manufacturing" [*Quality Engineering*, **16**(2), 269–282, 2003–2004] gave 20 measurements in a gauge repeatability and reproducibility study for each of the two operators and the averages were given for each operator. We will use the data for the first operator and that is given below.

Computing the Average

For the first operator, six of the numbers were 0.043, thirteen were 0.044, and one was 0.045. The average is obviously very close to 0.044, so that would be one choice for "a" in the coding equation. Since the sum of the deviations below 0.044 is 0.006 and the sum above 0.044 is 0.001, the average deviation is thus $-0.005/20 = -0.00025$. The average of the 20 numbers is thus $0.044 - 0.00025 = 0.04375$. With practice this can be done very quickly, in much less time than it takes to enter the numbers in a calculator.

Computing the Standard Deviation

The standard deviation or variance can be calculated almost as easily. It doesn't matter what number is subtracted from the original numbers since the standard deviation and variance are unaffected by that choice. So subtracting 0.044 from each number and then multiplying by 1,000, we obtain six numbers that are -1, thirteen that are 0, and one number that is $+1$. The variance of those numbers is $(7 - (-5)^2/20)/19 = 115/380 = 0.303$, so the variance of the original numbers is $0.303 \times 10^{-6} = 0.000000303$ and the standard deviation is the square root of that number. Although this requires more work than computing the average, it still can be done very quickly with practice.

Although coding greatly simplifies these computations, we would like to be able to compute these statistics with software as much as possible. Given below is the MINITAB output for the data in this example. The statistics are produced by using the DESC command with appropriate subcommands, with in this case the mean, standard deviation, variance, median, and interquartile range having been requested. (No subcommands would have been needed if the third and first quartiles would have been sufficient, as is indicated in Section 2.5, with the user then computing the interquartile range from the two numbers.)

```
Descriptive Statistics: C1

Variable      Mean      StDev      Variance      Median          IQR
C1        0.043750   0.000550   0.000000303    0.044000     0.001000
```

Another measure of variation, which is used for two variables such as two process characteristics, is *covariance*. This is a measure of how two random variables "co-vary." For variables X and Y, the sample covariance is given by

$$S_{xy} = \frac{\sum_{i=1}^{n}(X_i - \overline{X})(Y_i - \overline{Y})}{n - 1}$$

Note that if Y were replaced by X, this would produce the sample variance, S^2, given previously. If S_{xy} were divided by the square root of the products of the sample variances for X and Y, this would produce the sample *correlation coefficient*—a unit-free statistic whose value must lie between -1 and $+1$, inclusive. A value of $+1$ would result if all of the points could be connected by a straight line with a positive slope; a value of -1 would occur if all of the points could be connected by a straight line with a negative slope. Neither extreme case could be expected to occur in practice, however. Correlation is covered in more detail in Chapter 8.

There are a few other measures of variability that are occasionally used, but the range, interquartile range, variance, and standard deviation are the ones that have been used most frequently for a single variable, and these are the ones that will be used in this book, with emphasis on the variance and standard deviation. The correlation coefficient is frequently given when there are two variables, and covariance is used in Section 9.1, with correlation covered in Section 8.3.

2.3 ESTIMATING PARAMETERS FROM GROUPED DATA

In the previous sections it was assumed that the data were not grouped. Often, however, the raw data will not be available to everyone involved in using the data to make decisions, especially since there are now extremely large datafiles that are routinely being used in many applications of statistics. Assume that a histogram has been constructed from a frequency distribution (as discussed in Section 1.5.1), and it is desired to know (or estimate) the average of the data values. Without the raw data, it isn't possible to compute the average, but it can be estimated. This is done by using the midpoint of each class interval to represent all of the observations within the interval. Specifically,

$$\overline{x}^* = \frac{\sum\limits_{i=1}^{k} f_i (MP)_i}{\sum\limits_{i=1}^{k} f_i}$$

with $(MP)_i$ denoting the midpoint of the ith class and f_i denoting the corresponding frequency. This is actually a *weighted average*, with the weights being the class frequencies.

This estimate will work well if observations within each interval are approximately uniformly distributed in the population (the continuous uniform distribution is given in Section 3.4.2) and will also work well when the histogram is approximately symmetric relative to its center. If the histogram is skewed, however, skewness in class intervals on opposite sides of the mean will not necessarily be offsetting. Using the midpoint for an interval is similar to computing the median for the interval, and the mean and median will differ, perhaps considerably, if the data are at least moderately skewed within the interval.

Estimating the variance or standard deviation from grouped data presents an additional problem since the use of midpoints leads to an inflated estimate of the population variance. If the population from which the grouped data were obtained can be represented by a distribution that is continuous and has "tails" of at most moderate size (such as normal distributions given in Section 3.4.3), the following procedure should provide a reasonable approximation to what the calculation of the sample variance would be if the raw data were available. First, compute the sample variance using the midpoints as if they were the raw observations. Second, subtract from this result $w^2/12$, with w denoting the width of each interval. (Of course, this rule approximation can't be used if the last class is open-ended and/or there are classes of unequal widths.)

Thus, the approximated sample variance is

$$S^{2*} = \frac{\sum\limits_{i=1}^{k} (f_i (MP)_i - \overline{X}^*)^2}{n-1} - \frac{w^2}{12}$$

$$= \frac{\sum\limits_{i=1}^{k} (f_i (MP)_i)^2 - \left(\sum\limits_{i=1}^{k} f_i (MP)_i\right)^2 \Big/ n}{n-1} - \frac{w^2}{12}$$

with f_i, $(MP)_i$, and $\overline{X}^*$ as previously defined. This is known as Sheppard's correction. How well the approximation works will vary over datasets, and there is no guarantee that the approximation will be in the right direction for a particular dataset, as the reader is asked to show in Exercise 2.58. Overall, however, the approximation is considered to work well.

■ **EXAMPLE 2.4**

Consider the data in Example 1.2 and assume that you have been presented a histogram of the data, such as in Figure 1.8. The frequencies and the class midpoints for the six classes shown in that graph are given below.

Class midpoint	Frequency
12	3
15	18
18	12
21	7
24	5
27	5

Using this frequency distribution, the average value is then estimated as

$$\overline{X}^* = \frac{\sum\limits_{i=1}^{k} f_i (MP)_i}{\sum\limits_{i=1}^{k} f_i}$$

$$= \frac{3(12) + 18(15) + 12(18) + 7(21) + 5(24) + 5(27)}{50}$$

$$= \frac{924}{50}$$

$$= 18.48$$

and the sample variance is estimated as

$$S^{2*} = \frac{\sum\limits_{i=1}^{k} f_i (MP)_i^2 - \left(\sum\limits_{i=1}^{k} f_i (MP)_i\right)^2 \Big/ n}{n - 1} - \frac{w^2}{12}$$

$$= \frac{3(12)^2 + \cdots + 5(27)^2 - 924}{49} - \frac{3^2}{12}$$

$$= 18.50 - 0.75 = 17.75$$

■

2.4 ESTIMATES FROM A BOXPLOT

Since we may estimate μ and σ from grouped data, it is natural to ask whether we may estimate these parameters from other summarizations of data, such as a boxplot. Just as is the case with grouped data, we must also make assumptions about the population of data values that have been sampled in obtaining the data for the boxplot. If the midline on the plot is approximately halfway between the endpoints of the box, we could use the median as a rough estimate of μ. Since the difference between those endpoints is the interquartile range, we could use that difference in providing a rough estimate of σ. In doing so, however, we must make a particular distributional assumption, which might not be a plausible assumption. We return to this issue in Section 4.6.

2.5   COMPUTING SAMPLE STATISTICS WITH MINITAB

The basic statistics described in this chapter can be computed using the DESC(ribe) command in MINITAB or in menu mode by selecting "Basic Statistics" from the menu and then selecting "Display Descriptive Statistics." The output includes the mean, trimmed mean (with 5% trimmed off each end), median, standard deviation, maximum, minimum, 25th percentile (first quartile, Q_1), and 75th percentile (third quartile, Q_3). MINITAB does not provide estimates of the mean and standard deviation for grouped data.

2.6   SUMMARY

One way to characterize a probability distribution (covered in Chapter 3) is to give its parameters or, more realistically, to estimate those parameters since they will almost certainly be unknown. The population mean, variance, and standard deviation are generic parameters in the sense that their estimates can be used in describing any set of data. In one case (the normal distribution), μ and σ do appear explicitly in the function that represents the distribution, and these are the mean and standard deviation, respectively. Generally, this is not the case, however, with the mean and variance being functions of the parameters that appear explicitly in the probability function. In those cases, the objective is the same: to estimate the unknown parameters and to estimate the mean, variance, and/or standard deviation.

Although simple statistics such as $\bar{x}$ are the ones most frequently used, it is important to recognize that robust statistics such as the trimmed mean, trimmed variance, and trimmed standard deviation will generally be superior to their untrimmed counterparts in the presence of extreme observations. Estimation is discussed in more detail in Chapter 4.

We should also keep in mind that in this age of extremely large data sets and data mining, there are challenges in data summarization and data visualization that of course were not envisioned when simple statistics were used extensively and were deemed adequate. Clearly, we would not want to compute the average of 10 million observations, even if we have the capability of doing so as those 10 million observations would almost certainly come from a mixture of distributions with different means. Therefore, we should keep in mind the computing environment at the time and the objectives that exist in particular situations regarding parameter estimation.

REFERENCE

Rocke, D. M. (1989). Robust control charts. *Technometrics*, **31**, 173–184.

EXERCISES

2.1. The following are six measurements on a check standard, with the check standard measurements being resistivities at the center of a 100 ohm · cm wafer: 96.920, 97.118, 97.034, 97.047, 97.127, and 96.995.

(a) Compute the average and the standard deviation of the measurements.

(b) Could the average have easily been computed without having to enter the numbers as given in a computer or calculator (e.g., by entering integers in a calculator)? Explain.

(c) Similarly, assume that one wished to compute the variance without using the numbers as given. Could this be done?

2.2. Show that the sample variance and sample standard deviation are unaltered when 1000 is added to each of the numbers 10, 20, 30, 40, and 50. Explain why the variance and standard deviation are unchanged, supporting your argument by using the formulas for the variance and standard deviation.

2.3. A "batting average" was mentioned in Section 2.1 as being one type of average. What is being averaged when a batting average is computed? Is this what sportswriters and baseball fans think about being averaged to obtain a batting average? Explain.

2.4. A department of six faculty members each rank three job candidates with 1 being the highest rank. An average rank is computed for each applicant. Assume that there are no ties. What must be the average of the three average ranks?

2.5. Show by appropriate substitution in the expression for σ that the value of σ for heights in inches for some population is 12 times the value of σ when height is measured in feet (and fractions thereof).

2.6. Show that the value of σ is unchanged if a constant is added to every number in a population.

2.7. We can always compute an average of a group of numbers, but we also need to ask whether or not the average will make any sense. For example, the following numbers are provided by Western Mine Engineering, Inc. (www.westernmine.com); they represent the operating cost indices for surface mines in the United States from 1988 to 2000: 85.5, 90.4, 95.7, 97.7, 98.7, 99.0, 100.0, 102.4, 105.7, 107.0, 106.7, 108.7, 113.5.

(a) Compute the average of these 13 numbers.

(b) Does this number estimate some population parameter? If so, what is it?

(c) Compute the standard deviation of the numbers. What is the unit of measurement for the standard deviation?

2.8. Can the mean of the data represented by Figure 1.8 be estimated using only Figure 1.8? If so, obtain the estimate of the mean. Since the histogram is skewed, would you expect your estimate to be reasonably close to the actual mean? Explain.

2.9. Show that

$$\sum_{i=1}^{n} (X_i - \overline{X})^2 = \sum_{i=1}^{n} X_i^2 - \left(\sum_{i=1}^{n} X_i\right)^2 \Big/ n$$

2.10. In statistics compiled by the Major League Baseball Players Association for the year 2000, 126 starting pitchers (with 19 or more starts) had an average salary of $3,064,021, whereas 165 relief pitchers (10 or fewer starts and 25 or more relief appearances) had an average salary of $1,220,412. Assume that you want to compare the variability of salaries of starting pitchers versus relief pitchers. Could you simply compute their respective standard deviations and compare them? Explain.

2.11. For the data described in Exercise 2.10, assume that you want to construct a boxplot to summarize the salaries of starting pitchers. You are not given the raw data but you are given 10 summary statistics.

(a) Which statistics do you need in order to construct the boxplot?

(b) Would a boxplot be an appropriate choice for a graphical display given the number of observations and the nature of the data?

2.12. Find the mean, median, and range for the following numbers: 5, 8, 3, 4, 7, 2, 9.

2.13. Let k denote the standard deviation for a sample of 25 numbers.

(a) If each number is multiplied by 100, what is the standard deviation of the new set of numbers, as a function of k?

(b) Is the coefficient of variation affected by this multiplication? Explain.

2.14. Given the following data for X and Y, each expressed as a function of some number a, express the sample variance of Y as a function of the sample variance of X:

$$X : a, a + 1, \ a + 2, \ a + 3$$

$$Y : 2.5a + 2, \ 2.5a + 4.5, \ 2.5a + 7, \ 2.5a + 9.5$$

2.15. Assume that you work for a company with 500 employees and you have to fill out a report on which you will give the typical salary for an employee with the company. What statistic will you use for this purpose? Would your answer be any different if the company had 50 employees, or 5,000 employees?

2.16. Consider the following (small) population of numbers: 2, 5, 7, 8, 9. Compute the sample variance for each of the 10 different samples of size 3, then compute the variance for the population. Now compute the average of those 10 sample variances. Does the average equal the population variance? What adjustment must be made so that the average will be equal to the population variance? (Don't attempt the adjustment; simply state what must be done.)

2.17. Many instructors compute summary statistics on test scores and provide these to the class or post the results outside the instructor's office door. If the variance of the test scores is given, what is the unit for that measure? Would you recommend that an instructor use the variance or standard deviation as the measure of variability, or perhaps some other measure?

2.18. Assume that you have a sample of size 2 and the range is $\sqrt{8}$. Determine the standard deviation.

2.19. Assume that a random variable is measured in inches. Which of the following will *not* be in inches: (a) mean, (b) variance, and (c) standard deviation?

2.20. Compute the median for the following sample of 10 numbers: 23, 25, 31, 43, 32, 32, 19, 27, 41, and 22.

2.21. A sample of 50 observations is modified by dividing each number by 3 and then subtracting 75: that is, $Y = X/3 - 75$, with X representing the original observations. If $s_Y^2 = 1.2$, what is s_X^2?

2.22. If for a sample of size 20, $\sum_{i=1}^{20} (X_i - \overline{X})^2 < 19$, which will be larger, the standard deviation or the variance?

2.23. Purchasing shares of company stocks became affordable (in terms of commissions) for the average person some years ago with the advent and popularity of online brokerages. If an investor buys 200 shares of JBL at 28.40 in December, 100 shares at 27.64 in January, and 150 shares at 25.34 in March, (a) what was the total amount that the investor paid for the stock, and (b) what was the average price paid for the 450 shares that were purchased? If 200 additional shares were bought in June at that average price, would the average for the 650 shares change? Why or why not?

2.24. The shipping department of a leading high technology company ships out 234 units of a product on the first day of a 5-day work week. If the number of units shipped increases each day for the five days, the average number of units shipped per day over the five days must be at least equal to what number? If no more than 400 units could be shipped in a day, but shipments increased each day, what is the largest number that the average number of shipments could be?

2.25. Once when I was a graduate student, a fellow graduate student wanted to know the average grade he assigned for a particular semester. Assume that the grades were 14 A's, 24 B's, 26 C's, 10 D's, and 4 F's. (This was 30 years ago, in the days before grade inflation.) Computing the overall average in one's head, as I did, is a type of "lightning calculation," as it is called. Letting A, B, C, D, and F be designated by 4, 3, 2, 1, and 0, respectively, compute the average grade assigned in your head if you can do so; otherwise, use a calculator.

2.26. The following is (modified) MINITAB output for a sample of 50 observations:

```
Descriptive Statistics: C2

Variable   N  Average  Median  Minimum  Maximum      Q1      Q3
C2        50   20.560  21.000   10.000   30.000  15.750  26.000
```

(a) What does the relationship between the median and the average suggest about the symmetry of the data?
(b) What would the relationship between the median and mean have been if the data had been right skewed (i.e., the tail of the distribution were on the right)?

2.27. For a normal distribution (see Section 3.4.3), the variance of S^2 is $2\sigma^4/(n-1)$. Thus, if σ^2 is much greater than, say, 20, the variance of S^2 will be greater, for practical sample sizes, than the parameter that it is trying to estimate. What does this suggest about the choice of sample size when the population variance is believed to be large?

2.28. The summary data given by Western Mine Engineering, Inc. (www.westernmine.com) from its 2001 Annual Mine Cost Service U.S. Mine Labor Survey included the following. For the 117 non-union mines in the survey, the range for Paid Life Insurance Coverage was 5K–120K, with an average of 39K. Since the average differs greatly from the middle of the interval, would the IQR have given a better or worse picture of the variability of the coverages? Explain.

2.29. Assume two datasets, with the second one obtained by multiplying each number in the first dataset by 0.9. If the coefficient of variation of the first dataset is a, what is the coefficient of variation for the second dataset as a multiple of a?

2.30. Find the median of the following numbers: 1.94, 2.96, 4.9, 4.12, 6.87.

2.31. The following definition of a statistic is given at an Internet site: "any number calculated from sample data, describes a sample characteristic." Can we accept this definition and still speak of the variance of a statistic? Explain.

2.32. Consider the data in the file delta.dat, which contains data on number of minutes that Delta flights departing from Atlanta departed late on June 30, 2002. If you want to use a measure to represent the typical number of minutes that a Delta flight from Atlanta departed the gate late on June 30, 2002, would you use the average, median, or trimmed average? Or, since these three numbers differ considerably for this dataset, would you use the interquartile range or some other measure? Explain. If the average were used, what is the population parameter that it would estimate and thus what is the population? Or is there an obvious population? Explain.

2.33. An employee needs to perform a quick calculation to find the average of 50 numbers, all of which are of the form 6.32xx. To simplify the computation, she visually multiplies each number by 10,000 and then subtracts 63,200 from the transformed number, thus leaving the "xx" part in integer form. If the average of the transformed numbers is 42.8, what is the average of the original numbers?

2.34. Show that $\sum_{i=1}^{n}(x_i - a)^2$ is minimized when $a = \bar{x}$.

2.35. Assume that $n = 40$ and $\sum_{i=1}^{n}(x - a)^2 = 612$. If $a = \bar{x} + 3$, what is the numerical value of $\sum_{i=1}^{n}(x - \bar{x})^2$?

2.36. Find the average and the median of the following numbers: 5, 8, 3, 4, 7, 2, 9.

2.37. The sample variance (s^2) is found to be 35.6 for a sample of 25 observations. If each of the original numbers were multiplied by 20 and then 10 was subtracted from each of the transformed numbers, what would be the value of s^2 for the transformed set of 25 numbers?

2.38. Explain what is wrong, if anything, with the following statement: "The coefficient of variation for a population with a mean of 25 and a standard deviation of 5 is $5/25 = 0.2$." Is the statement correct? Explain.

2.39. Which would be more adversely affected by outliers, the sample standard deviation, s, or a statistic defined as $\sum_{i=1}^{n} |x_i - \bar{x}|/\sqrt{n}$? Explain. Which statistic would you recommend for general use?

2.40. Assume that we have a set of numbers $a_i, i = 1, 2, \ldots, 50$. Explain why the sample variance of these numbers is the same as the variance of the numbers that result after each number is multiplied by -1.

2.41. Assume that we have 100 numbers and subsequently put them in ascending order.
   (a) Explain how the median would be computed.
   (b) If each of these 100 ordered numbers were multiplied by 1.5, what would be the relationship between the median of these transformed numbers and the median of the original numbers?

2.42. Consider a set of ten observations measured in yards. What will be the unit of measurement for the sample variance?

2.43. Construct a sample of size 3 for which the sample variance exceeds the mean.

2.44. List three units of measurement commonly found in engineering for which the square of the measurement would not have any physical meaning. What measure of variability would you recommend for each of these cases?

2.45. A set of data is transformed to make the data easier to work with. If the sample variance for the original data was 125.67 and the sample variance of the transformed data was 1.2567, what transformation was used? Could there be more than one answer to this question? Explain.

2.46. Explain why a few very large values in a sample would cause the mean to be larger than the median.

2.47. Assume you are given a sample of six numbers that are deviations of each original number from the average of the numbers. One of the numbers was given with the wrong sign, however. If the numbers are 2.3, 3.1, 4.2, -5.2, and 1.8, which number has the wrong sign?

2.48. Consider a sample of 10 positive numbers. If each number is greater than the square of the number, the average of the numbers must be less than what number?

2.49. Determine the median from the following stem-and-leaf display (given in Section 1.4.1).

$$
\begin{array}{l|llllll}
6| & 4 & 5 \\
7| & 2 & 2 & 3 & 4 & 8 & 9 \\
8| & 2 & 3 & 5 & 7 & 7 & 8 \\
9| & 2 & 3 & 6 & 7
\end{array}
$$

2.50. If the numbers in Exercise 2.49 are all multiplied by -1, what is the numerical value of the third quartile for the new set of numbers?

2.51. If the standard deviation of a set of 50 numbers is 6.23, what will be the standard deviation of the transformed numbers if each number is multiplied by 100 and then 600 is subtracted from each number that results from the multiplication?

2.52. Assume that 17 is the value of Q_1 in a sample of 15 numbers and $Q_3 = 87$. If the distribution of these numbers is symmetric about the median and the data are transformed using the transformation $y^* = a + by$, with the statistics given above being for y, what is one possible combination of a and b if the median of y^* is 8?

2.53. Assume that an engineering statistics class consists of 25 men and 10 women. If the median height of the women is 65.4 inches and the median height of the men is 71.6 inches, can the median height for the men and women combined be determined? Why or why not?

2.54. How would you describe a sample of 25 observations if the median was much closer to Q_3 than to Q_1, with Q_3 not being much greater than the median?

2.55. Consider the following sample of 24 observations: 10, 12, 13, 15, 18, 21, 24, 27, 29, 32, 33, 35, 36, 38, 42, 44, 45, 46, 48, 49, 51, 52, 53, and 56. Compute the mean and the standard deviation of the ungrouped data. Then put the data in classes of 10–19, 20–29, 30–39, 40–49, and 50–59 and compute the estimated mean and estimated standard deviation from the grouped data. Compare the results and comment.

2.56. Assume that a manufacturing manager uses the coefficient of variation to essentially define an acceptable value of σ for each process by comparing the coefficient of variation for each process to the coefficient of variation of a standard, well-behaved process. If the latter has $\sigma/\mu = 0.42$, to what value must σ be reduced for a second process to meet this standard if the mean of 16 is considered desirable and the present value of σ is estimated to be 8.3?

2.57. Cornell [*ASQC Statistics Division Newsletter*, **14**(1), 11–12, 1994] explained, in a lighthearted article, how a person could maximize the octane rating of the gas in his or her tank for a given cost, or conversely, minimize the cost for a given octane rating. Sample three service stations in your neighborhood and for each service station determine the mixture of 87, 89, and 92 (or 93) octane that you should put in your car to maximize the octane rating for a $15 purchase. (Assume for the sake of the illustration that a mixture of different octanes won't harm your engine.) Then determine how to obtain an octane rating of 88 for the lowest cost. Note that the problem involves weighted averages.

2.58. Apply Sheppard's correction to the calculation of the sample variance for the following dataset of 223 numbers consisting of the integers 1–100 plus the integers 20–80, 30–70, and 40–60, with the respective integers placed in four classes of equal widths. Compare the corrected value with the actual value and comment. If you consider the corrected value to be inferior to the uncorrected value, do you consider the problem to be that too few classes are used? Comment.

2.59. The American Statistical Association conducts salary surveys of its members. The results are published and are also available at the ASA website `http://www.amstat.org/profession/salarysurvey.pdf`. Assume that you want to analyze the variability in the salary of full professors at research universities and you want to see how the variability changes over time (i.e., over surveys). The individual salaries are not given, however, so a variance or standard deviation could not be computed. Furthermore, the class intervals (i.e., for the number of years at the rank of full professor) are unequal and the last class is open-ended (e.g., 33 or more years). The frequencies of each class are given, in addition to, Q_1, Q_2, and Q_3. Given this limited information, how would you compare the variability in the salaries over time?

2.60. Assume a department of five employees. Three of them are above the average age for the department by 1, 4, and 2 years, respectively; one person is equal to the average and the other person is younger than the average.
  (a) Can the average age be determined? Explain.
  (b) What is the variance of their ages?

2.61. As an exercise, a student computes the average of three numbers. One number is two units above the average and another number is equal to the average. What is the standard deviation of the three numbers?

2.62. Assume that you wish to compare the month-to-month variability in advertising expenditures for a large company and a small company in a particular city. What statistic would you use for the comparison?

2.63. Obtain the most recent salary information for professionals in your major field and, assuming it is in the form of class intervals with corresponding class frequencies, describe the distribution. Is the distribution at least close to being symmetric in regard to years in the profession? Explain.

2.64. Compute the sample covariance between X and Y when $(X, Y) = (1, 3), (2, 7), (4, 8),$ and $(5, 2),$ and $(6, 4)$.

2.65. Consider the (ordered) sample observations: 2, 8, 23, 24, 28, 32, 34, 38, 42, 44, 45, 55, 58, 61, 65, 68, 71, 72, 98, and 99. Compute the sample variance with

10% trimming (from each end), and compare this number with the sample variance computed without trimming. Comment.

2.66. Grade summaries for the second semester of the 2001–2002 academic year at Purdue University were available online at the time of writing. Given below are the average grade-point averages (GPA) for each type of engineering major and the number of students with each major. Determine the overall average GPA for all engineering majors.

Major	Number	Average GPA
Freshman engineering	1924	2.613
Aero & Astro	370	2.868
Chemical	379	2.984
Civil	399	2.938
Construction	142	2.886
Land surveying	6	2.804
Industrial	460	2.844
Interdisciplinary	87	2.886
Materials	60	2.962
Mechanical	784	2.862
Nuclear	71	2.871

2.67. The average of 10 numbers is 100. If the numbers 79 and 89 are removed from the group of numbers and the average of the eight remaining numbers is computed, what is the average of those numbers?

2.68. A company computes its average advertising expenditure (in thousands) by averaging over its seven divisions. The average is 168. Two of the divisions subsequently report that they believe they may have reported erroneous figures. The figures that they reported were 162 and 174. Consequently, if the average is temporarily reported using only the other five divisions, explain why the previously reported average will be unchanged.

2.69. The following statistics were given at the website for the U.S. Geological Survey (USGS), with the numbers pertaining to the USGS External Server and the statistics being for one particular month during the last quarter of 2002.

Successful requests	14,744,979
Average successful requests per day	491,522
Successful requests for pages	1,471,649
Average successful requests for pages per day	49,057

(a) Can the month for which the statistics were given be determined? If so, what is the month? If the month cannot be determined, explain why not.

(b) The averages are obviously rounded to the nearest integer. What was the units digit in each number before they were each rounded, if this can be determined? If it cannot be determined, explain why.

2.70. You are given the following 15 observations: 6.2, 6.8, 6.7, 7.0, 7.3, 7.5, 7.1, 7.6, 7.7, 7.2, 7.5, 7.7, 7.8, 8.0, and 7.9.

   (a) Compute the moving averages of size 5 and the moving ranges of size 2. Is there any advantage to using the moving averages instead of the overall average as an estimate of "current" location for this dataset?

   (b) In general, how many moving averages of size k and how many moving ranges obtained from "moving" groups of size w will there be for n observations?

2.71. Can a sample of size 2 be constructed for which the standard deviation and the variance are the same? If so, construct the sample. If it is not possible, explain why it can't be done.

2.72. The interquartile range is used by many organizations. The U.S. Energy Information Administration in a report for the year 2000 (that at the time of writing is available at `http://www.eia.doe.gov/cneaf/coal/cia/html/d1p01p1.html`) gave the interquartile range and average mine price for coal (in dollars per short ton) by state and by mine type (underground and surface) for coal-producing states and regions. The following numbers were given for underground mines in Kentucky. [As stated at that website, average mine price is calculated by dividing the total free on board (f.o.b.) mine value of the coal produced by the total production.]

	Average Mine Price	Interquartile Range
Total	$24.31	$3.46
Eastern	25.32	2.09
Western	21.42	1.85

Explain why the interquartile range is so much greater for the entire state than it is for either the eastern or western parts of the state.

2.73. Construct a sample of 10 observations for which the range is twice the interquartile range.

2.74. If a symmetric distribution of quiz scores for a small class has a mean of 8 and an interquartile range of 2, what is the score above which 25% of the scores fell?

2.75. The average salary paid to all employees in a company is $50,000. The average annual salaries paid to male and female employees were $52,000 and $42,000, respectively. Determine the percentage of males and females employed by the company.

2.76. Show that the numerator of the expression for the covariance given in Section 2.2 is equivalent to $\sum XY - \left(\sum X \sum Y\right)/n$.

CHAPTER 3

Probability and Common Probability Distributions

In Chapter 1 we saw how the distribution of the observations in a sample could be captured with a histogram. It is also important to consider the distribution of population values, as inferential procedures are based on the distribution that is assumed. In this chapter we present the distributions that are frequently used in engineering applications.

Probability calculations and distributions of probability are an integral part of engineering-based systems, as explained and illustrated by, for example, Haimes (2004). The author makes the important point that the "risk of extreme and rare events is misrepresented when it is solely measured by the expected value of risk" (p. 34). Thus, what is needed is the distribution of probability over the possible outcomes, not just an indication of what should happen "on average." In software engineering, it is useful to be able to approximate the number of bugs per thousand lines of code, as illustrated by Haimes (2004, p. 614).

Efficient use of probability distributions is important in various types of engineering work. For example, Hess (2004) laments the fact that particle size distributions are often not well-presented and discusses remedies for the problem.

Statistical distributions (i.e., the distribution of one or more random variables) are also frequently referred to as *probability distributions*, so to understand these distributions and the inferential procedures on which they are based, we must first study probability. After that, many common continuous and discrete statistical distributions are presented. Both types of distributions are used extensively in engineering. For example, semiconductor technology development and manufacture are often concerned with continuous random variables (e.g., thin-film thickness and electrical performance of transistors), as well as discrete random variables (e.g., defect counts and yield).

3.1 PROBABILITY: FROM THE ETHEREAL TO THE CONCRETE

During a typical day many of us will hear statements that are probability-type statements, although the word "probability" might not be used. One example (statement 1) is a weather forecast in which a meteorologist states: "There is a 10% chance of rain tomorrow." What does such a statement actually mean? For one thing, it means that, in the opinion of the meteorologist, it is very unlikely that rain will fall tomorrow. Contrast that with the following statement (statement 2): "If a balanced coin is tossed, there is a 50% chance that a head will be observed." Disregarding for the moment that the coin could land on its edge (an unlikely possibility), there are two possible outcomes, a head or a tail, and they are equally likely to occur. Finally, consider the following statement (statement 3): "I found a slightly bent coin in the street. I am going to toss this coin 500 times and use the results of this experiment to determine an estimate of the likelihood of obtaining a head when the coin is tossed once. Thus, if I observe 261 heads during the 500 tosses, I estimate there is a 52.2% (=261/500) chance of observing a head when the coin is tossed once."

There are some important differences between these three statements. Statement 1 has to be at least somewhat subjective since it is not possible to repeat tomorrow 500 times and observe how many times it rains, nor is it possible to know the exact likelihood of rain as in the case of the balanced coin. The second and third statements are illustrative examples of the two approaches that will be discussed in succeeding chapters: (1) acting as if the assumption is valid (i.e., the coin is balanced) and (2) not making any assumption but rather collecting data and then drawing some conclusion from the analysis of that data (e.g., from the 500 tosses of the coin). The latter approach is obviously preferable if there is any question as to whether or not the assumption is valid, and if the consequences of making a false assumption are considerable. Unfortunately, statistical assumptions are often not checked and the consequences can be extreme. Methods for checking assumptions are illustrated in succeeding chapters. In general, assumptions should always be checked unless there is prior information to suggest that the assumption is valid.

The word *probability* has not as yet been used in this section; instead *percent chance* has been used. The two terms can be thought of as being virtually synonymous, however. The first statement that was given (the weather forecast) is essentially a subjective probability statement. Statement 2 could be expressed concisely as

$$P(\text{head}) = \frac{1}{2}$$

which is read as "the probability of a head equals 1/2" on a single toss of the balanced coin. Thus, a 50% chance is equivalent to a probability of one-half.

Just as percentages must range from 0 to 100, the probability of some arbitrary "event" (such as observing a head) must be between 0 and 1. An "impossible event" (such as rolling a seven on a single die) would be assigned a probability of zero. The converse is not true, however: if an event is assigned a probability of zero, it does not mean that the event is an impossible event. As mentioned previously, a coin *could* land on its edge (and most assuredly will if it is tossed enough times), but we customarily assign a probability of zero to that possible event.

With statement 3, no probability was assumed; instead, it was "estimated." In practice, this is customary since practical applications of probability go far beyond tossing a balanced coin or rolling a single die. In this instance, the probability of observing a head on a single toss of the misshapen coin was estimated by tossing the coin a large number of times and counting the number of heads that was observed. Are 500 tosses adequate? That depends on the degree of accuracy required in estimating the true probability. The idea of determining the number of trials in an experiment from a stated error of estimation will not be pursued here, but is discussed in subsequent chapters, especially Section 5.8. In general,

$$\frac{x}{n} \to p \quad \text{as } n \to \infty$$

with x denoting the number of times that the event in question occurs, n denoting the number of trials, and p being the true probability that the particular event will occur on a single trial. The symbol $\to$ should be read as "approaches" and $n \to \infty$ designates the number of trials becoming large without bound.

3.1.1 Manufacturing Applications

How can these concepts be applied in a manufacturing environment? Assume that a particular plant has just opened and we want to estimate the percentage of nonconforming units of a particular product that the process is producing. How many items should we inspect? We would certainly hope that the percentage of nonconforming units is quite small. If it is, and if we were to inspect only a very small number of units, we might not observe any nonconforming units. In that case our estimate of the percentage of nonconforming units produced would be zero (i.e., $x/n = 0$), which could certainly be very misleading. At the other extreme, we could inspect every unit that is produced for a particular week. This would be rather impractical, however, if the production item happened to be steel balls and thousands of them were produced every week. Consequently, a compromise would have to be struck so that a practical number of items would be inspected. For a reasonable number of items to be inspected, the percentage of nonconforming items that the process is producing could then be *estimated* by dividing the number of nonconforming units observed by the total number of items that were inspected (i.e., x/n). That is, we are using a *relative frequency approach*.

DEFINITION

In the *relative frequency approach* to probability, the probability of some event is estimated by the number of times that the event occurred in n trials, divided by n.

3.2 PROBABILITY CONCEPTS AND RULES

The preceding section contained a discussion of the different ways of viewing probability. In this section we consider basic probabilistic concepts and rules for determining probabilities. In the preceding section the word "event" was used in regard to the coin-tossing example. We compute probabilities for events, so events must obviously be clearly defined. But first we must have an *experiment*—such as tossing a coin. This leads to the *sample space*, which is the collection of all possible outcomes. For example, if the experiment is to toss a coin, there are two possible outcomes: head and tail and the sample space for the experiment could be denoted as $S = \{\text{head, tail}\}$.

DEFINITIONS

In general, the *sample space* for an experiment contains all possible values of a random variable, which can be listed only if the sample space is finite. An *event* contains one or more elements of the sample space.

Thus, if the experiment were to roll a single die and the event is an even number, then $P(\text{event}) = 3/6 = 1/2$. Thus, the event contains three elements of the sample space and is said to be a *composite event*. In both of these examples the *relative frequency* definition of probability was invoked. That is, the probability of the event is the number of times the event can be expected to occur "relative" to the set of possible outcomes.

There are rules for determining probabilities associated with multiple events. To use a simple illustration, assume that the experiment consists of tossing a coin and rolling a die and we want the probability of observing a 6 on the die and a head on the coin. Does the outcome of the coin toss have any influence on the outcome of the roll of the die? Obviously not, so the events are independent. That is, the outcome for one part of the experiment has no effect on the outcome for the other part. In general, with A and B denoting two independent events, $P(A \text{ and } B)$, which we can represent equivalently as $P(A \cap B)$ or $P(AB)$, is given by

$$P(A \text{ and } B) = P(A)\,P(B) \qquad (3.1)$$

DEFINITION

Two events, A and B, are *independent* if the occurrence or nonoccurrence of one of them has no effect on the probability of occurrence of the other one. Then

$$P(A \cap B) = P(A)\,P(B).$$

Thus, for the experiment consisting of rolling a die and tossing a coin, $P(\text{head on coin}$ and 6 on die$) = P(\text{head on coin}) \times P(6 \text{ on die}) = (1/2)(1/6) = 1/12$. If we listed the sample space for the experiment, we would see that there are 12 possible, equally likely outcomes, and we want the probability of one of those outcomes occurring. Thus, the probability is $1/12$. Of course, outcomes of an experiment will not always be equally likely, but when they are, the answer is both easy to compute and intuitive.

If we want the probability of at least one of the two events occurring, we obviously need a new formula. If two events cannot occur at the same time, it stands to reason that the probability of either one of them occurring should be the sum of their respective probabilities. That is, $P(A \text{ or } B)$, which is often written $P(A \cup B)$, is given by $P(A \text{ or } B) = P(A) + P(B)$. For example, if we toss a single coin and $A = $ coin is a head and $B = $ coin is a tail, clearly $P(A \text{ or } B) = P(A) + P(B) = 1/2 + 1/2 = 1$ since these are the two possible outcomes, and the probability associated with the entire sample space is 1. (Of course, a coin could land on its edge—and will eventually if we toss it an enormous number of times, as stated previously—but the probability of this occurrence is infinitesimally small and is therefore taken to be zero. This illustrates the fact that an event that is assigned a probability of zero is not necessarily an impossible event, as stated previously, but an impossible event, such as the probability of observing a 7 on a die, does have a probability of zero.)

When we set out to determine $P(A$ or $B)$, we start with the general expression given in the following definition.

DEFINITION

For any two events A and B, $P(A$ or $B) = P(A) + P(B) - P(A$ and $B)$ with $P(A$ and $B)$ determined by whether A and B are independent or dependent.

The numerical value of $P(A$ and $B)$ will be determined by the relationship between the two events. As indicated previously, if the two events cannot occur simultaneously, then the right side simplifies to $P(A) + P(B)$, which implies that $P(A$ and $B) = 0$. The latter is expressed in the following definition.

DEFINITION

If two events are *mutually exclusive* (i.e., they cannot occur simultaneously), the probability of their simultaneous occurrence is 0. Relative to independence, we may think of mutually exclusive as being a form of dependency.

If, however, the events can occur together—as in the example with a die and a coin—then $P(A$ and $B)$ is nonzero. For that experiment, $P(A$ and $B) = P(A) P(B)$ because the events were independent. *When the two events are independent, $P(A$ or $B)$ becomes*

$$P(A \text{ or } B) = P(A) + P(B) - P(A)P(B) \tag{3.2}$$

For example, $P(\text{head or a } 6) = P(\text{head}) + P(6) - P(\text{head and a } 6) = 1/2 + 1/6 - 1/12 = 7/12$. That is, if we count the number of elements in the sample space that have either a head or a 6, we will see that the number is 7: the six that occur with a head, including a 6 on the die, plus a 6 or the die occurring with a tail. Note that if we didn't subtract $P(A \cap B)$, we would be counting one point in the sample space twice, namely, {head, 6}, since the point is part of the sample space corresponding to both events A and B.

The reader will encounter the term *joint probability* in a few places in subsequent chapters (such as in Section 4.4.1), so we can mention that $P(\text{head and a } 6)$ in the preceding illustration is a joint probability in that it is the probability of the joint occurrence of two events. In general, a joint probability is the probability of the occurrence of two or more events.

Two events need not be either mutually exclusive or independent, with mutually exclusive being a form of dependence, as stated previously. Let $P(B|A)$ represent the probability that event B occurs given that event A has occurred. For mutually exclusive events, $P(B|A) = 0$, but the probability won't necessarily be zero when two events are dependent. For example, assume that we roll a die and it rolls out of sight. Someone locates the die and states that the outcome was at most a 4. Let B denote "3 on the die" and A denote "at most 4 on the die." Then it should be apparent that $P(B|A) = 1/4$. What has happened is the sample space has been reduced from six possible outcomes to four by conditioning on what was observed.

This is an example of a *conditional probability* problem, and for this example we can easily reason out the answer without having to formally compute it. This will not always be the case, however. Therefore, we need to consider how $P(B|A)$ is computed in such situations. Specifically,

$$P(B|A) = \frac{P(A \text{ and } B)}{P(A)} \tag{3.3}$$

To see that this represents a reduced sample space, we use the relative frequency definition of probability so that $P(A \text{ and } B) = n(A \text{ and } B)/N$, the number of elements in the sample space in the intersection of statements A and B divided by the number of elements in the sample space (N). If we define $P(A)$ similarly, we can see that the expression for $P(B|A)$ reduces to $n(A \text{ and } B)/n(A)$. Then the notion of a reduced sample space becomes readily apparent.

If we solve for the numerator in the fraction of Eq. (3.3), we obtain $P(A \text{ and } B) = P(A) P(B|A)$. This is often referred to as the *multiplication rule*, which is simply the way that we obtain $P(A \text{ and } B)$ when the events are neither independent nor mutually exclusive. It then follows that one applicable expression for $P(A \text{ or } B)$ is

$$P(A \text{ or } B) = P(A) + P(B) - P(A) P(B|A)$$

As an example, let's say that you as a student have two final exams on the same day. You would like to study at least 20 hours for each exam, but there is hardly enough time available to do so, and you will not begin studying for the second exam until you have finished studying for the first exam. Let $A = $ study 20 hours for first exam, let $B = $ study 20 hours for second exam, and $P(B|A) = .10$. If $P(A) = .60$ and $P(B) = .20$, then $P(A \text{ or } B) = .60 + .20 - (.60)(.10) = .74$. Of course, the logical question to ask is: "Can we determine $P(B)$ without knowing whether or not A has occurred?" We can always construct such examples, but do they make any sense? Clearly, there are instances in which it is logical to determine both $P(B)$ and $P(B|A)$ when we have a clearly defined sample space. But such problems exist more in textbooks than in real life. In this example, the student might be able to guess at $P(B)$ a priori, but it would obviously be easier to determine $P(B|A)$. Of course, there are ways to solve practical probability problems using prior data, without having to rely too heavily on judgment. Engineering problems of this nature are addressed in Section 3.2.1.1.

3.2.1 Extension to Multiple Events

Solving practical problems often involves working with more than two events. The rules for two events given in the preceding section can be extended to more than two events. For example, if we wanted to obtain $P(A \cup B \cup C)$, we would compute it as $P(A \cup B \cup C) = P(A) + P(B) + P(C) - P(A \cap B) - P(A \cap C) - P(B \cap C) + P(A \cap B \cap C)$. In general, for k events $A_1, A_2, \ldots, A_k$,

$$P(A_1 \cup A_2 \cdots \cup A_k) = \sum_{i=1}^{k} P(A_i) - \sum_{i<j}^{k} P(A_i \cap A_j) + \cdots$$

with the signs alternating. Similarly, if there is joint independence of the events, then

$$P(A_1 \cap A_2 \cdots \cap A_k) = P(A_1)P(A_2) \ldots P(A_k)$$

3.2.1.1   Law of Total Probability and Bayes' Theorem

The *law of total probability* is an important result that is used in Bayes' theorem, which in turn can be used to solve some important engineering problems.

We will illustrate both Bayes' theorem and the law of total probability with the following example.

■   **EXAMPLE 3.1**

Setting

Assume that there is a stage in a particular manufacturing process in which a unit of production can go through any one of three identical machines. Measurements are made at the next stage of the production process to determine if the unit is still in conformance with standards after the processing stage in which it passes through one of the three machines. Each of the machines has an operator, and although the machines are identical, the operators of course are not. A process engineer finds a nonconforming unit at the start of the morning shift. This is cause for alarm since nonconforming units are observed rather infrequently and especially would not be expected to occur at the start of a shift. In particular, nonconforming units have comprised only 0.015%, 0.022%, and 0.018% of the units produced by machines 1, 2, and 3, respectively. Since the operator for the second machine is relatively new, that machine produces only 28% of the total output, compared to 34% and 38% for machines 1 and 3, respectively. If a machine is not functioning properly, that should be detected as quickly as possible, but which machine should be checked first?

Determining Conditional and Unconditional Probabilities

The answer probably isn't obvious. (Actually, we can answer the question without performing any probability calculations, as will be explained later, but the direct way to solve the problem is to use a probabilistic approach.) We have observed a nonconforming unit, so we will let that event be represented by N, and will let M_1, M_2, and M_3 denote the three machines. We need to convert the given percentages into probabilities and we can do that as follows. We have $P(M_1) = .34$, $P(M_2) = .28$, and $P(M_3) = .38$ from the output percentages that were given for each machine. Furthermore, we also have $P(N|M_1) = .00015$, $P(N|M_2) = .00022$, $P(N|M_3) = .00018$. We need to determine $P(M_i|N)$ for $i = 1, 2$, and 3, and first check the machine that corresponds to the highest probability.

Note that what we need to determine is the reverse of what we are given, as we were given $P(N|M_i)$ and we are trying to find $P(M|N_i)$. Where do we begin? The logical starting point is to use what we have learned. That is, we can use Eq. (3.3), which leads to

$$P(M_i|N) = \frac{P(M_i \cap N)}{P(N)}$$

An obvious problem that we face at this point is that we were not given either of the two components in this fraction. Consequently, we have to solve for them. Using the

1, H	2, H	3, H
4, H	5, H	6, H
1, T	2, T	3, T
4, T	5, H	6, T

Figure 3.1   Illustration of law of total probability.

multiplication rule presented in Section 3.2, we have

$$P(M_i \cap N) = P(N|M_i)P(M_i)$$

and for each value of i we know the probability for each expression on the right-hand side because that information was given. Now we must determine $P(N)$, and the solution is not obvious. A graph may help, so consider Figure 3.1, which relates to the example with the die and coin.

The set H, which is represented by the first two rows of the table, would represent a head when the coin is tossed. Clearly if a head results, it must occur with either a 1, 2, 3, 4, 5, or 6 and if we add together the areas represented by $H \cap 1, H \cap 2, \dots, H \cap 6$, we obtain H.

When we convert this to a probability statement, we obviously have

$$P(H) = P(H \cap 1) + P(H \cap 2) + \cdots + P(H \cap 6)$$

This illustrates the *law of total probability*, which states that if an event must occur with one and only one of a set of mutually exclusive events, and one of these mutually exclusive events must occur, the sum of the probabilities of the intersections must equal the first event. This is both intuitively apparent and can be seen from Figure 3.1.

Computing the Bayes Probabilities

If we apply this result to the machine problem, we thus have that $P(N) = \sum_{i=1}^{3} P(N \cap M_i)$. Since $P(N \cap M_i) = P(N|M_i)P(M_i)$, when we put the pieces together we have

$$P(M_i|N) = \frac{P(N|M_i)P(M_i)}{\sum_{i=1}^{3} P(N|M_i)P(M_i)} \tag{3.4}$$

Performing the calculations using the probabilities given in the preceding section shows that the three probabilities, for $i = 1, 2,$ and 3, are .28, .34, and .38, respectively. Therefore, machine #3 is the one that should be checked first.

I stated previously that we could reach this conclusion rather quickly, and in fact without having to do any extensive calculations. It is clear that machine #3 is more likely than machine #1 since #3 has both a higher proportion of nonconforming units and accounts for more of the output. So it is between #2 and #3. The latter accounts for a higher fraction of the output but the former has a higher proportion of nonconforming units. But .38/.28 > .00022/.00018, so "clearly" the numbers point to machine #3. If $P(N|M_3)$ had been .000162 instead of .00018, then the two ratios would have been .38/.28 = .00022/.000162 = 1.36, and we would have had $P(M_3|N) = P(M_2|N) = .35$.

This is not to suggest that problems of this type should be solved in the ad hoc manner just described. Rather, the intent is to try to provide some insight into what is going on "behind the scenes." The formal way to work the problem is to work it the first way; that is, to use *Bayes' theorem*, which is what is given in Eq. (3.4). (The theorem was originally given in a paper by the Reverend Thomas Bayes, who presented it to the Royal Society in 1763.)

DEFINITION

Bayes' Theorem

Assume a collection of mutually exclusive and exhaustive events $A_i, i = 1, 2, \ldots, k$, such that all of these events have a probability greater than zero of occurring and one of the A_i must occur. Then for any other event B for which $P(B) > 0$,

$$P(A_i|B) = \frac{P(A_i \cap B)}{P(B)} = \frac{P(B|A_i)}{\sum_{i=1}^{k} P(B|A_i)P(A_i)} .$$

Bayes' theorem has considerable applicability to problems in engineering and science. Other applications include medical applications, in which one might want to know the probability that a person has a particular disease given that a person tests positive with a particular test, using information about the effectiveness of the test when a person does have the disease. In all problems in which Bayes' theorem is applicable, we are trying to determine the reverse of what is given, and we are usually going backward in time. For example, with the nonconforming units example, we were computing the probability that the unit *came* from a particular machine, *after* the unit had already been observed.

3.3 COMMON DISCRETE DISTRIBUTIONS

A discrete random variable was defined in Section 1.3. When we speak of the *distribution* of a discrete random variable, we are referring to how the total probability of 1.0 for any statistical distribution (also called a probability distribution) is distributed among the possible values of the random variable. That is, for each manner in which a random variable might be defined, there is a probability distribution for that random variable, which for a discrete random variable is called a *probability mass function (pmf)*.

DEFINITION

A *probability mass function (pmf)* is a function with nonnegative values that gives the probability for each of the possible values of a discrete random variable.

The *cumulative distribution function* (*cdf*) gives the cumulative probability up to a specified value of the random variable. (Note that this term is used rather than "cumulative mass function.") This concept will be illustrated later.

For example, when a coin is tossed, the total probability of one is distributed equally over the two possible outcomes, head and tail, so that $P(\text{head}) = P(\text{tail}) = .50$. If $X =$ outcome on the toss, with head $= 1$ and tail $= 0$, we have

$$f(x) = \frac{1}{2} \qquad x = 0, 1$$

as the probability mass function. This is the simplest probability distribution or probability mass function and is called the *discrete uniform distribution* (mass function). A random variable generally has more than two possible values, however, with the total probability of one being spread over all possible values in accordance with the likelihood of occurrence of each possible value.

■ **EXAMPLE 3.2**

A random variable assumes the values -1, 0, and 1 with probabilities $k/4$, k, and $k/2$, respectively. What is the value of k?

Solution   Since the probabilities must sum to one, it follows that $(7/4)k = 1$ so $k = 4/7$. Thus, the probabilities are $1/7$, $4/7$, and $2/7$, respectively.   ■

Now assume that we have

$$f(x) = \frac{1}{4} \qquad x = 1, 2, 3, 4$$

The cumulative distribution function, $F(x)$, would be obtained as $F(x) = \sum_1^x (1/4)$, so that $F(1) = 1/4$, $F(2) = 2/4$, $F(3) = 3/4$, and $F(4) = 1$.

In the sections that follow, we present the most common discrete distributions of the hundreds that exist. As you study these distributions, it is very important to keep in mind that these distributions serve only as models of reality; they do not capture reality exactly. Only in textbook examples can we even begin to contemplate the possibility that the distributions that are used do indeed capture reality. Even then, there are questions that must be addressed. For example, is $P(\text{head})$ really .5 even though the head and shoulders that are on a penny would seemingly cause it to be heavier than the tail side, which doesn't have anything that is raised? (For a dissenting opinion, see Gelman and Nolan (2002).) Furthermore, could we toss a penny exactly the same way each time so that the probabilities are what they are assumed to be? For nonconforming units, if one unit is nonconforming in coming off an assembly line, is the probability that the next unit is nonconforming likely to be unaffected by what has just been observed? Not likely, even though this is often assumed. If we decide that there is a dependency, what is the nature of the dependency and how do we model it?

The important point is that we can never assume that the distribution we are using is appropriate, even though it might seem as though it *must* be appropriate. When we select a distribution we are making an assumption, and assumptions in statistics *must almost always be checked*. This point cannot be overemphasized and indeed will be repeated often throughout the remainder of the book.

Accordingly, when a distribution is presented, methods of testing the adequacy of the fit (to a scenario or to data) will be discussed.

3.3.1 Expected Value and Variance

We often need to know the expected value and variance (and/or standard deviation) for a discrete random variable, as well as for a continuous random variable. We can think of expected value as being a long-term average, as well as being our best guess as to what should happen on a single outcome, if the expected value is a possible outcome.

To illustrate, if a basketball player is fouled and shoots two free throws, how many would we expect him to make? Since the events of shooting each free throw are independent (at least that is the logical assumption), we can use Eq. (3.1) to determine the probability that he makes both free throws, and also compute the probability that he misses both free throws. If he is a 70% free-throw shooter, it follows that the probability that he makes both free throws is $(.70)(.70) = .49$, and the probability that he misses both free throws is thus $(.30)(.30) = .09$. It follows then that the probability of him making one of the two is $1 - .09 - .49 = .42$. Thus, our best guess as to what would happen on a single pair of free throws is that he makes both of them.

But if we define the random variable as $X =$ number of free throws made out of the pair attempted, the expected value of X, written $E(X)$, is not 2. Rather, $E(X)$ is, for a discrete random variable, a weighted average of the possible outcomes of X, with the weights being the probabilities associated with the possible outcomes. Thus, for this example, $E(X) = 0(.09) + 1(.42) + 2(.49) = 1.40$. Thus, if this player were to be fouled in the act of shooting a very large number of times, we would expect that the average of the number of free throws made for every pair attempted would be approximately 1.40. Note that 1.40 is 70% of 2, which is as we would expect since he is a 70% free-throw shooter. That is, on average he scores 70% of the possible number of points when he shoots the pair of free throws.

DEFINITION

Formally, we obtain the *expected value of a discrete random variable* as

$$E(X) = \sum_x x P(x)$$

with the sum going over all possible values of X and $P(x)$ denoting the probability that x is observed. The $E(X)$ is denoted by μ.

In the discrete case, expected value should be very intuitive if we view it as a long-term (expected) average. If we roll a die and do it a million times, the average number of spots observed over the one million rolls will be virtually 3.5, which is also the average of the integers 1–6. If we toss two coins, we would expect to observe one head, this being both the expected value as well as what we would expect for a particular toss of the two coins since the expected value is, in this case, also a possible value.

■ **EXAMPLE 3.3**

The number of memory chips, M, that are needed in a personal computer depends on the number of applications programs, A, that the owner intends to run simultaneously. Assume that the probability distribution of A, $P(A)$, over all owners is given by

$$P(a) = 0.05(7.5 - a) \qquad a = 1, 2, 3, 4$$

What is the expected number of applications programs, with the expectation of course going over all owners?

Solution

$$E(A) = \sum_{a=1}^{4} a[0.05(7.5 - a)] = 0.375 \sum_{a=1}^{4} a - 0.05 \sum_{a=1}^{4} a^2$$

$$= 0.375(10) - 0.05(30) = 2.25$$

If 4 chips are required to run two programs simultaneously and 6 chips are required for three programs, a new computer owner who isn't sure what her needs will be would be wise to have at least 6 memory chips. ■

Expected value for a continuous random variable follows the same general principle, but not the same mechanics since the outcomes are infinite and thus cannot be enumerated. Expected value for continuous random variables is discussed in Section 3.4.1.

To find the *variance* of a random variable, denoted by σ^2, we would want to measure the spread of the possible values about μ. Accordingly, we define $\sigma^2 = E(X - \mu)^2$. For a discrete random variable, $E(\text{"anything"}) = \sum_x (\text{"anything"}) P(x)$. That is, whatever expression we are taking the expected value of becomes the expression before $P(x)$ in the summation. It can be shown that $E(X - \mu)^2 = E(X^2) - (E(X))^2$, as the reader is asked to show in Exercise 3.20. The second expression is generally easier to use since the mean doesn't have to be subtracted from each possible value of X.

The standard deviation, σ, is of course the square root of the variance.

DEFINITION

The *variance* of a random variable, either discrete or continuous, is given by $\sigma^2 = E(X - \mu)^2$. In the discrete case, this is computed as either $\sum (X - \mu)^2 P(x)$ or $\sum X^2 P(x) - (\sum X P(x))^2$, with each sum over all possible values of the random variable X. The *standard deviation*, σ, is the square root of the variance.

It is important to distinguish between the type of mean and variance discussed in Chapter 2, which were for a sample and a population, and the mean and variance of a probability distribution. If we had a finite population, there is the possibility, at least conceptually, of computing the population mean and variance. In this chapter we are concerned with the mean and variance of a random variable for an underlying probability distribution. We obviously cannot determine the mean and variance of a random variable that has a certain distribution without using the function that represents the distribution, whereas we can calculate a population mean or variance, if practical, without reference to *any* probability distribution. Thus, a sample mean, population mean, and the mean of a random variable that has a certain probability distribution are all quite different, and of course this is also true for a variance.

3.3.2 Binomial Distribution

This is one of the most frequently used discrete distributions. It is presented in detail in this section because of its wide applicability and also because it is used with control charts and other statistical procedures that are presented in later chapters.

The binomial distribution can be used when the following conditions are met.

1. When there are two possible outcomes (e.g., heads and tails). These outcomes are arbitrarily labeled *success* and *failure*, with the outcome that is labeled "success" being the one for which one or more probabilities are to be calculated. There is no intended connotation of good versus bad. For example, if probabilities for various numbers of nonconforming items are to be computed, "nonconforming" is labeled "success."

2. There are *n* trials (such as *n* tosses of a coin) and the trials are independent. The trials are called *Bernoulli trials*, being named for a prominent mathematician, James Bernoulli (1654–1705), who proposed his binomial distribution that was later published in 1713. The independence assumption means that the probability of an outcome for a particular trial is not influenced by the outcome of any preceding trials. Furthermore, the probability of a success on a single trial does not vary from trial to trial.

Clearly, a coin-tossing experiment conducted to determine the true probability of the coin landing on "heads" meets these requirements. Recall the discussion at the beginning of Section 3.3, which suggests that such an experiment might indeed be necessary, rather than just assuming that the probability is .50. Nevertheless, for the sake of illustration we shall assume that the probability is .50. The probabilities of observing 0, 1, and 2 heads will now be obtained in a somewhat heuristic manner. It should be apparent that whatever happens on the second toss of the coin is independent of the outcome on the first toss. As discussed in Section 3.2, when two events are independent, the probability of both of them occurring is equal to the product of their separate probabilities of occurrence. Thus, if H_1 represents a head on the first toss and H_2 represents a head on the second toss, then

$$P(H_1 \text{ and } H_2) = P(H_1)P(H_2)$$
$$= (1/2)(1/2)$$
$$= 1/4$$

Similarly, it could be shown that the probability of two tails (zero heads) equals 1/4. One head could be observed in one of two ways, either on the first toss (followed by a tail) or on the second toss (preceded by a tail). Since the probability for each sequence is 1/4, the probability of observing one head is equal to the sum of those two probabilities, which is 1/2.

If we define the random variable X as

$$X = \text{number of heads observed}$$

we can put these probabilities together to form the probability distribution of X. If we let $P(X)$ represent "the probability of X" (with X assuming the three different values), we then have the following:

X	$P(X)$
0	1/4
1	1/2
2	1/4

Thus, if this experiment were repeated an extremely large number of times, we would theoretically expect that 1 head would occur 50% of the time, 2 heads would occur 25% of the time, and 0 heads would also occur 25% of the time.

Assume that 16 Bernoulli trials were conducted and the observed numbers for 0, 1, and 2 heads were 5, 7, and 4, respectively. Obviously, these are close to the theoretical frequencies of 4, 8, and 4, respectively. Although the theoretical frequencies constitute our "best guess" of what should occur, we should not be surprised if the observed frequencies differ somewhat from the theoretical frequencies. In fact, it may be apparent that the observed frequencies could never equal the theoretical frequencies unless the number of experiments was a multiple of 4. (If the number of experiments is not a multiple of 4, the theoretical frequencies for 0 heads and 2 heads would not be an integer.) Even if the number of experiments is a multiple of 4, we should not expect the observed frequencies to be equal to the expected frequencies because of random variation. The important point, however, is that the difference between the observed and theoretical frequencies should become very small as the number of experiments becomes very large.

The way that the probabilities were found for the coin-tossing experiment would certainly be impractical if the number of tosses, n, was much larger than two. In virtually any practical application, n will be much larger than two, so there is clearly a need for a general formula that can be used to obtain binomial probabilities. The following symbols will be used.

$p = $ the probability of observing a success on a single trial
$1 - p = $ the probability of *not* observing a success on a single trial (i.e., a failure)
$n = $ the number of trials
$x = $ the number of successes for which the probability is to be calculated

Regardless of the size of n, it is easy to find $P(x)$ when $x = 0$ or $x = n$. There is only one way to observe either no successes or all successes. Therefore, since the trials are independent, $P(n)$ is simply p multiplied by itself n times (i.e., p^n). Similarly, $P(0) = (1 - p)^n$. It is by no means obvious, however, what $P(x)$ equals when x is equal to neither 0 nor n.

If we wanted the probability of x successes followed by $n - x$ failures, that would clearly be

$$p^x(1 - p)^{n-x} \tag{3.5}$$

If, instead, we just wanted the probability of x successes without regard to order, the answer would obviously be larger than what would be produced by Eq. (3.5). For example, if you tossed a coin 10 times, there are many different ways in which you could observe 4 heads and 6 tails, one of which would be 4 heads followed by 6 tails.

In general, the number of ways that x successes can be observed in n trials is $\binom{n}{x}$, which is defined as

$$\binom{n}{x} = \frac{n!}{x!(n-x)!} \tag{3.6}$$

where $n! = 1 \cdot 2 \cdot 3 \cdot 4 \cdots n$. For example, if $n = 5$ and $x = 3$, then

$$\binom{5}{3} = \frac{5!}{3!2!}$$
$$= \frac{120}{(6)(2)}$$
$$= 10$$

If such computation is to be done by hand (although hand computation is rather old-fashioned, in general), it is easier to first simplify the quotient by dividing the larger of the two numbers in the denominator into the numerator. Thus,

$$\frac{5!}{3!} = \frac{1 \cdot 2 \cdot 3 \cdot 4 \cdot 5}{1 \cdot 2 \cdot 3} = 4 \cdot 5 = 20$$

so that

$$\frac{5!}{3!2!} = \frac{20}{2} = 10$$

By putting Eq. (3.5) together with Eq. (3.6), we have the following general expression for the probability of x successes in n trials:

$$P(x) = \binom{n}{x} p^x (1-p)^{n-x} \tag{3.7}$$

Although it was easy to find the probabilities of observing 0, 1, and 2 heads without using Eq. (3.7), the direct approach would be to use the latter. Thus, we could have found the probability of observing one head as follows:

$$P(1) = \binom{2}{1} (.5)^1 (.5)^1$$
$$= 2(.5)(.5)$$
$$= .5$$

Note that if we were to attempt to determine the probability of 2 heads as

$$P(2) = \binom{2}{2}(.5)^2(.5)^0$$
$$= \frac{2}{2!0!}(.25)(1)$$

we would have to know what to do with $0!$. It is defined to be equal to one, and it should be apparent that if it were defined in any other way, we would obtain an incorrect answer for this example! Specifically, we *know* that $P(2) = (.5)(.5) = .25$; therefore, we must have

$$P(2) = \frac{2!}{2!0!}(.25)(1)$$
$$= 1(.25)(1)$$
$$= .25$$

which will result only if $0!$ is defined to be equal to one, as it is.

■ **EXAMPLE 3.4**

Problem Statement

A practical problem might be to determine the probability that a lot of 1000 capacitors contains no more than one nonconforming capacitor, if as in the very recent past, 1 out of every 100 capacitors produced is nonconforming. There are clearly two possible outcomes (conforming and nonconforming), but are the "trials" independent so that the probability of any particular capacitor being nonconforming does not depend on whether or not any of the previously produced capacitors were nonconforming? This is the type of question that must be addressed for any manufacturing application, and applications of the binomial distribution, in general.

Assumption, Model, and Computations

If this assumption is not valid, the binomial distribution would be of questionable value as a model for solving this problem. For this example we shall assume that the assumption is valid. The words *no more than one* indicate that we should focus attention upon 0 nonconforming units and 1 nonconforming unit and add the two probabilities together. Thus,

$$P(0) = \binom{1000}{0}(.01)^0(.99)^{1000}$$
$$= (.99)^{1000}$$
$$= .000043$$

and

$$P(1) = \binom{1000}{1}(.01)^1(.99)^{999}$$
$$= .000436$$

Therefore, the probability of no more than one nonconforming unit in a lot of 1000 capacitors is $.000043 + .000436 = .000479$.

Both of these individual probabilities are quite small, but that is due primarily to the fact that X (= the number of nonconforming capacitors) could be any one of 1001 numbers (0–1000). Therefore, we should not expect any one probability to be particularly high. ∎

If the binomial distribution is a probability distribution (as it is), all of the probabilities will add to one. It can be shown with some algebraic manipulation that

$$\sum_{x=0}^{n} \binom{n}{x} p^x (1 - p)^{n-x} = 1$$

for any combination of n and p. The key is recognizing that the sum represents the binomial expansion $[1 + (1 - p)]^n$, and of course this expression is obviously equal to 1. (You should recall that for the coin-tossing experiment with $n = 2$ and $p = 1/2$, the probabilities were 1/4, 1/2, and 1/4, which obviously do add to 1.) It should also be noted that the binomial distribution is a *discrete* probability distribution because the random variable X can assume only integer values.

Expected value for a binomial distribution was indirectly illustrated in Section 3.3.1, but we would like to obtain $E(X)$ without having to enumerate all of the possible values of X, as was done in that example. Therefore, we need a simpler way of determining $E(X)$ for the binomial distribution, and similarly for other distributions.

There are two commonly used ways of obtaining the expected value, one of which is quite easy to use. The Bernoulli trials that were mentioned previously correspond to the Bernoulli distribution. A Bernoulli random variable has two possible outcomes: 0 and 1. If we let 0 represent failure and 1 represent success, the expected value of the Bernoulli random variable is $(1)(p) + (0)(1 - p) = p$. Since the trials are independent, it follows that for the sum of n independent trials,

$$E(X) = np$$

For the coin-tossing experiment, $n = 2$ and $p = 1/2$, so, $np = 1$ and for the capacitor example, $n = 1000$ and $p = .01$, so $np = 10$. Thus, the theoretical results coincide with what common sense would tell us. If we consider a second capacitor example with $n = 500$ and $p = .0133$ (one divided by 75), then $np = 6.67$. This last result was probably not obvious. Note that $E(X)$ for the binomial distribution will not always be an integer, although X itself must always be an integer.

This should not seem incongruous, however, because $E(X)$ is simply the theoretical "average" value of X and should be very close to the actual average value of X if a binomial experiment were repeated a very large number of times.

The concept of *variance* of a random variable, as discussed in Section 3.3.1, is very important in statistics, in general, and particularly so for the control charts that will be presented in later chapters. You will recall that the sample variance, S^2, was introduced in Section 2.2. It was calculated from the data in a sample and is a *sample statistic*. The use of sample statistics to estimate population parameters is one of the primary uses of statistics. Such estimation will be done extensively in constructing control charts in later chapters, as well as in the other statistical procedures to be presented. [See Chapter 4 for various methods of obtaining (point) estimators.]

■ **EXAMPLE 3.5**

Problem Statement

A modem transmits one million bits. Each bit is 0 or 1 and the bit values are independent. Assume that the probability each bit is successfully transmitted is .999. What is the probability that at most 900 bits are not successfully transmitted?

Solution   In the not-so-distant past this type of problem would be worked by approximating the binomial probability using a normal distribution (see Section 3.4.3) because of the extremely large sample size. Today there is no need to do that as we can use MINITAB or other statistical software to obtain the answer directly, and using MINITAB we obtain .00069. So the probability is quite small and the probability that, say, at most 100 bits are not successfully transmitted is of course much smaller.   ■

The variance for a probability distribution is *somewhat* different, however, in that it is not calculated from data. Such a variance will henceforth be denoted by either $Var(X)$ or, as in Section 3.3.1, by σ^2, which is understood to represent σ_x^2 if X is the only random variable that is being discussed at the time.

We will introduce the concept of a covariance between two random variables. If X is a binomial random variable, we may obtain its variance in the following way, and relate it to one of the expressions previously given for a discrete random variable. Recall from Section 3.3.1 that for an arbitrary random variable, W, the variance of W is defined as $E(W^2) - (E(W))^2$, with $E(W^2) = \sum w^2 P(W = w)$ and $E(W) = \sum w P(W = w)$.

The *covariance* between two random variables, say, W and Y, is similarly defined as $E(WY) - E(W)E(Y)$. This is often written as $Cov(W, Y)$ or as σ_{wy}. Covariance measures the manner in which two random variables "co-vary," as was stated in Section 2.2 when the sample covariance was introduced. Note that if Y is replaced by W, we obtain the expression for $Var(W)$. Thus, we might say, perhaps somewhat trivially, that this is what we obtain when a variable "co-varies" with itself. A positive covariance means that as one variable increases (decreases), the other variable increases (decreases). With a negative covariance, one variable increases (decreases) while the other variable decreases (increases). Note that the sample covariance, s_{xy}, given in Section 2.2 is computed from *data* and is thus different from the covariance between two random variables.

Although correlation is discussed in some detail in Section 8.3, we will mention at this point that the population or theoretical correlation, ρ_{xw}, between two random variables X and W is defined as $\rho_{xw} = \sigma_{xw}/(\sigma_x \sigma_w)$. (Note that, if $W = X$, we obtain $\rho_{xw} = 1$ since $\sigma_{xx} = \sigma_x^2$. This means that a random variable is perfectly correlated with itself, a fact that should be intuitively obvious.)

For a Bernoulli random variable, it can easily be seen that $E(W^2) = p$, so the variance of a Bernoulli variable is $p - p^2 = p(1 - p)$. We can then find the variance of a binomial random variable by using the same general approach as was used for determining the mean. That is, the variance of a binomial random variable is n times the variance of the corresponding Bernoulli variable. Thus, for the binomial random variable we have $Var(X) = np(1 - p)$.

Note that this does not depend on any sample data, only on the sample size. It was shown previously that the $E(X)$ for the binomial distribution is quite intuitive, but the $Var(X)$ cannot be explained in a similar manner. The square root of the $Var(X)$ can, however, be explained in the following manner. If p is close to 1/2 and n is at least 15 or 20, then

$$P\left[E(X) - 3\sqrt{Var(X)} \le X \le E(X) + 3\sqrt{Var(X)}\right] = 0.99$$

In words, we would expect the value of X to fall between the two endpoints of the interval almost all of the time. Thus, the square root of the variance (i.e., the standard deviation) can be combined with the mean to determine an interval that should contain X with a high probability. The standard deviation for a probability distribution thus measures the spread of the *possible* values of a random variable, whereas the sample standard deviation, S, is a measure of the spread of the *actual* values of a random variable that are observed in a sample.

The binomial distribution has been used extensively in quality improvement work. For example, it is used in constructing p charts and np charts (i.e., control charts) that are presented in Chapter 11. There are many other types of applications, including the following.

■ EXAMPLE 3.6

Problem Statement

A memory module consists of nine chips. The device is designed with some redundancy so that it works even if one of its chips is defective, but not more than one. Assume that the chips function (or not) independently and that each chip contains four transistors and functions properly only when all of its transistors work. A transistor works with probability .995, independent of any other transistor. What is the probability that a chip works? What is the probability that the memory module works?

Solution   Since the transistors are independent, the probability that a chip works is the probability that all four transistors work, which is $(.995)^4 = .98$ because of the independence. The memory module will work if at most one chip is defective. The probability that no chip is defective is $(.98)^4 = .834$ since the chips function independently. The probability that one chip is defective is the binomial probability $\binom{9}{1}(.02)^1(.98)^8 = .153$. The probability that the memory module will work is thus $.834 + .153 = .987$.   ■

3.3.2.1   *Testing for the Appropriateness of the Binomial Model*

Recall the discussion in the preceding section regarding observed and expected frequencies. If the binomial model is inappropriate, we would expect the two sets of frequencies to differ considerably. Accordingly, one test for the appropriateness of the binomial model would be to compare the observed and expected frequencies. Just because a random variable has two possible outcomes does not mean that the binomial distribution is an appropriate model. In particular, the true variance might be greater than the binomial variance, which can occur when observations are correlated. A simple way to check this is to compare $n\bar{p}(1 - \bar{p})$ with $\sum (x - \bar{x})^2 / n$, with the two computations performed over at least a few dozen samples. (Here $\bar{p}$ denotes the average of the sample proportions, averaged over the samples, and $\bar{x}$ denotes the average number of nonconforming units, also averaged over the samples.)

The two values will almost certainly not be equal but should not differ by an order of magnitude.

3.3.3 Hypergeometric Distribution

This distribution, like the binomial distribution, has been used extensively in sampling inspection work. The two distributions are similar in that both assume two possible outcomes. They differ, however, in that the hypergeometric distribution is applicable when sampling is performed without replacement. The following example is used for illustration.

■ **EXAMPLE 3.7**

Problem Statement

Assume it is known that a lot of 1000 condensers contains 12 nonconforming condensers. What is the probability of observing at most one nonconforming condenser when a random sample of 50 is obtained from this lot?

Solution   Why can't the binomial distribution be used to solve this problem, using, $p = 12/1000 = .012$, $n = 50$, and $x \leq 1$? Note the subtle difference between this example and the example with the 1000 capacitors that was used to illustrate the binomial distribution. For the latter there was no sampling from a stated finite population; it was simply a matter of determining the probability of observing at most one nonconforming unit in 1000 when it is known (or assumed) that $p = .01$. For the current problem, however, we cannot say that the probability is .012 that any condenser in the random sample of 50 condensers is nonconforming. This is because the probability of any particular condenser being nonconforming depends on whether or not the previously selected condenser(s) were nonconforming. For example, the probability that the second condenser selected is nonconforming is 12/999 if the first condenser selected is conforming, and 11/999 if the first condenser is nonconforming.

Therefore, the binomial distribution cannot be used since p is not constant. We can, however, deduce the answer using the same type of heuristic reasoning as was used for the coin-tossing problem in the section on the binomial distribution. First, how many different samples of size 50 are possible out of 1000? Of that number, how many will contain at most one nonconforming unit? The answer to the first question is

$$\binom{1000}{50} = \frac{1000!}{50! \, 950!}$$

This is analogous to the earlier example concerning the number of possible ways of obtaining one head in two tosses of a balanced coin, which was

$$\binom{2}{1} = \frac{2!}{1! \, 1!}$$

Thus, the two tosses were, in essence, partitioned into one for a head and one for a tail. For the current problem, the 1000 condensers are partitioned into 50 that will be in the sample and 950 that will not be in the sample.

How many of these $\binom{1000}{50}$ samples will contain at most one nonconforming unit? There are $\binom{12}{0}\binom{988}{50}$ ways of obtaining zero nonconforming units and $\binom{12}{1}\binom{988}{49}$ ways of obtaining exactly one nonconforming unit. Therefore, the probability of having at most one nonconforming condenser in the sample of 50 is

$$\frac{\binom{12}{0}\binom{988}{50} + \binom{12}{1}\binom{988}{49}}{\binom{1000}{50}} = .88254$$

The combinatorics in the numerator [such as $\binom{12}{0}$ and $\binom{988}{50}$] are multiplied together because of a counting rule that states that if one stage of a procedure can be performed in M ways and another stage can be performed in N ways, the two-stage procedure can be performed in MN ways. For this example we can think of the number of ways of obtaining 0 nonconforming units out of 12 as constituting one stage, and the number of ways of obtaining 50 good items out of 988 as the other stage. Of course, the sample is not collected in two stages, but the sample can be viewed in this manner so as to determine the number of ways that zero nonconforming units can be obtained. (Note the analogy with two independent events, whose probability of simultaneous occurrence is equal to the product of the probabilities of the occurrence of each one.) ∎

The general form for the hypergeometric distribution is as follows:

$$P(x) = \frac{\binom{D}{x}\binom{N-D}{n-x}}{\binom{N}{n}} \qquad x = \max(0, n - N + D), 1, 2, \ldots, \min(n, D)$$

where N represents the number of items in the finite population of interest, D represents the number of items in the population of the type for which a probability is to be calculated (nonconforming units in the previous example), and $N - D$ represents the number of items of the other type (e.g., conforming units) that are in the population. The sample size is represented by n, of which x are of the type for which the probability is to be calculated, and $n - x$ are of the other type. D must be at least as large as x, otherwise $\binom{D}{x}$ would be undefined, and, obviously, x cannot exceed n. Thus, x cannot exceed the minimum of n and D [i.e., $\min(n, D)$]. Similarly, x cannot be negative, but the smallest possible value of x is not zero because $N - D$ must not be less than $n - x$. Thus, $N - D \geq n - x$, so that $x \geq n - N + D$. Of course, $n - N + D$ could be negative for a given combination of n, N, and D, so the lower bound on x is thus $x \geq \max(0, n - N + D)$.

3.3.4 Poisson Distribution

This distribution can be used when dealing with rare events. One application in semiconductor manufacturing would be to examine the number of chips that fail on the first day of operation. The distribution has also been used in quality improvement work to construct control charts for "nonconformities," which have also been called defects. (A product can have one or more nonconformities without being proclaimed a "nonconforming unit," so the terms are not synonymous. Specifically, a very minor nonconformity might be considered almost inconsequential. See also the discussion in Section 11.7.2.) Other applications include queueing theory and queues (i.e., waiting lines) in various places, including doctors' waiting rooms. There is a long list of other applications, including incoming calls at a switchboard, counts of the numbers of victims of specific diseases, the distribution of plants and animals in space, and number of misprints in a text.

The Poisson distribution is similar to the binomial distribution in that it is also a discrete distribution. In fact, it can be used to approximate binomial probabilities when n is large and p is small.

Specifically, if we again consider the binomial distribution

$$P(x) = \binom{n}{x} p^x (1 - p)^{n-x}$$

and let $n \to \infty$ and $p \to 0$ in such a way that np remains constant ($=\lambda$, say), it can be shown that

$$\lim_{n \to \infty} P(x) = \frac{e^{-\lambda}\lambda^x}{x!} \qquad x = 0, 1, 2, \dots \qquad (3.8)$$

with "lim" being short for "limit." The letter e represents the nonrepeating and nonterminating mathematical constant $2.71828\ldots$, and the right-hand side of Eq. (3.8) is the Poisson distribution, with λ representing the mean of X. Since this distribution can be obtained as a limiting form of the binomial distribution as $n \to \infty$ and $p \to 0$, it stands to reason that it should be possible to use the Poisson distribution to approximate binomial probabilities, when n is large and p is small, and obtain a reasonably good approximation (if a computer is not available to produce the exact answer, which would be the simplest and most direct approach).

The Poisson distribution is a valuable distribution in its own right; it is not just for approximating binomial probabilities. The distribution is named for Simeon Poisson (1781–1840), who presented it as a limit of the binomial distribution in 1837. It was stated earlier that the Poisson distribution is used as a basis for constructing control charts for nonconformities. Before the distribution can be applied in a physical setting, however, it must be determined whether or not the assumptions for the distribution are at least approximately met. These assumptions will be illustrated with the following example.

■ **EXAMPLE 3.8**

Problem Statement

Assume that sheets of steel are being produced in a manufacturing process and the random variable X is the number of surface scratches per square yard. Before it can be claimed that X is a Poisson random variable, the following questions must be addressed.

1. Do the scratches occur randomly and independently of each other?
2. Is the possible number of scratches per square yard quite large (theoretically it should be infinite)?

The second question might be answered in the affirmative, but perhaps not the first one. If there are quite a few surface scratches on one section of steel, we might expect to observe a sizable number of scratches on adjacent sections. Or perhaps not. One important point to keep in mind about modeling nonconformities, however, is that clustering may occur. If so, then the Poisson distribution is inappropriate and some type of modified Poisson distribution must be used. The need for such a modification has been increasingly discussed in the statistical literature during the past ten years. See also the discussion in Fang (2003), for example.

In any event, for this problem and for other practical problems our objective should be to determine whether or not a particular distribution can logically serve as a *model* (and nothing more) for physical phenomena under investigation. We can rarely expect all of the assumptions to be met exactly; we can only hope that they are approximately satisfied. We should be concerned, however, if there is evidence of a radical departure from the assumptions, and our concern should lead us to seek alternative approaches to the problem.   ■

■ **EXAMPLE 3.9**

Problem Statement

A disk server receives requests from many client machines, with the probability that the server receives k requests within a 10-millisecond interval modeled by a Poisson distribution with a mean of 0.9. What is the probability that three requests are received during a 10-millisecond interval?

Solution

$$P(X = 3) = \exp(-0.9)(0.9)^3/3! = .049.$$ ■

3.3.4.1 Testing for the Appropriateness of the Poisson Model

As with the binomial distribution and other distributions, we should test the appropriateness of the Poisson distribution whenever its use as a model is contemplated. As with the binomial distribution, the Poisson variance may understate the variance of the random variable that is being modeled. Oftentimes this occurs because of clustering.

A simple, but somewhat crude, method for comparing the observed sample variance with the Poisson variance was given by Snedecor and Cochran (1974, p. 198). Using a goodness-of-fit approach, covered formally in Section 15.3, $Y = \sum_{i=1}^{k} (c_i - \bar{c})^2/\bar{c}$ is distributed approximately as chi-square with $(k - 1)$ degrees of freedom, where k denotes the number of independent samples, c_i represents the count in each sample, and $\bar{c}$ denotes the average of those counts. (The chi-square distribution is covered in Section 3.4.5.1.) Since the sample variance of c is $s_c^2 = \sum_{i=1}^{k} (c_i - \bar{c})^2/(k - 1)$, and $\bar{c}$ is the estimator of both the Poisson mean and Poisson variance, it follows that $Y = (k - 1)(\text{sample variance})/(\text{estimator}$ of Poisson variance). To illustrate, if a sample of $k = 10$ counts is 5, 8, 7, 2, 4, 5, 3, 2, 5, and 6, then $\bar{c} = 4.70$, $s_c^2 = 4.01$, and $Y = 9(4.01)/(4.70) = 7.68$. Since $P(Y < 7.68) = .4332$ using the chi-square approximation, there is no reason to doubt that the sample variance can be represented by the Poisson variance. (Our conclusion would have been different if the probability had been quite small, say, less than .05.)

How good is this approximation? Well, the approximation is based on the assumed adequacy of the normal approximation to the Poisson distribution. That approximation is generally assumed to be adequate when the Poisson mean is at least 5, although it is shown in Chapter 11 that for control chart purposes this rule of thumb is inadequate. Here the mean is only 4.70, but since the mean and variance differ very little, a formal test seems unnecessary.

3.3.5 Geometric Distribution

In recent years this distribution has been used as an alternative to the binomial distribution, in particular, and also as an alternative to the Poisson distribution. The need for the geometric distribution is discussed and illustrated in Section 11.3.

The geometric distribution is frequently referred to as a "waiting time" distribution. This can be illustrated by contrasting the geometric distribution with the binomial distribution when a single success is considered. With the binomial distribution, the single success could occur on the first trial, the nth trial, or any trial in between.

The geometric distribution is the appropriate distribution when the probability is obtained of the single success occurring on the nth trial. (If there is interest in, say, the probability of the kth success occurring on the nth trial, with k greater than one, the appropriate distribution is the negative binomial distribution, which we will not present here.)

In order for the first success to occur on the nth trial, the first $n - 1$ trials must obviously result in failure. The probability of that occurring is $(1 - p)^{n-1}$. Therefore, we multiply this component by the probability of obtaining a success so as to obtain the probability mass function as

$$P(n) = (1 - p)^{n-1} p \qquad n = 1, 2, \ldots$$

with n denoting the number of trials that occur before the first success is observed (on the nth trial). Thus, n is the random variable for the geometric distribution. The distribution for n depends solely on p. If p is small, the mean of n is large, and conversely. It can be shown that the mean is $1/p$ and the variance is $(1 - p)/p^2$.

■ EXAMPLE 3.10

Problem Statement

Assume that a quality engineer oversees the inspection of a certain product that a company has had great difficulty producing defect-free. One hundred items are inspected at the beginning of each shift and a worker reports finding a bad unit on the 12th unit that is inspected. The engineer suspects a problem because only 3% of the units had been declared bad and thus unacceptable in recent weeks.

Analysis

If the probability of an unacceptable unit were truly .03, the probability that the first such unit were observed on the 12th item inspected is

$$P(12) = (0.97)^{11}(.03) = .0215$$

This is a small probability but the engineer would probably be more interested in the probability of observing a bad unit *by* the 12th unit inspected. That probability can be shown to be .306 by computing the probabilities for each possible value of n and summing all of the probabilities. Since .306 is not a small probability, the value of p might be .03, but a quick check of the equipment would probably be desirable. ∎

3.4 COMMON CONTINUOUS DISTRIBUTIONS

A continuous probability distribution is such that the random variable can assume any value along a continuum, and such a distribution is called either a *probability density function* or a *probability distribution function*. Either way, the abbreviation is *pdf*. Analogous to the discrete case, the corresponding cumulative distribution function (*cdf*) is obtained as $F(x) = \int_{-\infty}^{x} f(x)dx$ with $f(x)$ denoting the *pdf*.

For example, if X = height of an adult human being, then X can assume any value between, say, 24 inches and 108 inches (Goliath's estimated height). Although a person might actually be 68.1374136 inches tall, that person is not likely to ever have his or her height recorded as such. Thus, random variables that are continuous are, in essence, "discretized" by the use of measuring instruments. Height, for example, is usually not recorded to an accuracy greater than one-fourth of an inch.

3.4.1 Expected Value and Variance

The concepts of expected value and variance also apply to continuous distributions, but the expressions are different. Specifically, summation is replaced by integration, so that $E(X) = \int x\, f(x)\,dx$, where $f(x)$ denotes the function that represents the probability distribution, and the limits of integration are over the range of X for which the distribution is defined. Other expected value expressions, such as $E(X^2)$, are defined analogously. For example, $E(X^2) = \int x^2 f(x)\,dx$. Recall from Section 3.3.1 that $E(\text{"anything"}) = \sum_x (\text{"anything"})P(x)$. (Similarly, for a continuous random variable we have $E(\text{"anything"}) = \int (\text{"anything"})f(x)\,dx$.

In particular, let $g(x)$ denote an arbitrary function of x. Then $E(g(x)) = \int g(x) f(x)\,dx$. If $g(x)$ is of the form $g(x) = a + bx$, then certain general results can be obtained, as the reader is asked to show in Exercise 3.21. These results apply whether a distribution is continuous or discrete.

Although the median of a probability distribution is used less often than the mean, the median, x_0, is defined such that $\int_{-\infty}^{x_0} f(x)\,dx = 0.5$.

3.4.2 Determining Probabilities for Continuous Random Variables

Probabilities for a discrete distribution can be obtained simply by inserting the value of the random variable for which the probability is desired into the probability mass function, if the probability is not actually given by the *pmf*, which occurs when the latter is a constant. A continuous random variable does not have a finite number of possible values, however, so a nonzero probability cannot be obtained for any one value.

Therefore, probabilities can be obtained only for an interval of possible values.

> **DEFINITION**
>
> Probabilities for a continuous variable are obtained by integrating the probability density function, $f(x)$, over the range given in a probability statement. That is, $P(a \leq X \leq b) = \int_a^b f(x)\,dx$.

For example, for a *continuous uniform distribution* defined on the interval $(0, 1)$, $f(x) = 1$ and thus $P(0.25 \leq X \leq 0.50) = \int_{0.25}^{0.50} (1)\,dx = 0.50 - 0.25 = .25$. This answer should be intuitively apparent since the probability statement covers one-fourth of the range of possible values of the random variable and the probability is uniform over that interval.

3.4.3 Normal Distribution

It is somewhat unfortunate that any statistical distribution is called "normal," since this could easily create the false impression that this distribution is "typical" and the other hundreds of distributions are "atypical."

As Geary (1947) pointed out decades ago, there is no such thing as a normal distribution *in practice*, but a normal distribution (also sometimes called a Gaussian distribution) is often used to approximate the *actual* (unknown) distribution of many random variables. The equation for the distribution is given by

$$f(x) = \frac{1}{\sigma\sqrt{2\pi}} e^{-(x-\mu)^2/2\sigma^2} \qquad -\infty < x < \infty$$

where μ denotes the mean of the distribution, σ represents the standard deviation, π is the mathematical constant $3.14159\ldots$, and $-\infty < x < \infty$ indicates that the random variable X can assume any real number. The latter should tell you that we are simply using a model of reality because no real-life random variable is going to assume values that approach plus infinity or minus infinity.

The fact that measurements, aptitude test scores, and so on have a lower bound greater than zero does not mean that a normal distribution cannot serve as an adequate model, however.

It should be noted that there is actually not just one normal distribution since there is a different normal distribution for each combination of μ and σ, so there is theoretically an infinite number of distinct normal distributions. The value of σ determines the shape of the distribution, and the value of μ determines the location. This is illustrated in Figures 3.2

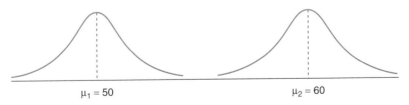

$\mu_1 = 50$ $\mu_2 = 60$

Figure 3.2 Two normal distributions with different means.

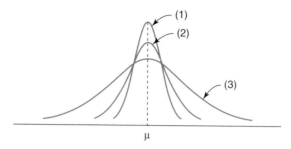

Figure 3.3 Three normal distributions with the same mean and different variances.

and 3.3. [The height of the curve at any point x is given by $f(x)$, but this is usually not of any practical interest.]

In practice, μ and σ are seldom known and must be estimated. A normal distribution has some important properties, as illustrated in Figure 3.4. The number in each section of the curve denotes the area under the curve in that section. For example, 0.34134 represents the area under the curve between μ and $\mu + \sigma$, which is also the area between μ and $\mu - \sigma$ since the curve is symmetric with respect to μ. That is, the shape of the distribution above μ is the mirror image of the shape of the distribution below μ, as stated previously. Stated differently, if we could fold one half of the distribution on top of the other half, the halves would coincide. The total area under the curve equals one, which corresponds to a total probability of one, with 0.5 on each side of μ.

The reader should make note of the fact that the area between $\mu - 3\sigma$ and $\mu + 3\sigma$ is $0.9973 [= 2(0.34134 + 0.13591 + 0.02140)]$. Thus, the probability is only $1 - .9973 = .0027$ that a value of the random variable will lie outside this interval. This relates to the "3-sigma limits" on control charts, which are discussed in Chapter 11.

The areas given in Figure 3.4 can be determined from Table B in the Appendix of Statistical Tables. That table is for a *particular* normal distribution specifically, the distribution with $\mu = 0$ and $\sigma = 1$. This distribution results when the transformation

$$Z = \frac{X - \mu}{\sigma}$$

is used with $X \sim N(\mu, \sigma^2)$ and "$\sim$" is read as "has," meaning that the random variable X has the indicated distribution, which in this case is a normal (N) distribution with the indicated parameters.

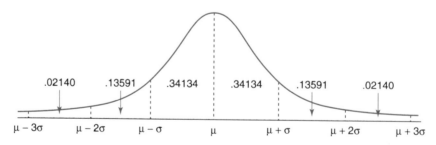

Figure 3.4 Areas under a normal curve.

Subtracting the mean of a random variable and dividing by its standard deviation is the way that a random variable is *standardized*. That is, if the distribution of X were known, then the distribution of the standardized random variable would have a distribution of the same name except that the mean would be zero and the standard deviation would be one. Thus, in the present context, the transformation produces the *standard normal distribution* with $\mu = 0$ and $\sigma = 1$, with the value of Z being the number of standard deviations that the value of the random variable X is from μ. The transformation must be used before probabilities for any normal distribution can be determined (unless some computing device is used).

■ **EXAMPLE 3.11**

Assumption and Problem Statement

To illustrate, suppose that a shaft diameter has (approximately) a normal distribution with $\mu = 0.625$ (feet) and $\sigma = 0.01$. If the diameter has an upper specification limit of 0.65, we might wish to estimate the percentage of shafts whose diameters will exceed 0.65, in which case a nonconformity will result.

Solution   Since $0.65 = \mu + 2.5\sigma$, we can see from Figure 3.4 that the percentage will be less than 2.275% since this would be the percentage for the area under the curve beyond $\mu + 2\sigma$. Specifically, we need to determine the shaded area given in Figure 3.5.

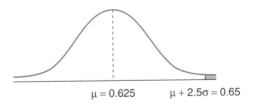

$$\mu = 0.625 \qquad \mu + 2.5\sigma = 0.65$$

Figure 3.5   Distribution of shaft diameters.

By using the transformation

$$Z = \frac{X - \mu}{\sigma}$$

we obtain

$$z = \frac{0.65 - 0.625}{0.01} = 2.5$$

Note that this z-value (which is given in lowercase to differentiate it from the random variable Z) is the number of standard deviations that the value of X is from μ. This will always be the case, as indicated previously.

We can now say that the probability of observing a z-value greater than 2.5 is the same as the probability of observing a shaft diameter in excess of 0.65. Thus, we need only determine the shaded area in Figure 3.6.

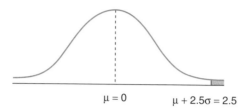

$$\mu = 0 \qquad\qquad \mu + 2.5\sigma = 2.5$$

Figure 3.6   Standard normal distribution.

Since this shaded area does not cover $\mu = 0$, we must look up the area for $z = 2.5$ in Table B and subtract it from 0.5000. We thus obtain $0.5000 - 0.49379 = 0.00621$. Putting all of this together we obtain

$$P(X > 0.65) = P(Z > 2.5) = .00621$$

Thus, we expect approximately 0.621% of the shafts to not be in conformance with the diameter specification. (The probability is approximate rather than exact since a distribution was assumed that was approximately normal.) In subsequent chapters we will use expressions such as $z_{\alpha/2}$ to denote the standard normal distribution with an upper tail area of $\alpha/2$. Therefore, in this example $z_{.00621} = 2.5$.   ■

Of course, we could also obtain this numerical result using statistical software, as illustrated for this problem in Section 3.4.3.1.

If there had been a lower specification limit, another z-value would have been calculated and the area obtained from that value would have been added to 0.00621 to produce the total percentage of nonconforming shaft diameters. For example, if the lower specification limit were 0.61, this would lead to a z-value of -1.5 and an area of 0.06681. The total percentage would then be $6.681\% + 0.621\% = 7.302\%$.

In determining areas under the z-curve, it is generally desirable (such as on class tests) to shade in the appropriate region(s) before going to Table B. This will lessen the chances of making an error such as *subtracting* a number from 0.5000 for a problem in which the number should instead be *added* to 0.5000.

The determination of probabilities using the normal distribution should be accomplished in accordance with the following step-by-step procedure:

1. Transform the probability statement on X to the equivalent statement in terms of Z by using the transformation

$$Z = \frac{X - \mu}{\sigma}$$

2. Shade in the appropriate region(s) under the z-curve as determined from the probability statement on Z.

3. Find the area(s) in Table B and obtain the answer in accordance with the following:

It should also be noted that $P(Z = a) = 0$ for any value of a. Thus, it is not possible to determine the probability that a shaft will have a diameter of, say, 0.640 feet; a probability can be determined only for an interval. *This is true for any continuous distribution.*

General Form of the Probability Statement		Action to Be Taken
1. $P(a < Z < b)$	$(a < 0, b > 0)$	Look up the area in the table for $z = -a$ and $z = b$ and add the two areas together to obtain the answer
2. $P(a < Z < b)$	$(0 < a < b)$	Look up the area for $z = a$ and subtract it from the area for $z = b$.
3. $P(a < Z < b)$	$(a < b < 0)$	Look up the area for $z = -b$ and subtract it from the area for $z = -a$.
4. $P(Z > a)$	$(a > 0)$	Look up the area for $z = a$ and subtract it from 0.5
5. $P(Z > a)$	$(a < 0)$	Look up the area for $z = -a$ and add it to 0.5.
6. $P(Z < a)$	$(a > 0)$	Look up the area for $z = a$ and add it to 0.5.
7. $P(Z < a)$	$(a < 0)$	Look up the area for $z = -a$ and subtract it from 0.5.
8. $P(Z > a$ or $Z < b)$	$(a > 0, b < 0)$	Look up the area for $z = a$ and $z = -b$, add the two areas together and subtract the sum from 1.0.

3.4.3.1 Software-Aided Normal Probability Computations
Although it is easy to use a normal table to determine probabilities, it is of course easier to use statistical software. Any statistical software can be used to determine normal probabilities. One advantage in using software is that it isn't necessary to first convert to the standard normal distribution.

For example, the problem posed in Section 3.4.1 of determining the probability of a shaft exceeding 0.65 feet when normality is assumed with $\mu = 0.625$ and $\sigma = 0.01$ can be determined using standard statistical software such as MINITAB.

3.4.3.2 Testing the Normality Assumption
The most commonly used method of testing for normality is to construct a normal probability plot. This is essentially a plot of the observed frequencies against the expected frequencies under the assumption of normality. The use of a normal probability plot is illustrated in Section 6.1.2. Histograms are also frequently used, but as stated in Section 1.5.1, a histogram is not a reliable indicator of the shape of a population distribution, as the shape of a histogram is strongly dependent on the number of classes that are used, with the optimum number of classes dependent on the shape of the (unknown) distribution.

3.4.4 t-Distribution

Consider the transformation

$$Z = \frac{\overline{X} - \mu}{\sigma/\sqrt{n}} \tag{3.9}$$

which is the standardization of the statistic $\overline{X}$, and which is what we would use in computing a probability for $\overline{X}$ as compared with computing a probability for X, under the assumption of normality, as was done in Section 3.4.1. That is, Z as given in Eq. (3.9) also has the standard normal distribution when X has a normal distribution.

If σ were unknown, however, which is the usual case, and we use s instead, we then have a statistic and its distribution that have not yet been discussed.

That distribution is the t-distribution, which is often referred to as *Student's t-distribution*. "Student" was the pseudonym used by W. S. Gosset (1876–1937) for the statistical papers that he wrote while employed as a brewer at St. James' Gate Brewery in Dublin, Ireland. According to one version of the story, the brewery had a rule that prohibited its employees from publishing papers under their own names, hence the need for a pseudonym.

Gosset worked with small samples, often in situations in which it was unreasonable to assume that σ was known. He was also concerned with probability statements on $\overline{X}$ instead of on X. Thus, he was more interested in, say, $P(\overline{X} > a)$ than $P(X > a)$. At the time of his work (circa 1900), it was well known that if X has a normal distribution with mean μ and variance σ^2, then Z defined in Eq. (3.9) has a normal distribution with $\mu = 0$ and $\sigma = 1$. This stems from the fact that $\overline{X} \sim N(\mu, \sigma^2/n)$ when $X \sim N(\mu, \sigma^2)$. Thus, $\overline{X}$ is "standardized" in Eq. (3.9) by first subtracting its mean (which is the same as the mean of X), and then dividing by its standard deviation. This is the same type of standardization that was used for X in obtaining $Z = (X - \mu)/\sigma$. The fact that the mean of $\overline{X}$ is the same as the mean of X is a theoretical result that can easily be established by taking the expected value of $\overline{X}$. Specifically,

$$E\left(\sum X/n\right) = (1/n)\sum E(X) = (1/n)(n\mu) = \mu$$

This result can easily be demonstrated for a finite population, as in the following example. Assume that a (small) population consists of the numbers 1, 2, 3, and 4 and we want to find the average of the sample averages in which the averaging is performed over all possible samples of size two. We would then obtain the results given in Table 3.1.

There are six possible samples of size 2 and the average of the six values of $\overline{X}$ is 2.5. Notice that this is also the average of the numbers 1, 2, 3, and 4. This same result would be obtained for samples of any other size, as well as for finite populations of any other size.

It should also be observed that there is slightly less variability in the $\overline{X}$ values than in the X values. For infinite populations (or approximately for finite populations that are extremely large)

$$\sigma_{\overline{x}} = \frac{\sigma_x}{\sqrt{n}}$$

with n denoting the sample size.

TABLE 3.1 **Sample Averages**

Sample		$\overline{x}$
1	2	1.5
1	3	2.0
1	4	2.5
2	3	2.5
2	4	3.0
3	4	3.5

When n is large (say, $n \geq 30$), the sample standard deviation, s, can be used as a substitute for σ in Eq. (3.9) so as to produce

$$Z = \frac{\overline{X} - \mu}{s/\sqrt{n}} \tag{3.10}$$

and Z will then be approximately normally distributed with $\mu = 0$ and $\sigma^2 = 1$. This will not be true when n is small, however, so the distribution of the right-hand side of Eq. (3.10) must be addressed for small-to-moderate sample sizes.

Gosset's 1908 paper, entitled "The Probable Error of a Mean," led to the t-distribution in which

$$t = \frac{\overline{X} - \mu}{s/\sqrt{n}} \tag{3.11}$$

although his paper did not exactly give Eq. (3.11). The equation for the distribution is

$$f(t) = \frac{1}{\sqrt{(n-1)\pi}} \frac{\Gamma(n/2)}{\Gamma[(n-1)/2]} \left(1 + \frac{t^2}{n-1}\right)^{-n/2} \qquad -\infty < t < \infty$$

although the equation is not generally needed and will not be used in this book. (Γ refers to the gamma function.)

What will be needed (and used) in subsequent sections of this book is the fact that it can be shown that a t-distribution results when a $N(0, 1)$ random variable is divided by a chi-square random variable (see Section 3.4.5.1) that has been divided by its degrees of freedom, with the t-statistic having the same degrees of freedom as the chi-square random variable. That is

$$t_v = \frac{N(0, 1)}{\sqrt{\chi_v^2/v}} \tag{3.12}$$

with χ_v^2 denoting a chi-square random variable with v degrees of freedom. Unlike the standard normal distribution and other normal distributions, the shape of the t-distribution depends on the sample size, or more specifically on the degrees of freedom. The dependence of the shape on the sample size is illustrated in Figure 3.7, and it can be shown mathematically that the t-distribution approaches the standard normal distribution as $n \to \infty$. There is very little difference in the two distributions when $n > 30$, so Eq. (3.10) might be used in place of Eq. (3.11).

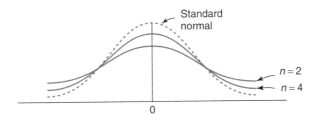

Figure 3.7 Student's t-distribution for various n.

Table C is used for the t-distribution. Unlike the table for the standard normal distribution (Table B), Table C cannot be used to determine probabilities for the t-distribution. Its uses will be illustrated in later chapters.

It is worth noting that the t-distribution arises not only when averages are used, but also for other statistics. In general,

$$ t = \frac{\widehat{\theta} - \theta}{s_{\widehat{\theta}}} $$

where θ denotes an arbitrary parameter to be estimated, $\widehat{\theta}$ is the estimator based on a sample, and $s_{\widehat{\theta}}$ is the sample estimator of $\sigma_{\widehat{\theta}}$. Of course, it follows from what was stated earlier that, strictly speaking, $\widehat{\theta}$ must have a normal distribution in order for the t-statistic to have a t-distribution, but in practice this works out okay as long as $\widehat{\theta}$ is approximately normally distributed.

The *degrees of freedom* for t are determined by the degrees of freedom for $s_{\widehat{\theta}}$. Loosely speaking, a degree of freedom is used whenever a parameter is estimated. The t-statistic in Eq. (3.11) has $n - 1$ degrees of freedom; other t-statistics can have fewer degrees of freedom for fixed n. In subsequent chapters, we will use expressions such as to $t_{\alpha/2, \nu}$ denote a t-variate with ν degrees of freedom and a tail area of $\alpha/2$. Thus, $t_{18, .025}$ denotes a variate from a t-distribution with 18 degrees of freedom and a tail area of 0.025.

3.4.5 Gamma Distribution

The gamma distribution is actually a family of distributions represented by the equation

$$ f(x) = \frac{1}{\beta^{\alpha} \Gamma(\alpha)} x^{\alpha - 1} e^{-x/\beta} \qquad x > 0 \qquad (3.13) $$

for which $\alpha > 0$ and $\beta > 0$ are parameters of the distribution and Γ refers to the gamma function. Almost all distributions do not have the mean and/or standard deviation given as parameters (unlike the "normal" distribution), and the gamma distribution is typical in this regard. The mean of a gamma random variable is $\alpha\beta$ and the variance is $\alpha\beta^2$.

The gamma distribution is used in various applications, including queuing applications and birth–death processes. It is also often used as a prior distribution in Bayesian applications, including Bayesian reliability estimation. (Reliability analysis is covered in Chapter 14.) Special cases of the gamma distribution are used more extensively, however, and those special cases are considered in the next two sections.

3.4.5.1 Chi-Square Distribution

This distribution is a special case of the gamma distribution. Specifically, the chi-square distribution is obtained by letting $\alpha = r/2$ and $\beta = 2$ in Eq. (3.13), with r denoting the degrees of freedom of the chi-square distribution. That is, the distribution is given by

$$ f(x) = \frac{1}{2^{r/2} \Gamma(r/2)} x^{r/2 - 1} e^{-x/2} \qquad x > 0 $$

This distribution plays a central role in statistics because, for example, the square of a standard normal random variable has a chi-square distribution with one degree of freedom,

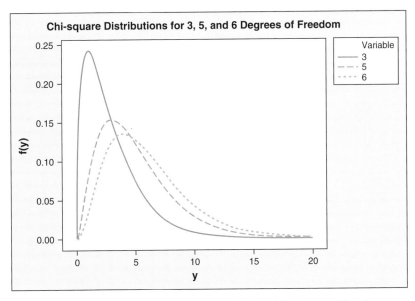

Figure 3.8 Chi-square distributions for 3, 5, and 6 degrees of freedom.

and the sum of the squares of k independent normal random variables has a chi-square distribution with degrees k of freedom. (Degrees of freedom are lost when parameters are estimated, as indicated previously, so estimating the mean of a normal distribution will cause a chi-square distribution with k degrees of freedom to have $k - 1$ degrees of freedom when such estimation is used.)

The chi-square distribution will be involved in the methods that are presented in subsequent chapters.

Since the chi-square distribution is a gamma distribution, the mean must be $\alpha\beta$, which simplifies to r, the degrees of freedom. Similarly, the variance of $\alpha\beta^2$ simplifies to $2r$. The shape of a chi-square distribution depends on r, as shown in Figure 3.8, with the distribution approaching a normal distribution as r increases.

3.4.5.2 Exponential Distribution

If we let $\alpha = 1$ in Eq. (3.13), we obtain the exponential distribution, which historically has been used extensively in the fields of life testing and reliability, although it does not play quite so dominant a role in those areas today. It does, however, serve as a model in life testing and reliability for items subjected to random fatal shocks and for the time to failure for a complex system of nonredundant components. In queuing problems, it serves as a model for the waiting time for the arrival of the next customer or the time between arrivals, and in time-and-motion studies it can serve as a model for the time to complete a task.

The role of the exponential distribution in reliability analysis is discussed further and illustrated in Chapter 14.

The equation for the distribution is

$$f(x) = \frac{1}{\beta}e^{-x/\beta} \qquad x > 0 \tag{3.14}$$

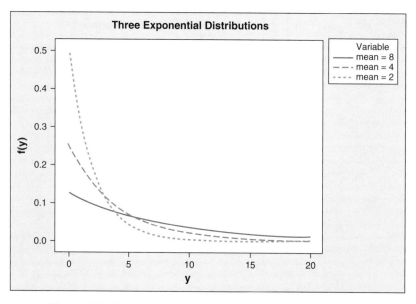

Figure 3.9 Exponential distributions with means of 2, 4, and 8.

with β denoting the mean of the distribution because, of course, $\alpha\beta = (1)\beta = \beta$. Similarly, the variance is $\alpha\beta^2 = (1)\beta^2 = \beta^2$. The shape of the distribution depends on the mean, β, as shown in Figure 3.9.

When the mean is small, such as 2, most of the probability will be concentrated near zero. This ceases to be true as the mean increases, as illustrated by Figure 3.9.

3.4.6 Weibull Distribution

Like an exponential distribution, Weibull distributions have been used extensively in life testing and reliability. In particular, a Weibull distribution is often used to model the tensile strength of materials, especially brittle materials such as boron and carbon. The equation for the (two-parameter) distribution is

$$f(x) = \beta/\alpha(x/\alpha)^{\beta-1}e^{-(x/\alpha)^\beta} \qquad x > 0$$

with $\alpha > 0$ and $\beta > 0$ denoting the parameters of the distribution; the former is the scale parameter and the latter is the shape parameter. (There is also a three-parameter Weibull distribution when a location parameter is used.) As with other distributions, the shape of a Weibull distribution depends on the values of the parameters. When $\beta = 1$, a Weibull distribution reduces to an exponential distribution. The mean and the variance of the Weibull distribution are given by $(\alpha)\Gamma[(\beta + 1)/\beta]$ and $(\alpha^2)\{\Gamma[(\beta + 2)/\beta] - (\Gamma[(\beta + 1)/\beta])^2\}$, respectively, with $\Gamma(\cdot)$ denoting the gamma function, as stated previously.

As with certain other distributions, the Weibull distribution has been written in an alternative form in the literature, namely,

$$f(x) = \alpha\beta(\alpha x)^{\beta-1}e^{-(\alpha x)^\beta} \qquad x > 0$$

References on how a Weibull distribution can be used in quality improvement work include Berrettoni (1964), and the role that the Weibull distribution plays in reliability analysis is discussed in Chapter 14. See also the applications in Murthy, Xie, and Jiang (2004).

■ **EXAMPLE 3.12**

It is important to realize that some of these distributions, such as the Weibull distribution, are not just of academic interest. Lee and Matzo (1998) give leakage data from 94 randomly selected fuel injectors. Eighty-five of the injectors have leakage, and these are to be used in determining the leakage distribution. A histogram of the data using the power-of-2 rule to determine the number of classes is given in Figure 3.10. We can see that the histogram is highly skewed.

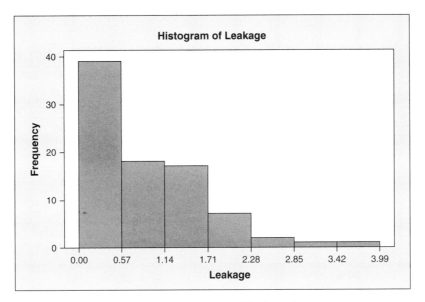

Figure 3.10 Histogram of leakage data.

■

Other distributions that are used less frequently in life testing and reliability are presented and discussed briefly in Chapter 14, in addition to the distribution that is covered in the next section.

3.4.7 Smallest Extreme Value Distribution

This is a less common distribution that is included in this chapter because it is mentioned in Chapter 14 and because it has considerable utility as a reliability distribution. It is also covered here because it is related to the Weibull distribution and has often been used to analyze Weibull data. An extreme value of course can be either large or small, and it is the smallest extreme value distribution that is useful in reliability applications

(Chapter 14), such as the electrical strength of materials, as the largest extreme value distribution is seldom used in failure data analysis (Chapter 14).

The *pdf* is

$$f(y) = \frac{1}{\delta} \exp[(y - \lambda)/\delta] \cdot \exp\{-\exp[(y - \lambda)/\delta]\} \qquad -\infty \leq y \leq \infty$$

where λ is the location parameter that can assume any value, and δ is the scale parameter, which must be positive. It can be shown that $\log(Y)$ has a Weibull distribution.

The mean and variance of the extreme value distribution are $\lambda - 0.5772\delta$ and $1.645\delta^2$, respectively.

3.4.8 Lognormal Distribution

This is another distribution that has considerable utility in reliability analysis and which is accordingly discussed in that context in Chapter 14. The *pdf* is given by

$$f(y) = \{1/[(2\pi)^{1/2} y\sigma]\} \exp\{-[\log(y) - \mu]^2/(2\sigma^2)\} \qquad y > 0$$

with μ denoting the mean of $\log(Y)$ and σ denoting the standard deviation of $\log(Y)$. Thus, these are *not* the parameters of Y. The mean and variance of Y are $\exp(\mu + \sigma^2/2)$ and $\exp(2\mu + \sigma^2)[\exp(\sigma^2) - 1]$, respectively. (This is the two-parameter lognormal distribution; there is also a three-parameter lognormal distribution.)

The lognormal and normal distributions are related in that $\log(Y)$ has a normal distribution when Y has a lognormal distribution. The distribution will have different appearances for different combinations of μ and σ. Figure 3.11 shows three lognormal distributions. The distribution with the highest peak has $\mu = 1$ and $\sigma = 1$; the distribution with the next highest peak has $\mu = 2$ and $\sigma = 2$; and the other distribution has $\mu = 3$ and $\sigma = 2$. Notice that the distributions differ considerably even though the parameter values do not differ greatly.

3.4.9 *F* Distribution

This is another distribution that plays a prominent role in statistical theory and applications. If two independent chi-square random variables are each divided by their respective degrees of freedom, and a fraction is formed from these two fractions, the result is a random variable that has an F distribution. That is,

$$F_{v_1 v_2} = \frac{x^2_{v_1}/v_1}{x^2_{v_2}/v_2} \qquad (0 < F < \infty)$$

with v_1 and v_2 denoting the degrees of freedom for each of the chi-square random variables, which are also the numerator and denominator degrees of freedom, respectively, for the random variable denoted here by F (which has an F distribution).

The shape of the distribution depends on v_1 and v_2, so, as with the other distributions discussed in this chapter, there is not a single F distribution, but rather a family of such distributions. Three of these distributions are given in Figure 3.12, with the distribution with the highest peak having $(v_1, v_2) = (5, 30)$; with the second highest peak being $(3, 20)$,

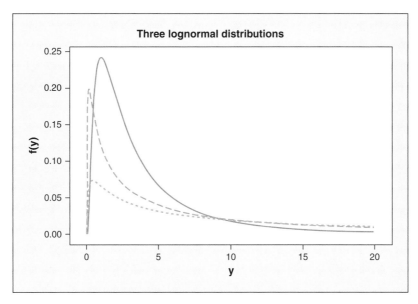

Figure 3.11 Three lognormal distributions.

and the other being (4, 15). Notice that small changes in v_1 and moderate changes in v_2 have very little effect on the distribution.

Table D at the end of the book gives the values of F for different combinations of v_1 and v_2 as well as for $\alpha = 0.01$ and 0.05, with α denoting the upper tail area (the upper tail is the only one that is used in most statistical procedures).

It is worth noting the relationship between the t-distribution and the F-distribution, as this relationship is used in Section 12.2.1.2. Consider the square of the t-statistic in Eq. (3.12):

$$t_v^2 = \frac{(N(0, 1))^2}{\chi_v^2/v} \tag{3.15}$$

We recognize that the denominator is in the form of the denominator of the F-statistic, so we need only show (or accept) that the numerator has a chi-square distribution with one degree of freedom. As stated in Section 3.4.5.1, the square of a standard normal random variable is a chi-square random variable with one degree of freedom. (The proof of this result is straightforward but is beyond the intended scope of this book.) Since the numerator is a chi-square random variable with one degree of freedom, it follows that t_v^2 is $F_{1,v}$; that is, the square of the t-statistic is an F-statistic that has an distribution with one degree of freedom for the numerator and v degrees of freedom for the denominator.

A result that is sometimes useful is

$$F_{\alpha,v_1,v_2} = \frac{1}{F_{1-\alpha,v_2,v_1}}$$

Using this result, lower-tail F-values can easily be obtained from upper-tail F-values. For example, assume that we need $F_{.95,4,26}$. This is equal to $1/F_{.05,26,4} = 1/5.7635 = 0.1735$,

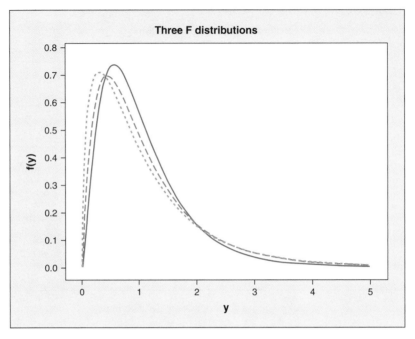

Figure 3.12　Three F distributions.

with α denoting the upper tail area. Of course, computational aids such as this are not as important in today's computing environment as they were decades ago, but they can still be useful on occasion.

3.5　GENERAL DISTRIBUTION FITTING

The fitting of probability distributions is a lost art, and consequently the best-known distributions are apt to be fit to data, or even worse, simply assumed to be adequate, when a distribution that is "between" two common distributions might be the best choice.

A continuous distribution is usually characterized by its first four moments, with moments being, by definition, either central moments or noncentral moments. The rth noncentral population moment is defined as $E(X^r)$, with the corresponding central moment defined as $E(X - \mu)^r$. Thus, "central" means moments about the mean, with noncentral moments being moments about zero. Thus, $E(X) = \mu$ is the first noncentral moment and $E(X - \mu)^2 = \sigma^2$ is the second central moment.

The third and fourth (central) moments are $E(X - \mu)^3$ and $E(X - \mu)^4$, respectively. These measure the skewness and the kurtosis, respectively, although they are generally made into unit-free number expressions by dividing the third moment by $(\sigma^2)^{3/2}$ and dividing the fourth moment by $(\sigma^2)^2$. These expressions are generally denoted by $\sqrt{\beta_1}$ and β_2, respectively. The former is a measure of the extent to which the distribution departs from a symmetric distribution, and the latter is a measure of how "peaked" a distribution is, although the measure is also a function of whether the tails of the distribution are heavy or not. For a normal distribution, $\sqrt{\beta_1} = 0$ and $\beta_2 = 3.0$.

The moments for a distribution can be determined most efficiently by using the moment generating function for a distribution, provided that the moment generating function exists. The function is obtained, for both discrete and continuous distributions, by computing $E(e^{tX})$. Moment generating functions are covered in books on mathematical statistics but will not be pursued here.

The first two sample moments are the sample mean and the sample variance, respectively. Although the skewness and kurtosis sample statistics are defined as $\sum (x - \bar{x})^3/n$ and $\sum (x - \bar{x})^4/n$, respectively, they, like their population counterparts, are made into unit-free numbers by dividing by $(s^2)^{3/2}$ and $(s^2)^2$, respectively. (Note that the divisor for each of these is n and not the $n - 1$ that is used for the sample variance. In general, the divisor for sample moments is n; the sample variance being the exception.). Values of these statistics that are, for a large sample, close to 0 and 3.0, respectively, would suggest fitting a distribution that is close to a normal distribution.

The selection of a distribution must necessarily be done with the aid of a computer, and there are software packages available for this purpose. It is better to use a computer-aided distribution fitting approach than to restrict one's attention to commonly used distributions that might not provide a good fit to a particular set of data.

After a distribution is selected by matching sample moments with population moments, confirmation can be made by using a *probability plot*, which was mentioned previously. The plot, which is illustrated in Section 6.1.2 in testing for a normal distribution, is a graphical way of determining if the sample data match up well with what would be expected if the data had come from the hypothesized distribution. There are various ways of constructing a probability plot, but the general idea is to see if the sample data are close to the percentiles of the distribution that is under consideration. Probability paper was used decades ago, with the points plotted by hand on the paper. The percentiles of the distribution on the vertical axis would be appropriately scaled and if the ordered sample data matched up well with those percentiles, the data would plot almost as a straight line. Now we generally use computers in place of probability paper and Figures 6.1 and 6.2 are computer-generated normal probability plots that are of the probability paper type. Probability plots can also easily be constructed for many other distributions, and MINITAB provides the capability for constructing such a plot for 13 other distributions.

3.6 HOW TO SELECT A DISTRIBUTION

There are certain guiding principles that can be followed in selecting a probability distribution in a particular application, and for a student to determine which distribution to use on a test problem.

A distinction must be made between when to use the binomial distribution and when to use the Poisson distribution. One main difference is that the largest possible value of the binomial random variable is the sample size, whereas there is no upper bound on the largest possible value of the Poisson random variable. A binomial distribution can be used when there are Bernoulli trials with two possible outcomes for each trial, and although we can use the Poisson distribution to model, say, the number of defects, we do not have an inspection unit that is declared to either have or not have a defect. So there is no Bernoulli trial.

Among the other discrete distributions, it is also necessary to distinguish between the hypergeometric distribution and the binomial distribution. There are two possible outcomes in practical applications of each, but with the former, sampling is done without replacement from a finite population, which is not the case with the binomial distribution. There is also a subtle difference between the geometric distribution and the binomial distribution, with the former being a waiting time distribution. Clearly, there is a difference between the probability of observing the first defective capacitor on the fourth capacitor examined (a geometric distribution problem), and the probability of observing one defective capacitor in the first four that are examined.

The picture is somewhat different with continuous distributions, as some distributions come into play primarily from a theoretical perspective as being distributions of functions of sample statistics, whereas other distributions are used for modeling. Whereas the wording of a problem or objective can determine which discrete distribution to use, it is somewhat different with continuous distributions. Subject-matter knowledge will often guide the selection of a distribution, as in reliability work. Although the term "normal distribution" suggests that this distribution is used more than any other continuous distribution, which is undoubtedly true as normality or approximate normality is assumed when many statistical procedures are used, there are many situations, including reliability work, when a skewed distribution is required. The t-distribution results when normality is assumed for individual observations, so it is tied-in with the normal distribution.

3.7 SUMMARY

The choice of which distribution to select for the purpose of determining probabilities such as those given in this chapter and for other purposes depends to a great extent on how much accuracy is required as well as the availability of tables and/or software.

Many statistical methods are based on the assumption of a normal distribution for individual observations and in the case of regression analysis (Chapters 8 and 9), the error term in a regression model is assumed to have a normal distribution. Similarly, all of the control charts presented in Chapter 11 have an implicit assumption of either a normal distribution (approximately) or the adequacy of the approximation of the normal distribution to another distribution. The consequences of making such an assumption when the assumption is untenable will vary from chart to chart, and in general from statistical method to statistical method. This is discussed in subsequent chapters.

In general, distribution fitting is quite important, although it is somewhat of a lost art. Using as many as the first four sample moments (mean, variance, and standardized skewness and kurtosis), one or more distributions could be fit to the data and tested for the quality of the fit. The testing can be performed using numerical measures such as goodness-of-fit tests (see Chapter 16) and by constructing probability plots, as in Chapter 6.

Many probability distributions were presented in this chapter. These are not simply of academic interest as some of them are used extensively in engineering. For example, even extreme value distributions (see Section 3.4.7), which are generally not presented in introductory statistics texts, are used frequently in practice. An example is the paper by J.-H. Lin and C. C. Wen "Probability Analysis of Seismic Pounding of Adjacent Buildings" (*Earthquake Engineering and Structural Dynamics*, **30**, 1339–1357, 2001). One of the major findings of their study is that the Type I extreme value distribution provides a good fit to the probability distribution of the required separation distance to avoid pounding under certain earthquake motions. (There is one Type I extreme value distribution for the largest value and one for the smallest value.)

REFERENCES

Berrettoni, J. N. (1964). Practical applications of the Weibull distribution. *Industrial Quality Control*, **21**(1), 71–79.

Fang, Y. (2003). C-charts, X-charts and the Katz family of distributions. *Journal of Quality Technology*, **35**, 104–114.

Geary, R. C. (1947). Testing for normality. *Biometricka*, **34**, 209–242.

Gelman, A. and D. Nolan (2002). You can load a die, but you can't bias a coin. *The American Statistician*, **56**(4), 308–311.

Haimes, Y. Y. (2004). *Risk Modeling, Assessment, and Management*. Hoboken, NJ: Wiley.

Hess, W. (2004). Representation of particle size distributions in practice. *Chemical Engineering and Technology*, **27**(6), 624–629.

Lee, C. and G. A. D. Matzo (1998). An evaluation of process capability for a fuel injector process using Monte Carlo simulation. In *Statistical Case Studies: A Collaboration Between Academe and Industry* (R. Peck, L. D. Haugh, and A. Goodman, eds.). Philadelphia, PA: Society for Industrial and Applied Mathematics and Alexandria, VA: American Statistical Association.

Murthy, D. N. P., M. Xie, and R. Jiang (2004). *Weibull Models*. Hoboken, NJ: Wiley.

Snedecor, G. W. and W. G. Cochran (1974). *Statistical Methods*, 7th ed. Ames, IA: Iowa State University Press.

EXERCISES

3.1. What is the numerical value of $P(B|A)$ when A and B are mutually exclusive events?

3.2. Show algebraically why $P(M_3|N)$ had to be equal to $P(M_2|N)$ in the modified example in Section 3.2.1.1 due to the fact that $P(N|M_2)/P(N|M_3) = P(M_3)/P(M_2)$.

3.3. A survey shows that 40% of the residents of a particular town read the morning paper, 65% read the afternoon paper, and 10% read both.

    (a) What is the probability that a person selected at random reads the morning or the afternoon paper, or both?

    (b) Now assume that the morning and afternoon papers are subsequently combined, as occurred with the Atlanta papers in February 2002, with the paper delivered in the morning. If the paper doesn't lose (or gain) any subscribers, can the probability that a person selected at random receives the single paper be determined? If so, what is the probability? If not, explain why not.

3.4. A system is comprised partly of three parallel circuits. The system will function at a certain point if at least two of the circuits are functioning. If the circuits function independently and the probability that each one is functioning at any point in time is .9, what is the probability that the system is functioning (relative to those circuits)?

3.5. Given $P(A) = 0.3$, $P(B) = 0.5$, and $P(A|B) = 0.4$, find $P(B|A)$.

3.6. A die is rolled and a coin is tossed. Find the probability of obtaining a 1, 3, 4, or 6, and a head or a tail.

3.7. Give an example of a discrete random variable and indicate its possible values. Then assign probabilities so as to create a probability mass function.

3.8. A company produces 150,000 units of a certain product each day. Think of this as a sample from a population for which 0.4% of the units are nonconforming. Letting X denote the number of nonconforming units per day, what is the numerical value of σ_x^2? (Assume that the units are independent, but as a practical matter does it seem as though one could automatically assume this? Why or why not?)

3.9. Given the probability density function $f(x) = 4x, 1 < x < \sqrt{1.5}$, determine $P(1.0 < X < 1.1)$.

3.10. Assume a normal distribution with a known standard deviation, with 34% of the observations contained in the interval $\mu \pm a$. If the standard deviation is reduced by 50% and the mean is unchanged, what percentage of observations will lie in the interval $\mu \pm a$?

3.11. Assume that the amount of time a customer spends waiting in a particular supermarket checkout lane is a random variable with a mean of 8.2 minutes and a standard deviation of 1.5 minutes. Suppose that a random sample of 100 customers is observed.

    (a) Determine the approximate probability that the average waiting time for the 100 customers in the sample is at least 8.4 minutes.

    (b) Explain why only an approximate answer can be given.

3.12. A random variable X has a normal distribution with a mean of 25 and a variance of 15. Determine a such that $P(X < a) = .9850$.

3.13. A balanced coin is tossed six times. What is the probability that at least one head is observed?

3.14. Let X_1, X_2 be a random sample of size $n = 2$ from a discrete uniform distribution with $f(x) = 1/4, x = 1, 2, 3, 4$. Let $Y = X_1 - X_2$. Determine the mean and variance of Y.

3.15. Explain how you might proceed to determine a continuous distribution that provides a good fit to your set of data.

3.16. The probability that a customer's order is shipped on time is .90. A particular customer places three orders, and the orders are placed far enough apart in time that they can be considered to be independent events. What is the probability that exactly two of the three orders are shipped on time?

3.17. Explain what is wrong with the following statement: "The possible values of a binomial random variable are 0 and 1."

3.18. Given $P(A|B) = .4$, $P(B|A) = .3$, and $P(A) = .2$, determine $P(B)$.

3.19. Quality improvement results when variation is reduced. Assume a normal distribution with a known standard deviation, with 34% of the observations contained in the interval $\mu \pm a$. If the standard deviation is reduced by 20% but the mean is not kept on target and increases by 10% from μ, obtain if possible the percentage of observations that will lie in the interval $\mu \pm a$? If it is not possible, explain why the number cannot be obtained.

3.20. Show that $E(X - \mu)^2 = E(X^2) - (E(X))^2$.

3.21. Show that $E(a + bX) = a + bE(X)$ and $Var(a + bX) = b^2 Var(X)$.

3.22. Given: $f(x) = 3x^2 \quad 0 < x < 1$.
  (a) Graph the function.
  (b) On the basis of your graph, should $P(0 < X < 0.5)$ be less than .5, equal to .5, or greater than .5? Explain.
  (c) Answer the same question for $E(X)$.
  (d) Compute $P(0 < X < 0.5)$.
  (e) Determine $E(X)$.

3.23. A company employs three different levels of computer technicians, which it designates with classifications of A, B, and C. Thirty percent are in classifications A and B, and 40% are in C. Only 20% of the "A" technicians are women, whereas the percentages for B and C are 30% and 35%, respectively. Assume that one technician has been selected randomly to serve as an instructor for a group of trainees. If the person selected is a man, what is the probability that he has a B classification?

3.24. Assume that flaws per sheet of glass can be represented by a Poisson distribution, with an average of 0.7 flaws per sheet. What is the probability that a randomly selected sheet of glass has more than one flaw?

3.25. Assume that you use a random number generator to generate an integer from the set of integers 1–4 and the selection is truly random. Let X represent the value of the selected integer and compute $Var(X)$.

3.26. Explain, mathematically, why the variance of a random variable is nonnegative when the possible values of a random variable (either discrete or continuous) are all negative. [An example would be $f(x) = 1/5, x = -6, -4, -3, -2, -1$.] What would be the nature of the distribution of a random variable that has a zero variance?

3.27. If a die is rolled three times, give the probability distribution for the number of twos that are observed.

3.28. In a throw of 12 dice, let X_j denote the number of dice that show j. Find $E(X_1)$.

3.29. You are contemplating investing in a particular stock tomorrow and one of your brokers believes there is a 40% chance that the stock will increase over the next few weeks, a 20% chance that the price will be essentially unchanged, and a 40% chance that the price will decline. You ask another broker and she gives you 35%, 30%, and 35%, respectively, for the three possible outcomes. Assume that you accept these percentages. If $X =$ the outcome under the first broker's judgment and $Y =$ the outcome under the second broker's judgment, which random variable has the smaller variance? What course of action would you follow?

3.30. Critique the following statement. "I don't see how we can assume that any random variable in an engineering application has a normal distribution because the range of any practical engineering variable will not be minus infinity to plus infinity, yet this is the range of any normal distribution."

3.31. Assume that time to fill an order for condensers of all types at a particular company is assumed to have (roughly) a normal distribution. As a result of an influx of new employees, it is claimed that the mean time has increased but not the variance, and the distribution is still approximately normal. Critique the following statement: "The distribution of times may appear to have the same variance, but if the distribution is normal, the variance must have decreased because the increase in time has moved the distribution closer to a conceptual upper boundary on time to fill an order (say, 2 months), so that the range of possible times has decreased and so has the variance."

3.32. Let the probability mass function (*pmf*) of X be given by

x	$P(X = x)$
5	1/4
6	1/4
7	1/4
8	1/4

Let $Y = X + 6$. Give the *pmf* of Y and determine $E(Y)$.

3.33. If X and Y are independent random variables, $\sigma_x^2 = 14$, $\sigma_y^2 = 20$, $E(X) = 20$, and $E(Y) = 40$, what must be the numerical value of $E(XY)$?

3.34. A balanced coin is flipped n times. If the probability of observing zero heads is 1/64, what is the probability of observing n heads?

3.35. Assume that $X \sim N(\mu = 50, \sigma^2 = 16)$.
   (a) Determine $P(X > 55)$.
   (b) Determine $P(X = 48)$.
   (c) Determine the lower bound on how large the sample size must have been if $P(\overline{X} > 51) < .06681$.

3.36. Critique the following statement: "People always say that aptitude test scores, such as SAT scores, have a normal distribution, but that can't be right because test scores are discrete and the normal distribution is a continuous distribution."

3.37. The following question was posed on an Internet message board: "If I have three machines in a department that have the potential to run for 450 minutes per day, but on average each one runs for only 64 minutes, what is the probability of them all running at the same time?" The person wanted to know if one of the machines could be replaced, this being the motivation for the question. Can this answer be determined, and will the computed probability address the central issue?

3.38. Three construction engineers are asked how long a particular construction project should take to complete. If we assign equal weight to their estimates, and those estimates are 256 days, 280 days, and 245 days, respectively, can a probability distribution be formed from this information? If so, what is the name of the distribution? In particular, is the distribution discrete or continuous? If X denotes the time to complete the project, what is $E(X)$?

3.39. The term "engineering data" is often used in the field of engineering to represent relationships, such as the following relationship: tensile strength (ksi) $= 0.10 + 0.36$(CMMH), with CMMH denoting "composite matrix microhardness." If we wanted to model tensile strength in a particular application by using an appropriate statistical distribution, would the relationship between tensile strength and CMMH likely be of any value in selecting a distribution? What information must be known, in general, before one could attempt to select a model?

3.40. A two-stage decision-making process is in effect. In the first stage, one of two possible courses of action (A_1 and A_2) is to be taken, and in the second stage there are three possible courses of action (B_1, B_2, and B_3). If each of the two possible courses of action has an equal chance of occurring in the first stage and B_3 has a probability of .40 of being selected with the other two being equally likely, what is the probability of A_1 and either B_1 or B_3 being selected if the course of action in the second stage is independent of the selection in the first stage?

3.41. Which one of the four moments discussed in the chapter is in the original unit of measurement?

3.42. What is always the numerical value of the first population moment about the mean?

3.43. If $P(A|B) = P(A)$, show that $P(B|A)$ must equal $P(B)$.

3.44. Determine the second moment about the mean for the continuous uniform distribution defined on the unit square in Section 3.4.2.

3.45. Given: $f(x) = 1/4 \quad 0 < x < 4$.
   (a) Show that this is a probability density function.
   (b) Sketch the graph of the function. Is this one of the distribution types covered in the chapter? Explain.
   (c) Determine $Var(X)$.

3.46. Explain why moments for a normal distribution can be computed directly using integration, but probabilities under a normal curve cannot, in general, also be computed using integration.

3.47. Given the following *pdf*, determine $E(X)$ and the variance:

X	2	4	6	7
P(X)	1/4	1/3	1/6	1/4

3.48. Assume an arbitrary *pmf*, $f(x)$. Show that $Var(x)$ cannot be negative.

3.49. Assume that you are given (most of) a *pmf*. If you were given the following, explain why the listed probabilities could not be correct:

X	1	2	3	4	5
P(X)	1/8	1/4	1/3	1/2	?

3.50. With $Z \sim N(0, 1)$, find z_0 such that $P(Z > z_0) = .8849$.

3.51. Assume that in a very large town, 20% of the voters favor a particular candidate for political office. Assume that five voters are selected at random for a television interview and asked for their preferences. If they express their preferences independently, what is the probability that at most one of them will indicate a preference for the other candidate?

3.52. Assume a binomial distribution and let a denote the probability of zero successes in a particular application. Express the probability of one success as a function of a.

3.53. Consider a binomial distribution problem with $P(X = n) = 1/128$. What is $P(X = 0)$? Can there be more than one answer? Explain.

3.54. Explain the conditions under which the binomial distribution would be inappropriate even when there are only two possible outcomes.

3.55. Given the following probability mass function—$f(x) = a, 2 \le x \le 8$—determine the numerical value of a.

3.56. Why does the standard normal distribution have shorter tails than the t-distribution? Stated differently, why does $Z = (\overline{X} - \mu)/(\sigma/\sqrt{n})$ have less variability than $t = (\overline{X} - \mu)/(s/\sqrt{n})$?

3.57. Assume that $X \sim N(\mu = 20, \sigma = 3)$. If $Y = 2X - 8$:
   (a) What are the mean and variance of Y?
   (b) Would your answers be different if you weren't asked to assume normality? Explain.

3.58. Critique the following statement: "For a random sample of size n, the possible values of a Poisson random variable are 0 through n."

3.59. Given: $f(x) = 4x^3, 0 < x < a$:
   (a) What must be the value of a for the function to be a *pdf*?
   (b) Explain why $E(X)$ is not midway between 0 and a.

3.60. When a variable is standardized, what will be the mean of the standardized variable?

3.61. The August 2, 2000 edition of *People's Daily* stated that, according to a national survey of 4,000 people, 91.6% of Chinese are in favor of amending the existing Marriage Law. Approximate the probability that out of 5 randomly selected Chinese from the same population to which the survey was applied, at least 4 will favor an amendment. Why is this an approximation and not an exact value?

3.62. Western Mine Engineering, Inc. conducts surveys to obtain data on the operations of mines in the United States and elsewhere. In their 2002 survey of 180 U.S. mines, all 82 of the union mines indicated that workers' medical insurance was either partially or completely paid by the company, with 97 of the non-union mines indicating that the company paid the insurance.

  (a) If you obtained a random sample of five of the 180 U.S. mines that paid all or part of their workers' medical insurance, what is the probability that at least three of the mines in your sample were non-union mines?

  (b) If you randomly selected one of the 180 U.S. mines, what is the probability that it was a non-union mine?

  (c) If you randomly select two mines (without replacement from the 180 mines), what is the probability that one is a union mine and the other is a non-union mine?

  (d) If you randomly selected one of the 180 mines and discovered that you have obtained one that pays all or part of the medical insurance, what is the probability that the mine you selected was a non-union mine?

3.63. Acceptance sampling methods are used in industry to make decisions about whether to accept or reject a shipment, for example. Assume that you have a shipment of 25 condensers and you decide to inspect 5, without replacement. You will reject the shipment if at least one of the five is defective.

  (a) In the absence of any information regarding past or current quality, does this appear to be a good decision rule?

  (b) What is the probability of rejecting the shipment if the latter contains one bad condenser?

3.64. Give two reasons why the function $f(x) = x$, $-1 < x < 1$, could not be a probability density function (*pdf*).

3.65. Consider the function $f(x) = x^2$ with the range of the function as defined in Exercise 3.64. Is this a *pdf*? Explain.

3.66. Explain why a sample size does not come into play when computing Poisson probabilities, unlike the case with binomial probabilities.

3.67. There is a 25% chance that a particular company will go with contractor A for certain work that it needs to have done, and a 75% chance that it will go with contractor B. Contractor A finishes 90% of his work on time and contractor B finishes 88% of his work on time. The work that the company wanted done was completed on time, but the identity of the contractor was not widely announced.

What is the probability that the contractor selected for the job was contractor *A*?

3.68. Assume that two random numbers are generated from the standard normal distribution.

   (a) What is the probability that they both exceeded 1.60?

   (b) What is the probability that at most one of the two numbers was less than 1.87?

3.69. Assume that one of the eight members of the customer service department of a particular retailer is to be randomly selected to perform a particular unsavory task. Numbers are to be generated from the unit uniform distribution [i.e., $f(x) = 1$, $0 < x < 1$]. The employees are assigned numbers 1–8 and the first random number is assigned to employee #1, the second random number to employee #2, and so on. The first employee whose assigned random number is greater than 0.5 is assigned the task.

   (a) What is the probability that employee #4 is picked for the job?

   (b) What is the probability that no employee is selected and the process has to be repeated?

3.70. Engineers frequently define their data interval as the mean plus and minus the standard deviation.

   (a) Under the assumption of normality, what percentage of values in the population would this cover *if* the reference had been to μ and σ?

   (b) Since the reference is actually to $\bar{x} \pm s$, what is the probability that for a sample of 30 observations the interval given by $\bar{x} \pm s$ will be exactly the same as the interval given by $\mu \pm \sigma$?

3.71. Bayes' theorem is often covered in engineering statistics courses. A student is skeptical of its value in engineering, however, and makes the following statement after class one day: "In working a Bayes problem, one is generally going backward in time. I am taking another class that is covering the contributions of W. Edwards Deming. He claimed that 'statistics is prediction,' which obviously means that we are moving forward in time. I don't see how any statistical tool that takes us back in time can be of any value whatsoever."

   (a) How would you respond to this student?

   (b) If possible, give an application of Bayes' theorem from your field of engineering or science. If you are unable to do so, do you agree with the student? Explain.

3.72. Consider the *pdf* $f(x) = cx + d$, $-1 < x < 1$.

   (a) Is it necessary to know the value of c in order to solve for d? If so, explain why; if not, solve for the value of d.

   (b) If $P(0.2 < X < 0.8)$ can be determined from the information that is given, obtain the probability. If not, explain why the probability cannot be obtained.

   (c) Obtain $E(X)$ as a function of c.

3.73. Assume $X \sim N(\mu, \sigma^2)$. Obtain $E(X^2)$ as a function of μ and σ^2.

3.74. In answering questions on a multiple-choice exam, a student either knows the answer or guesses, with the probability of guessing correctly on each question being 0.25 since each question has four alternatives. Let p denote the probability that the student knows the answer to a particular question.

   (a) If a student answered the question correctly, what is the probability that the student guessed, as a function of p?

   (b) Show that, even though p is unknown, the probability cannot exceed one.

3.75. Assume that 10 individuals are to be tested for a certain type of contagious illness and a blood test is to be performed, initially for the entire group. That is, the blood samples will be combined and tested, as it is suspected that no one has the illness, so there will be a cost saving by performing a combined test. If that test is positive, then each individual will be tested. If the probability that a person has an illness is .006 for each of the 10 individuals, can the probability that the test shows a positive result be determined from the information given? If not, what additional information is needed?

3.76. A computer program is used to randomly select a point from an interval of length M, starting from 0 ($M > 0$).

   (a) (Harder problem) Let the point divide the interval into two segments and determine the probability that the ratio of the shorter to the longer segment is less than .20.

   (b) Let X denote the value that will result when the computer program is used and now assume that the interval begins at A and ends at B. Determine $Var(X)$ as a function of A and B.

3.77. Becoming proficient at statistical thinking is more important than the mastery of any particular statistical technique. Consider the following. There are certain organizations that have limited their memberships to males and have not had female members. A person applies for membership with his or her height indicated, but gender not filled in. Since the name "Kim" can signify either a man or a woman, the membership committee is unsure of the gender of an applicant with this first name. The person's height is given as 5ft $7\frac{1}{2}$in. The mean heights of all men and all women are of course known (approximately), and assume that the standard deviations are also known approximately. Assume that heights are approximately normally distributed for each population and explain how the methods of this chapter might be used to determine whether the applicant is more likely a woman or a man.

3.78. A department has six senior faculty members and three junior faculty members. The membership of a departmental committee of size 3 is to be determined randomly. What is the probability that at most one junior faculty member is selected to be on the committee?

3.79. Service calls come to a maintenance center in accordance with a Poisson process at a rate of $\lambda = 2.5$ calls every 5 minutes. What is the probability that at least one call comes in 1 minute?

3.80. A random variable has a normal distribution with a mean of 0. If $P(X > 10) = .119$, what is the numerical value of σ?

3.81. The lifetime of a device has an exponential distribution with a mean of 100 hours. What is the probability that the device fails before 100 hours have passed? Explain why this probability differs considerably from .5 even though the mean is 100.

3.82. Assume that a particular radio contains five transistors and two of them are defective. Three transistors are selected at random, removed from the radio, and inspected. Let X denote the number of nonconforming transistors observed and determine the probability mass function of X.

3.83. Assume that X has an exponential distribution with a mean of 80. Determine the median of the distribution.

3.84. (Harder problem) Demonstrate the memoryless property of an exponential distribution by showing that the probability that a capacitor will function for at least 100 more hours does not depend on how long it has been functional.

3.85. Assume that all units of a particular item are inspected as they are manufactured. What is the probability that the first nonconforming unit occurs on the 12th item inspected if the probability that each unit is nonconforming is .01?

3.86. Assume that a particular automotive repair facility doubles as an emission testing station, and the manager believes that arrivals for emission testing essentially follow a Poisson distribution with a mean of 3 arrivals per hour. The station can handle this many arrivals, but it cannot, for example, handle 6 arrivals in an hour. What is the probability of the latter occurring?

3.87. Consider a particular experiment with two outcomes for which the binomial distribution seemed to be an appropriate model, with $p = 1/8$. The probability of $n - 1$ successes is equal to 350 times the probability of n successes. What is the numerical value of n?

3.88. What is the probability that a 6 is observed on the eighth roll of a die when a die is rolled multiple times?

3.89. (Harder problem) Two students in your engineering statistics class decide to test their knowledge of probability, which has very recently been covered in class. They each use MINITAB or other statistical software to generate a number that has a continuous uniform distribution over a specified interval. The first student uses the interval 0–3 and the second student uses the interval 0–1. Before the random numbers are obtained, the students compute the probability that the first student's number will be at least twice the second student's number.
(a) What is that probability?
(b) What is the probability after the random numbers have been obtained? Explain.
(c) What is the probability that the maximum of the two numbers is at most 1/2?
(d) What is the probability that the minimum of the two numbers is at least 1/4?

3.90. A power network involves substations A, B, and C. Overloads at any of these substations might result in a blackout of the entire system. Records over a very long period of time indicate that if substation A alone experiences an overload, there is a 1% chance of a network blackout, with the percentages for B and C being 2% and 3%, respectively. During a heat wave there is a 55% chance that substation A alone will experience an overload, with the percentages for stations B and C being 30% and 15%, respectively. (Although two or more substations have failed, this has happened so rarely that we may use 0% as the percentage of the time that it has occurred, as that is the number to the nearest integer.) A blackout due to an overload occurred during a particular heat wave. What is the probability that the blackout occurred at substation A?

3.91. You notice that only 1 out of 20 cars that are parked in a particular tow-away zone for at least 10 minutes is actually towed. Because of very bad weather, you are motivated to park your car as close as possible to a building in which you have short, 10–15-minute meetings for two consecutive days. You decide to take a chance and park in the tow-away zone for your meetings on those two days. You decide, however, that if your car is towed on the first day that you won't park there on the second day. What is the probability that your car will be towed? (Have you made an assumption in working the problem that might be unrealistic? Explain.)

3.92. Phone messages come to your desk at a rate of 3 per hour. Your voice mail isn't working properly and since your secretary is off for the day, you are afraid that you will miss a call if you step out of your office for 15 minutes.
   (a) Can the probability of this occurring be determined from the information that is given? Why or why not?
   (b) Would it be possible to assume that the calls have a binomial distribution? Why or why not?
   (c) Work the problem under the assumption of a Poisson distribution with a mean of 2 calls per hour. Is this a reasonable assumption?

3.93. Two students in a statistics class are told that their standard scores (i.e., z-scores as in Section 3.4.3 except that $\bar{x}$ and s are used instead of μ and σ, respectively) on a test were 0.8 and -0.2, respectively. If their grades were 88 and 68, what were the numerical values of $\bar{x}$ and s?

3.94. Consider the value of $t_{11,.025}$. Determine, using software if appropriate, the value of α in $z_{\alpha/2}$ such that the z-value is equal to $t_{11,.025}$.

3.95. Consider the probability statement $P(2z_{\alpha/2} > z_0) = .236$, with $z \sim N(0, 1)$. Determine the numerical value of z_0. What is the numerical value of α?

3.96. A probability problem that gained nationwide attention several years ago was related to the game show "Let's Make a Deal." The following question was posed in the "Ask Marilyn" column in *Parade* magazine in 1990:

> Suppose you are on a game show, and you're given the choice of three doors. Behind one door is a car; behind the others, goats. You pick a door, say, #1, and the host, who knows what's behind the doors, opens another door, say, #3, which has a goat. He says to you "Do you want to pick door #2?" Is it to your advantage to switch choice of doors?

The question can be answered using probability, but the question confounded a large number of people who initially thought that they knew the answer, but in fact they did not. What is your answer? If you wish to compare your answer and reasoning with the "final word" on the matter that appeared in the statistical literature, you are invited to read "Let's Make a Deal: The Player's Dilemma" (with commentary), by J. P. Morgan, N. G. Chaganty, R. C. Dahiya, and M. J. Doviak, *The American Statistician*, **45**(4), 284–289, 1991.

3.97. A department with seven people is to have a meeting and they will all be seated at a round table with seven chairs. Two members of the department recently had a disagreement, so it is desirable that they not be seated next to each other. If this requirement is met but the seating is otherwise assigned randomly, how many possible seating arrangements are there and what is the probability associated with any one of those arrangements?

3.98. An engineering consulting firm has two projects (which we will label as A and B) that it wants to work on in a forthcoming week, and because of limited resources it can work on only one project at a time. Project A will require more time than project B, and it is possible that neither project will be completed in a week. Let $P(B|A)$ denote the probability that project B is completed within a week given that project A has been completed, and similarly define $P(A|B)$. The head of the firm does not understand conditional probability and simply wants to know the chances of both projects being completed within the week.

(a) Can that be determined from what is given? Why or why not? Explain.
(b) Can this be determined if either $P(A)$ or $P(B)$ is known, without conditioning on the occurrence or nonoccurrence of the other event? Explain.

3.99. Given that $P(X = 1) = 2P(X = 4)$, complete the following table and determine the expected value of X.

x	1	4	6
$P(x)$	—	—	1/6

3.100. Assume that $X \sim N(\mu, \sigma^2)$. If $\mu = 19$ and $P(X < 22.5) = .63683$, determine σ^2.

CHAPTER 4

Point Estimation

The remaining chapters are devoted to statistical inference, that is, making inferences about population parameters from sample statistics. We begin our discussion of inference with point estimation and then proceed into interval estimation and hypothesis tests, which are covered in subsequent chapters.

4.1 POINT ESTIMATORS AND POINT ESTIMATES

We begin by making the important distinction between *point estimators* and *point estimates*. The former are statistics whereas the latter are the numerical values of statistics. That is, $\overline{X}$ is a point estimator of μ, and if a sample of 100 observations is taken and $\overline{X}$ is computed to be 86.4, then the latter is the point estimate, which is often denoted by a lowercase letter (e.g., $\overline{x}$). It is important to distinguish between the two terms because the expression "properties of point estimates" doesn't make any sense because a number cannot have any statistical properties, such as having a variance.

4.2 DESIRABLE PROPERTIES OF POINT ESTIMATORS

A point estimate will almost certainly not be equal to the parameter value that it serves to estimate, but we would hope that it would not deviate greatly from that number. Furthermore, although in most applications of statistics we usually take only one sample within a short time period, if we were to take a second sample we would hope that the second estimate would not differ greatly from the first estimate. This can be accomplished by taking a large sample.

4.2.1 Unbiasedness and Consistency

We would also hope that, from a purely conceptual standpoint, if we kept taking samples of increasing size (50, 100, 500, etc.), the difference between the point estimate and the population value would approach zero. If so, then the estimator is *consistent*. Obviously, this is a desirable property.

Modern Engineering Statistics   By Thomas P. Ryan
Copyright © 2007 John Wiley & Sons, Inc.

Although multiple samples are sometimes taken, such as when control charts are used (see Chapter 11), we need properties that apply directly to a single sample. One such property is *unbiasedness*. Consider an arbitrary estimator, $\widehat{\theta}$, of some parameter θ (mean, variance, proportion, etc.). If the expected value of $\widehat{\theta}$ [i.e., $E(\widehat{\theta})$] is equal to θ, then $\widehat{\theta}$ is an *unbiased estimator* of θ. (Expected value was illustrated in Sections 3.3.1 and 3.4.1.) Unbiasedness can also be viewed as a limiting result in that if the number of samples of a fixed size becomes large, then the average of the $\widehat{\theta}$ values becomes arbitrarily close to θ. For example, let's assume that our population consists of the first ten positive integers. We select a sample of five numbers from this population, with repeat numbers allowed, and compute the average of the five numbers. We then repeat this process a very large number of times and compute the average of the averages. That number should be extremely close to 5.5, the average of the ten numbers in the population. (Recall this type of illustration in Section 3.4.4.) Again, we are not particularly interested in what happens when many samples are taken if we are just going to take a single sample, but unbiasedness does apply to a single sample. Specifically, if we take only one sample, our best guess as to what the average of the five numbers will be is the same 5.5, although the actual average may differ more than slightly from 5.5.

Although unbiasedness is a desirable property, there will be instances in which other properties should receive greater consideration, especially when the bias is small.

Bias is even ignored when some statistical methods are used, and in such cases the bias is usually small. For example, under the assumption of a normal distribution, s^2 is an unbiased estimator of σ^2, but s is a biased estimator of σ. When confidence intervals are computed (e.g., see Chapter 5), the bias is ignored. The bias is adjusted for, however, when control chart methods (Chapter 11) are used, since if $E(s) = a\sigma$, it then follows that $E(s/a) = \sigma$. If the value of a is very close to 1, the bias might be ignored, although in typical control chart applications a is not very close to 1. Strictly speaking, the bias adjustment is made under the assumption that one has the right model for the observations, such as using a normal distribution as a model. Similarly, one speaks of unbiased estimators in regression analysis (Chapter 8), but the estimators are unbiased in general only if the postulated model is correct, which isn't likely to happen.

Therefore, although we may speak conceptually of unbiased and biased estimators, from a practical standpoint estimators in statistics are often biased simply because the postulated model is incorrect, and such estimators cannot be made unbiased in the absence of strong prior information.

4.2.2 Minimum Variance

There will often be multiple estimators of a parameter that are unbiased. To give a simple illustration, assume that we are to take a sample of four observations from an unknown population for the purpose of estimating the population mean. Consider two estimators, $\overline{X}$ and X^*:

$$\overline{X} = \frac{X_1 + X_2 + X_3 + X_4}{4} \qquad X^* = \frac{X_1 + 2X_2 + 2X_3 + X_4}{6}$$

with X_1, X_2, X_3, and X_4 denoting the random variables whose realizations are the four numbers in the sample.

Since $E(X) = \mu$, the expected value of a sum is the sum of the expected values [i.e., $E(\sum x_i) = \sum E(x_i)$], and the expected value of a constant times a random variable is the constant times the expected value of the random variable, as the reader was asked to show in Exercise 3.21 [e.g., $E(\sum x_i/n) = (1/n)E(\sum x_i)$], it follows that $E(\overline{X}) = \mu$. By applying the same rules for expected values, we can also show that $E(X^*) = \mu$.

These results are summarized below.

RULES FOR EXPECTED VALUES

1. $E(X) = \mu$, the population mean.
2. $E(aX) = aE(X)$.
3. $E(\text{constant}) = \text{constant}$.
4. $E(\sum_{i=1}^{n} X_i) = \sum_{i=1}^{n} E(X_i)$.

Since both estimators are unbiased, which one is better? Of course, on the surface X^* seems to be a rather nonsensical estimator since the four random variables have unequal weights, but in the use of robust statistics (mentioned in Section 2.2), observations will typically have unequal weights. Such weighting is generally performed *after* the data have been obtained, however.

If the X_i are independent, as they will be when a random sample is obtained, $Var(\sum X_i) = \sum Var(X_i)$. Combining this result with $Var(aW) = a^2 Var(W)$ for any random variable W (as the reader was asked to show in Exercise 3.21), it follows that $Var(\overline{X}) = \sigma^2/4$ and $Var(X^*) = 5\sigma^2/18$, with σ^2 denoting $Var(X)$. Thus, $\overline{X}$ has a slightly smaller variance and is the better estimator (as we would have logically guessed). The general result is $Var(\overline{X}) = \sigma^2/n$.

It should be noted that this result is actually an "infinite population" result. If a sample is obtained from a finite population and the population is small (not a common occurrence), it is necessary to multiply this result by a *finite population correction factor (fpc)*. Specifically, the *fpc* is given by $\sqrt{(N - n)/(N - 1)}$, with N denoting the population size and n denoting the sample size. Note that the numerical value of the *fpc* will be close to 1.0 (and thus have no effect), when n is small relative to N. The usual recommendation is to not use the *fpc* whenever $n/N < 0.05$. Since N is generally quite large (and often unknown), this requirement is usually met.

These results (for the infinite or large sample case) are summarized below.

RULES FOR VARIANCES

Let $X_1, X_2, \ldots, X_n$ represent a sequence of *independent* random variables, such as the random variables corresponding to the observations in a sample of size n.

Using the following rules, we can obtain the variance of any function of a random variable:

1. $Var(aX_i) = a^2 Var(X_i)$.
2. $Var(\sum_{i=1}^{n} X_i) = \sum_{i=1}^{n} Var(X_i)$.
3. $Var(\text{constant}) = 0$.
4. $Var(\text{constant} + X) = Var(X)$.

We may note that the right-hand side of the equation in Rule #2 has an additional term, $2\sum_{i=1}^{j-1}\sum_{j=2}^{n} Cov(X_i, X_j)$, when the X_i are not independent.

4.2.3   Estimators Whose Properties Depend on the Assumed Distribution

For this simple example we didn't need to assume a distribution in the process of determining which estimator had the smaller variance. That won't be the case, in general, however. For example, let's assume that we wanted to compare the sample median with the sample mean as an estimator of the population mean. In general, if we choose to use the sample median as a statistic, we would assume that this estimates the population median, which is the same as the population mean in the case of a symmetric distribution. The sample mean and sample median are both unbiased for a symmetric distribution, but the sample median is a biased estimator of the population mean for asymmetric distributions. This logically follows from the fact that the population mean and median are different for such distributions.

Since it is well known that the sample mean is the minimum variance unbiased estimator (MVUE) for the mean of a population that is normally distributed, it follows that the sample median must have a larger variance than the sample mean for that distribution. It is not true, however, that we would always prefer the mean over the median for an arbitrary symmetric distribution, as with a heavy-tailed, symmetric distribution extreme values could easily occur that would have a large effect on the sample mean, especially for a small sample, and cause it to be a poor estimate of the population mean. So the sample median could be a better choice for an arbitrary heavy-tailed distribution.

4.2.4   Comparing Biased and Unbiased Estimators

In Section 4.2.2 we compared two unbiased estimators in terms of their respective variances. In such a situation we will want to use the estimator with the smaller variance, but what if we have to choose between a biased estimator and an unbiased estimator? Unless the biased estimator has a larger variance than the unbiased estimator, the choice may not be obvious. Consequently, there is a need for a comparison yardstick that incorporates both bias and variance. That yardstick is the *mean squared error* of an estimator, defined as $E(\widehat{\theta} - \theta)^2$ for an arbitrary estimator $\widehat{\theta}$. With some algebra we obtain

$$E(\widehat{\theta} - \theta)^2 = E(\widehat{\theta} - E(\widehat{\theta}) + E(\widehat{\theta}) - \theta)^2$$
$$= E([\widehat{\theta} - E(\widehat{\theta})]^2 + 2[E(\widehat{\theta}) - \theta][\widehat{\theta} - E(\widehat{\theta})] + [E(\widehat{\theta}) - \theta]^2)$$

which results from squaring the expression on the right-hand side as a binomial. Then, when the expected value operator is applied to each of the three terms, the middle term drops out because the factor $[E(\widehat{\theta}) - \theta]$ is a constant, and $E(\widehat{\theta} - E(\widehat{\theta})) = E(\widehat{\theta}) - E(\widehat{\theta}) = 0$. Thus, we have

$$E(\widehat{\theta} - \theta)^2 = E(\widehat{\theta} - E(\widehat{\theta}))^2 + [E(\widehat{\theta}) - \theta]^2$$
$$= Var(\widehat{\theta}) + [Bias(\widehat{\theta})]^2$$
$$= MSE(\widehat{\theta})$$

with *MSE* denoting "mean squared error."

If an estimator is unbiased, the bias term disappears and the mean squared error is then simply the variance. Since expected value means "average," $E(\widehat{\theta} - E(\widehat{\theta}))^2$ is thus the average squared deviation of $\widehat{\theta}$ from its average.

There is obviously no need to use the *MSE* criterion when comparing two unbiased estimators, as that would reduce to comparing their variances. Instead, the criterion should be used for comparing two biased estimators and for comparing a biased estimator with an unbiased estimator.

As the reader is asked to show in Exercise 4.2, if the divisor of X^* as defined in Section 4.2.2 is changed so that the divisor is 10 instead of 6, then X^* has a smaller variance than $\overline{X}$, but that advantage is more than offset by the large squared bias when, for example, $\mu = 10$ and $\sigma^2 = 4$. Therefore, $\overline{X}$ has the smaller mean squared error.

4.3 DISTRIBUTIONS OF SAMPLING STATISTICS

Statistical distributions were covered in Chapter 3. These give the distribution of the total probability over the possible values of the random variable, which is the same as the distribution of the random variable for a sample of size one.

Of course, in practice we don't take samples of size one, so we need to consider the distributions of sample statistics for arbitrary sample sizes. Knowledge of these distributions is needed more for later chapters than for this chapter, but the concept is introduced here because the variability of sample statistics is inherent in material that is presented in later sections of this chapter.

Assume that $X \sim N(\mu, \sigma^2)$. Then what is the distribution of $\overline{X}$? If the sample observations are independent, it can be shown that $\overline{X} \sim N(\mu, \sigma^2/n)$, where the result for the mean and the variance was established previously in Section 3.4.4 and can also be obtained using the rules in Section 4.2.2, but not for the distribution of the statistic. The fact that $\overline{X}$ has a normal distribution can be shown using moment generating techniques (beyond the scope of this book), whereas the mean and variance can be obtained using the rules in Section 4.2.2, as the reader is asked to show in Exercise 4.1.

An important point of this result is that the variance of $\overline{X}$ decreases as the sample size increases. Of course, this is desired as we don't want any estimator that we use to have a large variance. We do have to pay a price for high precision as sampling costs money and the larger the sample, the higher the cost.

The estimator of the standard deviation of $\overline{X}$ is generally termed the *standard error*, which is the term applied to the estimated standard deviation of *any* statistic. Such usage is not universal, however, and some authors refer to such an estimator as the *estimated standard error*. We will follow convention and not use the latter term. This result is stated formally in Section 4.5. The concept of a point estimator and its standard error are central to statistical inference.

4.3.1 Central Limit Theorem

As has been stated repeatedly, we generally will not know the distribution of the random variable with which we are concerned, and if we don't know the distribution of a random variable X, we don't know the distribution of $\overline{X}$ either. This is more of a problem in interval estimation and hypothesis testing than it is in point estimation, but it can also create problems in point estimation. This is because, as discussed later in Sections 4.4.1 and 4.4.2, in using a method for parameter estimation such as the method of maximum likelihood, we have to be able to specify the distribution of X in order to use the method.

It was indicated initially in the section on the t-distribution (Section 3.3.4) that $\overline{X} \sim N(\mu, \sigma^2/n)$ when $X \sim N(\mu, \sigma^2)$. When the distribution of X is unknown (the usual case), the distribution of $\overline{X}$ is, of course, also unknown, as stated previously. When the sample size is large, however, the distribution of $\overline{X}$ will be approximately normal. How large must the sample size be? That depends on the shape of the distribution of X. If the distribution differs very little from a normal distribution (e.g., a chi-square distribution with a moderate number of degrees of freedom), a sample size of 15 or 20 may be sufficient. At the other extreme, for distributions that differ greatly from a normal distribution (e.g., an exponential distribution), sample sizes in excess of 100 will generally be required.

Stated formally, if $X_1, X_2, \ldots, X_n$ constitute a sequence of independent random variables (not necessarily identically distributed) with means $\mu_1, \mu_2, \ldots, \mu_n$ and variances $\sigma_1^2, \sigma_2^2, \ldots, \sigma_n^2$, then

$$Z = \frac{\sum_{i=1}^{n} X_i - \sum_{i=1}^{n} \mu_i}{\sqrt{\sum_{i=1}^{n} \sigma_i^2}}$$

approaches the standard normal distribution [i.e., $N(0, 1)$] as n approaches infinity. We may alternatively express the result in terms of the sample average.

Let's first assume that the X_i are not identically distributed. The theorem leads to the assumption of approximate normality for at least a moderate sample size if there are no dominant effects among the X_i. This often happens in quality improvement work, in which a particular effect may result from the summation, loosely speaking, of many small effects. When we take a random sample, the X_i will all have the same distribution (and of course there are no dominant effects).

4.3.1.1 Illustration of Central Limit Theorem

We will illustrate the Central Limit Theorem in the following way. We need a random variable with a known, nonnormal distribution so that we can see how normality is approached as the sample size increases. Assume that we want to simulate the roll of a single die 1000 times. Since each outcome is equally likely, the appropriate distribution is the discrete uniform distribution (see Section 3.3). Simulating 1000 tosses of the die produces the empirical distribution shown in Figure 4.1.

The distribution starts becoming bell-shaped when the average outcome of only two tosses of the die is computed and this is done 1000 times, as shown in Figure 4.2, and the normal distribution shape starts coming into view in Figure 4.3 when the average of ten tosses is computed, and so on. The student is asked to similarly see the Central Limit

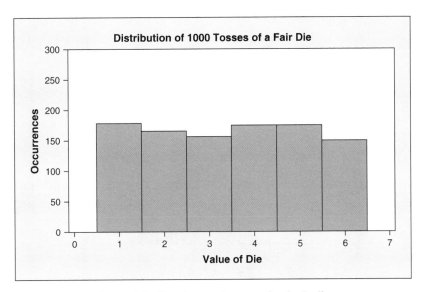

Figure 4.1   One thousand tosses of a single die.

Theorem in action in Exercise 4.22 by using the Java applet that is indicated, or by using the MINITAB macro that will produce various graphs, including the three that are given here.

The Central Limit Theorem has unfortunately been misused in practice: one example is that it has been cited as forming the underlying foundation for many of the control charts that are presented in Chapter 11. It is shown in that chapter, however, that the "normal approximations" that are used for determining the control limits for several

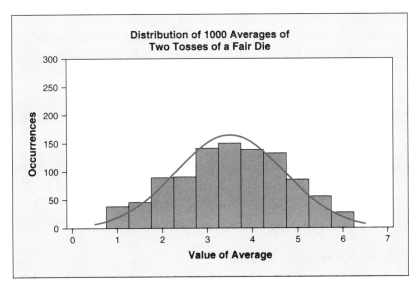

Figure 4.2   The average of two tosses of the die, performed 1,000 times.

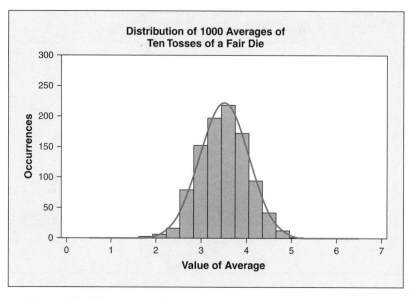

Figure 4.3 The average of ten tosses of the die, performed 1,000 times.

charts will often be inadequate because the sample sizes that are typically used are too small.

4.3.2 Statistics with Nonnormal Sampling Distributions

Although the distributions of many statistics approach a normal distribution as n approaches infinity, the rate of convergence could be sufficiently slow that "asymptotic normality" (i.e., approaching normality as n goes to infinity) really isn't relevant. In particular, the sampling distribution of the standard deviation or variance of a sample from a normal population will not be particularly close to normality for any practical sample size. This problem is even more acute (and perhaps intractable) for such measures of variability when sampling from nonnormal populations. Accordingly, the appropriate distribution will have to be used for statistical intervals and hypothesis tests such as those given in subsequent chapters. When the distribution is unknown, bootstrap methods are often used (see Section 5.7).

4.4 METHODS OF OBTAINING ESTIMATORS

Three commonly used methods of obtaining point estimators are presented in each of the next three subsections.

4.4.1 Method of Maximum Likelihood

One method of obtaining point estimators is to obtain them in such a way as to maximize the probability of observing the set of data that has been observed in the sample. Although this may sound like a play on words, the concept is intuitively appealing. Therefore, it is not surprising that Casella and Berger (1990, p. 289) stated. "The method of maximum likelihood is, by far, the most popular technique for deriving estimators." Other prominent statisticians have made similar statements.

To understand what a maximum likelihood estimator is, we first need to understand the concept of a *likelihood function*. The latter is simply a joint probability (as discussed briefly in Section 3.2). For example, assume that we want to obtain the maximum likelihood estimator of λ in the Poisson distribution. If we let $X_1, X_2, \ldots, X_n$ denote the random variables whose realizations are $x_1, x_2, \ldots, x_n$, the joint probability $P(X_1 = x_1, X_2 = x_2, \ldots, X_n = x_n)$ is given by

$$P(X_1 = x_1, X_2 = x_2, \ldots, X_n = x_n) = \prod_{i=1}^{n} f(x_i)$$
$$= \prod_{i=1}^{n} \frac{e^{-\lambda} \lambda^{x_i}}{(x_i)!}$$

which is the likelihood function. It is generally easier to work with the natural logarithm (log) of the likelihood function. Since the log of a product is the sum of the logs, we thus have the log(likelihood function) given by

$$\sum_{i=1}^{n} \log \left(\frac{e^{-\lambda} \lambda^{x_i}}{(x_i)!} \right) = \sum_{i=1}^{n} \left[\log(e^{-\lambda}) + \log(\lambda)x_i - \log(x_i!) \right]$$

Taking the derivative of this expression with respect to λ and setting it equal to zero, we obtain

$$-n + \frac{1}{\lambda} \sum_{i=1}^{n} x_i = 0$$

Solving this equation for λ, we then obtain $\widehat{\lambda} = \left(\sum_{i=1}^{n} x_i \right) / n = \bar{x}$. This result is unsurprising because λ is the mean of the Poisson distribution and we are thus estimating the population (distribution) mean by the sample mean.

■ **EXAMPLE 4.1**

Sometimes maximum likelihood estimators cannot be obtained by using calculus. To illustrate, consider the continuous uniform distribution defined on α and β, which is given by

$$f(x) = \frac{1}{\beta - \alpha} \qquad \alpha < x < \beta$$

The likelihood function is

$$P(X_1 = x_1, X_2 = x_2, \ldots, X_n = x_n) = \prod_{i=1}^{n} f(x_i)$$
$$= \frac{1}{(\beta - \alpha)^n}$$

Since the likelihood function is not a function of data, it is obvious that we cannot obtain the maximum likelihood estimators of α and β by direct means. We can see that the likelihood function will be maximized when $\beta - \alpha$ is made as small as possible. By definition of the *pdf*, α can be no larger than the smallest observation in a sample of size n and similarly β cannot be smaller than the largest observation. Therefore, the maximum likelihood estimators are $\widehat{\beta} = x_{\max}$ and $\widehat{\alpha} = x_{\min}$. ∎

Maximum likelihood estimators do not always exist (i.e., the likelihood function may not have a finite maximum), and when they do exist the estimator may be biased, as the reader will observe in Exercise 4.3. It is generally straightforward to convert such estimators to unbiased estimators, however. (Also see Exercise 4.3 for an example of this.) It should also be kept in mind that maximum likelihood estimators are "distribution dependent." That is, we cannot simply obtain the maximum likelihood estimator of μ, the mean for an arbitrary distribution, as we must specify the distribution and for some distributions the mean is a product of parameters (such as the gamma distribution). So we are limited to obtaining maximum likelihood estimators for parameters in the distribution. Furthermore, if we specify the distribution to be a normal distribution, for example, we are immediately starting out "wrong" since the actual distribution is almost certainly not going to be the same as the specified distribution. Good and Hardin (2003) decry the extensive use of maximum likelihood. The bottom line is that the method of maximum likelihood should be used with caution.

4.4.2 Method of Moments

Another point estimation method is the *method of moments*. The name is derived from the fact that with this approach the sample moments are equated to the corresponding population moments. These are "moments about the origin," with the mth population moment defined as $E(X^m)$. The corresponding sample moments are defined as $\sum X^m / n$ for a sample of size n. Thus, for equating first moments we have $\sum X / n = E(X) = \mu$. Therefore, the method of moments estimator of μ is $\overline{X} = \sum X / n$. Note that this result does not depend on the specification of a particular probability distribution. For a normal distribution the mean of course is denoted by μ, whereas for a Poisson distribution, for example, the mean is λ.

We generally want to estimate at least the mean and variance of a distribution (and also the third and fourth moments if we are trying to fit a continuous distribution, in particular, to a set of data). When we go beyond the first moment, the estimation becomes nontrivial. For example, let's assume that we wish to estimate σ^2. We need $E(X^2)$. By definition, $E(X - \mu)^2 = \sigma^2$, and using the algebraic simplification that the reader was asked to derive in Exercise 3.20, we obtain $E(X^2) = \mu^2 + \sigma^2$. Equating this to the second sample moment and using $\overline{X}$ as the estimator of μ, as shown above, we obtain $\sum X^2 / n = \overline{X}^2 + \sigma^2$. Substituting $\sum X / n$ for $\overline{X}$ and solving for σ^2, we obtain (after slight algebraic simplification),

$$\widehat{\sigma}^2 = \frac{\sum (X - \overline{X})^2}{n}$$

We observe that this is $(n - 1)S^2/n$, with S^2 denoting the sample variance. Note that we obtain this result without making any distributional assumption. If we wish to determine the properties of this estimator, we of course have to specify a distribution, which in turn specifies what σ^2 represents in terms of the parameters of the distribution.

For example, if we assume a normal distribution, we can show, as the reader is asked to do in Exercise 4.29, that S^2 is an unbiased estimator of σ^2, so that the method of moments estimator is thus biased and in fact is the same as the maximum likelihood estimator. Of course, an unbiased estimator can easily be obtained from this estimator simply by multiplying $\widehat{\sigma}^2$ by $n/(n - 1)$.

If we assume a Poisson distribution, the method of moments estimator of λ is $\overline{X}$, but problems ensue if we try to estimate the variance, as the reader will observe in Exercise 4.4, since the mean and the variance are the same but the method doesn't take that into account.

The method of moments is a viable alternative to maximum likelihood when maximum likelihood is intractable, and it has been used extensively in various areas of engineering, such as control theory, and in other fields, including econometrics. For example, Rao and Hamed (2000) provide extensive coverage of the method of moments and maximum likelihood as being important tools in statistical hydrology for use by hydrologists and engineers.

4.4.3 Method of Least Squares

A frequently used estimation method is the *method of least squares*. This is the standard method of estimating the parameters in a regression model (see Section 8.2.2). In general, with this method parameters are estimated such that the sum of the squared deviations of the observations from the parameter that is being estimated is minimized.

■ **EXAMPLE 4.2**

For example, in estimating a mean we would minimize $\sum (x - \mu)^2$. We might guess that $\overline{X}$ would be the least squares estimator of μ simply because it is the customarily used estimator. It is easy to show that it is indeed the least squares estimator. Specifically, we start with $\sum (x - \overline{X} + \overline{X} - \mu)^2$, which is obviously equivalent to the original expression. Then we group the four terms into $(x - \overline{X})$ and $(\overline{X} - \mu)$ and square the newly created expression as a binomial. The middle term vanishes because $\sum (x - \overline{X})$ $(\overline{X} - \mu) = (\overline{X} - \mu) \sum (x - \overline{X}) = 0$ since the sum of deviations about the mean is zero, as was shown in Section 2.2. The expression thus simplifies to $\sum (x - \overline{X})^2 + n(\overline{X} - \mu)^2$. Since the last term is nonnegative, the smallest possible value for the sum is $\sum (x - \overline{X})^2$, which results if μ is replaced by its estimator, $\overline{X}$. ■

Least squares estimators in regression analysis, which is one of the most widely used statistical techniques, are discussed in detail in Chapter 8.

4.5 ESTIMATING $\sigma_{\widehat{\theta}}$

Consider an arbitrary estimator $\widehat{\theta}$ and an arbitrary distribution. For common distributions, it isn't difficult to obtain closed-form expressions for $\widehat{\theta}$ and for its standard deviation, the estimate of which is called the *standard error*, as stated previously.

DEFINITION

The estimator of the standard deviation of a statistic is generally referred to as the *standard error*. This definition is not universal, however, and some authors use the term *estimated standard error* to represent what herein is designated as the standard error.

A standard error has an interpretation that is similar to the interpretation of a sample standard deviation: it is an estimate of the variability of the estimator $\widehat{\theta}$ in repeated sampling from a fixed population. In many, if not most, applications the population is not fixed, however, and interpretation of a standard error is difficult if there are major, short-term changes in a population. Good and Hardin (2003, p. 95) discuss the standard error, concluding that it is not a meaningful measure unless the statistic for which the standard error is being calculated has a normal distribution.

For many distributions a closed-form expression is not possible, however. Under such conditions, standard errors are often obtained using *bootstrapping*, which is a resampling procedure. Specifically, one acts as if the sample is really the population and obtains repeated samples from the original sample. An estimator of interest, $\widehat{\theta}$, is then used for each sample, producing a series of estimates and an empirical distribution of $\widehat{\theta}$, from which a standard error can be estimated.

Although bootstrapping is a popular procedure, we should keep in mind that in using the procedure one is trying to generate new data from the original data, something that, strictly speaking, is impossible. For this reason bootstrapping is somewhat controversial. Nevertheless, bootstrapping often provides good results, depending on the purpose and the sample size.

As an example of when bootstrapping won't work, consider a sample of size 10 and the objective of estimating the first percentile of the distribution for the population from which the data were obtained. We can obtain as many samples of a given size, such as 5, as we want to, but we should not expect to be able to obtain a good estimate of the first percentile. More specifically, if we recognize that the smallest observation in a sample of size 10 essentially gives us the 10th percentile of the sample, how can we expect to obtain subsamples from this sample that has such a minimum value and expect to be able to estimate a much smaller percentile (the first) of a population?

In general, bootstrapping won't work unless the sample observations line up in such a way as to be representative of the population *for the stated objective*. We will generally need about a thousand observations in order to be able to obtain a good estimate of the first percentile, so only an extremely unusual sample of size 10 would give us any chance of obtaining a good estimate when sampling from that sample.

See Section 5.7 for additional information and an example on bootstrapping. See also Efron and Gong (1983) and a more recent, but slightly more rigorous article by Efron (2002), as well as the critical view found in Young (1994) and the accompanying discussion.

4.6 ESTIMATING PARAMETERS *WITHOUT* DATA

Occasionally, there is a need to estimate a parameter *before* a sample is taken, perhaps to make a certain decision or judgment. For example, we may need an estimate of σ. If the distribution were symmetric and we knew that virtually all observations should be contained in the interval $\mu \pm k\sigma$, then if we had a good idea of the endpoints of the interval—the difference of which we will call $L = X_{max} - X_{min}$—then we could estimate σ as $\hat{\sigma} = L/2k$. Note that we don't need to know μ in order to obtain such an estimate. We simply need to know the largest and smallest values that the random variable will generally be, and to have a good idea of the value of k. Normal distributions were covered in Section 3.4.3 and for such distributions $k = 3$.

There are various other ways to estimate σ, such as from the interquartile range and from probability plots.

■ **EXAMPLE 4.3**

Recall the data that were used in Section 1.4.4 (which are in Exercise 1.7). The SAT score values for Q_1 and Q_3 are reported by colleges and universities and are published in an annual issue of *U.S. News and World Report*. These score percentiles could be used to estimate σ for a normal distribution if we were willing to assume that the scores are approximately normally distributed (which would seem to be a reasonable assumption). The 75th percentile for a normal distribution is at $\mu + 0.6745\sigma$ and the 25th percentile is at $\mu - 0.6745\sigma$. It follows that the interquartile range, $Q_3 - Q_1$, equals $\mu + 0.6745\sigma - (\mu - 0.6745\sigma) = 1.3490\sigma$. Therefore, it follows that an estimate of σ is given by $(Q_3 - Q_1)/1.3490$, or as $0.713(Q_3 - Q_1)$. For example, if $Q_1 = 1160$ and $Q_3 = 1280$, σ would be estimated as $(1280 - 1160)/1.3490 = 88.96$. Is this really a large sample approximation since we are acting as if the sample interquartile range is equal to the population interquartile range? That depends on how we define the population. If our interest centers only on a given year, then we *have* the population, assuming that the numbers were reported accurately.

Similarly, σ could be estimated using other percentiles, if known, of a normal distribution, and in a similar way σ_x could be estimated in general when X does not have a normal distribution if similar percentile information were known.   ■

4.7 SUMMARY

Parameter estimation is an important part of statistics. Various methods are used to estimate parameters, with least squares and maximum likelihood used extensively. The appeal of the former is best seen in the context of regression analysis (as will be observed in Chapter 8), and maximum likelihood is appealing because maximum likelihood estimators maximize the probability of observing the set of data that, in fact, was obtained in the sample. For whatever estimation method is used, it is desirable to have estimators with small variances.

The best way to make the variance small is to use a large sample size, but cost considerations will generally impose restrictions on the sample size. Even when data are very inexpensive, a large sample size could actually be harmful as it could result in a "significant" result in a hypothesis test that is of no practical significance. This is discussed further in Section 5.9. In general, however, large sample sizes are preferable.

REFERENCES

Casella, G. and R. L. Berger (1990). *Statistical Inference*. Pacific Grove, CA: Brooks/Cole. (The current edition is the second edition, 2001.)

Efron, B. (2002). The bootstrap and modern statistics. In *Statistics in the 21st Century* (A. E. Raftery, M. A. Tanner, and M. T. Wells, eds.). New York: Chapman Hall/CRC and Alexandria, VA: American Statistical Association.

Efron, B. and G. Gong (1983). A leisurely look at the bootstrap, jackknife, and cross-validation. *The American Statistician*, **37**, 36–48.

Good, P. I. and J. W. Hardin (2003). *Common Errors in Statistics (and How to Avoid Them)*. Hoboken, NJ: Wiley.

Rao, A. R. and K. H. Hamed (2000). *Flood Frequency Analysis*. Boca Raton, FL: CRC Press.

Young, A. (1994). Bootstrap: more than a stab in the dark? (with discussion). *Statistical Science*, **9**, 382–415.

EXERCISES

4.1. Show that the mean and variance of $\overline{X}$ are μ and σ^2/n, respectively, with μ and σ^2 denoting the mean and variance, respectively, of X. Does the result involving the mean depend on whether or not the observations in the sample are independent? Explain. Is this also true of the variance? Is there a distributional result that must be met for these results to hold, or is the result independent of the relevant probability distribution?

4.2. Show that $\overline{X}$ as given in Section 4.2.2 has a smaller mean squared error than $X^* = (X_1 + 2X_2 + 2X_3 + X_4)/10$ when $\mu = 10$ and $\sigma^2 = 4$. Assume that the variance of individual observations is the variance of the yield for a manufacturing process and the plant manager believes that the variance can be reduced. Assume that the variance is subsequently reduced to 2 but the mean is also reduced. For what values of the mean would $\overline{X}$ still have a smaller mean squared error? Comment.

4.3. Critique the following statement: "I see an expression for a maximum likelihood estimator for a particular parameter of a distribution that I believe will serve as an adequate model in an application I am studying, but I don't see why a scientist should be concerned with such expressions as I assume that statistical software can be used to obtain the point estimates."

4.4. Obtain the method of moments estimator for the variance of a Poisson random variable. Comment.

4.5. Assume that $X \sim N(\mu = 20, \sigma^2 = 10)$ and consider the following two estimators of μ, each computed from the same random sample of size 10:

$$\widehat{\mu}_1 = \sum_{i=1}^{10} c_i x_i \text{ with } x_1, x_2, \ldots, x_{10} \text{ denoting the observations in the sample, and}$$

$$c_i = 0.2, \quad i = 1, 2, \ldots, 5$$
$$= 0, \quad i = 6, 7, \ldots, 10$$

and $\widehat{\mu}_2 = \sum_{i=1}^{10} k_i x_i$ with the x_i as previously defined and

$$k_i = 0.10 \quad \text{for} \quad i = 1, 2, \ldots, 9$$
$$= 0 \quad \text{for} \quad i = 10$$

(a) Determine the variance of each estimator.
(b) Is each estimator biased or unbiased? Explain.
(c) Is there a better estimator? If so, give the estimator. If not, explain why a better estimator cannot exist.

4.6. The amount of time that a customer spends in line at a particular post office is a random variable with a mean of 9.2 minutes and a standard deviation of 1.5 minutes. Suppose that a random sample of 100 customers is observed.

(a) Determine the approximate probability that the average waiting time for the 100 customers in the sample is at least 9.4 minutes.
(b) Explain why only an approximate answer can be given.

4.7. Assume that $X \sim N(10, \sigma^2)$.

(a) Obtain the maximum likelihood estimator for σ^2.
(b) Is the estimator unbiased? Explain. If biased, what must be done to convert it to an unbiased estimator?

4.8. In Exercise 4.7, the mean of a normal distribution was assumed known, for simplicity. Generally, neither the variance nor the mean will be known. Accordingly, obtain the maximum likelihood estimators for both the mean and the variance for a normal distribution.

4.9. Assume that a sample of 10 observations has been obtained from a population whose distribution is given by $f(x) = 1/5, x = 1, 2, 3, 4, 5$. If possible, obtain the maximum likelihood estimator of the mean of this distribution. If it is not possible, explain why it isn't possible.

4.10. Obtain the maximum likelihood estimator for the variance of the Poisson distribution. Is the estimator biased or unbiased? Explain. If biased, can an unbiased estimator be obtained? If so, how would that be accomplished?

4.11. Assume that a random variable has a distribution with considerable right skewness (i.e., the tail of the distribution is on the right). Describe as best you can what the distribution of the sample mean will look like relative to the distribution of X.

4.12. What is the variance of the point estimate 25.4 obtained from a sample of $n = 100$? Suppose that an average is being updated by sequentially adding observations to the original sample of size 100. What is the variance of the average of the estimate 25.4 and the average of the next 100 observations?

4.13. Assume that a random variable has a distribution that is approximately normal. It would be rare for an observation from the population that has this distribution to be either larger than 120 or smaller than 30. Using only this information, what would be your estimate of σ^2?

4.14. Consider a random sample of size 2 from an arbitrary population and let $L = a_1 x_1 + a_2 x_2$.

(a) Give one set of possible values for a_1 and a_2 such that L will be an unbiased estimator of μ.

(b) Among the class of unbiased estimators, what choice of a_1 and a_2 will minimize the variance of L?

4.15. Given $f(x) = 2x, 0 < x < 1$, determine the method of moments estimator of μ if possible. If it isn't possible, explain why.

4.16. Is the sample mean a consistent estimator of the population mean? Explain.

4.17. Assume that $X \sim N(50, 25)$. If 1,000 samples are produced in a simulation exercise and the range of the $\overline{X}$ values is 47.3 to 52.8 but the sample size was not recorded, what would be your "estimate" of the sample size?

4.18. Explain what an unbiased estimator is and also explain why unbiasedness is a desirable property of an estimator.

4.19. Explain the difference between an estimate and an estimator.

4.20. Under the assumption of a normal distribution, $E(s^2) = \sigma^2$, but $E(s) \neq \sigma$. A student offers the following explanation for this phenomenon: "This is easy to explain because the expected value operator is a linear operator and you can't take the square root of a linear operator and still expect to have a linear operator. In fact, it is not even possible to take the square root of a linear operator." Do you agree? Explain.

4.21. Consider the standard normal distribution in Section 3.4.3. Obtain a random sample of 25 from that distribution, using MINITAB or other software. We know that $\sigma_{\overline{x}} = 0.2$. Obtain 1,000 (bootstrap) samples of size 10 and estimate $\sigma_{\overline{x}}$. (It will be necessary to write a macro in MINITAB to accomplish this.) Comment on your results relative to the known value.

4.22. The best way to see the Central Limit Theorem at work is to use one of the Java applets that are available on the Web. There are many such applets and as of this writing one of the best can be found at http://www.maths.soton.ac.uk/ teaching/units/ma1c6/links/samplingapplet/samplingapplet .html. Use all of the options that are available at that site to produce histograms based on different-sized samples from each of the three distributions: normal, exponential, and uniform. Note how the shape of the histogram changes as the sample size changes for a given distribution. At what size sample would you say that the distribution of the sample mean appears to be approximately normal for each distribution? Do you agree with the number of classes that were used in constructing each of the histograms? If you disagree, do you believe that your conclusions about the minimum sample sizes necessary for approximate normality would have been different if a different number of classes had been used? (If for some reason this particular applet is no longer available, do the exercise on an applet that is available.)

4.23. Can the maximum likelihood estimator of β for the exponential distribution given in Section 3.4.5.2 be obtained using the approach illustrated in Section 4.4.1? If so, obtain the estimator. If not, explain why not and determine if the estimator can otherwise be determined.

4.24. Assume it is known that certain capacitor measurements hardly ever exceed 2.6 inches or are less than 1.8 inches. If it can be further assumed that the measurements are approximately normally distributed, what would be an estimate of σ^2?

4.25. Is it possible to obtain the method of moments estimators for the mean and variance of the continuous uniform distribution (see Section 3.4.2) defined on the interval $(0,1)$? If so, obtain the estimators; if not, explain why it isn't possible.

4.26. It is believed that averages of five observations from a particular manufacturing process rarely exceed 6.5 or are less than 3.4.
   (a) Could we use this information to determine the corresponding extremes for individual observations? Explain.
   (b) Could this determination be made if approximate normality were assumed for the individual observations? Explain.
   (c) If the individual observations were approximately normally distributed, what would be the approximate value of μ?

4.27. A store decides to conduct a survey regarding the distribution of men and women that visit the store on Saturdays. A particular Saturday is chosen and an employee is told to record a "1" whenever a man enters the store and a "2" whenever a woman enters. Let X denote the number of women who enter the store. Can $Var(X)$ be determined based on this information? If so, what is the variance? If not, explain why the variance cannot be obtained.

4.28. Suppose you are told that for some estimator $\widehat{\lambda}$ of the Poisson parameter λ, $E(\widehat{\lambda}) = n\lambda/(n-1)$.
   (a) Is $\widehat{\lambda}$ a biased or unbiased estimator of λ? Why or why not?
   (b) What can be said when n is large?

4.29. Show that S^2 is an unbiased estimator of σ^2 when a normal distribution is assumed.

4.30. Explain why there is no maximum likelihood estimator for the parameters of the pdf $f(x) = x/8, 0 < x < 4$. Does the maximum likelihood estimator exist for μ_x? Why or why not?

4.31. Show that $\sum_{i=1}^{n} (x_i - a)^2$ is minimized when $a = \bar{x}$. Does this mean that $\bar{x}$ is thus the least squares estimator of μ_x? Explain.

4.32. Consider the following computer output, which was used in Exercise 2.26:

```
Descriptive Statistics: X
Variable   N Average Median Minimum Maximum    Q1      Q3
X         50  20.560 21.000 10.000   30.000  15.750 26.000
```

If this is all you were given and you needed an estimate of σ, could you obtain an estimate from this information? If so, how good do you think the estimate would be? If it isn't possible to obtain the estimate, what minimal additional information would you need in order to be able to produce an estimate?

4.33. Consider the information given in Exercise 1.63.
  (a) What must be assumed in order to estimate σ for the fraternity GPA using any method given in this chapter? Does such an assumption seem plausible? If so, what would be the estimate? If the assumption seems implausible, explain why.
  (b) Could the same approach be used to estimate the standard deviation of the sorority GPA? Explain.

4.34. Generate a sample of size 20 from the discrete uniform distribution, with the distribution defined on the integers 1–10. Then generate 10,000 bootstrap samples of size 10 from the "population" of observations in the sample of size 20. Compute the mean for each of the 10,000 samples and determine the standard error of the mean. Compare this result with the actual standard deviation of the mean (i.e., $\sigma_{\bar{x}}$), which you will need to determine. How does the standard error of the mean compare to the standard deviation of the mean?

4.35. Explain why the maximum likelihood estimator of the mean of a normal distribution has the same variance as the method of moments estimator.

4.36. Suppose that X and Y are random variables that represent an insurer's income from two different types of life insurance policies, with X and Y being independent. If $\mu_x = 55,000$, $\mu_y = 41,000$, $\sigma_x = 400$, and $\sigma_y = 300$, what is the variance of the insurer's average income for the two policies?

4.37. A manager is considering giving employees in her division a 4% cost-of-living raise plus a flat $600 bonus in addition to the raise. Assume that the employees in her division currently have a mean salary of $43,000 with a standard deviation of $2,000. What will be the new mean and the new standard deviation after both the raise and the bonus?

4.38. At the end of Section 4.7 it was stated that large samples are preferable. Accordingly, critique the following statement: "I don't see why experimenters don't always use very large samples in an effort to minimize the variance of their estimators."

4.39. Explain why there is no maximum likelihood estimator for the *pdf* $f(x) = 1$, $0 < x < 1$.

4.40. Consider the *pdf* $f(x) = (\alpha + 1)x^{\alpha}$, $0 < x < 1, \alpha > 0$. Obtain the maximum likelihood estimator of α if possible. If this is not possible, explain why not.

4.41. Assume that you are a supermarket manager and you want to obtain some idea of the variability in the amount of time required to check out a customer in the express lane once the customer reaches the front of the line compared to the amount of time for customers in the lanes that do not have a limit on the number of items. You know from past data that customers in the other lanes have an average of 29.8 items, whereas customers in the express lane have an average of 8.9 items. If the estimate of the standard deviation for the other lanes is 2.2 minutes, what would be the estimate of the standard deviation for the express lane, or is it possible to even obtain an estimate from this information? Explain.

4.42. A sample of 35 items is obtained and it is believed that the Poisson distribution is an adequate model for the data. If the sum of the data values is 234.6, what is the estimate of the variance for this Poisson distribution?

4.43. Data in a population database that are in inches are converted to centimeters. What is the variance of the new data as a function of the variance of the old data?

4.44. Can we speak of the standard error of an estimator without specifying a probability distribution? Explain.

4.45. Information given at the Energy Information Administration website (www.eia.doe.gov) showed that in 2000 the interquartile range for underground mines in Alabama was \$22.54 per short ton. Assuming a normal distribution, could σ be estimated from this information? If so, what would be the estimate? If not, explain why an estimate cannot be obtained.

CHAPTER 5

Confidence Intervals and Hypothesis Tests—One Sample

Point estimators of parameters and the standard errors of those estimators were covered in Chapter 4. In this chapter we use point estimators and their standard errors in constructing confidence intervals and hypothesis tests.

By a confidence interval we mean an interval that will very likely contain the unknown value of the parameter for which the interval has been constructed. This is termed an interval estimate, as contrasted with a point estimate. By a hypothesis test we mean a test of an unknown, hypothesized value of a parameter or a test of a distributional assumption.

It is important to see the direct connection between each type of interval estimate and the corresponding hypothesis test, so that relationship will be illustrated and discussed throughout the chapter. It is also important to recognize the various incorrect ways in which the results can be stated. For example, we can't speak of the probability that a hypothesis is true because what we are hypothesizing is fixed, such as a distributional assumption, a value for a parameter, or the equality of two parameters. In classical (non-Bayesian) statistics, as is presented in this book, probability statements are made only about random variables. Improper statements about hypothesis testing and other types of misstatements are discussed by Gunst (2002), and the reader is referred to that paper for additional reading.

5.1 CONFIDENCE INTERVAL FOR μ: NORMAL DISTRIBUTION, σ NOT ESTIMATED FROM SAMPLE DATA

We will begin with the simplest case: a confidence interval for the mean of a normal distribution with σ assumed known. Assume that we want to construct a 95% confidence interval. What does this provide? It provides the following: if we were to construct 100 such intervals, our best guess as to the number we would expect to contain the mean is 95. (If we performed such an experiment once, the number probably wouldn't be 95, but if we performed the experiment many times and averaged the results, the average of the outcomes would be very close to 95, and might be, say, 94.9.) Figure 5.1 shows what the outcomes might be for part of one such experiment.

Modern Engineering Statistics   By Thomas P. Ryan
Copyright © 2007 John Wiley & Sons, Inc.

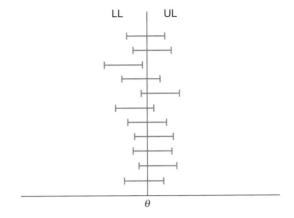

Figure 5.1 Confidence intervals (random) and the mean (fixed).

Of course, we are going to construct only *one* interval, so can we say that we are 95% confident that our interval will contain the mean? We can make such a statement as long as we haven't already obtained a sample and constructed the interval, because once we have the interval it either contains the mean or it does not. That is, loosely speaking, it is either a "0% interval or a 100% interval." We don't know which it is because we don't know the value of the mean, and if we did know that value we of course would not have to construct an interval estimate! Thus, a 95% confidence interval *does not provide* an interval that will contain the mean 95% of the time.

This leads to the following definition of a confidence interval.

DEFINITION

A $100(1 - \alpha)\%$ *confidence interval* for some parameter θ is an interval, often of the general form $\widehat{\theta} \pm (Z \text{ or } t)s_{\widehat{\theta}}$, which before the data are collected and the interval is constructed, has a probability of $1 - \alpha$ of containing θ, provided that the assumptions are true.

Here "Z or t" means that we will be using either the standard normal distribution or the t-distribution. As stated, we will initially assume that $X \sim N(\mu, \sigma^2)$ and that we desire to construct a confidence interval for μ. The corresponding point estimator is $\overline{X}$, as stated previously, and we will use $\overline{X}$ in constructing a confidence interval. We will also assume that σ is known. Such an assumption is usually unrealistic (as is the normality assumption, of course) because it is unlikely that we would know the population standard deviation but not know the population mean, but this provides a starting point and we will later relax these assumptions.

From Section 3.4.4 we know that

$$Z = \frac{\overline{X} - \mu}{\sigma/\sqrt{n}}$$

has the standard normal distribution. Therefore, it follows that

$$P\left(-Z_{\alpha/2} \le \frac{\overline{X} - \mu}{\sigma/\sqrt{n}} \le Z_{\alpha/2}\right) = 1 - \alpha \tag{5.1}$$

with $Z_{\alpha/2}$ denoting the standard normal variate such that $\alpha/2$ is the area under the standard normal curve to the right of the value of $Z_{\alpha/2}$, for a selected value of α. In order to obtain a confidence interval for μ, we simply have to perform the appropriate algebra on the double inequality so that μ is in the center of the interval. Multiplying across the double inequality by $\sigma/\sqrt{n}$, then subtracting $\overline{X}$, and finally dividing by -1 (which of course reverses the direction of each inequality) produces

$$P(\overline{X} - Z_{\alpha/2}\sigma/\sqrt{n} \le \mu \le \overline{X} + Z_{\alpha/2}\sigma/\sqrt{n}) = 1 - \alpha \tag{5.2}$$

Thus, the lower limit, which will be written $L.L.$, is at the left of the double inequality, and the upper limit, which will be written $U.L.$, is on the right.

A relatively narrow interval would obviously be desirable, and a very wide interval would be of little value. The width, W, of the interval is defined as $W = U.L. - L.L. = 2Z_{\alpha/2}\sigma/\sqrt{n}$, as the reader can observe from Eq. (5.2). With σ assumed to be known and fixed in this section, the width is thus a function of Z and n. The higher the degree of confidence, the larger the value of Z. Specifically, if we want a 95% confidence interval, then $\alpha = .05$ and $\alpha/2 = .025$. From Table B we can see that $Z_{\alpha/2} = 1.96$. Similarly, if $\alpha = .01$, $Z_{\alpha/2} = 2.575$; and if $\alpha = .10$, $Z_{\alpha/2} = 1.645$. These results are summarized as follows:

Degree of Confidence (%)	Z
90	1.645
95	1.96
99	2.575

Of course, using the smallest of these Z-values so as to minimize the width of a confidence interval wouldn't really accomplish anything: we would then have the lowest degree of confidence that the interval we will construct will contain the parameter value. A small interval width could be created in a non-artificial way by using a large sample size, but taking samples costs money, and usually the larger the sample, the greater the cost, as stated in previous chapters.

5.1.1 Sample Size Determination

One approach would be to specify a desired width for the interval, then solve for n and decide whether or not that value of n is financially feasible. Since $W = 2Z_{\alpha/2}\sigma/\sqrt{n}$, it follows that $n = (2Z_{\alpha/2}\sigma/W)^2$.

Alternatively, we might wish to construct a confidence interval such that we have a maximum error of estimation of the population parameter, with the specified probability. Since $\overline{X}$ is in the center of the interval, it follows that the maximum error of estimation is $U.L. - \overline{X} = \overline{X} - L.L.$, and we will represent this difference by E. (We say the *maximum* error of estimation because the value of E is clearly the greatest possible difference between

$\overline{X}$ and μ with the stated probability since the endpoints of the interval form our interval estimate for μ.) Thus, $E = W/2$, so we could equivalently solve for n in terms of E, which produces $n = (Z_{\alpha/2}\sigma/E)^2$.

5.1.2 Interpretation and Use

Proper interpretation of a confidence interval is important, as well as proper use of confidence intervals in *statistical inference*. The latter can be defined in the following manner.

DEFINITION

Statistical inference refers to the use of sample data to make statements and decisions regarding population parameters, which are generally unknown.

When we construct a confidence interval, we are using one form of statistical inference, as we use a probability statement that involves a population parameter in making a statement about the parameter. We know that before we take a sample and construct the interval, the probability is $(1 - \alpha)$ that the interval we are about to construct will contain the population value, *provided that our assumptions are true*. This is the proper way to explain a confidence interval because, as stated previously, once we have constructed the interval, the interval either contains the parameter or it does not. That is, we can no longer speak of $100(1 - \alpha)\%$ confidence, as our "degree of confidence" is either 0% or 100%, and we don't know which it is.

From a practical standpoint, we will never be able to construct an exact $100(1 - \alpha)\%$ confidence interval because our assumptions will *almost never* be true. That may sound like a profound statement, but its validity can easily be established. As was explained in Section 3.4.3, there is no such thing as a normal distribution in practice, so by assuming normality we are guaranteeing that we will not have a $100(1 - \alpha)\%$ confidence interval right off the bat. Furthermore, even if σ is assumed to be known in a particular application, it is not going to be known exactly, so this further erodes our "confidence."

Another possible problem concerns the manner in which the sampling is performed. The construction of a confidence interval assumes that a *random sample* has been obtained, and there is the matter of how the population is defined and viewed.

Consider the following example, which is motivated by a practical, and undoubtedly common, problem that I encountered.

■ **EXAMPLE 5.1**

Problem Statement

A common manufacturing problem is that the tops of bottles are put on too tightly during the manufacturing process so that elderly people, in particular, may have great difficulty removing the tops. In about 1996, a well-known manufacturer of apple juice (which would probably prefer not to have its name mentioned here!) experimented with its manufacturing process, with one end result being that the author of this book, a competitive weightlifter, was unable to remove the top without virtually having to knock it off. This experience was reported to the company, and the public relations person who handled the complaint mentioned the change in the manufacturing process. I was "rewarded" with three coupons that enabled me to purchase three more bottles of the product at a reduced price. That did little good, however, because if I couldn't remove the top of one bottle, I probably wasn't going to be able to remove the top of each of three more bottles!

How was this problem handled at the manufacturing end? That isn't known to this writer, but in time the top became easier to remove. We might surmise how the company would approach the problem.

What should be the first step? Take a sample, right? How many observations? Let X denote the amount of pressure required to remove the top. (*Note*: Torque may be a more appropriate measure, but we will use pressure as a simplification.) If the company has never addressed this problem before, and additionally does not know anything about the distribution of X over the bottles that it manufactures, it would be very difficult to construct a confidence interval for μ, the average amount of pressure required to remove the top. We might take a sample of $n = 50$ bottles and appeal to the Central Limit Theorem (Section 4.3.1), but 50 could be well short of what is needed if the distribution of X is highly skewed.

Take a Sample

Let's assume that the company decides to play it safe and takes a sample of 100 bottles. If we are going to construct a confidence interval, we need to have a random sample. That is, each possible sample of size 100 should have the same probability of being selected. However, in order to take a random sample, we need to enumerate the elements of the population, which forces us to first define the population. If our sample is obtained on one particular day, the population would logically be defined as all of the bottles produced on that day. But could a list of those items (a *frame*, in sampling terminology) be constructed? If constructing such a list would be impossible or at least inconvenient, a random sample could not be taken. Strictly speaking, a sample that is not a random sample would invalidate the probability statement that was the starting point in obtaining the general expression for a confidence interval, which has already been invalidated by the fact that we don't know the distribution of X!

Distributional Assumption

We can extricate ourselves from this predicament by rationalizing that the distribution of $\overline{X}$ for samples of size 100 is probably very close to a normal distribution, and by trying to take a sample that is "almost as good as a random sample." Specifically, if we take a *convenience sample*, such as selecting 10 items at the beginning of each hour for 5 hours, we would hope that our *convenience sample* is almost as good as a random sample in terms of the values of X being randomly determined.

Under certain conditions that will not be the case, however, as observations made close together in time will often be correlated, depending on what is being measured. Furthermore, if we begin sampling without a frame and are sampling from a population that isn't finite, we technically would not be obtaining a random sample from a single population if we continued the sampling for very long as populations change over time. Thus, we would eventually be sampling from two or more populations with a sampling scheme that could consequently have some undesirable properties.

Analysis

Assume that the process engineers believed that $\sigma = 3$ (approximately), and for the sample of size 100 it is found that $\overline{X} = 21.84$ (pounds of pressure). Assuming approximate normality (we don't have enough observations to have a very powerful test for it, anyway), we thus have the following for a 95% confidence interval:

$$
\begin{aligned}
L.L. &= \overline{X} - Z_{\alpha/2}\sigma/\sqrt{n} \\
&= 21.84 - 1.96(3)/\sqrt{100} \\
&= 21.25 \\
U.L. &= \overline{X} + Z_{\alpha/2}\sigma/\sqrt{n} \\
&= 21.84 + 1.96(3)/\sqrt{100} \\
&= 22.43
\end{aligned}
$$

(Note that $U.L. + L.L. = 2\overline{X}$. This serves as a check that the numbers have been listed correctly, but does not guarantee that both values are correct. It simply tells us that either both numbers are incorrect or they are both correct.)

Interpretation of Confidence Interval

What is the proper interpretation of this interval? For the population that was sampled (on the selected day), we believe that the mean for all bottles on that day is somewhere between 21.25 and 22.43. If the machines are supposed to be set so that 20 pounds of pressure is needed to remove the tops, then the process is (apparently) out of control because the interval does not cover 20. When we reach such a conclusion, we are using a confidence interval to test a hypothesis, and as stated at the beginning of the chapter, the relationship between confidence intervals and hypothesis tests will be stressed throughout this chapter. ∎

We will return to this example later in the chapter and discuss it in the context of hypothesis testing.

5.1.3 General Form of Confidence Intervals

Confidence intervals that are constructed using either Z or t are all of the same general form, and recognizing this makes the study of confidence intervals easier.

Two-sided confidence intervals that utilize either the t-distribution or the standard normal distribution are always of the form

$$
\widehat{\theta} \pm t(\text{or } Z)\widehat{\sigma}_{\widehat{\theta}}(\text{or } \sigma_{\widehat{\theta}})
$$

where θ denotes an arbitrary parameter to be estimated, $\widehat{\theta}$ is the corresponding point estimator of that parameter, and $s_{\widehat{\theta}}$ is the estimated standard deviation of the point estimator. The confidence interval that was constructed in Example 5.1 was of this form since $\theta = \mu$, $\widehat{\theta} = \overline{X}$, and $\sigma_{\widehat{\theta}} = \sigma/\sqrt{n}$. (Although prior knowledge might be used in estimating σ, as in that example, σ will generally have to be estimated from recent data, so that $\widehat{\sigma}_{\widehat{\theta}} = \widehat{\sigma}/\sqrt{n} = s/\sqrt{n}$ will be the typical case.)

Confidence intervals are not always symmetric about $\widehat{\theta}$, however. One example is a confidence interval for σ^2 (using the chi-square distribution) with s^2 not being in the middle of the interval. Such intervals will be illustrated later.

5.2   CONFIDENCE INTERVAL FOR μ: NORMAL DISTRIBUTION, σ ESTIMATED FROM SAMPLE DATA

Although prior information to suggest an estimate of σ is often available, it is more common for σ to be estimated from sample data. When that occurs, the appropriate distribution is no longer the standard normal distribution. Rather, the t-distribution is used, as indicated in Section 3.4.4. Thus, relative to Section 5.1.3, the "t or Z" part is t. Therefore, a confidence interval for the mean is of the form

$$\overline{X} \pm t_{\alpha/2,n-1}s/\sqrt{n}$$

■   **EXAMPLE 5.2**

Objective and Assumption

Assume that we wish to construct a 99% confidence interval for μ, and we believe that the population from which we will obtain a sample is approximately normal.

Sample and Confidence Interval

We obtain a sample of size 25, with $\overline{x} = 28.6$ and $s = 4.2$. The confidence interval is then $28.6 \pm 2.7969(4.2/\sqrt{25}) = 28.6 \pm 2.2375$, so $L.L. = 26.36$ and $U.L. = 30.84$.   ■

Although it is certainly not imperative that software be used for such simple computations, it is nevertheless somewhat easier. The general recommendation is to use Z in place of t for large sample sizes, but there is no need to do that when software is available.

5.2.1   Sample Size Determination

In general, the width of the interval, $W = 2\,t_{\alpha/2,n-1}s/\sqrt{n}$, depends on the sample size in two different ways, as n appears in two places. (This expression for W is obtained in the same general algebraic way as the expression for W using Z that was given in Section 5.1.) This creates a problem in trying to solve for n because the solution is a function of t, which itself is a function of n. There is not a simple solution to this problem—unless we use software. This will be illustrated shortly.

One suggested approach is to take a small pilot sample, compute s, then use trial-and-error to try to arrive at a value for n as a function of t. That is a risky approach, however, because the variance of s is a function of n, so s will have a large variance if a small pilot sample is used. More specifically, the value of s that might be used to solve for n could differ considerably from the value of s that results when the sample of size n is obtained. If the latter is much larger than the pilot sample value of s, this means that a larger sample should have been taken. In general, problems with sample size determination for the case when the t-statistic should be used make such a determination by hand calculation rather impractical.

Fortunately, we can avoid such headaches because there are readily available "Internet calculators" that can be used for this purpose. For example, for the sake of simplicity let's assume that $\sigma = 1$ and we want a 95% confidence interval with the (maximum) error of estimation to be 0.5. Russ Lenth has developed a Java applet that can be used for this purpose, which is available at http://www.stat.uiowa.edu/~rlenth/Power/. (Select

"CI for the mean.") Other applets for sample size determination may also be available on the Web.

The use of the applet produces $n = 17.76$. We generally round sample sizes up, regardless of the value of the decimal fraction, so that the interval will not have less than the desired degree of confidence. Accordingly, $n = 17.76$ rounds up to 18 and the degrees of freedom for t is thus 17. Then $t_{.025,17} = 2.1098$, and solving for $n = (t\sigma/E)^2 = [(2.1089)(1)/0.5]^2 = 17.805$, which differs only slightly from what the applet provided.

Sample size/power determinations can also be made, of course, using various software, although in JMP and MINITAB, for example, sample size is determined in accordance with a specified power to detect a change of a specified amount from a hypothesized parameter value.

To reiterate what was indicated in Section 3.4.4, the fact that we are using the value of a t-variate in constructing the interval doesn't mean that the normal distribution is not involved. Rather, the individual observations must be assumed to have come from a population with a normal distribution, and in general the random variable that is in the numerator of the t-statistic must have a normal distribution. If that is not the case, the coverage probability for the confidence interval (such as .95) will not be what is assumed. This is because $\sqrt{n}(\overline{X} - \mu)/s$ does not have a t-distribution when X does not have a normal distribution. This is not a serious problem, however, as long as there is not more than a slight deviation from normality. See, for example, the discussion in Moore and McCabe (2002) and Pearson and Please (1975).

5.3 HYPOTHESIS TESTS FOR μ: USING Z AND t

Historically, hypothesis tests and confidence intervals have almost always been presented in different chapters in statistics textbooks. That is, different types of confidence intervals are presented together, as are different types of hypothesis tests. A different approach is taken in this text because confidence intervals and hypothesis tests are related. In fact, they are almost like different sides of a coin. Therefore, it is important for readers to understand this relationship, which would be harder to portray if confidence intervals and hypothesis tests were presented in different chapters.

5.3.1 Null Hypotheses Always False?

The reader should understand at the outset that hypothesis tests have some glaring weaknesses, and these should be understood. Indeed, various writers have claimed that null hypotheses are always false, so testing a hypothesis is nonsensical. (A null hypothesis is the hypothesis that is being tested. Except when a hypothesis of a specific probability distribution is being tested, the null hypothesis is what we doubt to be true, which is why we test it.)

Let's examine this claim. Assume that we have some normal distribution and we wish to test the hypothesis that $\mu = 50.25$. Since the random variable can assume any real number, how likely is it that the average of the numbers in the population is exactly 50.25, or is exactly *any* number? Thus, we know that our hypothesis is false for almost any number that we might select, unless the hypothesized value is recorded only to the nearest integer (e.g., nearest inch) or in general to the nearest whole number. If we know that the hypothesized value is false, why would we bother to test it?

Obviously, we would not have the same problem if we hypothesized that the mean was at most 50, or that it was at least 50, so that we weren't specifying a single value. We do get into some trouble when we use a null hypothesis that specifies more than a single value for the parameter being tested, however, as is explained in Section 5.3.2.

So it isn't true that all null hypotheses are false, but they will be false when testing distributional assumptions, and they will almost certainly be false whenever population parameters are not rounded to the nearest integer.

If the null hypothesis is a specified distribution, the hypothesis will almost always be wrong, since "all models are wrong, but some are useful," as stated by G. E. P. Box.

Of course, if we are going to construct a confidence interval or hypothesis test, we should check our assumption of a normal distribution (or whatever distribution is assumed). But, as has been stated previously in this chapter and in Section 3.4.3, there is no such thing as a normal distribution in practice, so why should we bother to perform such a test?

Actually, the apparent absurdity of performing certain hypothesis tests is tempered somewhat by the realization that we need some idea of how close our hypothesis is to being met, and a hypothesis test will more or less gives us this information. Thus, if the distribution is relatively close to a normal distribution, a test for normality will likely not lead to rejection of the hypothesis (unless the sample size is extremely large—more about this later), and of course no great harm will occur in most applications of statistical tests if the actual distribution is close to the assumed normal distribution when normality is assumed.

Therefore, although hypothesis tests have been in ill-repute for many years [e.g., see Tukey (1991) and Nester (1996)], they are not likely to go away anytime soon. Consequently, we have to live with them, but we need to understand that they do have serious limitations, and frequently a different approach will be preferred. For example, Reeve and Giesbrecht (1998) stated "Many questions that are answered with hypothesis testing could be better answered using an equivalence approach." (The latter is used with dissolution tests.)

5.3.2 Basic Hypothesis Testing Concepts

The process of hypothesis testing begins with the hypothesis that is to be tested, and of course this must be determined before a sample is taken. This is a statement, such as a statement about a parameter, that we doubt, as stated previously. Therefore, we wish to test it. We do not test statements that we believe to be true since hypothesis testing can never be used to prove a hypothesis. Therefore, if we fail to reject the null hypothesis, we do not say that we accept it. Rather, we say that we failed to reject it. This is a subtle difference, but an important one.

We can only "disprove" a null hypothesis, with a specified (small) probability of being wrong when we reject it. However, when we test the assumption of a normal distribution, we hope that we *don't* reject the hypothesis for that would force us to use a different inferential approach, such as a nonparametric (distribution-free) test. Nonparametric tests are covered in Chapter 16.

The hypothesis that we test is termed the *null hypothesis* (denoted as H_0), and the statement that we believe to be true is termed the *alternative hypothesis* (denoted as H_a). Thus, if we reject the null hypothesis, we accept the statement that we believe to be true when we constructed the test. For example, universities that report high SAT scores for entering freshmen while simultaneously reporting a moderately high acceptance rate might be suspected of inflating the former. Assume that Greenwood University (fictitious), a private university, reports that their average SAT score is 1306. A publication that ranks universities has reason to believe that this number is too high, and so it dispatches certain members of its staff to look into the matter. There would be no point in the "investigators" contacting the Registrar's Office and asking for the average SAT score because if the university is in fact inflating the number, the Registrar's Office would probably just give that number.

Therefore, the investigators decide to take a sample, and they try their best to take a relatively random sample to test the null hypothesis, with H_0: $\mu = 1306$ and $H_a : \mu < 1306$. This would be called a "lower-tailed" test because H_0 would logically be rejected in favor of H_a only if the value of the test statistic, which will be illustrated shortly, was sufficiently far into the lower (left) tail of the distribution of the test statistic such that it would be unlikely to observe a value as small or smaller than what was observed if H_0 were true. If the alternative hypothesis had been $\mu > 1306$, then the test would have been an upper-tailed test and a large value of the test statistic would lead to rejection of H_0. This should become clear with the examples that follow.

They select 100 current students for a survey and one of the questions is the student's SAT score. Assuming that all of the answers to the SAT question were truthful, the results were $\overline{X} = 1293$ and $s = 36$.

Which statistic do we use? We don't know σ, so technically we should use t, provided that we can assume that SAT scores are approximately normally distributed. In general, we would expect aptitude test scores to be approximately normally distributed, so the assumption of approximate normality in this case seems quite reasonable.

Tables of the t-distribution don't generally go as high as 99 degrees of freedom, however, which is what we have here since $n - 1 = 99$. Of course, we could use computer software and if we did so we would find only a small difference between the t-values and the corresponding Z-values. Therefore, we could use Z as a (satisfactory) approximation and proceed as follows. We want to compute $P(\overline{X} < 1293)$ because the alternative hypothesis is that $\mu < 1306$ (i.e., the inequality signs are the same).

Pictorially, we have Figure 5.2.

We want to convert this to the corresponding graph for Z with Z defined as $Z = (\overline{X} - \mu)/(s/\sqrt{n})$. Then we have approximately the curve in Figure 5.3, with the -3.61 resulting from the following computation:

$$Z = \frac{\overline{X} - \mu}{s/\sqrt{n}}$$
$$= \frac{1293 - 1306}{36/100}$$
$$= -3.61$$

That is, -3.61 on the Z-curve corresponds to approximately 1293 on the curve for $\overline{X}$, with this being approximate because the distribution of X was not specified, so the distribution of $\overline{X}$ is also unknown but is probably close to a normal distribution because of the large sample size.

If we computed $P(Z < -3.61)$, since the alternative hypothesis is "less than," we would be obtaining the probability under the assumption that the null hypothesis is true, since we are using 1306 in place of μ. That is, we would be actually computing the conditional probability $P(Z < -3.61|\mu = 1306)$, which equals .0002, using Table B, or .000153 if a

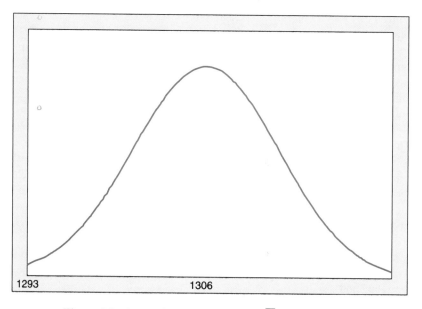

Figure 5.2   Approximate distribution of $\overline{X}$ with $\mu = 1306$.

computer is used. If we used a computer and used the *t*-distribution, we would also obtain
.0002 (actually .000242). Thus, the difference is very small in absolute terms and is also
inconsequential relative to the decision that would be made.

To summarize what we have done to this point, we have stated the null and alternative
hypotheses, decided on the test statistic (*Z*), and computed the value of the test statistic

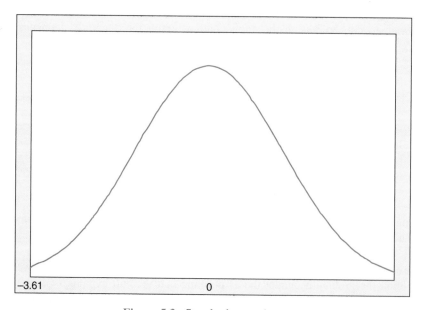

Figure 5.3   Standard normal curve.

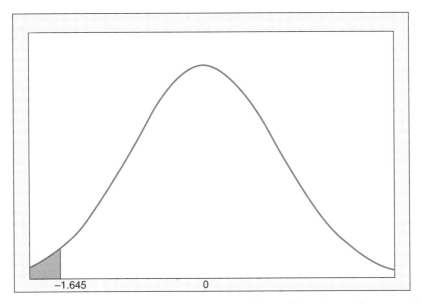

Figure 5.4 Rejection region for one-sided test with $\alpha = .05$ and normality assumed.

using the data in the sample. We haven't attempted to reach a decision yet, however, and that is the next step. The conditional probability of .0002 means that there is only a 2 in 10,000 chance of obtaining a Z-value this small or smaller when the null hypothesis is true.

This conditional probability is also called the p-value for the test, which is defined and discussed extensively in Section 5.8.1. For this example, there is strong evidence, in the form of a small *p-value*, that favors the alternative hypothesis. (*Note*: In this type of application the null hypothesis could conceivably be true because we should regard the average, which almost certainly would not be an integer, as being rounded to the nearest integer. This then creates a finite number of possible values for the mean, one of which is the correct value.)

In order to reach the type of decision that was reached in the preceding example, the decision-maker must have some threshold value in mind for the conditional probability. That is, we have to decide when the conditional probability is small enough to lead us to reject the null hypothesis. The old approach to hypothesis testing was to state explicitly the threshold value as either a probability or as a value of the test statistic, and then either compare the conditional probability to the threshold probability, or compare the threshold value of the test statistic to the observed value of the test statistic (or even state the threshold value in the unit of measurement).

Since we have a directional alternative hypothesis (meaning either less than or greater than), we might reject H_0 when the conditional probability is less than, say, .05, or equivalently when the value of the Z-statistic is less than -1.645. That is, a value of the Z-statistic less than -1.645 would fall in the *rejection region*, which would be defined by this upper bound, as shown in Figure 5.4.

The .05, which is the area under the normal curve to the left of -1.645, is the (approximate) probability of observing a value of the Z-statistic that is less than -1.645 when the null hypothesis is true. That is, 5% of the time the null hypothesis would be incorrectly rejected. These concepts are discussed in more detail in Section 5.8.

If H_a had been greater than rather than less than, the threshold value for the Z-statistic would have been $+1.645$ rather than -1.645, again assuming that the threshold value for the conditional probability is .05.

If the alternative hypothesis had been H_a: $\mu \neq 1306$ (i.e., a two-sided test so that H_0 would be rejected for either a small value of Z or a large value of Z), the critical values for Z would have been ± 1.96, assuming a threshold probability, which is generally denoted as α, of .05. (The numerical value of α is the probability of rejecting the null hypothesis when it is true; see Section 5.8.) Relationships of this type are related to the short table given in Section 5.1 since a two-sided hypothesis test with a specified value of α relates directly to the corresponding confidence interval with the same value of α, and, for example, a one-sided hypothesis test with $\alpha = .05$ corresponds to a 90% confidence interval.

Consider the null hypothesis being in the form $H_0 : \mu \leq \mu_0$. Can we then determine, for example, $P(Z > 2.21 \mid \mu \leq 100)$? No, because in order to compute the probability we must condition on a value of μ, and we can't do that if we are not hypothesizing a particular value of μ. (For example, we can't compute a value of Z by writing "$X - \leq 100$" in the numerator of the expression for Z.) Obviously, the only way we can obtain one number for Z is to use a specific value for μ. Doing so means that we are conditioning on that value of μ in the probability statement, not on a range of values for μ.

Of course, an extreme value of Z relative to $\mu = 100$ would be even more extreme if μ were less than 100. So although we technically can't speak of a specific value of α when the null hypothesis contains an interval (and would thus have to speak of a maximum or minimum value), we can still carry out the test as if we had a point value rather than an upper bound or a lower bound.

5.3.3 Two-Sided Hypothesis Tests Vis-à-Vis Confidence Intervals

It was stated at the beginning of the chapter that the relationship between confidence intervals and hypothesis tests would be stressed and illustrated throughout the chapter. Therefore, we now show that we could have tested the hypothesis in the preceding example by constructing a confidence interval.

Assume for the sake of illustration that σ is known and consider Eq. (5.1). Since this was the starting point for constructing the $100(1 - \alpha)\%$ confidence interval, if we have a two-sided hypothesis test (i.e., the alternative hypothesis is "$\neq$") with a significance level of α, we would suspect that the confidence interval will not contain the hypothesized value if the value of the test statistic does not fall between $-Z_{\alpha/2}$ and $Z_{\alpha/2}$. That is, the inequality that is used in constructing the confidence interval is immediately violated. Let Z_0 and μ_0 denote the numerical value of the test statistic and the hypothesized mean, respectively, and assume that $Z_0 > Z_{\alpha/2}$ so that the null hypothesis is rejected "on the high side." Then by solving the inequality

$$\frac{\overline{X} - \mu_0}{\sigma/\sqrt{n}} > Z_{\alpha/2}$$

for μ_0, we obtain

$$\mu_0 < \overline{X} - Z_{\alpha/2}\frac{\sigma}{\sqrt{n}}$$

Thus, if we reject H_0 on the high side (i.e., a positive value for the test statistic), the hypothesized value of the mean lies below the lower limit of the confidence interval. If H_0 had been rejected on the low side (i.e., a negative value for the test statistic), μ_0 would have been above the upper limit of the confidence interval, as the reader is asked to show in Exercise 5.2.

Similarly, if the null hypothesis is *not* rejected, the confidence interval must contain the hypothesized value, as the reader is asked to show in Exercise 5.3, and if the confidence interval contained the hypothesized value, then the null hypothesis would not be rejected.

Because of this direct relationship between hypothesis tests and confidence intervals, which holds for almost all hypothesis tests (as will be seen later in the chapter), we could test a hypothesis by constructing the corresponding confidence interval. Indeed, this would be the preferred way of doing it because a confidence interval provides more information than the hypothesis test, in addition to providing the same information that a hypothesis test provides.

■ EXAMPLE 5.3

Sample and Assumption

A machine part has a nominal value of 50 millimeters for its diameter. Fifty parts are inspected and the average diameter is 50.7 millimeters. It has generally been assumed that $\sigma = 0.3$ millimeter and it is understood that the distribution of diameters is approximately normal.

Confidence Interval

A 95% confidence interval on the mean diameter is $50.7 \pm 1.96(0.3)/\sqrt{50} = 50.7 \pm 0.08 = (50.62, 50.78)$. Since the interval does not include 50, we would conclude that the mean has changed and this would have been the conclusion had a two-sided hypothesis test been performed with $\alpha = .05$.   ■

5.3.4 One-Sided Hypothesis Tests Vis-à-Vis One-Sided Confidence Intervals

The same type of relationship exists between a one-sided (i.e., directional) hypothesis test and a one-sided confidence interval. The latter has not been presented to this point in the chapter and might be better termed a one-sided confidence bound or limit since either the upper limit will be infinity or the lower limit will be minus infinity depending on whether it is a lower bound or an upper bound, respectively. Thus, since only one limit will be a finite number, it would be best not to refer to it as an interval.

There are many scenarios for which a one-sided confidence bound is needed rather than a two-sided interval. For example, a person would likely be interested in the average miles per gallon, averaged over different types of driving for, say, the first five years of driving a new car. A confidence interval for that average (mean) would be useful, but obviously it would be the lower limit that is of interest, as a mean near the upper limit would probably be viewed as a "bound." Using a simulator to obtain the mileage data, one might thus construct the lower bound, which would be of the form

$$\mu > \overline{X} - Z_\alpha \frac{\sigma}{\sqrt{n}}$$

(Note that Z_α is used rather than $Z_{\alpha/2}$ since this is a one-sided bound.)

A $100(1 - \alpha)\%$ one-sided confidence bound for μ with σ assumed known corresponds to a one-sided hypothesis test with either $H_0: \mu \geq \mu_0$ and $H_a: \mu < \mu_0$, $H_0: \mu \leq \mu_0$ and $H_a: \mu > \mu_0$, or $H_0: \mu = \mu_0$ and either $H_a: \mu < \mu_0$ or $H_a: \mu > \mu_0$. Note the possible forms for H_0, with and without an equal sign (only). The form of the hypothesis test and the conclusion that is reached do not depend on whether the null hypothesis contains an inequality or not. If an old car is taken in for an emissions test as required by a particular state, we might view the "null hypothesis" in that case as the maximum allowable value, such as the upper bound (i.e., "$\leq$") on nitric oxide emission. If the car fails the test, the "alternative hypothesis" of "$>$" is accepted. Strictly speaking, this is not a classical example of a hypothesis test since only one reading is obtained, but it does have some of the elements of a hypothesis test since testing apparatus are often in need of repair and the level of nitric oxide depends on various factors, including how hot the motor was at the time of the test. So there is an element of variation and thus the possibility of reaching the wrong conclusion, just as there is with a classical test.

We should not use "$<$" or "$>$" alone in the form of the null hypothesis. This is because we cannot, for example, substitute "<50" in a test statistic; we obviously must substitute a number, not an inequality.

Consider a hypothesis test of the form $H_0: \mu \geq \mu_0$ and $H_a: \mu < \mu_0$ and we do *not* reject H_0. This would correspond to an upper confidence bound, as we would expect. To see this, we observe that

$$\frac{\overline{X} - \mu_0}{\sigma/\sqrt{n}} \geq -Z_\alpha$$

when the null hypothesis is not rejected. Solving this inequality for μ_0 produces $\mu_0 \leq \overline{X} + Z_\alpha \sigma/\sqrt{n}$. Thus, μ_0 is less than the upper bound on μ and is therefore an "acceptable" value of μ, which corresponds to the null hypothesis being "accepted" (or properly stated, "not rejected," as explained earlier).

Conversely, if this null hypothesis had been rejected, then "$\geq$" in the Z-statistic inequality would be replaced by "$<$" and the second inequality would be $\mu_0 > \overline{X} + Z_\alpha \sigma/\sqrt{n}$. That is, μ_0 would be an unacceptable value for μ because it exceeded the upper bound for μ. Thus, μ_0 would be rejected by both the hypothesis test and the confidence interval, and of course the signals must be the same.

The illustration of the other combination of the null and alternative hypotheses would proceed similarly, and this combination would correspond to the miles per gallon example. Specifically, the alternative hypothesis would be $H_a: \mu > \mu_0$ and rejecting the null hypothesis in favor of this alternative hypothesis would result in $\mu_0 < \overline{X} - Z_\alpha \sigma/\sqrt{n}$. That is, the lower confidence limit would exceed μ_0, so the latter would not be a possible value for μ (with the stated probability, of course). Conversely, if the null hypothesis were not rejected, this would result in $\mu_0 > \overline{X} - Z_\alpha \sigma/\sqrt{n}$. That is, μ_0 would exceed the lower limit and thus be an acceptable value since it is within the "interval" (lower limit, infinity).

Therefore, for the miles per gallon example, testing H_0: $\mu \geq 25$ and *not* rejecting it would be the same as concluding that 25 is greater than the lower limit, since $\mu_0 = 25$. If we reject the null hypothesis, we are saying that 25 is less than the lower limit, so 25 is not a likely possible value. If 25 is the smallest acceptable value by manufacturing specifications, then we don't want to reject this null hypothesis.

Of course, our estimate of μ is $\overline{X}$, so when we go from $\overline{X}$ to $\overline{X} - Z_\alpha \sigma/\sqrt{n}$, we are simply going below $\overline{X}$ by an amount that reflects the sampling variability of $\overline{X}$ on the side of $\overline{X}$ in which we are interested, and similarly for the upper limit.

5.3.5 Relationships When the *t*-Distribution is Used

The relationships between confidence intervals and hypothesis tests that exist when Z is used also exist when t is used. That is, for two-sided hypothesis tests and two-sided confidence intervals, $Z_{\alpha/2}$ is replaced by $t_{\alpha/2, n-1}$, and for one-sided tests and one-sided intervals, Z_α is replaced by $t_{\alpha, n-1}$. [*Note*: The degrees of freedom for t is not always $n-1$, as the degrees of freedom for t is the degrees of freedom on which the estimate of $\sigma_{\text{statistic}}$ is based, with the latter defined for the specific procedure that is used and the statistic (not the test statistic) that is employed.]

5.3.6 When to Use *t* or Z (or Neither)?

This is a question that students and practitioners must address. Some instructors (such as this author) have been known to not give any partial credit on tests when the wrong statistic is used, as this is somewhat analogous to starting on a trip by driving in the wrong direction, which would obviously prevent one from reaching the desired destination unless the misdirection were corrected.

Table 5.1, which applies to a confidence interval or hypothesis test for μ, should prove helpful.

The first two cases are clear-cut, although unrealistic. *If* the distribution were normal and σ were known, then the standard normal distribution is the exact distribution. For the next two cases, the proper statistic is t but Z could be used as a substitute if n were large. This is not a robustness issue (i.e., not a matter of being sensitive or insensitive to departures from assumptions) and so the user could determine when large is "large enough." For example, let's assume that $n = 125$, σ is unknown, and a 95% confidence interval is desired. We

TABLE 5.1 **Choice of Test Statistic**

Distribution	Sample Size	σ	Test Statistic
Normal	Large	Known	Z
Normal	Small	Known	Z
Normal	Large	Unknown	Z or t
Normal	Small	Unknown	t
Unknown	Large	Known	Z
Unknown	Large	Unknown	Z
Unknown	Small	Known	Neither
Unknown	Small	Unknown	Neither

know that if we use Z, the value of Z we will use is 1.96. The following MINITAB output shows that the corresponding t-value is 1.9793.

```
Inverse Cumulative Distribution Function
Student's t distribution with 124 DF

P (X <= x)         x
   0.9750       1.9793
```

If a user were then to compute $2(1.9793 - 1.96)s_{\bar{x}}$, this would provide the difference of the widths of the confidence intervals, with the interval using t being the wider interval. In most cases this difference is likely to be inconsequential.

The next two cases *do* involve robustness considerations. If we don't know the distribution (the usual case), then we don't know what test statistic to use. If σ were known, then the Central Limit Theorem applies when the sample size is "large." Of course, what is sufficiently large depends on the shape of the parent population. See the robustness results given in Moore and McCabe (2002) and Pearson and Please (1975). When σ is unknown, we should obtain a good estimate of it with a large sample. Here the Central Limit Theorem does not directly apply since σ is unknown, but we could still invoke it with the understanding that it applies only approximately.

The last two cases in Table 5.1 will frequently occur, especially when sampling is expensive. If we have a small sample and we know nothing about the distribution, then we can't use either t or Z. Instead, we must use a *corresponding nonparametric approach* (see Section 16.2.3), or perhaps a bootstrap approach (Section 5.7). The former, which is also called a distribution-free approach, is the more conservative approach. The latter might provide better results in a particular application but could also be riskier, depending on the size of the initial sample and the shape of the parent population.

5.3.7 Additional Example

We will illustrate the connection between hypothesis tests and confidence intervals with the following example.

■ EXAMPLE 5.4

Sample

Light emitting diodes (LEDs) are expected to revolutionize the lighting industry by greatly increasing the lifespan of various lighting devices. Assume that accelerated testing (Section 14.3) has produced lifetimes of 100 lighting displays of a particular type with $\overline{X} = 76.5$ and $s = 6.25$ (units are 1,000 hours).

Hypothesis Test and Assumption

It is desired that the average lifetime be greater than 75 (000), so the manufacturer uses the data that have been collected to test this hypothesis, with $H_0: \mu \leq 75$ and $H_a: \mu > 75$. It is decided to initially assume that the distribution of $\overline{X}$ is probably reasonably close to a normal distribution because of the large sample size, and then proceed on that assumption.

Computations

Invoking the large-sample approximation, the test statistic is thus

$$
\begin{aligned}
Z &= \frac{\overline{X} - \mu_0}{s/\sqrt{n}} \\
&= \frac{76.5 - 75}{6.25/\sqrt{100}} \\
&= 2.40
\end{aligned}
$$

Conclusion

The company would like to see clear evidence that the average is greater than 75, and it would prefer to see a p-value less than .01. The p-value can easily be shown (such as by using Table B) to be .0082. Thus, the evidence would be deemed to be sufficient.

Alternative Approach

A more useful approach would be to construct a lower bound on the lifespan, and the desired upper bound on the p-value as stated by the company corresponds to a 99% lower bound on the lifespan. We "know" that the lower bound will slightly exceed 75 because the p-value was slightly less than .01. Numerically, the lower bound is $76.5 - 2.326(6.25/\sqrt{100}) = 75.05$.

Thus, the result is essentially what we expected. The hypothesis test and confidence interval gave us the same message, but the confidence interval gave us the message in the units that were of interest, rather than in terms of a probability. ∎

5.4 CONFIDENCE INTERVALS AND HYPOTHESIS TESTS FOR A PROPORTION

In this section we consider confidence intervals and hypothesis tests for a proportion, such as the proportion of transistors produced by a company that are out of tolerance. It was shown in Section 3.3.2 that the variance of the binomial random variable is $np(1 - p)$. This variance is typically used in constructing confidence intervals and hypothesis tests. Implicit in the derivation of the variance is the assumption of a constant value of p, in addition to the assumption of independent observations. Both assumptions may be questionable in a practical application and should therefore be checked. Obviously, the first assumption could be checked only if multiple samples are obtained, but generally only one sample is obtained when a confidence interval or hypothesis test is constructed.

A simple, heuristic way of checking the first assumption would be to plot the observations (successes and failures, ones and zeros) in a single sample over time. Time might then be blocked into equal-sized intervals and p computed within each interval. If the values of p differ greatly, then p is probably not constant. Similarly, if the zeros and ones do not appear to be occurring randomly, the observations are probably not independent. A simple test for the adequacy of the binomial distribution assumption in a particular application was given in Section 3.3.2.1. Tests such as that one (or even simple visual tests) should be employed rather than just assuming that the binomial distribution is an adequate model in a particular application.

For the sake of simplicity, however, we shall assume in this section that the binomial assumption is adequate. As stated previously, when we construct a confidence interval using Z or t, the confidence interval for some parameter θ is always of the form (when $\sigma_{\hat{\theta}}$ is not assumed known) $\hat{\theta} \pm (Z \text{ or } t)\hat{\sigma}_{\hat{\theta}}$.

The use of either Z or t implies that normality in some form is being assumed: either a normal distribution or the adequacy of a normal approximation. The normal approximation to the binomial distribution can be written $A = (X - np)/\sqrt{np(1-p)}$, which equals $(\hat{p} - p)/\sqrt{p(1-p)/n}$ if we divide numerator and denominator by n. (*Note:* If binomial probabilities were approximated, $+0.5$ and/or -0.5 would be added to X, depending on the form of the probability statement.) *If* the normal approximation approach were applied to confidence intervals, a confidence interval for p would be obtained as

$$\hat{p} \pm Z_{\alpha/2}\sqrt{\hat{p}(1 - \hat{p})/n} \tag{5.3}$$

(Note that we must assume the use of $\hat{\sigma}_{\hat{\theta}}$ instead of $\sigma_{\hat{\theta}}$ because the latter is a function of p and if we knew p we wouldn't need to construct a confidence interval for it.)

This approach using Eq. (5.3) is the most frequently used approach, but only when the parameter is close to 0.5 is the interval likely to be adequate. Rules of thumb are often given in terms of the numerical values of np and $n(1-p)$, with the idea that the interval will be adequate when each of the two products exceeds either 5 or 10. The value of p more strongly determines the adequacy of the interval than does either np or $n(1-p)$, however.

5.4.1   Approximate Versus Exact Confidence Interval for a Proportion

So is there a better way to construct the confidence interval? Clopper and Pearson (1934) presented an exact approach for obtaining the interval, but Agresti and Coull (1998) showed in their article, "Approximate Is Better than "Exact" for Interval Estimation of Binomial Proportions," that the exact Clopper–Pearson approach isn't the best way to construct the interval. Specifically, they showed that, for $n \leq 100$, with the exact approach the coverage probability may be considerably higher than $1 - \alpha$. For example, an "exact" 95% confidence interval is guaranteed to be at least 95%, but might actually be a 98% interval if n is very small (e.g., 15), and can be noticeably larger than 95%—such as 96.5%—even when $n = 100$. Certainly, this sounds paradoxical, but this "error" in the exact approach is due in part to the fact that the binomial distribution is discrete rather than continuous, with cumulative probabilities taking large jumps when n is small, and smaller jumps when n is large.

As is being stressed in this chapter and in the next chapter, there is a direct connection between confidence intervals and hypothesis tests, so it would be logical to use a confidence interval for p that corresponds to the test of $H_0: p = p_0$ versus $H_a: p \neq p_0$. The test statistic that is generally used to test H_0 is

$$Z = \frac{\hat{p} - p_0}{\sqrt{p_0(1 - p_0)/n}}$$

with the value of Z compared with $\pm Z_{\alpha/2}$. This might seem to be as objectionable as using Z to construct the interval because $\hat{p}$ does not have a normal distribution. It is approximately normally distributed when n is large, however. But what about when n is small; should t be used instead of Z? There is no "t-test" for a proportion because the individual observations must have a normal distribution, at least approximately, for a t-test to be used, but here the individual observations have a binomial distribution.

The confidence interval given earlier in this section does not correspond to this hypothesis test because the estimated standard deviation of $\widehat{p}$ is used in the confidence interval, whereas the standard deviation under H_0 is used in the hypothesis test. Thus, this is a rare exception in which we do not have a direct correspondence between the confidence interval and the corresponding hypothesis test, and this is due to the fact that the standard deviation of the estimator of the parameter is a function of the parameter. The hypothesis test, by definition, must be a function of the hypothesized value of the parameter, whereas the confidence interval utilizes an estimate of the parameter obtained from a sample.

The method recommended by Agresti and Coull (1998) and by Brown, Cai, and DasGupta (2001), and originally given by Wilson (1927), is to use the form of the confidence interval that corresponds to this hypothesis test. That is, solve for the two values of p_0 (say, p_{upper} and p_{lower}) that result from setting $Z = Z_{\alpha/2}$ and solving for $p_0 = p_{upper}$, and then setting $Z = -Z_{\alpha/2}$ and solving for $p_0 = p_{lower}$. Although this might sound complicated, the appropriate expressions can be obtained by straightforward but slightly tedious algebra. Such algebraic manipulation isn't necessary, however, as the appropriate expressions are given in various sources. Specifically, we have

$$U.L. = \frac{\widehat{p} + z_{\alpha/2}^2/2n + z_{\alpha/2}\sqrt{\widehat{p}(1 - \widehat{p})/n + z_{\alpha/2}^2/4n^2}}{1 + z_{\alpha/2}^2/n} \tag{5.4a}$$

$$L.L. = \frac{\widehat{p} + z_{\alpha/2}^2/2n - z_{\alpha/2}\sqrt{\widehat{p}(1 - \widehat{p})/n + z_{\alpha/2}^2/4n^2}}{1 + z_{\alpha/2}^2/n} \tag{5.4b}$$

This approach can be substantiated on the grounds that it corresponds to the large-sample hypothesis test and is also supported by the research of Agresti and Coull (1998). We illustrate it with the following example.

■ **EXAMPLE 5.5**

A company has been having serious problems with scrap and rework so one of its quality engineers decides to investigate a particular process. A sample of 150 items is taken on a particular day and an alarmingly high percentage, 16%, is found to be nonconforming (i.e., defective). The engineer decides to construct a 95% confidence interval on the true proportion of defective units at that time, realizing that point estimates do vary. The computations proceed as follows.

$$
\begin{aligned}
U.L. &= \frac{\widehat{p} + z_{\alpha/2}^2/2n + z_{\alpha/2}\sqrt{\widehat{p}(1 - \widehat{p})/n + z_{\alpha/2}^2/4n^2}}{1 + z_{\alpha/2}^2/n} \\[2mm]
&= \frac{.16 + (1.96)^2/2(150) + 1.96\sqrt{.16(.84)/150 + (1.96)^2/4(150)^2}}{1 + (1.96)^2/150} \\[2mm]
&= \frac{0.2457}{1.0256} = 0.2395
\end{aligned}
$$

$$L.L. = \frac{\widehat{p} + z^2_{\alpha/2}/2n - z_{\alpha/2}\sqrt{\widehat{p}(1-\widehat{p})/n + z^2_{\alpha/2}/4n^2}}{1 + z^2_{\alpha/2}/n}$$

$$= \frac{.16 + (1.96)^2/2(150) - 1.96\sqrt{.16(.84)/150 + (1.96)^2/4(150)^2}}{1 + (1.96)^2/150}$$

$$= \frac{0.1256}{1.0256} = 0.1224$$

The 16% defective rate was very troublesome and the upper limit of the confidence interval was even more so, thus necessitating corrective action. ■

One advantage of this procedure is that its worth does not depend strongly on the value of n and/or p. In addition to the fact that the standard approach is unsatisfactory from the standpoint of coverage probabilities, there is a distinct possibility that the lower limit may not exist. Clearly, a proportion cannot be negative, and a lower limit of zero would also be meaningless since that is the lower limit by definition. The computed lower limit with the standard approach given in Eq. (5.3) will not be positive, however, unless $\widehat{p} > 1.96/(n + 1.96^2)$. Notice that the larger n is, the more likely the inequality will be satisfied, in general. However, if n were only, say, 30, then $\widehat{p} > .0579$ is required in order to have a lower limit. In many applications, such as quality improvement applications, we would expect p to be small (.01 or less), so we might need a very large sample in order to have a positive lower limit with the standard approach. Obtaining a negative computed lower limit for something that cannot be negative should cause us to question the approach that is being used.

We do much better with the improved approach, as the computed lower limit can never be negative, regardless of the value of n, as the reader is asked to show in Exercise 5.17.

As stated previously, if we wanted to construct a one-sided confidence limit, such as a lower limit, we would simply replace $Z_{\alpha/2}$ by Z_α in the expression for the upper or lower limit, whichever is desired. For example, we might want an upper bound on the proportion of time that one or more components of an electrical system fail to function properly, or an upper bound on the percentage of nonconforming units of a certain type, or an upper bound on the percentage of units that have one or more nonconformities.

Since the confidence interval given by Eqs. (5.4a) and (5.4b) stems from the large-sample hypothesis test, we should keep in mind that it can perform poorly in situations for which the normal approximation will have difficulty. In particular, estimating tail probabilities for a nonnormal distribution can be difficult, yet when we work with hypothesis test endpoints we are in the tails of the distribution, especially when a 99% confidence interval is desired.

We observe that the width of the confidence interval, which we will not write explicitly here, is a function of $\widehat{p}$, which of course we will not have until we have taken a sample. Thus, as with almost all of the other confidence intervals in this chapter, it is not particularly practical to solve for n. A crude approach would be to substitute .5 for $\widehat{p}$, which will maximize n.

5.5 CONFIDENCE INTERVALS AND HYPOTHESIS TESTS FOR σ^2 AND σ

The confidence intervals presented for μ thus far in the chapter (using Z or using t) are symmetric about the point estimator of the parameter. This is also true for the standard way of obtaining a confidence interval for p but is not true for the improved approach that was given in Section 5.4.1. The reason that almost all of these intervals are symmetric is that the distribution that is used in producing the intervals is itself symmetric. Certainly, whenever the form of the interval is $\hat{\theta} \pm W_{\alpha/2} S_{\hat{\theta}}$, with W here denoting an arbitrary variate, the confidence interval will obviously be symmetric about $\hat{\theta}$ because of the "$\pm$".

We cannot, however, simply choose to use this form when constructing *any* confidence interval because the form will depend on the relevant distribution theory. To illustrate, consider a confidence interval for σ^2. To obtain the confidence interval we use the chi-square distribution because

$$\frac{(n-1)S^2}{\sigma^2} \sim \chi^2_{n-1}$$

as the reader is asked to show in Exercise 5.59.

Therefore, it follows that

$$P\left(\chi^2_{n-1,\alpha/2} \leq \frac{(n-1)S^2}{\sigma^2} \leq \chi^2_{n-1,1-\alpha/2}\right) = 1 - \alpha$$

From this starting point, we simply need to manipulate the form so that σ^2 will be in the middle of the double inequality. Doing the appropriate algebra produces

$$P\left(\frac{(n-1)S^2}{\chi^2_{n-1,1-\alpha/2}} \leq \sigma^2 \leq \frac{(n-1)S^2}{\chi^2_{n-1,\alpha/2}}\right) = 1 - \alpha \tag{5.5}$$

so that the expression on the left gives the lower limit and the expression on the right gives the upper limit. Here the values of χ^2 are the values of the variate with a cumulative probability of $\alpha/2$ and $1 - \alpha/2$, respectively.

The width of the interval of course is given by the difference between the two fractions in Eq. (5.5). Note, however, that if we tried to solve for n for a specified width, we would have the same type of problem as encountered previously for other types of intervals: the χ^2 value depends on n and we won't have a value for s until we have taken a sample. Thus, solving for n so as to have a confidence interval with a specified width is not practical.

We will use the following example to illustrate the construction of the interval.

■ **EXAMPLE 5.6**

Objective and Sample

A leading personal computer manufacturer is interested in the variance of the time that its customer service personnel take to respond to customers at the company's "Internet chat" facility at its website. So a sample of 100 records is obtained with the objective of constructing a confidence interval for the variance of the time, as a high variance would suggest that perhaps some customer service personnel are not responding promptly.

Sample, Assumption, and Computation

The 100 response times have a variance of 123.9 seconds squared. If we assume that the response times are approximately normally distributed, we can construct the 95% confidence interval as

$$L.L. = \frac{(n-1)S^2}{\chi^2_{n-1,\,1-\alpha/2}} = \frac{99(123.9)}{128.422} = 95.5$$

$$U.L. = \frac{(n-1)S^2}{\chi^2_{n-1,\,\alpha/2}} = \frac{99(123.9)}{73.361} = 167.2$$

Alternative Approach

Now assume that the company decides it would rather have a confidence interval for σ instead of for σ^2, as the company would prefer not to lose the unit of measurement. How should the company proceed?

The relevant distribution is then the chi distribution, as the square root of a random variable that has a chi-squared distribution has a chi distribution, as we might expect. We need not be concerned with this distribution, however, as the endpoints of the confidence interval for σ would simply be the square roots of the endpoints of the interval for Eq. (5.5). Therefore, for this example we have $L.L. = 9.77$ and $U.L. = 12.93$.   ■

■ EXAMPLE 5.7

Consider a circuit that constructs the sum of two binary numbers (i.e., a full adder). The time to stabilization of the sum after inputs change and the carries propagate from the least significant bit to the most significant bit is a random variable because the inputs are variable and the capacitances will also vary. Assume that the time can be modeled by a normal distribution. Five measurements are taken and the values are: 47.2, 50.9, 46.3, 45.1, and 47.4 nanoseconds. Would it be possible or practical to construct a confidence interval or at least obtain a point estimate for the variance based on the information given?

Solution   Since normality was stated, a confidence interval for σ^2 can be obtained using the chi-square distribution. The sample variance can be shown to be 4.697. It follows that the 95% confidence interval on the population variance is

$$L.L. = \frac{(n-1)S^2}{\chi^2_{n-1,\,1-\alpha/2}} = \frac{4(4.697)}{11.1433} = 1.686$$

$$U.L. = \frac{(n-1)S^2}{\chi^2_{n-1,\,\alpha/2}} = \frac{4(4.697)}{0.4844} = 38.786$$

Notice that this is an extremely wide interval relative to the magnitude of the sample variance, so the practical value of a confidence interval of this width might have to be questioned. Of course, the culprit is the small sample size, so a larger sample might have to be taken to obtain a confidence interval that is of acceptable width. ∎

5.5.1 Hypothesis Tests for σ^2 and σ

If we wanted to test $H_0: \sigma^2 = \sigma_0^2$, we would compute

$$\chi^2 = \frac{(n-1)S^2}{\sigma_0^2} \tag{5.6}$$

and compare the computed value against the appropriate value(s) of χ_{n-1}^2, which, as with other distributions, depends on the value of α and whether or not the test is one-sided, and of course additionally depends on the value of n. Note that Eq. (5.6) was the starting point for deriving the confidence interval for σ^2.

To illustrate, assume that the variability of a measured characteristic in a manufacturing process has purportedly been improved, but the plant manager would like to see hard evidence of that. There are two ways in which one might proceed. If a future comparison of the old process versus the new process had been planned, a sample of items from the old process might have been taken, for comparison with a sample from the new process. The methodology for this scenario would then be that given in Section 6.4. If the old process had not been sampled, however, then the best that could be done would be to take a sample from the new process, compute S^2, and compare that with records that suggest what the variance was under the old process. The latter would then be taken as the hypothesized value and we would hope that we could reject that value with a one-sided hypothesis test, which would suggest that the variability of the process had indeed been reduced.

Assume that records indicate that σ^2 was approximately 2.86, but no sample was taken from the previous process. A sample of 100 items is obtained from the new process and S^2 = 2.43. Could the difference between 2.86 and 2.43 be due to sampling variability, or does this likely constitute evidence that the process has indeed been improved? Thus, we want to test $H_0: \sigma^2 = 2.86$ versus $H_a: \sigma^2 < 2.86$.

As with the other test scenarios, we could indirectly test this hypothesis by constructing a one-sided confidence bound, or directly by performing the hypothesis test. Since confidence intervals are more informative, we will construct the confidence bound. We want an upper bound on σ^2, and we hope that the upper bound is below 2.86. The form of the upper bound is

$$U.L.\ \sigma^2 < \frac{(n-1)S^2}{\chi_{n-1,\alpha}^2}$$

Although we might not be interested in doing so in a particular situation, a variation of what has been presented so far in this chapter regarding the relationship between confidence intervals and hypothesis tests would be to solve for the degree of confidence (i.e., solve for α) such that the upper bound on the confidence interval is 2.86. This value of α would also be the p-value for the hypothesis test. This would be an indirect way of solving for the p-value.

Solving for $\chi^2_{n-1,\alpha}$ involves only algebra, but then software would have to be used to determine α. The degree of confidence is then $100(1 - \alpha)\%$. Setting the upper bound equal to 2.86 and solving for $\chi^2_{n-1,\alpha}$, we thus obtain

$$\frac{(n - 1)S^2}{\chi^2_{n-1,\alpha}} = 2.86$$

so that

$$\begin{aligned}
\chi^2_{n-1,\alpha} &= \frac{(n - 1)S^2}{2.86} \\
&= \frac{99(2.43)}{2.86} \\
&= 84.1154
\end{aligned}$$

We can solve for α using MINITAB, for example, by specifying this cumulative probability for a chi-square distribution with 99 degrees of freedom. Doing so gives $\alpha = .1428$, so $1 - \alpha = .8572$. Thus, the 85.72% upper bound would be 2.86. Of course, we generally use a higher degree of confidence than this, so our upper bound with a higher degree of confidence would exceed the hypothesized value, which we would thus not be able to reject with one of the commonly used values of α. Therefore, we cannot reject the null hypothesis.

If we had performed a conventional hypothesis test with, say, $\alpha = .05$, we would have computed the value of the χ^2 test statistic, which the reader can observe is the 84.1154 that we have already obtained and compared that with $\chi^2_{n-1,.05}$, which can be shown to be 77.0463.

Since $\chi^2_{\text{calculated}} > \chi^2_{\text{critical}}$ and is thus not in the rejection region, we would fail to reject the null hypothesis. If we want to use a p-value approach, the p-value is .1428 (which of course we already obtained), so we would fail to reject H_0 since this number is greater than .05. Thus, we must conclude that there is no statistical evidence that the process variability has been reduced.

Again, the reason for going through all of these computations (not all of which were necessary in working the problem) is to show that a hypothesis test and the corresponding confidence interval are essentially different sides of a coin, as stated previously.

5.6   CONFIDENCE INTERVALS AND HYPOTHESIS TESTS FOR THE POISSON MEAN

Let X denote the number of nonconformities, for example, for a specified area of opportunity, and let λ denote the corresponding mean. The confidence interval that is probably the most widely used is

$$X \pm Z_{\alpha/2}\sqrt{X} \tag{5.7}$$

which is of the general form $\widehat{\theta} \pm Z s_{\widehat{\theta}}$ that was mentioned previously. An argument can certainly be made, however, that if we are going to use the large-sample hypothesis test that is of the form

$$Z = \frac{X - \lambda_0}{\lambda_0}$$

then at least in the case of large samples we should invert this hypothesis test to obtain the corresponding confidence interval, just as was done in the binomial case. Doing so produces the interval

$$\frac{2X + Z_{\alpha/2}^2 \pm Z_{\alpha/2}\sqrt{Z_{\alpha/2}^2 + 4X}}{2} \tag{5.8}$$

The simplest types of confidence intervals are the ones that are going to be used the most often, and although the interval given by Eq. (5.8) is not as simple as the interval given by Eq. (5.7), it is easier to justify. Barker (2002) considered the Eq. (5.8) interval in addition to eight other confidence interval forms in a comparison study for the case when $\lambda \leq 5$ and found that in the case of 95% confidence intervals, this form for the interval comes closer to maintaining the 95% coverage probability than does any other closed-form interval.

Therefore, this interval might be used for both large and small values of λ. See also the method given by Schwertman and Martinez (1994) and the methods that were reviewed by Sahai and Khurshid (1993).

■ **EXAMPLE 5.8**

Problem Statement and Sample

A company's vice president for quality improvement is concerned about the sudden appearance of surface blemishes on the company's flat panel displays. Accordingly, she asks that a sample of 500 such displays be obtained during March, and it is found that 23 have surface blemishes. This provides her with some idea of the magnitude of the problem, but she asks that a confidence interval additionally be constructed.

Confidence Interval and Conclusion

Using Eq. (5.8), the 95% confidence interval is (3.83, 46.03). If the sample taken was a representative sample, this provides an indication of the number of surface blemishes that can be expected per 500 units. If much of this interval would be considered unacceptable by management, then corrective action should be taken.

Notice that the interval obtained using Eq. (5.8) is symmetric about $X + Z_{\alpha/2}^2/2$, not symmetric about X. ■

5.7 CONFIDENCE INTERVALS AND HYPOTHESIS TESTS WHEN STANDARD ERROR EXPRESSIONS ARE NOT AVAILABLE

Let θ denote an arbitrary parameter, with $\widehat{\theta}^*$ denoting an arbitrary estimator. A closed-form expression for $s_{\widehat{\theta}^*}$ is necessary in order to calculate confidence intervals and perform hypothesis tests using an equation, such as those presented in previous sections. As stated in Section 4.5, such expressions are easily obtained for estimators presented in textbooks, but in many applications it will not be possible to obtain the necessary expression. Consequently, *bootstrap methods* have become popular (in some quarters) during the past ten to fifteen years.

There are different types of bootstrap approaches, but the general idea is as follows. A sample is taken and a thousand or more subsamples of the same size as the original sample are taken from this sample, with replacement, acting as if the sample were the population. For k subsamples, $\widehat{\theta}^*$ is computed for each subsample and the standard error of $\widehat{\theta}^*$ is then computed as

$$s_{\widehat{\theta}^*} = \sqrt{\frac{\sum_{i=1}^{k} [\widehat{\theta}_i^* - \text{ave.}(\widehat{\theta}^*)]^2}{k-1}}$$

There is no set number for k that is agreed on, but the U.S. Food and Drug Administration, for example, has recommended $k = 2{,}000$ (see `http://www.fda.gov/cder/bioequivdata/statproc.htm`) as part of its bioequivalence guidelines.

For a $100(1 - \alpha)\%$ confidence interval, the endpoints of the interval are given by the $\alpha/2$ and $1 - \alpha/2$ percentiles of the distribution of the $\widehat{\theta}_i^*$ that has been generated from the bootstrapping. This is a nonparametric bootstrap approach as no distribution assumption is involved. There is also a parametric bootstrap and various alternative approaches, as stated previously.

■ **EXAMPLE 5.9**

Statistical Approach

In his paper, "Confidence Intervals for the Mean of Non-normal Data" (*Quality and Reliability Engineering International*, **17**, pp. 257–267, 2001), F. K. Wang gave an example of a scenario where a bootstrap confidence interval would be warranted, although Wang did not construct the interval(s). The following data were stated as being two random samples from an accounting firm in Taiwan.

```
Accounts receivable (unit = NT$1,000):

510, 2684, 2907, 3165, 3553, 3640, 7267, 13571

Accounts payable (unit = NT$1,000):

819, 1272, 4857, 5047, 6617, 9324, 14597, 21904,
38824, 42409, 48056, 51818
```

Transformation

Wang tested each sample for normality and concluded "these two data [sic] are non-normal." He then proceeded to construct a confidence interval for each mean using a normal theory approach, a Box–Cox transformation (to normality) approach, and a bootstrap approach. He concluded that the bootstrap approach produced the superior results because the confidence intervals using that approach had the smallest width among the three intervals constructed for each of the two population parameters.

Before we proceed to construct a bootstrap confidence interval for the accounts receivable population mean and the accounts payable population mean, a few comments are in order. First, it isn't practical to test for normality with such small samples, as any test for normality will have low power to detect a departure from normality because of the small sample sizes. Furthermore, *any* small sample from a normal population is going to look nonnormal simply because there aren't enough observations for a histogram of the data to be mound-shaped. Of course, we know that (exact) normality doesn't exist in practice anyway, as has been emphasized.

That said, the degree of departure from normality for these populations undoubtedly depends on the mixture of large, medium, and small firms and more weight should be placed on that than the information from small samples.

With these issues in mind, we will proceed to construct the bootstrap confidence intervals and compare the results with those given by Wang (2001).

Computations

Using 1,000 bootstrap samples, we obtain 2505.88 as the 2.5 percentile of the averages and 7503.88 as the 97.5 percentile, so that the width of the interval is 4998. Note that the midpoint of the interval, 5004.88, is considerably above the average value of 4662. This shows the influence of an extreme observation, 13,571, as that value is greater than the average by a much larger amount than the smallest value, 510, is below the average. In general, bootstrapping will fail when outliers are present. We don't know whether or not the extreme value is a valid data point. If it is a valid data point, then it is suggestive of a skewed distribution.

Wang constructed the confidence interval as the average of the bootstrap averages plus and minus 1.96 times the standard error of the averages, thus acting as if the bootstrap averages were normally distributed. The distribution is skewed, however, because of the outlier. Consequently, Wang's confidence interval (2831.6, 8168.4) differs considerably from the interval given in the preceding paragraph.

For the other sample, the 95% bootstrap confidence interval is (10,358.3, 31,567.0) and the distribution of the bootstrap averages is reasonably close to exhibiting perfect symmetry. The lower limit differs considerably from the lower limit in Wang's interval, however, as his interval was (11,326.2, 32,069.6). ∎

The success of the bootstrap approach obviously depends on how well the initial sample represents the population. Obviously, the larger the sample, the more likely that it will be reasonably representative of the population. Not surprisingly, the only theoretical support for bootstrapping methods are asymptotic results (i.e., as the size of the initial sample approaches infinity).

In essence, the user of bootstrap methods is acting as if the original sample contains more information than it actually does contain. It is not possible to increase the amount of information in a sample, but in essence that is what the user is trying to do. Thus, although bootstrap methods have been used successfully in many applications, they do have a somewhat shaky foundation and should be used cautiously, except in certain applications where they are known to fail and thus should not be used at all. In particular, they generally fail when the population distribution is markedly skewed. See Davison and Hinkley (1997) and Efron and Tibshirani (1993) for an introduction to bootstrap methods and bootstrap confidence intervals. See also Chernick (1999) for, in particular, a discussion of six common myths about bootstrapping. The references given in Section 4.5 of this text may also be of interest.

5.8 TYPE I AND TYPE II ERRORS

When we test a hypothesis and reach a conclusion, we will have made either one of two possible correct decisions, or one of two possible incorrect decisions, depending on the decision that we made. Specifically, there are two possible decisions, reject or not reject the null hypothesis, and the null hypothesis could be either true or false, although as indicated in Section 5.3.1, null hypotheses that are true will not be encountered very often. There are thus four possible combinations of decisions and outcomes, as is indicated in the table below.

		Decision	
		Not Reject	Reject
Null Hypothesis	True	Correct	Incorrect
	False	Incorrect	Correct

This characterization of the four possibilities, although standard, is almost too simplistic since, as stated previously, null hypotheses will generally be false. If we fail to reject the null hypothesis, that may mean that it is "almost true," or it may mean that we simply used a sample size that was too small. Sample size determination was discussed in Section 5.1, for example, but it has not been discussed in this chapter in the context of hypothesis testing. In general, the sample size used for hypothesis testing should be sufficient to detect, with a stated probability, the smallest difference that is considered to be consequential between the true state of nature and what is hypothesized.

For example, assume that our null hypothesis is H_0: $\mu = 50$ but in reality, $\mu = 50.3$. The fact that we are testing $\mu = 50$ means that we doubt that value, but if μ were 50.3 we might not care that we failed to reject 50! That is, the difference between 50.0 and 50.3 might not be of practical significance. Assume, however, that the difference between 50 and 51 *is* of practical significance, so that if the mean were really 51, we would want to reject 50 and thus detect this difference with a specified probability. Let's assume that we want to detect the difference with probability .90. How large a sample should we take to accomplish this?

The question that we are asking relates to the *power of the test*. That is, how powerful is our test at detecting specified differences from what we have hypothesized? For simplicity, let's assume that we have a one-sided test with $\alpha = .05$, and we will further assume

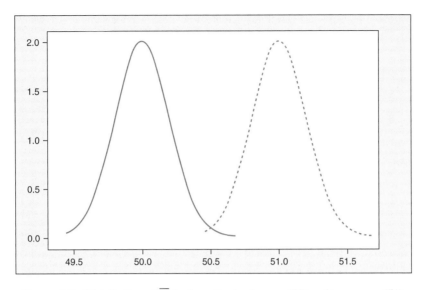

Figure 5.5 Distribution of $\overline{X}$ for hypothesized mean (50) and true mean (51).

normality and that $\sigma = 2$. Then the test statistic will be

$$
\begin{aligned}
Z &= \frac{\overline{X} - \mu}{\sigma / \sqrt{n}} \\
&= \frac{\overline{X} - 50}{2 / \sqrt{n}}
\end{aligned}
$$

We then want to solve for n such that $P(Z > 1.645 \mid \mu = 51) = .90$. It is important to note that we are conditioning on $\mu = 51$ because we are assuming this to be the true value. Before we continue further, it will be helpful to use a graph as a reference point, and the relevant graph is given in Figure 5.5.

Using this graph as a guide to allow us to see the null distribution and the assumed distribution, we can solve for n (the derivation is given in the chapter Appendix) as

$$
\begin{aligned}
n &= \left(\frac{\sigma(Z_\alpha + Z_\beta)}{\mu_1 - \mu_0} \right)^2 \\
&= \left(\frac{2(1.645 + 1.28155)}{51 - 50} \right)^2 \\
&= 34.26
\end{aligned}
$$

which we would round up to 35 so as to have *at least* the power that we desired. It is helpful if the reader can understand the mechanics, but we don't need to perform calculations such as these every time we want a power calculation. Instead, there are software packages we can use for these computations. For example, the following is the MINITAB output with the sequence of commands that produced the output given at the textbook website. Notice that the probability ("power") is slightly greater than .90 since the sample size was rounded up.

```
Power and Sample Size

1-Sample Z Test

Testing mean = null (versus > null)
Calculating power for mean = null + difference
Alpha = 0.05 Assumed standard deviation = 2

Difference Sample Size Target Power Actual Power
         1          35          0.9      0.905440
```

The reason the probability is labeled "power" in the last section of the output is because this is in reference to the *power of the test*. We clearly want to have a test and sample size that will detect a significant departure from the hypothesized value with a high probability, which means high or good power. The symbol used to denote power is generally $1 - \beta$. Since this is the probability of rejecting the null hypothesis when it is false, β is the probability of failing to do so and failing to reject a false null hypothesis is termed a Type II error. The other type of error, rejecting the null hypothesis when it is true, is termed a Type I error. The probability of the latter occurring is equal to the significance level, α, that is selected by the user if a classical hypothesis testing approach is used. If not, then a "p-value," as discussed previously and in more detail in the next section, is computed, and the user decides whether or not the value is small enough to reject the null hypothesis.

5.8.1   *p*-Values

DEFINITION

A *p-value* is the probability that the value of the test statistic is at least as extreme, relative to the null hypothesis, as the value that was observed, when the null hypothesis is true.

For example, assume that we test H_0: $\mu = 50$ versus H_a: $\mu \neq 50$, we assume normality, and the value of the Z-statistic is 2.01. The $P(Z > 2.01 \mid \mu = 50) = .0222$. This is a two-sided test and if we used $\alpha = .05$, we would reject H_0 if $Z > 1.96$ or $Z < -1.96$. Since the calculated Z-value is close to one of these two numbers, we would expect the p-value to be close to .05. When we have a two-sided test, the words "at least as extreme" in the definition of a p-value refer to either a large or small (extreme) value of the test statistic. Accordingly, the probability given above is doubled because $Z = -1.96$ is as extreme as $Z = 1.96$. Therefore, the p-value for this example is .0444 and we reject H_0. (If the need to double the probability to obtain the p-value isn't clear, note that if we didn't double the probability, we would have to compare it with $\alpha/2$ instead of with α. Otherwise, we would reach the wrong conclusion if, for example, α were .05 and the probability were .03.)

We should note that a p-value is a random variable because it is determined by the value of the test statistic, which is a random variable. If we obtained a second sample, we would almost certainly have a different p-value, just as the second sample will almost certainly differ from the first sample. The extent to which p-values would vary over successive samples would depend on the extent to which the values of the test statistics would vary over those samples, which will in turn depend on the sample size that is used. This is something that should be kept in mind when p-values are reported and decisions are made based on p-values.

We should also note that a p-value can exceed .50 for a one-sided test, as the value of the test statistic could be in the opposite direction from the alternative hypothesis. For example, consider the hypotheses $H_0: \mu = 50$ versus $H_a: \mu < 50$. If the sample mean exceeds 50, the p-value will exceed .50. Of course, a p-value for a two-sided test will be much larger than .50 if the sample mean is close to 50.

When a decision is made based on the p-value, the user must obviously have some threshold value in mind, and that value might indeed be .05. The advantage of using a p-value is that the classification of a p-value as "large" or "small" does not depend on what test was performed or what test statistic was used. The use of p-values is well established, especially on computer output.

A p-value is frequently defined as "the smallest significance level that will result in the null hypothesis being rejected." Such a definition is potentially confusing, however, as a p-value and a significance level are unrelated concepts, and a person can (and probably should) discuss one without discussing the other. Clearly, one could take a classical approach to hypothesis testing and make a decision without ever computing a p-value. Similarly, a decision can be made from a p-value without specifying a significance level. Furthermore, a p-value is a function of the data; a significance level is specified by an experimenter and is independent of the data. Since the two concepts are essentially unrelated, it is best not to relate them.

Nevertheless, if one insists on relating the two concepts, the following statement, which is a (necessary) modification of a statement that has appeared in the literature can be made: "If one repeatedly applies a statistical procedure at a specific significance level to distinct samples taken from the same population and if the null hypothesis were true and all assumptions were satisfied, then the p-value will be less than or equal to the significance level with the proportion of time that this occurs given by the significance level."

There is some confusion among users of p-values as to what it represents. It does *not* represent the probability of the null hypothesis being true or false because in the frequentist approach to statistics (the standard approach), no probabilities are assigned to possible values of a parameter.

Although there is no direct mathematical relationship between α and β (i.e., we cannot solve for one given the other), in general, there is a monotonic relationship between them. That is, if one is increased, the other will decrease. Consider the example given at the start of this section. The value of $1 - \beta$, the power, was specified as .90 for $\mu = 51$, so the probability of committing a Type II error is $1 - .90 = .10$. Thus, the probabilities for the two types of errors are different for this example, as they will be for virtually every example. The probabilities will be very close if the true parameter value is extremely close to the hypothesized value, as $1 - \beta$ will "converge" to α as we move the true parameter value closer to the hypothesized value. This should be clear because the probability of rejecting a true hypothesis is $1 - \alpha$, so the probability of rejecting an "almost true" hypothesis is then very close to $1 - \alpha$. Thus, failing to reject means that β is very close to $1 - (1 - \alpha) = \alpha$.

What would the probability of a Type II error have been if α had been .01 instead of .05? Would the probability be higher or lower? Consider the two distributions for $\overline{X}$ given in Figure 5.2. If α were decreased, the Z-value that would be used in determining the dividing line for rejecting or not rejecting the null hypothesis in terms of $\overline{X}$ would increase. This would then cause the cutoff value in terms of $\overline{X}$ to increase and thus move closer to the true value of $\mu = 51$. Thus, there would be a smaller area under the curve for the true distribution that is above the cutoff value, so β would increase and $1 - \beta$ would thus decrease.

This is an example of what happens in general with Type I and Type II error probabilities, as they move in opposite directions. For a fixed value of n, if one is increased (decreased), the other will decrease (increase). For a fixed value of α, we can increase the value of $1 - \beta$ by increasing the sample size, as the reader is asked to show in Exercise 5.4.

5.8.2 Trade-off Between Error Risks

This raises an interesting question: Which type of error should we try hardest to avoid making? Consider a courtroom setting in which a person is charged with committing a crime. The null hypothesis is that the person on trial is innocent (recall that the null hypothesis is what we doubt, in general, and of course a person is considered to be innocent until proven guilty). It would be unthinkable to incarcerate an innocent person (a Type I error), but it would also be highly undesirable to fail to convict a guilty person (a Type II error). Obviously, α is not going to be specified for such a scenario because we don't have a statistical hypothesis. However, it should be apparent that if we require extremely strong evidence before we convict someone ("beyond a shadow of a doubt") then we run the risk of failing to convict a person who is guilty. So the same general type of relationship between the Type I and Type II errors exists for this scenario as when a statistical hypothesis is used. [Readers interested in discussions of Type I and Type II errors in the judicial process are referred to Feinberg (1971) and Friedman (1972).]

5.9 PRACTICAL SIGNIFICANCE AND NARROW INTERVALS: THE ROLE OF n

Solving for n so as to have a confidence interval of a specified width was discussed and illustrated in Section 5.2.1, and the impracticality of trying to do it for certain other types of confidence intervals was discussed in subsequent sections.

In general, considerable care should be exercised in the selection of n in hypothesis testing. If n is too large, a hypothesis test will generally be too sensitive in that the null hypothesis will be rejected when the true value of the parameter differs only slightly, and immaterially, from the hypothesized value. Thus, statistical significance can result in the absence of practical significance. Consequently, sample size might be determined in such a way as to essentially equate the two, as has been advocated.

Another way to address this issue is to provide some built-in protection with the form of the null hypothesis. For example, assume that we are virtually certain that a new manufacturing process is superior to the old process in terms of process yield, but we suspect that the difference might be small. To guard against the possibility of rejecting the null hypothesis when the difference is small, the null hypothesis could be constructed to have the general form H_0: $\mu \leq \mu_0 + \delta$, so that the alternative hypothesis is H_a: $\mu > \mu_0 + \delta$, with δ suitably chosen. This same idea would apply when two samples are involved, representing perhaps data from a new process and a standard process, with the onus on the new process to be better than the standard process by an amount δ in terms of, say, process yield. This is discussed at the beginning of Chapter 6. Some would argue that this is the way *every* null hypothesis should be constructed, with the value of δ chosen by the scientist/practitioner. If H_0: $\mu \leq \mu_0 + \delta$ were used, $\mu_0 + \delta$ would of course replace μ_0 in the numerator of the test statistic.

The situation is different in confidence interval construction, however. The narrower the interval, the more valuable is the interval. We would love to be able to estimate parameters with little or no error, and we can achieve a small error of estimation with a large sample size. An interval that is narrow solely because n is large would be problematic when a confidence interval is being used to test a hypothesis, however, as then we have exactly the same type of problem as in hypothesis testing. We see the narrowness of the interval when we construct it, but if we used a regular hypothesis testing approach we would not as easily ascertain that the hypothesis was rejected because of the large sample size.

Of course, a narrow interval comes with a price, as the larger the sample size, the greater is the cost of obtaining the sample.

5.10 OTHER TYPES OF CONFIDENCE INTERVALS

The emphasis in this chapter has been mostly on showing the connection between one-sided confidence bounds and one-sided hypothesis tests for a single sample, as the scenario in most of the examples motivated the use of a one-sided bound. Commonly used confidence intervals have been covered, but not some lesser known but potentially important intervals. Accordingly, in this section we briefly survey other important types of confidence intervals.

One such confidence interval is an interval on the probability that a critical value of some sort is exceeded, such as a specified number of equipment malfunctions during a certain time period, or a confidence interval on the probability that the diameter of a ball bearing exceeds a certain value. This is often referred to as a confidence interval on an exceedance probability. The latter are often used in engineering.

Such an interval relates somewhat to the idea of constructing a confidence interval for a process capability index (discussed in Chapter 11), which is essentially a confidence interval on the distance that the closest of two tolerance limits is to μ (assuming that there are two limits), with observations outside the tolerance limits being unacceptable. Tolerance intervals are discussed in Chapter 7.

An approach that is conceptually similar is to obtain a confidence bound for a probability. For example, assume that a person has an old car and he isn't sure if it will pass the annual emissions test that it must undergo. He wonders whether he should have it checked out by his mechanic before he takes it to a testing station, so as to perhaps save himself the embarrassment of having his vehicle fail the test. He is concerned about one particular component of the test, so he does some research and discovers that 84% of the cars that are of the type that he owns pass the test on the first attempt. Since this is a reasonable percentage, he is tempted to just go ahead and have the test; then see his mechanic later, if necessary. He realizes that he has a point estimate and having had a statistics course or two, he wonders how much uncertainly is built into this number. Therefore, he would like to see a lower confidence bound on the percentage, and then take his chances if the bound is at least 75%.

It might seem as though this is nothing more than a lower confidence bound on the binomial parameter, and indeed that would be one way to approach the problem. It wouldn't be the best way, however, because when we use that approach, we are ignoring the possibility that continuous data might have been dichotomized to obtain the binary data. Since the original data would contain much more information than the binary data, it would be better to use the original data.

To illustrate the difference, we can consider the simple example of tossing a coin and computing the proportion of heads, or consider a scenario of the proportion of patients who are free from heart disease for the five years after exhibiting strong signs of susceptibility to heart disease. In each case we have two possible outcomes. Similarly, we have two possible outcomes if we are interested in the proportion of work days that a particular employee arrives late. The difference is that in the last two scenarios there is an underlying continuous distribution, namely, the distribution of time that patients in aggregate are free from heart disease and the distribution of arrival times at work, respectively, whereas with the first example there is no underlying distribution.

The exceedance probability can be computed simply by using whatever distribution is assumed. For example, a surface temperature of 200 degrees Fahrenheit for a particular manufactured device might be viewed as critical, and so we are interested in the proportion of units for which the temperature exceeds this value, hoping that the proportion is small. *If we took a very large sample, we could substitute $\bar{x}$ for μ and s for σ, and then approximate the exceedance probability using a Z-statistic as in Section 3.4.3, provided that degrees Fahrenheit can be assumed to be (approximately) normally distributed, as we will assume for the sake of illustration.

For example, if $\bar{x} = 186$, $s = 5.1$, and $n = 350$, we would approximate the exceedance probability as $\Phi[(200 - 186)/5.1)] = \Phi(2.745) = .9970$, with $\Phi(\cdot)$ denoting the cumulative standard normal probability for the indicated value. If we wanted, say, a lower confidence bound on the proportion, it would be necessary to use special tables that were given in Odeh and Owen (1980).

With 350 observations we should at least be able to obtain a general idea of the shape of the distribution of population values. If we had a much smaller sample size (say, 85), it would be unwise to try to make any distributional assumption based on the distribution of the 85 numbers. In that case, Eqs. (5.4a) and (5.4b) could be used, with this being essentially a nonparametric approach because modeling of the 85 sample observations is not being attempted.

5.11   ABSTRACT OF MAIN PROCEDURES

Population Mean.   When a confidence interval for a population mean is to be constructed or a hypothesis test performed, the user/student must decide whether t or Z should be used. The conditions under which one or the other or neither should be used were summarized in Table 5.1. Of course, approximate normality of the observations should exist before t or Z is used, with closeness to normality that is needed dependent on the sample size.

Population Proportion. The choice between alternative approaches is different when a confidence interval or hypothesis test for a population proportion is used, as t is not used under any conditions. Instead, there are various options that are available and the recommended approach for a confidence interval is to use Eqs. (5.4a) and (5.4b), with the form of the hypothesis test given near the beginning of Section 5.4.1.

These methods are based on the assumption that the binomial distribution is an appropriate model. This means, in particular, that the variability in the random variable is adequately represented by the binomial variance.

Population Standard Deviation and Variance. Confidence intervals and hypothesis tests for a population standard deviation or variance are obtained using the chi-square distribution. Again, approximate normality must be assumed, with the assumption being more critical than when inference is made regarding the population mean since any departure from normality is squared, loosely speaking. The confidence interval for σ^2 is obtained using Eq. (5.5), and the hypothesis test is performed using Eq. (5.6). Of course, the endpoints of a confidence interval for σ are obtained simply by taking the square root of the endpoints of the interval for σ^2.

Poisson Mean. The confidence interval is obtained by using Eq. (5.8) and the hypothesis test is performed using the Z-statistic that immediately precedes it. Of course, the Poisson distribution assumption should be checked before proceeding.

5.12 SUMMARY

It is important to note the direct relationship between hypothesis tests and confidence intervals, and also to understand the general form for these tests and intervals whenever t or Z (or something else) is used. Without this understanding, the many different types of confidence intervals and hypothesis tests presented herein and in other textbooks might appear to be simply an endless stream of unrelated methods.

It is also desirable to understand the weaknesses of hypothesis testing so that these tests can be viewed in the proper light. Because of the direct relationship between confidence intervals and hypothesis tests, it might seem that these weaknesses must also be shared by confidence intervals. That isn't true, however. A decision about a hypothesis is reached whenever a hypothesis test is used, but this is not necessarily the case when a confidence interval is constructed, unless the interval is constructed for the express purpose of testing the hypothesis. That is, the confidence interval approach gives both the interval and the result of a test, if desired. Even when used for hypothesis testing, a confidence interval is superior to a hypothesis test because the interval is in the original units and the effect of a possibly too large sample can be seen in the narrowness of the interval.

Despite the considerable amount of criticism that hypothesis tests have received in recent years, they are not likely to go away since p-values are ingrained in statistical output and whenever a p-value is observed, a decision is apt to be made that is based at least partly on that value.

Confidence intervals should definitely receive preference, however, and there are various types of confidence intervals that have important applications in engineering and in other disciplines beyond the standard types of confidence intervals that are presented in textbooks. Some of these types of intervals were discussed briefly in Section 5.10; other types can be found in Hahn and Meeker (1991) and in the statistics literature.

APPENDIX: DERIVATION

The sample size formula given in Section 5.8 can be derived as follows. The power in detecting the true mean of 51 is to be .90, so that $\beta = .10$. Since this is a one-sided (upper-tailed) test, we thus have the following probability statement:

$$P(\overline{X} < \mu_0 + Z_\alpha \sigma/\sqrt{n} \,|\, \mu = \mu') = .10$$

so that the appropriate algebra produces

$$P\left(\frac{\overline{X} - \mu'}{\sigma/\sqrt{n}} < Z_\alpha + \frac{\mu_0 - \mu'}{\sigma/\sqrt{n}} \,\Big|\, \mu = \mu'\right) = .10$$

$$\Phi\left(Z_\alpha + \frac{\mu_0 - \mu'}{\sigma/\sqrt{n}}\right) = .10$$

with $\Phi(\cdot)$ denoting the cumulative normal *pdf*. Thus,

$$Z_\alpha + \frac{\mu_0 - \mu'}{\sigma/\sqrt{n}} = \Phi^{-1}(.10) = \Phi^{-1}(\beta) = Z_\beta$$

Solving the equation

$$Z_\alpha + \frac{\mu_0 - \mu'}{\sigma/\sqrt{n}} = Z_\beta$$

for n produces the equation that was given in Section 5.8.

REFERENCES

Agresti, A. and B. A. Coull (1998). Approximate is better than "exact" for interval estimation of binomial proportions. *The American Statistician*, **52**(2), 119–126.

Barker, L. (2002). A comparison of nine confidence intervals for a Poisson parameter when the expected number of events is ≤ 5. *The American Statistician*, **56**(2), 85–89.

Brown, L. D., T. T. Cai, and A. DasGupta (2001). Interval estimation for a binomial proportion. *Statistical Science*, **16**(2), 101–133.

Chernick, M. R. (1999). *Bootstrap Methods: A Practitioner's Guide*. New York: Wiley.

Clopper, C. J. and E. S. Pearson (1934). The use of confidence or fiducial limits illustrated in the case of the binomial. *Biometrika*, **26**, 404–413.

Davison, A. C. and D. V. Hinkley (1997). *Bootstrap Methods and Their Application*. Cambridge, UK: Cambridge University Press.

Efron, B. and R. J. Tibshirani (1993). *An Introduction to the Bootstrap*. New York: Chapman and Hall.

Feinberg, W. E. (1971). Teaching the Type I and Type II errors: The judicial process. *The American Statistician*, **25**, 30–32.

Friedman, H. (1972). Trial by jury: Criteria for convictions, jury size and Type I and Type II errors. *The American Statistician*, **26**, 21–23.

Gunst, R. F. (2002). Finding confidence in statistical significance. *Quality Progress*, **35** (10), 107–108.

Hahn, G. J. and W. Q. Meeker (1991). *Statistical Intervals: A Guide for Practitioners*. New York: Wiley.

Moore, D. S. and G. P. McCabe (2002). *Introduction to the Practice of Statistics*. New York: W. H. Freeman.

Nester, M. R. (1996). An applied statistician's creed. *Applied Statistics*, **45**, 401–410.

Odeh, R. E. and D. B. Owen (1980). *Tables for Normal Tolerance Limits, Sampling Plans and Screening*. New York: Marcel Dekker.

Pearson, E. S. and N. W. Please (1975). Relation between the shape of population distribution and the robustness of four simple test statistics. *Biometrika*, **62**, 223–241.

Reeve, R. and F. Giesbrecht (1998). Dissolution method equivalence. In *Statistical Case Studies: A Collaboration Between Academe and Industry* (R. Peck, L. D. Haugh, and A. Goodman, eds.). Alexandria, VA: American Statistical Association and Philadelphia, PA: Society for Industrial and Applied Mathematics.

Sahai, H. and A. Khurshid (1994). Confidence intervals for the mean of a Poisson distribution: A review. *Biometrical Journal*, **35**, 857–867.

Schwertman, N. C. and R. A. Martinez (1994). Approximate Poisson confidence limits. *Communications in Statistics—Theory and Methods*, **23**, 1507–1529.

Tukey, J. W. (1991). The philosophy of multiple comparisons. *Statistical Science*, **6**, 100–116.

Wilson, E. B. (1927). Probable inference, the law of succession, and statistical inference. *Journal of the American Statistical Association*, **22**, 209–212.

EXERCISES

5.1. Show that the form of a confidence interval for μ with σ estimated from a small sample is as was given in Section 5.2 by using the same general starting point as was used in Section 5.1 and then deriving the expression for the interval.

5.2. Show that if a two-sided hypothesis test for H_0: $\mu = \mu_0$ with significance level α, normality, and an assumed known σ is rejected and the value of the test statistic is negative, μ_0 must lie above the upper limit of a $100(1 - \alpha)\%$ two-sided confidence interval.

5.3. Show that if a two-sided hypothesis test for H_0: $\mu = \mu_0$ with significance level α and normality and an assumed known σ is *not* rejected, then the corresponding $100(1 - \alpha)\%$ confidence interval *must* contain μ_0.

5.4. Consider a two-sided hypothesis for the mean of a normal distribution and show that for a fixed value of α and a fixed true mean that is different from the hypothesized mean, the value of $1 - \beta$ will increase as n increases.

5.5. Assume that a sample of 64 observations will be obtained from a population that has a normal distribution. Even though σ is unknown, the practitioner decides to obtain a 95% confidence interval for μ using Z (i.e., using 1.96). What is the actual probability that the interval will contain μ? (*Note*: You may find it necessary to use appropriate software in working this problem.)

5.6. In constructing confidence intervals, it is important to know what the population is and to know whether or not we have a random sample from some (stable) population. Accordingly, assume that a person uses the salaries for all second basemen on major league teams, computing the average and standard deviation of those salaries. If a 95% confidence interval were constructed, what would it be a confidence interval for, if anything? In particular, what is the population?

5.7. One of my professors believed that students should take four times as long to finish one of his exams as it took him to complete it. He completes a particular exam in 15 minutes, so he reasons that the students should complete the exam in 60 minutes. (The students are allowed 70 minutes.) Assume that each of 40 students in a class writes down the length of time that it took to complete the first test. The times are as follows, to the nearest minute: 55, 61, 58, 64, 66, 51, 53, 59, 49, 60, 48, 52, 54, 53, 46, 64, 57, 68, 49, 52, 69, 60, 70, 51 50, 64, 58, 44, 63, 62, 59, 53, 48, 60, 66, 63, 45, 64, 67, and 58. A student complains that the test was too long and didn't provide her sufficient time to look over her work. (Assume that 5 minutes should be sufficient for doing so, and this time was now included in the professor's estimate.) Perform the hypothesis test and state your conclusion.

5.8. Why do we say that we "fail to reject H_0" rather than saying that we "accept H_0"? If we don't reject it, then what else are we going to do other than accept H_0 since only two hypotheses, the null and alternative hypotheses, are under consideration?

5.9. An experimenter constructs a 95% two-sided confidence interval for μ, using a sample size of 16. Normality of the individual observations is assumed and the population standard deviation is unknown. The limits are: lower limit $= 14$, upper limit $= 24$. If instead of constructing the confidence interval, the experimenter had tested H_0: $\mu = 10$ against H_a: $\mu \neq 10$ using $\alpha = .05$, what was the numerical value of the test statistic?

5.10. Complete the following sentence. A p-value is the probability of _____ (more than one word).

5.11. Given a particular data set, would a null hypothesis for a two-sided test that is rejected by one data analyst using $\alpha = .01$ also be rejected by another analyst using $\alpha = .05$ if each person used the same test procedure so that the only difference was α? (Assume that no errors are made.) Now assume that the analyst who used $\alpha = .01$ did *not* reject the null hypothesis. Can the outcome be determined for the other analyst who used $\alpha = .05$? Explain.

5.12. Assume that a student constructs, using simulated data from a known distribution with known parameter values, two hundred 99% two-sided confidence intervals for some parameter θ. If this experiment is then repeated 1,000 times, what number would be our best estimate of the number of intervals that contain θ? Would you expect to actually see this number?

5.13. Assume that H_0: $\mu = 25$ is rejected in favor of H_a: $\mu \neq 25$ using $\alpha = .01$. Assume further that a 99% two-sided confidence interval is subsequently constructed using the same set of data that was used to test the hypothesis. If the value of the appropriate test statistic was negative, the upper limit of the confidence interval must have been _____. (Use either "greater than" or "less than" as part of your answer.)

5.14. Can a sample variance ever be at least approximately in the center of a confidence interval for σ^2? Explain.

5.15. Assume that your driving time to school/work is (approximately) normally distributed and you wish to construct a confidence interval for your (long-term) average driving time. So you record the time for 10 consecutive days and find that the average is 23.4 minutes and the standard deviation is 1.2 minutes. If you wish to construct a 99% confidence interval for what your average driving time would be for a very long time period, what assumption will you have to make? With that assumption, what is the interval?

5.16. Show that the lower limit given by Eq. (5.4b) can never be negative. (*Hint*: The computed lower limit is nonnegative if $\hat{p} + z_{\alpha/2}^2/2n > z_{\alpha/2}\sqrt{\hat{p}(1-\hat{p})/n + z_{\alpha/2}^2/4n^2}$. Start from this point and then perform the appropriate algebra.)

5.17. An experimenter performs a hypothesis test of $H_0: \mu = \mu_0$ versus $H_a: \mu > \mu_0$, using a sample of size 100, and obtains a p-value of .05. Since this is a borderline value, the test is repeated and the absolute value of the test statistic is greater than with the first test. Knowing only this information, what course of action would you take/suggest?

5.18. An engineer sought to reduce the variability of roundness measurements over work pieces and believes that he has succeeded. State one hypothesis test that could be used, giving both the null and alternative hypotheses. Must any assumption be made in order to perform the test? If so, state the assumption(s) and indicate how to test the assumption(s).

5.19. A process engineer decides to estimate the percentage of units with unacceptable flatness measurements. A sample of size 100 is obtained and a 95% confidence interval is obtained using the standard approach: $\hat{p} \pm Z_{\alpha/2}\sqrt{\hat{p}(1-\hat{p})/n}$. The computed lower limit is negative, which causes the engineer to proclaim that constructing such an interval is a waste of time since a proportion cannot be negative. Explain to the engineer why the lower limit is negative and suggest a superior approach, if one is available.

5.20. An experimenter performs a two-sided test of $H_0: \mu = \mu_0$ and obtains $Z = 1.87$. Determine the p-value. What would the p-value have been if the test were one-sided?

5.21. Explain why we cannot state, for example, that "we are 95% confident that μ is between 45.2 and 67.8."

5.22. Critique the following statement: "Why should we settle for a 95% confidence interval instead of a 100% confidence interval. The latter would remove all doubt and would surely be more useful."

5.23. An experimenter decides to construct a 95% confidence interval for μ and starts to use t because σ is unknown. Someone else in her department informs her that she could use Z because the sample size is $n = 36$. What will be the numerical difference between the widths of the two confidence intervals if $s = 12.1$?

5.24. The following data have been simulated from a nonnormal distribution:

```
  2.30071    1.58192  -0.42940  -0.11641  -0.19029  -1.28538    1.09995
 -1.21649    0.05247   0.10919   0.21804  -0.58013   0.70508    1.23839
  1.94917    2.41419  -0.38660   0.10442  -0.49846   1.81957  -0.73201
  0.14199  -0.11403   0.24233   0.45994   0.74652   2.02368  -1.96190
  1.65069  -1.82210
```

Does a test for nonnormality impart the correct message? (This can be performed in MINITAB, for example, by constructing a normal probability plot.) Could a confidence interval for μ still be constructed, using methodology given in this chapter, despite the nonnormality? If so, construct the 95% interval. Would such an interval be a 95% confidence interval? Explain.

5.25. A 99% confidence interval for p is desired that has width 0.04. Can such an interval be constructed? Explain.

5.26. A 90% confidence interval for μ is to be constructed with a specified width. If σ is unknown, can this be accomplished? Since σ is generally unknown, what would you recommend to practitioners who want to construct confidence intervals of specific widths?

5.27. Assume that a 95% confidence interval is to be constructed for μ. If $n = 41$, $s = 10$, and normality is assumed, we know that we could use either Z (as an approximation) or t in constructing the interval. If Z is used and the width of the confidence interval is a, what would be the width of the interval, as a multiple of a, if t had been used instead of Z?

5.28. Given a particular data set, if a hypothesis is rejected by one data analyst using $\alpha = .01$ and a two-sided test, would it also be rejected by another analyst using $\alpha = .05$ and a one-sided test if each person used the same test procedure so that the only difference was α? (Assume that no errors are made.) Explain.

5.29. I once had a student who could not understand the concept of a p-value, despite several different ways in which I tried to explain it. Consider one of your relatives who has never taken a statistics course (assuming that you have such a relative), and try to give a more intuitive definition than was given in Section 5.8.

5.30. Critique the following statement. "I am going to perform a study in which I compare computer chips produced by Intel with those produced by Applied Micro Devices (AMD). I will prove with a hypothesis test that the two companies produce the same percentage of defective chips."

5.31. The planet Saturn has a relative mass that is 95 times the relative mass of the planet Earth. Does this mean that the difference in relative masses is statistically significant? In general, what does "statistically significant" mean in the context of hypothesis testing?

5.32. Assume that we fail to reject $H_0: \mu = 28$ for a two-tailed test when $\bar{x} = 32$ and $\sigma = 9$ and approximate normality can be assumed.

   (a) Did this likely occur because the sample size was too small? Explain.
   (b) What is the largest value that n could have been if a decision was made to reject H_0 only if the p-value was less than .05?
   (c) What would you recommend to the experimenter about sample size choice?

5.33. If 3000 samples of $n = 30$ and $n = 99$ are generated from a particular right-skewed distribution, what relationship should be observed for:

   (a) The distribution of the sample averages for the two sample sizes.
   (b) The *average* of the 3000 sample averages for the two sample sizes.

5.34. Consider a 99% confidence interval for λ using Eq. (5.7). For what values of X will the lower limit be negative? Will a 99% confidence interval constructed using Eq. (5.8) have a lower limit that is negative for all or part of the range of X values for which the use of Eq. (5.7) produces a negative lower limit? Explain. Since λ is positive, by definition, what would your analysis suggest to the user who wants to avoid a useless lower limit?

5.35. Consider Example 5.1 in Section 5.1.2. Explain why a normal distribution is not automatically ruled out as a possible model for the amount of pressure required to remove the bottle top just because no one will be exerting a negative amount of pressure to remove the top, but negative values are included in the range of possible values for any normal distribution.

5.36. In the Winter 1999 issue of *Chance* magazine (http://www.amstat.org/ pressroom/cookies.pdf), Brad Warner and Jim Rutledge described an experiment performed by students in an introductory statistics course in answer to a challenge posed by Nabisco, Inc., with the company asking (perhaps as a marketing/advertising ploy) for the most creative way of determining that there are at least 1,000 chips in every 18-ounce bag of their chocolate chip cookies. There are various ways in which this problem can be tackled (including assuming a particular distribution for the number of chips in each bag versus not making any distributional assumption), and one approach would be to construct a confidence interval on an exceedance probability, as discussed in Section 5.10. Read the article and write a report of the analysis that was performed. Do you agree with the method of analysis? If not, what would you have done differently?

5.37. An important aspect of the use of statistical methods is selecting the appropriate tool for a stated objective. Assume that a company has to meet federal requirements regarding the stated weight of its breakfast cereal. If that weight is 24 ounces, why would we not want to take a sample and test $H_0: \mu = 24$? Instead of this, what would you suggest that the company do in terms of methodology presented in this chapter?

5.38. Assume that we are testing $H_0: \sigma^2 = 20$ versus $H_a: \sigma^2 > 20$ and $s^2 = 24$, with $n = 65$. Obtain the p-value.

5.39. Assume that the width of a 95% small-sample confidence interval for μ is 6.2 for some sample size. Several months later the same population is sampled and another 95% confidence interval for μ is constructed, using the same sample size, and the width is found to be 4.3. What is the ratio of the earlier sample variance to the more recent sample variance?

5.40. The importance of processes being in control when designed experiments are used is stressed in Chapter 12. Assume that a 99% confidence interval for μ with σ assumed to be known has been constructed, but then, unbeknownst to the experimenters, there is a 10% increase in the mean. If this had been known shortly after the confidence interval had been constructed, how would the endpoints of the interval have to be adjusted in order to keep the degree of confidence at 99%?

5.41. Assume that a sample of $n = 100$ observations was obtained, with $\bar{x} = 25$ and $s^2 = 10$. Based solely on this information, would it be practical to construct a 90% confidence interval for μ? Why or why not?

5.42. A sample of $n = 100$ items is obtained with the following results: $\bar{x} = 100$ and $s^2 = 16$. What would be the numerical value of the estimate of $\sigma_{\bar{x}}$?

5.43. Two practitioners each decide to take a sample and construct a confidence interval for μ, using the t-distribution. If one constructs a 95% interval and the other constructs a 99% interval, explain how the widths of the two intervals could be the same.

5.44. An experimenter intends to construct a 95% confidence interval for μ that has a width of 2. If a sample of the necessary size is obtained and $\bar{x} = 50$:

    (a) What will be the upper limit of the interval if $\sigma = 4$ and this is used in constructing the interval?

    (b) Based on the information that is given, will the interval be exactly a 95% confidence interval? Would it be exactly a 95% confidence interval if normality were assumed?

    (c) *If* the interval turned out to be (49, 51), would the probability be .95 that this interval contains μ if normality is assumed?

5.45. Assume that we computer-generate 10,000 lower 95% confidence bounds for p.

    (a) What would be the best guess of the number of bounds that do exceed p?

    (b) Would the expected number of bounds that exceed p be more likely to equal the actual number if only 1000 intervals had been constructed? Why or why not?

5.46. A company produces 150,000 units of a certain product each day. Think of this as a sample from a population for which 0.4% of the units are nonconforming.

   (a) Letting X denote the number of nonconforming units per day, what is the numerical value of σ_x^2? (Assume that the units are independent.)
   (b) Is it practical to construct a confidence interval for the true proportion of nonconforming units since 0.4% is quite small? If so, construct the interval (if possible). If not, explain why it is not practical or possible to construct the interval.

5.47. If H_0: $\mu = 40$ was rejected in favor of H_a: $\mu < 40$ with the random variable X assumed to have (approximately) a normal distribution, $n = 16$, $s = 5$, and $\alpha = .05$, the value of the test statistic (t or Z) must have been less than _____.

5.48. An experimenter wishes to estimate μ (with $\overline{X}$) so as to have a maximum possible error of estimation equal to 3 with probability .9544. (The odd number is for mathematical simplicity.) If $\sigma = 15$, how large must the sample be?

5.49. The following explanation of a confidence interval is given on the Web (`http://onlineethics.org/edu/ncases/EE18.html`): "If we performed a very large number of tests, 95% of the outcomes would lie in the indicated 95% bounded range." Do you consider this to be an acceptable explanation of a confidence interval? Explain. If not, how would you modify the wording?

5.50. Assume that a sample of 16 observations is obtained and the software that is used shows that $\overline{X} = 8$ and $s = 2$.

   (a) Assuming approximate normality, construct a 95% confidence interval for μ.
   (b) If the scientist whose data you used later contended that the interval is too wide and asked that the interval be constructed to be exactly one unit in width, how would you reconstruct the interval if you had to use the same data?
   (c) How would you proceed if resources allowed you to take a new sample?

5.51. Assume that a 99% confidence interval is constructed for μ with an assumed known value of $\sigma = 2$. What is the variance of the width of the interval?

5.52. Assume that H_0: $\mu = 50$ is tested against H_a: $\mu < 50$ and a sample of size 100 has $\overline{X} = 53.8$. If $s = 3.8$ and you were asked to write a report of the result of the hypothesis test, what would you state?

5.53. State three hypotheses that would be useful to test in your area of engineering or science and, if possible, test one of those hypotheses by taking a sample and performing the test.

5.54. What would you recommend to an experimenter who wished to determine how large a sample to take to construct a confidence interval for μ of a specified width but had no idea as to the shape of the population from which the sample will come?

5.55. A company report shows a 95% confidence interval for p with the limits not equidistant from $\widehat{p}$. A manager objects to the numbers, claiming (at least) one of them must be wrong. Do you agree? Explain. If not, what would be your explanation to the manager?

5.56. Consider the following exercise. You are tutoring a student in your class who doesn't understand confidence intervals. To help drive home exactly what a confidence interval is and how it works, you decide to generate samples from the standard normal distribution and construct 95% confidence intervals for the mean, which of course is zero.

   (a) Which of the following would you expect to come closest to having 95% of the intervals contain the mean of zero—1000 samples of size 10 or 100 samples of size 100?

   (b) Which will have the greatest average width—the 1000 intervals with a sample size of 10 or the 100 intervals with a sample size of 100?

5.57. As stated in Section 5.1, a confidence interval can be constructed to have a maximum error of estimation with a stated probability, with this maximum error of estimation being the half-width of the interval.

   (a) Explain, however, why a confidence interval for a normal mean can be constructed to have a specified half-width without the use of software or trial-and-error only if σ is known.

   (b) Explain how the expected half-width would be determined when σ is unknown and $n = 16$.

5.58. As stated by Hahn and Meeker (1991, p. viii), the wrong type of interval is often used in practice. At a consultant's website one finds the words "confidence intervals of targeted materials categories." Explain why the type of intervals alluded to probably aren't confidence intervals.

5.59. Show that $(n - 1)s^2/\sigma^2$ has a chi-square statistic.

5.60. A political pollster surveys 50 registered Republicans in Cobb County in Georgia and, among other things, she asks them their age. The average age for people in her sample is 43, with a standard deviation of 12.

   (a) If the pollster were to use this information to construct a confidence interval, what population parameter would she be estimating? Be specific.

   (b) Would you question the wisdom of constructing an interval based solely on the information stated in this problem?

5.61. The heat involved in calories per gram of a cement mixture is approximately normally distributed. The mean is supposed to be 90 and the standard deviation is known to be (approximately) 2. If a two-sided test were performed, using $n = 100$, what value of $\bar{x}$ would result in a p-value for the test of .242? What decision should be reached if this p-value resulted from the test?

5.62. Suppose it is desired to test that the melting point of an alloy is 1200 degrees Celsius. If the melting point differs from this by more than 20 degrees, the composition will have to be changed. Assume normality and that σ is approximately 15. What decision would you reach if a sample of 100 alloys were obtained and the average melting point were 1214?

5.63. A machine is supposed to produce a particular part whose diameter is in the interval (0.0491 inch, 0.0509 inch). To achieve this interval, it is desired that the standard deviation be at most 0.003 inch. It is obvious that some machines cannot meet this requirement and will be replaced. It is not clear whether one particular machine should be replaced or not as a sample of 100 observations had a standard deviation of 0.0035 inch. This is higher than the target value but the difference could be due to sampling variation. Test the null hypothesis that the standard deviation for this machine is at most 0.003, bearing in mind that management has decided that a *p*-value less than .01 will motivate them to replace the machine, and assuming that evidence suggests that the part diameters are approximately normally distributed.

5.64. There are various Internet scams, including various versions of the "Nigeria scam," with people in the United States being promised large amounts of money if they assist the party, who is ostensibly in Nigeria, to move millions of dollars out of that country into the United States. Let's say that you want to construct a confidence interval for the percentage of people who are duped by this scam.

(a) Does this seem like a workable project? In particular, how would you go about obtaining the data?

(b) Would you expect Eq. (5.3) to be adequate for this purpose, or would the use of Eqs. (5.4a) and (5.4b) probably be necessary?

5.65. Assume that a 95% confidence interval for μ with σ assumed known is of width W. What would be the width of a 99% confidence interval for μ, using the same data, as a function of W?

5.66. A motorist wishes to estimate the average gas mileage of her car using a 95% confidence interval with the confidence interval to be 2 units wide. If the standard deviation is known to be approximately 3:

(a) Explain in words what the standard deviation of 3 means in this context.

(b) Determine how many tanks of gas must be used in constructing the confidence interval.

(c) What assumption(s) did you make in working part (b)?

5.67. An experimenter performs a hypothesis test for the mean of an assumed normal distribution, using a one-sided test (greater than). If σ is assumed to be 4.6, the mean of a sample of 100 observations is 22.8, and the *p*-value for the test is .06681, what is the sign of $(\bar{x} - \mu)$ and what was the magnitude of the difference?

5.68. A city near a government weapons clean-up site is concerned about the level of radioactive by-products in its water supply. Assume that the EPA limit on naturally occurring radiation is 5 picocuries per liter of water. A random sample of 25 water specimens has a mean of 4.75 and a standard deviation of 0.82. We would like to be able to show that the mean level for the population is actually below 5. If we assume approximate normality, do these data allow us to reject the null hypothesis in favor of this alternative hypothesis? Explain. If we don't reject the null hypothesis, is that really a problem for this scenario? Explain.

5.69. Assume that a confidence interval for the mean is to be constructed and the t-value should be used instead of the Z-value because σ is unknown. Consider the difference between the t-value and the Z-value for a 95% confidence interval versus the difference between the two for a 90% interval. Which difference will be greater? For which will the percentage difference, relative to the t-value, be the larger? What does this suggest, if anything, about using Z-values as approximations for t-values when confidence intervals are constructed?

5.70. The Rockwell hardness index of steel is determined by pressing a diamond point into the steel and measuring the depth of penetration. Assume that for a certain type of steel, the standard deviation of the Rockwell hardness index is 7.4. A manufacturer claims that its steel has an average hardness index of at least 65. A sample of 64 units of steel is obtained and the average Rockwell hardness value is found to be 61.8. Assume normality of the index measurements.
  (a) If the manufacturer's claim is tested by constructing a one-sided confidence bound, should it be an upper bound or a lower bound? Explain.
  (b) For the appropriate type of confidence bound, what is the degree of confidence such that the bound is 65?

5.71. The specifications for a certain type of surveillance system state that the system will function for more than 12,000 hours with probability of at least .90. Fifty such systems were checked and eight were found to have failed before 12,000 hours. These sample results cast some doubt on the claim for the system.
  (a) State H_0 and H_a and perform the appropriate test after first stating what the assumptions are for the test, if any.
  (b) Construct the confidence bound that corresponds to this test and comment on the results.

5.72. The diameter of extruded plastic pipe varies about a mean value that is determined by a machine setting. If the standard deviation of the diameter is known to be about 0.08 inch, how large a sample must be taken so that the point estimator of the population mean will differ from the population mean by at most 0.02 inch with probability .99?

5.73. A 95% confidence interval is constructed for μ as a means of testing a null hypothesis. If the hypothesized value was exactly in the center of the interval, what would have been the value of the test statistic if a hypothesis test had been performed?

5.74. A large-sample test of H_0: $\mu = 50$ resulted in $Z = 1$. If a 95% confidence interval for μ had been constructed instead, what would have been the distance, in units of $\sigma_{\bar{x}}$, of the upper limit from the hypothesized value?

5.75. Several years ago D. J. Gochnour and L. E. Mess of Micron Technology, Inc. patented a method for reducing warpage during application and curing of encapsulated materials on a printed circuit board. Assume that the company has found that the warpage with the standard method has had a mean of 0.065 inch and before the method was patented the scientists wanted to show that their method has a smaller mean. They decide to construct a one-sided confidence bound, hoping that the bound will be less than 0.065. They took a sample of 50 observations and found that $\overline{X} = 0.061$, with $s = 0.0018$.

(a) What assumption(s), if any, must you make before you can construct the bound? Since assumptions must be tested, can the assumption(s) be tested with 50 observations? Explain.

(b) Make the necessary assumption(s), construct the bound, and draw a conclusion.

5.76. A company produces a resistor that is designed to have a mean of 100 ohms and a standard deviation of 2 ohms. There is some evidence that suggests the mean is not being met, although the target value for the standard deviation appears to be very close to the actual deviation. A two-sided confidence interval for the mean is to be constructed, but one member of the team who designed the resistor wants to be able to estimate the mean with a maximum error of estimation of 0.5 ohm with probability .95. Determine the number of observations that should be used in constructing the 95% confidence interval.

5.77. Indoor swimming pools are known to have poor acoustic properties. The goal is to design a pool in such a way that the mean time that it takes a low-frequency sound to die down is at most 1.3 seconds with a standard deviation of at most 0.6 second. Data from a study to test this hypothesis were given by Hughes and Johnson's article, "Acoustic Design in Nanatoriums" (*The Sound Engineering Magazine*, pp. 34–36, April 1983). The simulated data were as follows:

1.8	2.8	4.6	5.3	6.6
3.7	5.6	0.3	4.3	7.9
5.0	5.3	6.1	0.5	5.9
2.5	3.9	3.6	2.7	1.3
2.1	2.7	3.8	4.4	2.3
3.3	5.9	4.6	7.1	3.3

Analyze the data relative to the stated objective.

5.78. Assume that you encounter a confidence interval for the population mean of a random variable such as a trace contaminant, which of course cannot be negative, but the lower limit of the interval is well below zero. What is the likely cause of the problem and what would you do to correct it?

5.79. Consider Example 5.4 in Section 5.3.7. Would you have proceeded differently in working that problem? In particular, would you have made the distributional assumption that was made in that example or would you have tested that assumption? Would you have proceeded differently in terms of the distributional assumption if the objective had been to construct a confidence interval or bound on σ^2 rather than a confidence bound on the mean? Explain.

5.80. Explain why a confidence interval for a population mean would not be the appropriate interval to construct to show compliance to government regulations regarding the extent to which a pollutant level can deviate from a nominal value. Is there any type of confidence interval that would be suitable for that objective? Explain.

5.81. Explain why a p-value is *not* the probability that the null hypothesis is correct.

5.82. Using appropriate software, simulate 100 observations from the standard normal distribution [i.e., $N(0, 1)$ distribution], and construct a 95% confidence interval for the known mean of 0. Do this 1000 times. (It is easy to write code to do this.) What percentage of the intervals contained zero? Repeat nine more times. Did any of the percentages equal 95, exactly? If not, does this mean that when we construct a 95% confidence interval, it really isn't a 95% interval? Explain.

5.83. In a random sample of 400 industrial accidents for a large company over a period of time, it was found that 231 accidents were due to unsafe working conditions. Construct a 99% confidence interval for the true proportion of accidents due to unsafe conditions for that time period. Address the validity of this interval if these 400 accidents occurred over a period of, say, 25 years.

5.84. Consider the data that were given in Exercise 5.24. What would be the outcome if we tested the hypothesis that the data have come from a population with a mean of zero, using $\alpha = .05$? If we fail to reject the hypothesis, what is the largest value of the hypothesized mean for a two-sided test that would result in rejection of the null hypothesis with a positive value of the test statistic? Similarly, what is the smallest value of the hypothesized mean that would lead to rejection of the null hypothesis on the low end? Answer these same two questions if a one-sided test had been used (i.e., greater than for the first question and less than for the second question). Assume α is the same for each of these questions.

5.85. Assume that a large-sample test of H_0: $\mu = 35$ against H_a: $\mu < 35$ is performed and $Z = 1.35$. What is the p-value?

5.86. Assume that two engineers jointly conduct an experiment to determine if a particular process change affects the average process yield. One prefers to specify a significance level and selects .05, while the other prefers to take a more modern approach and reach a decision based on the p-value.
   (a) If this experiment were conducted a large number of times and if the null hypothesis were true, what proportion of times would the p-value be less than the specified value for the significance level?
   (b) Of course, the experiment will be conducted only once, however, and now assume that the null hypothesis is false. Can the probability that the p-value is less than the significance level of .05 be determined? If so, what is the probability? If not, explain why the determination cannot be made.

CHAPTER 6

Confidence Intervals and Hypothesis Tests—Two Samples

There are many instances in which there is a need to use a two-sample procedure. For example, a new and improved manufacturing process might be compared with the standard process in terms of process yield or in terms of the percentage of nonconforming units; or the variances of the two processes might be compared to see if process variability has been reduced; or the lifetimes of a new, improved light bulb may be compared with a standard light bulb in a reliability test; or the respective strengths of two materials may be compared; and so on.

As in the case of a single sample, we could construct a confidence interval or equivalently test a hypothesis. As in the case of single-sample hypothesis tests, we need to adopt a realistic view. That is, when we test whether two process yields are equivalent, we know almost certainly that they are not equal before we even obtain a sample from each process. Therefore, what we are really trying to see is whether there is more than a small difference. If so, then we will likely be able to reject the hypothesis of equality.

Therefore, it makes sense to test whether one parameter is greater than another parameter by a certain amount. For example, we might be interested in testing whether the yield of a new process is greater than the yield of the standard process by a specified amount, δ. That is, we might test $H_0: \mu_2 - \mu_1 = \delta$ versus $H_1: \mu_2 - \mu_1 > \delta$, with μ_2 being the mean for the new process. Since the alternative hypothesis is equivalent to $\mu_2 > \mu_1 + \delta$, if we reject H_0 in favor of H_1, we would conclude that the new process yield exceeds the old process yield by at least some desired amount. Since test statistics are constructed under the assumption that the null hypothesis is true, δ would be used in the numerator of the statistic. This is illustrated in Section 6.1.1.

6.1 CONFIDENCE INTERVALS AND HYPOTHESIS TESTS FOR MEANS: INDEPENDENT SAMPLES

The discussion in Chapter 5 regarding the mechanics of confidence intervals and hypothesis tests for the mean of one population extends in a natural way to samples from each of two populations. That is, we can use a confidence interval to test a hypothesis; we have to decide whether to use t or Z, and so on. There are a few additional wrinkles, however, as will be explained in subsequent sections.

6.1.1 Using Z

The confidence intervals and hypothesis tests when Z is the test statistic have the same general form as was emphasized in Chapter 5. That is, the confidence intervals are of the form $\widehat{\theta} \pm Z_{\alpha/2}s_{\widehat{\theta}}$ (or $\sigma_{\widehat{\theta}}$) and the hypothesis tests are of the form $Z = (\widehat{\theta} - \theta_0)/s_{\widehat{\theta}}$ (or $\sigma_{\widehat{\theta}}$), with θ_0 denoting the hypothesized value of θ.

Since two samples are involved, θ will obviously be defined differently from the way it is defined in the one-sample case. The logical starting point is to define θ and $\widehat{\theta}$ and then everything will follow logically from there. If the means, μ_1 and μ_2, of each of two populations with normal distributions were equal, then $\mu_1 - \mu_2 = 0$. The logical choice then is to let $\theta = \mu_1 - \mu_2$, and under the null hypothesis of equal population means, $\theta = 0$ so that $\theta_0 = 0$ is the hypothesized value. Then $\widehat{\theta} = \widehat{\mu}_1 - \widehat{\mu}_2$. With samples taken from two separate populations, the samples will be independent. [A method for nonindependent (paired) samples is given in Section 6.2.] It then follows that $Var(\widehat{\mu}_1 - \widehat{\mu}_2) = Var(\widehat{\mu}_1) + Var(\widehat{\mu}_2) = Var(\overline{x}_1) + Var(\overline{x}_2) = \sigma_1^2/n_1 + \sigma_2^2/n_2$ since the population means are of course estimated by the respective sample means.

If the population variances were known, the test statistic would be

$$Z = \frac{\widehat{\theta} - \theta_0}{\sigma_{\widehat{\theta}}}$$

$$= \frac{\overline{x}_1 - \overline{x}_2 - 0}{\sqrt{\sigma_1^2/n_1 + \sigma_2^2/n_2}}$$

After the Z-statistic is computed, the subsequent steps are the same as in the one-sample case. That is, the p-value could be computed and a decision could be based on that value.

■ EXAMPLE 6.1

Hypotheses and Data

To illustrate, assume that $H_0: \mu_1 - \mu_2 = 3$ and $H_a: \mu_1 - \mu_2 < 3$. Samples of 65 and 75 have been taken from the two populations, and the variances are assumed to be $\sigma_1^2 = 13.2$ and $\sigma_2^2 = 15.4$. The corresponding sample means are $\overline{x}_1 = 71.2$ and $\overline{x}_2 = 69.6$, respectively.

Computations

The value of the test statistic is then obtained as

$$Z = \frac{(\overline{x}_1 - \overline{x}_2) - 3}{\sqrt{\sigma_1^2/n_1 + \sigma_2^2/n_2}}$$

$$= \frac{(71.2 - 69.6) - 3}{\sqrt{13.2/65 + 15.4/75}}$$

$$= -2.19$$

Conclusion

Since $P(Z < -2.19 \,|\, \mu_1 - \mu_2 = 3) = .014$, the sample data provide moderately strong evidence that seems to contradict the null hypothesis, so we conclude that the mean of the first population is less than the mean of the second population plus 3. ∎

In this example the data were in the "right direction" relative to the alternative hypothesis, in that the Z-statistic was negative instead of positive. What if the Z-statistic had been positive? If it were more than slightly greater than zero, that would not be surprising as this would simply support the null hypothesis. But a Z-statistic that is significantly greater than zero should be cause for concern as this could mean that there could be some bad data or outliers that are good data, as experimenters would be expected to know the proper inequality relationship between two population means if they differed considerably. In any event, the null hypothesis would not be rejected because the numerical value of the test statistic would have the wrong sign from what would be needed to reject the null hypothesis.

If the inequality relationship between the sample means was totally unexpected, could the alternative hypothesis be changed before the analysis is performed? The reader is asked to explain why this cannot be done in Exercise 6.23.

From past discussions, the form of the corresponding confidence interval should be apparent. As in the one-sample case, it is of the form emphasized in Chapter 5. That is, the confidence intervals are of the form $\widehat{\theta} \pm Z_{\alpha/2}\, s_{\widehat{\theta}}$ (or $\sigma_{\widehat{\theta}}$), as stated at the beginning of this section. So with normality assumed for each population, the variances assumed known, and the sample sizes either small or large, the confidence interval for $\mu_1 - \mu_2$ would have the form

$$(\overline{x}_1 - \overline{x}_2) \pm Z_{\alpha/2}\sqrt{\sigma_1^2/n_1 + \sigma_2^2/n_2} \tag{6.1}$$

As in the one-sample case, we should recognize that the variances will generally be unknown. The question then arises as to whether or not Z can still be used, and if so, under what conditions it can be used. Recall Table 5.1 in Chapter 5, which indicated when Z should be used. If we extend the results in that table to two samples, then we would use Z *if* we knew that the sample data were from two normal populations and the variances were known, as in this section, and also in the case when we have (approximate) normality and unknown variances, but two large samples. In the latter case we would use

$$(\overline{x}_1 - \overline{x}_2) \pm Z_{\alpha/2}\sqrt{s_1^2/n_1 + s_2^2/n_2} \tag{6.2}$$

Consider the other two cases in Table 5.1 when Z would be used. If the population distributions are unknown, but the variances are known (an unlikely scenario), we would use Eq. (6.1) if we had large samples (because of the Central Limit Theorem), and we would use Eq. (6.2) if we had large samples but the variances are unknown. As in Table 5.1, the minimum sample sizes for the samples to be declared "large" depends on the shape of the distributions for the two populations, so a simple rule of thumb cannot be given, beyond stating that the oft-mentioned minimum sample size of 30 for each sample should be adequate as long as there is no more than a small-to-moderate departure from normality for the two populations.

6.1.2   Using t

If the two populations have (approximately) normal distributions, the variances are un-known, and the sample sizes are not large, then we would use the t-distribution. Unlike the one-sample case, however, we have to choose between the *exact* (also called "pooled" and independent sample) t-test and the *approximate* t-test. More specifically, the statistic

$$t^* = \frac{(\bar{x}_1 - \bar{x}_2) - 0}{\sqrt{s_1^2/n_1 + s_2^2/n_2}}$$

does have not a t-distribution, and for that reason it is denoted by t^* instead of t, and the test is called an approximate t-test. In order to use an exact t-test, we have to assume that the variances are equal, or that the ratio of the variances is known. (The latter is not likely to be of any real help, however, because if we don't know the variances, we are not likely to know their ratio either.)

The statistic t^* has approximately a t-distribution with degrees of freedom calculated as

$$d.f. = \frac{(s_1^2/n_1 + s_2^2/n_2)^2}{(s_1^2/n_1)^2/(n_1 - 1) + (s_2^2/n_2)^2/(n_2 - 1)}$$

Of course, the variances are almost certainly not going to be equal, but fortunately they need not be equal in order to use the exact t-test, although the test statistic is constructed under the assumption that the variances are equal. The statistic is given by

$$t = \frac{(\bar{x}_1 - \bar{x}_2) - 0}{s_{\bar{x}_1 - \bar{x}_2}}$$

$$= \frac{(\bar{x}_1 - \bar{x}_2) - 0}{s_p\sqrt{1/n_1 + 1/n_2}}$$

with

$$s_p^2 = \frac{(n_1 - 1)s_1^2 + (n_2 - 1)s_2^2}{n_1 + n_2 - 2}$$

denoting the "pooled" variance, which is obviously a weighted average of the two sample variances, with each variance weighted by the respective degrees of freedom. Thus, $s_p^2 = \hat{\sigma}^2$ is the estimator of the assumed common population variance $\sigma_1^2 = \sigma_2^2 = \sigma^2$.

Both of these tests are based on the assumption that observations *within* each sample are independent, as well as on independence between the two samples.

The approximate test is, as we would expect, more conservative than the exact test as a larger difference between the sample means is needed to produce a significant result.

The exact test can be used under certain conditions. Although normality of the two population distributions is assumed, the test is not undermined by a slight-to-moderate departure from normality. Good and Hardin (2003) state in their book, which is recommended reading for anyone who uses statistical methods, that nonnormality is generally not a problem with which to be concerned as long as the sample sizes are at least 12. Of course, major departures from normality will require larger sample sizes.

The assumption of equal variances can be more troublesome, however, as the exact t-test is undermined when the variances are unequal *and* the sample sizes differ considerably, especially when one of the sample sizes is small.

The logic behind this recommendation becomes apparent when we consider the conditions that would cause t and t^* and their respective degrees of freedom to differ. If $n_1 = n_2$, then $t = t^*$ and their respective degrees of freedom ($d.f.$) would then be equal only if $s_1^2 = s_2^2$. Thus, if we had the unlikely scenario that the sample variances were the same and additionally the sample sizes were the same (more likely), the tests are equivalent.

Depending on the difference between the sample variances and the difference, if any, in the sample sizes, it will often be desirable to test for equal variances. The selection of a particular test must be carefully made, however, because some tests for equal variances are highly sensitive to nonnormality, whereas the exact t-test itself is not sensitive to nonnormality. The hypothesis test of equal variances might be rejected when one of these sensitive tests is used, but the reason for the rejection could simply be moderate nonnormality. If the population variances were approximately equal and the nonnormality were not severe enough to undermine the exact t-test, the user could be needlessly pointed in the wrong direction.

Therefore, a test for equal variances that is not overly sensitive to nonnormality should be considered, if a test is used at all. There are several such tests, including the test proposed by Layard (1973). Another well-known test is one given by Levene (1960), which is used later in this section. Cressie and Whitford (1986) gave recommendations on the use of two-sample t-tests in the presence of unequal variances, concluding that the latter is generally not a problem unless the variances differ considerably and the sample sizes are also considerably different. See also Posten, Yeh, and Owen (1982).

Since the approximate t-test is also undermined by more than moderate nonnormality because the t-distribution is still being used, it is necessary to check for (approximate) normality of the two populations using each of the two samples.

■ **EXAMPLE 6.2**

Problem Statement

We will use the camshaft dataset that comes with the MINITAB software to illustrate these concepts. The file is CAMSHAFT.MTW and contains sample camshaft length measurements from each of two suppliers and from all of the camshafts used at an automobile assembly plant. Since the company was concerned about quality control, we might test that the average length of camshafts provided by the two suppliers is the same. Of course, we would perform this test only if we had reason to believe that the average lengths were not the same, since we can never prove the null hypothesis in hypothesis testing, as stated previously. Thus, as emphasized in Chapter 5, the null hypothesis is a statement that we doubt, except when testing a distributional assumption.

Preliminary Computations

The plant measured five camshafts of each type in each of four daily shifts and this was done for five work days. Thus, there are 100 measurements for each supplier, and preliminary calculations yield the following:

```
Variable    N     Mean     Variance
  Supp1     100   599.55    0.3835
  Supp2     100   600.23    3.5114
```

Objectives

The company might be interested in determining if there is a difference in variability of the camshaft lengths for the two suppliers, in addition to testing if the mean lengths differ.

Let's assume that the company had both objectives. What should be the starting point? Since comparing variability for the two suppliers is one of the objectives, we might as well look at that issue first. Since each variance is computed from 100 observations and the ratio of the sample variances is about 9:1, it seems apparent that the population variances differ more than slightly.

Tests for Equality of Variances

We can test this in MINITAB with the commands to do so given at the text website. The output is

```
F-Test (normal distribution)

Test Statistic:       0.109
P-Value:              0.000

Levene's Test (any continuous distribution)

Test Statistic:       74.752
P-Value:              0.000
```

The F-test is simply $F = s_1^2/s_2^2 = 0.109$, and the p-value is given as 0.000 because of the very small ratio. Recall from Section 3.4.7 that an F random variable results from the ratio of two independent chi-square random variables (i.e., independent samples), with each divided by its degrees of freedom. As was stated in Section 5.5, $(n - 1)s^2/\sigma^2$ has a chi-square distribution, assuming a normal distribution for the population from which the sample was obtained and s^2 was computed. It follows then that

$$\frac{\left[(n_1 - 1)s_1^2/\sigma_1^2\right]/(n_1 - 1)}{\left[(n_2 - 1)s_2^2/\sigma_2^2\right]/(n_2 - 1)} \tag{6.3}$$

which simplifies to s_1^2/s_2^2 under: $H_0: \sigma_1^2 = \sigma_2^2$ and has an F distribution with $(n_1 - 1)$ degrees of freedom for the numerator and $(n_2 - 1)$ degrees of freedom for the denominator.

It is often best not to use this test, however, because it is sensitive to nonnormality. The form of Levene's test used by MINITAB is actually the modification given by Brown and Forsythe (1974). Levene's test uses the absolute deviations of each observation from the sample mean. The Brown–Forsythe (1974) modification of Levene's test replaced the means by the medians. (This is explained in detail in Section 1.3.5.10 of the *e-Handbook of Statistical Methods*; see `http://www.itl.nist.gov/div898/handbook/eda/section3/eda35a.htm`.)

Of course, since these tests are the equivalent of *t*-tests, one must still assume equal variances: with Levene's test the variances of the absolute deviations from the means in the two populations are assumed equal, and with the Brown–Forsythe modification the variances of the absolute deviations from the medians in the two populations are assumed equal. With both methods, approximate normality of the difference of the averages of the absolute deviations must also be assumed and this becomes a shaky assumption because, for example, the absolute value of the standard normal random variable is a half-normal random variable, and a half-normal distribution is highly skewed. Therefore, very large sample sizes may be needed to "correct" for this skewness. What may still be unknown, however (was unknown as of 1996) is how well these tests perform when the variance assumption is not met and when only small-to-moderate sample sizes are used.

Obviously, these uncertainties make the approximate *t*-test an attractive alternative when the sample variances differ more than slightly, the sample sizes also differ considerably, and the variances of the two absolute deviations from the respective means or medians also differ more than slightly.

For the data in this example the signal is the same with each of the variance tests. This suggests that t^* should be used instead of t. We should bear in mind, however, that $t^* = t$ in this case because $n_1 = n_2$. Furthermore, when n_1 and n_2 are large, it won't make much difference which statistic we use. This is because the *d.f.* for both t^* and t will be large. Specifically, the *d.f.* for t^* can be written as $(n-1) + \left[(n-1)2s_1^2 s_2^2/(s_1^4 + s_2^4)\right]$. Thus, the *d.f.* must exceed the common sample size and in fact is 120 for this example

Recall from Section 3.4.4 that the *t*-distribution approaches the standard normal distribution as $d.f. \to \infty$. If the common sample size is large, both degrees of freedom will be large and will be between n and $2n$, with n denoting the common sample size. The result of this is that the critical values for the t and t^* statistics will be about the same. For example, assume that a two-sided test is used with a significance level of .05, or equivalently that this is viewed as the dividing line that separates small and not small *p*-values. The critical value for t for this example is 1.9720 and it is 1.9799 for t^*. There is hardly any difference in these two numbers, which is due to the fact that the equal sample sizes are large.

Test for Normality

In order to properly use either t or t^* for this problem, however, we must see evidence of something less than extreme nonnormality for the two populations. The normal probability plots for supplier #1 and supplier #2 are given in Figures 6.1 and 6.2, respectively.

We observe that use of the Anderson–Darling test for the supplier #1 data leads to rejection of the hypothesis of normality if we use a .05 significance level, but the data for supplier #2 appear to be close to normal. Notice in particular the difference of the *p-values*. This difference may seem surprising since the graphs are similar. Bear in mind, however, that we have large samples, which will thus have considerable power for detecting relatively small departures from normality. Furthermore, the plot for supplier #1 with the smallest points above the line and the largest points below the line does suggest a distribution with heavier tails than a normal distribution. There is also a pronounced absence of granularity

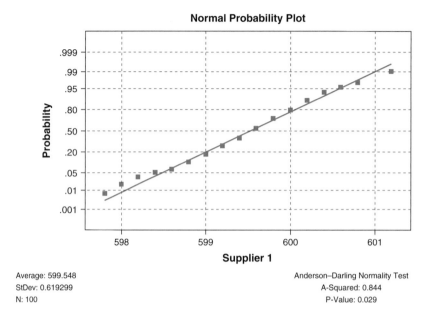

Figure 6.1   Normal probability plot for supplier #1.

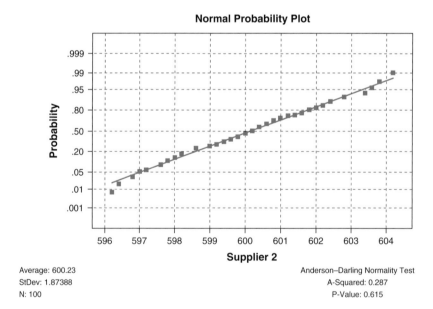

Figure 6.2   Normal probability plot for supplier #2.

in the data as 20 of the 100 observations have the same value, and there are similarly repeats of many other numbers.

Test for Equality of Means

Nevertheless, the nonnormality for supplier #1 is not severe enough to cause us any concern. Using MINITAB or other software, we obtain $t^* = -3.46$, which is far below the critical value of -1.9799 that was given earlier.

Conclusion

Thus, we conclude that the suppliers differ both in terms of the variability of the lengths of the camshafts that they supply and in terms of the mean lengths. ∎

Note that we could have used the hypothesis test counterpart to Eq. (6.2) since the sample sizes were large, and in effect we practically did so anyway because computing that form of Z is the same as computing t^*, as they are equivalent expressions. The only difference is that the critical values for each are not the same, but they hardly differ as the critical values for Z are ± 1.96 (assuming a two-sided test and $\alpha = .05$), which differs very little from ± 1.9799. Thus, as in the corresponding one-sample scenario in Table 5.1, this is a "t or Z case."

If a confidence interval rather than a hypothesis test had been desired, the interval based on the exact t-test would be

$$(\bar{x}_1 - \bar{x}_2) \pm t_{\alpha/2,\, n_1+n_2-2}\, s_p \sqrt{1/n_1 + 1/n_2}$$

6.1.3 Using Neither t nor Z

It was safe to use t^* in the preceding example, but that won't always be the case. When data are severely nonnormal (but continuous), the Wilcoxon two-sample test described in Section 16.3.1 should be used, provided that its assumptions are satisfied.

6.2 CONFIDENCE INTERVALS AND HYPOTHESIS TESTS FOR MEANS: DEPENDENT SAMPLES

We won't always have independent samples. In particular, if measurements are made on the same units, such as people, the samples will be dependent. For example, before and after readings might be obtained on a group of 20 patients who agree to participate in a clinical experiment to test the effectiveness of a new blood pressure drug.

Recall that with the independent-sample t-test the independence must be both within the two samples and between the samples. Clearly, the latter assumption is violated when there are repeat measurements on the same subjects. When there is a natural pairing, the obvious thing to do is to analyze the data as paired data: for example, compute "after minus before" values or the reverse, whichever is appropriate. This will have the effect of collapsing the two samples into one sample; then the single sample would be analyzed in the same way as when one starts with a single sample: that is, the methodology and assumptions of Chapter 5 apply.

Specifically, let X_2 = the "after" measurements and let X_1 = the "before" measurements. Assume that $d = X_2 - X_1$ is the appropriate differencing for a particular problem, and we believe that the differences should be positive. Then we would test $H_0 : \mu_d = 0$ against $H_a : \mu_d > 0$ and proceed as we would with a one-sample test. That is, the guidelines given in Table 5.1 and elsewhere apply; the only difference is the notation. Although $\mu_d = \mu_2 - \mu_1$, so that we are testing that the two population means are equal, as when the samples are independent, it would be better to think of a "population of differences" since the two samples are collapsed to form a sample of differences. Of course, this difference could be constructed as $X_2 - X_1$ or $X_1 - X_2$, depending on which is appropriate.

The test statistic is

$$t = \frac{\bar{d} - 0}{s_{\bar{d}}}$$

$$= \frac{\bar{d} - 0}{s_d / \sqrt{n}}$$

with $\bar{d}$ denoting the average of the differences, which of course is also the difference of the averages; that is, $\bar{d} = \bar{X}_2 - \bar{X}_1$. The denominator of course is the standard deviation of the average of the differences. The degrees of freedom for the t-statistic is $n - 1$. Thus, everything is the same as in the one-sample case once the differences have been formed, with the two sets of values that have been differenced not used directly in the calculations, with only the differences being used.

This test is most often called the *paired-t* test since the data are (inherently) paired.

■ **EXAMPLE 6.3**

Background

The following is a modification of a problem that has appeared on the Web and for which the intention was to use the paired-t test. We will examine whether or not that seems practical.

Data

The following are leading-edge and trailing-edge dimensions measured on five wood pieces. The two dimensions should directly correspond in order for the piece to fit properly in a subsequent assembly operation. The data are as follows:

Piece Number	Leading Edge (in.)	Trailing Edge (in.)
1	0.168	0.169
2	0.170	0.168
3	0.165	0.168
4	0.165	0.168
5	0.170	0.169

Obviously, this is a very small number of observations. The data clearly have a natural pairing since there are two measurements made on each piece of wood. With only four degrees of freedom for the paired-t test statistic, there would be very little power for detecting a departure from $H_0 : \mu_d = 0$. A more serious problem is the fact that it is almost impossible to gain any insight into the distributions of the two dimensions and/or their difference with such small sample sizes. So external information regarding the distributions of the measurements would have to be relied on. The reader is asked to consider this problem further in Exercise 6.8. ■

Recommendations are often given as to when the paired test should be used instead of the pooled test. Strictly speaking, if the data are paired, then the paired test should be used. The reason this is subject to some debate, however, is the degrees of freedom for the paired test is exactly half of the degrees of freedom for the pooled test, which makes the paired test a less sensitive test. In general, $Var(\bar{x}_1 - \bar{x}_2) = Var(\bar{x}_1) + Var(\bar{x}_2) - 2Cov(\bar{x}_1, \bar{x}_2)$ (see Section 4.2.2). This last term is incorporated implicitly when the paired test is used but of course is not used with the pooled test. If data are paired, then the covariance is positive, so the variance of the difference of the means will be overestimated when the samples are dependent/paired but the pooled test is used. This reduces the sensitivity of the pooled test when it is thus used improperly. So there is somewhat of a tradeoff and one might argue that the pooled test is preferable if the covariance term is small relative to what is gained with twice the degrees of freedom. Some would say that this would constitute improper use of statistical methodology, but using methods based on normality when normality doesn't exist is also "improper." Since t-tests are often performed with a small amount of data, such as in Example 6.3, it will frequently be necessary to think about whether or not the variables should be strongly related from subject matter considerations. Since data that are naturally paired will generally be strongly related, a good strategy would be to use the paired test routinely and to only infrequently consider the possible use of the pooled test with paired data

■ **EXAMPLE 6.4**

Data and Objective

The dataset SUGAR.MTW that comes with MINITAB has measurements on 28 wine samples for which the sugar content is measured both in the field and in the lab, with the lab measurements made using a new technique that is more reliable than a field measurement, but is also more expensive. The objective is to determine if there is a relationship between the lab and field measurements so that precision is not sacrificed by relying solely on the field measurement.

These are paired data because two measurements are made on each sample. A first step might be to test the hypothesis that there is no difference between the means, after testing the normality assumption, as there would be no point in trying to determine a relationship between the measurements if we can't reject the hypothesis of equal means. If the hypothesis cannot be rejected, the workers might as well just use the less expensive field measurement.

Test for Normality

The normality assumption is rejected, as shown by Figure 6.3.

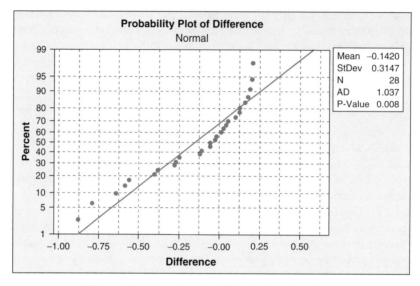

Figure 6.3   Normal probability plot for wine data.

Hypothesis Test

Of course, it is the distribution of $\overline{d}$ that is relevant but that distribution cannot be evaluated directly since there is only one $\overline{d}$. (There are methods for assessing this issue but those are beyond the scope of this book.) The value of the t-statistic for testing $H_0 : \mu_d = 0$ is -2.39, and the p-value for a two-sided test is .024. Thus, we reject the hypothesis of equal means if we use a significance level of .05 as the benchmark; but this result is somewhat shaky because of the nonnormality.

Conclusion

At this point there is some evidence that the means are different, so it would be worthwhile to determine if a reasonably strong relationship of some sort exists between the two measurements, so that the lab measurement (if made) can be estimated from the field measurement. Accordingly, we return to this problem in Exercise 8.67 in Chapter 8.   ∎

6.3   CONFIDENCE INTERVALS AND HYPOTHESIS TESTS FOR TWO PROPORTIONS

Testing the equality of two proportions is frequently of interest. For example, interest may center upon determining if an improved manufacturing process is superior to the standard process in terms of nonconforming units. So the hypothesis $H_0 : p_1 = p_2$ would be tested against $H_a : p_1 > p_2$, with p_1 denoting the proportion nonconforming from the standard process and p_2 denoting the proportion nonconforming from the improved process. As was true in the single-sample case, there is no such thing as a t-test for proportions, so we know that we must use Z. Thus, the statistic will have the general form $Z = (\widehat{\theta} - \theta_0)/\sigma_{\widehat{\theta}}$.

In order to see how we obtain that form in this case, we need to write H_0 equivalently as $H_0: p_1 - p_2 = 0$. Then it should be apparent that $\hat{\theta}$ is $\hat{p}_1 - \hat{p}_2$ and $\theta_0 = 0$. As in the single-sample case, the standard deviation is what is specified by the null hypothesis. The starting point in understanding this is to write $\sigma_{\hat{\theta}} = \sigma_{\hat{p}_1 - \hat{p}_2} = \sqrt{p_1(1 - p_1)/n_1 + p_2(1 - p_2)/n_2}$, with n_1 and n_2 denoting the size of each sample. The proportions are equal under the null hypothesis, so the standard deviation becomes $\sqrt{p(1 - p)(1/n_1 + 1/n_2)}$. There is then the question of how to estimate the hypothesized common value of p, since no value is implied by the null hypothesis. (Note how this differs from the single proportion hypothesis test, as in that case the value used for p is the hypothesized value.)

The only logical approach is to pool the data from the two samples to estimate p since a common value of p is being hypothesized. That is, $\hat{p} = (X_1 + X_2)/(n_1 + n_2)$. Therefore, with this estimate of p, we have

$$Z = \frac{\hat{p}_1 - \hat{p}_2 - 0}{\sqrt{\hat{p}(1 - \hat{p})(1/n_1 + 1/n_2)}} \tag{6.4}$$

Since a normal approximation approach is being used, the two sample proportions should not be extremely close to either 0 or 1, and preferably they should not differ greatly from .5.

■ **EXAMPLE 6.5**

Problem Statement

A company that is concerned about the quality of its manufactured products creates the position of Vice President of Quality Improvement and the occupant of the position sets out to improve the quality of the company's manufacturing processes. The first process that was addressed was the one that had caused the most problems, with one manufactured item becoming 35% nonconforming, by strict standards, when it went through the finishing stage. This of course was totally unacceptable, so various actions were taken to improve the finishing stage operation. This caused a reduction to 23%, still far too high, but a noticeable improvement. If each of the two percentages were computed from 100 items, does the appropriate one-sided hypothesis test show that there has been improvement?

Computations and Conclusion

We obtain

$$Z = \frac{\hat{p}_1 - \hat{p}_2 - 0}{\sqrt{\hat{p}(1 - \hat{p})(1/n_1 + 1/n_2)}}$$

$$= \frac{.35 - .23 - 0}{\sqrt{.29(1 - .29)(1/100 + 1/100)}}$$

$$= 1.87$$

This is not a very large Z-value as it produces a p-value of .031, which although less than .05, does not provide evidence of improvement as strong as might be desired (as would a p-value less than .01, for example). ■

6.3.1 Confidence Interval

Recall that in the one-sample proportion case given in Section 5.4.1 it is possible to derive a confidence interval for p that corresponds directly to the form of the hypothesis test. We can also do that in the two-sample case if we replace 0 in Eq. (6.4) by what it is the hypothesized value for: $p_1 - p_2$. Doing so is much easier than in the one-sample case, and we can immediately see that we obtain

$$(\widehat{p}_1 - \widehat{p}_2) \pm Z_{\alpha/2}\sqrt{\widehat{p}_1(1 - \widehat{p}_1)(1/n_1 + 1/n_2)} \tag{6.5}$$

This differs from the usual textbook expression of

$$(\widehat{p}_1 - \widehat{p}_2) \pm Z_{\alpha/2}\sqrt{\frac{\widehat{p}_1(1 - \widehat{p}_1)}{n_1} + \frac{\widehat{p}_2(1 - \widehat{p}_2)}{n_2}} \tag{6.6}$$

Which form should be used? It is easier to recommend the latter rather than the former, even though the former corresponds to the hypothesis test. The latter is clearly preferred when $p_1 \neq p_2$. Therefore, if the hypothesis were rejected, we clearly should use Eq. (6.6). If the hypothesis is not rejected, then it shouldn't make much difference which approach is used. In particular, the expressions are equivalent if $\widehat{p}_1 = \widehat{p}_2$, so the expressions will not differ greatly when the proportions differ only slightly. Thus, Eq. (6.6) should be used routinely.

Whereas MINITAB provides an option for how the hypothesis test is to be performed (pooling the data or not), the confidence interval is computed in accordance with Eq. (6.6). The command that produces both the confidence interval and the hypothesis test is PTWO.

■   **EXAMPLE 6.6**

Consider Example 6.5. If we had constructed a confidence interval for the difference of the two proportions, the interval would not have come close to containing 0 for any reasonable degree of confidence. The reader is asked to pursue this further in Exercise 6.35.   ■

6.4 CONFIDENCE INTERVALS AND HYPOTHESIS TESTS FOR TWO VARIANCES

There is often a need to compare two variances, such as when a manufacturing (or other) process has purportedly been improved in terms of a reduction in process variability and there is a subsequent desire to test the claim that improvement has been achieved.

Of course, methods for testing $H_0 : \sigma_1^2 = \sigma_2^2$ were discussed in Section 6.1.2 in the context of t-tests, and it was recommended that the approximate t-test be strongly considered because of the uncertainties associated with the tests for equal variances.

But if our focal point is testing equality of variances rather than testing equality of means, we must face the issue of how to test the variances, and similarly if we want a confidence interval for the ratio of two variances.

Consider the following example.

■ **EXAMPLE 6.7**

Problem Statement

A company was having a problem with one of its manufacturing processes (inconsistent performance) so a team was formed to try to improve the process. The variance of a particular product characteristic was estimated at 16.2 (based on 100 measurements) and that was considered unacceptable. After four weeks of work, the team was able to identify a few problem areas and the end result was that the variance was reduced to 15.0 (also based on 100 measurements). The question to be addressed is whether or not this apparent improvement is really just an illusion caused by random variation.

Test for Equality of Variances

One way to address the question would be to test $H_0 : \sigma_1^2 = \sigma_2^2$ using either an F-test or, preferably, Levene's test, as illustrated in Section 6.1.2. If the F-test that was given in Section 6.1.2 were used, this produces $F = s_1^2/s_2^2 = 16.2/15.0 = 1.08$ for this example. The p-value for a one-sided test of H_0 against $H_a : \sigma_1^2 > \sigma_2^2$ (with "1" denoting the old process) is .3345, so we would not reject H_0. Of course, before we used this test we would want to see evidence of approximate normality. (*Note*: There is a temptation to place the larger of the two sample variances in the numerator so that the ratio will be greater than one, but doing so creates a statistic that does not have an F-distribution.)

Confidence Interval on Ratio of Variances

Let's assume that the normality assumption does seem to be approximately met and that we instead prefer to construct a confidence interval on the ratio σ_1^2/σ_2^2. (We will look at a one-sided bound later.) As with the construction of other confidence intervals, this is accomplished by performing the necessary algebra on the appropriate expression so that σ_1^2/σ_2^2 will be in the middle of the double inequality. This is accomplished in this case by starting with Eq. (6.3). Thus,

$$P\left(F_{1-\alpha/2,n_1-1,n_2-1} \le \frac{s_1^2/\sigma_1^2}{s_2^2/\sigma_2^2} \le F_{\alpha/2,n_1-1,n_2-1}\right) = 1 - \alpha$$

Using the result from Section 3.4.7 that $F_{\alpha,v_1,v_2} = 1/F_{1-\alpha,v_2,v_1}$ (which is needed here so that we can obtain the interval for the desired ratio), we obtain

$$P\left(F_{\alpha/2,n_2-1,n_1-1} \le \frac{s_2^2/\sigma_2^2}{s_1^2/\sigma_1^2} \le F_{1-\alpha/2,n_2-1,n_1-1}\right) = 1 - \alpha$$

so that

$$P\left(F_{\alpha/2,n_2-1,n_1-1}\frac{s_1^2}{s_2^2} \le \frac{\sigma_1^2}{\sigma_2^2} \le \frac{s_1^2}{s_2^2}F_{1-\alpha/2,n_2-1,n_1-1}\right) = 1 - \alpha$$

Thus, the lower limit is the expression on the left of the interval and the upper limit is on the right of the interval.

For the present example we obtain 0.73 as the lower limit and 1.60 as the upper limit of a 95% confidence interval, as the reader is asked to verify in Exercise 6.38. These results can also be obtained using the sequence of MINITAB commands that is given at the textbook website.

Conclusion

A one-sided lower bound, which would correspond to the hypothesis test performed previously, would be obtained as $\sigma_1^2/\sigma_2^2 \geq F_{\alpha,n_2-1,n_1-1}s_1^2/s_2^2$. Making the appropriate substitutions produces 0.77. Since this number does not exceed 1.0, we cannot conclude that the new production process resulted in a smaller variance. Of course, this result was expected since the p-value for the one-sided hypothesis test was greater than .05.

As with the hypothesis test, we need to be able to assume at most only a small departure from a normal distribution before the interval should be used.

It is not possible to construct a confidence interval for σ_1^2/σ_2^2 using either the Levene approach or the Brown–Forsythe approach because variances of the raw observations are not in any way a part of these methods. Instead, one would have to construct a confidence interval on the difference between the two averages of absolute deviations since equality of those averages of absolute deviations is the null hypothesis that is implied when the Brown–Forsythe test is used.   ∎

6.5 ABSTRACT OF PROCEDURES

Population Means: Independent Samples. The student/user must first decide whether the samples are independent or dependent. If independent, then the user must decide whether to use t or Z, analogous to the decision that must be made in the one-sample case. Unlike the one-sample case, however, there are two t-statistics to chose from when there are two samples, and these were denoted as t^* and t, respectively, in Section 6.1.2. Assuming two populations that don't differ greatly from normality, it will be safe to use t as long as the sample sizes do not differ greatly if there is a considerable difference in the variances. Large percentage differences in both should motivate a user to select t^*.

Table 5.1 can be used as a guide as to when Z or one of the t-statistics should be used, with the obvious modification to make it applicable to two samples instead of one.

Population Means: Dependent Samples. If the data are naturally paired, such as when two observations are made on each person, the paired-t test given in Section 6.2 should be employed. It is worth remembering, however, that there cannot be dependence *within* each sample.

Population Proportions. This case is rather straightforward as Eq. (6.6) is used to produce a confidence interval of the difference of the two population proportions and Eq. (6.4) is used to test the equality of the proportions. Since a standard normal variate is being used for each, the two sample proportions should not be close to either zero or one.

Population Variances. Although the F-test for the equality of two variances was presented in Section 6.4, it is preferable to use a test that is not adversely affected by nonnormality. Such options do not exist when a confidence interval on the ratio of two variances is constructed, however, so if the interval given in Section 6.4 is to be used, there should be approximate normality for each of the two populations.

6.6 SUMMARY

We have covered testing the equality of two variances, two means, and two proportions, as well as the corresponding confidence intervals. These tests are frequently needed, and a test for the equality of more than two variances is needed for Chapter 12. Care must be exercised in selecting a test for the equality of variances since most tests are sensitive to nonnormality.

We should also bear in mind that null hypotheses are generally false simply because two populations are not likely to have *exactly* the same means, or the same proportions, or the same variances. Consequently, from a practical standpoint, we are actually testing to see if the means, proportions, or variances differ enough to lead us to reject the null hypothesis, keeping in mind that the likelihood of rejection is a function of the sample size and the actual difference of the population values.

REFERENCES

Brown, M. B. and A. B. Forsythe (1974). Robust tests for the equality of variances. *Journal of the American Statistical Association*, **69**, 364–367.

Cressie, N. A. C. and H. J. Whitford (1986). How to use the two sample *T*-test. *Biometrical Journal*, **28**(2), 131–148.

Good, P. I. and J. W. Hardin (2003). *Common Errors in Statistics (and How to Avoid Them)*. Hoboken, NJ: Wiley.

Layard, M. W. J. (1973). Robust large-sample tests for homogeneity of variances. *Journal of the American Statistical Association*, **68**, 195–198.

Levene, H. (1960). In *Contributions to Probability and Statistics: Essays in Honor of Harold Hotelling* (I. Olkin et al., eds). Stanford, CA: Stanford University Press, pp. 278–292.

Posten, H. O., H. C. Yeh, and D. B. Owen (1982). Robustness of the two-sample *t*-test under violations of the homogeneity of variance assumption. *Communications in Statistics, Part A—Theory and Methods*, **11**(2), 109–126.

EXERCISES

6.1. Show that t and t^*, as discussed in Section 6.1.2, are equal when the sample sizes are equal. Does this mean that the tests are equivalent when the sample sizes are equal? Explain.

6.2. Show that the equation for $\hat{p}$ given in Section 6.3 is a weighted average of the two sample proportions, with the weights being the respective sample sizes. Why are the data from the two samples pooled to obtain a common estimate?

6.3. Western Mine Engineering, Inc. provides considerable data on mining operations at its website. For example, for 2001 there was sample data on 194 U.S. mines, 77 of which were union mines. Of the 58 union mines at which there was a wage increase, the average increase was 2.4%. Of the 70 non-union mines at which there was a wage increase, the average increase was 2.7%. Would it be possible to test whether the wage increase was equal for 2001 for the two types of mines using one of the methods given in this chapter? Why or why not? If possible, perform the test and draw a conclusion.

6.4. The following data were given in 2002 at the Western Mine Engineering, Inc. website.

Type of Worker	Union		Non-union	
	Number of Mines	Average Salary	Salary Range	Average Salary
Mechanic—surface mine	82	19.25	98	16.99

Assume that the standard deviations, which were not given, are the same for the union and non-union salaries. What is the largest value of the common standard deviation that would result in rejection of the hypothesis of an equal mean hourly wage, using a two-sided test with $\alpha = .05$? What additional assumption(s) did you make when you performed the analysis? Can you check those assumptions with only the data that are given here? Explain.

6.5. A two-sample t-test is used to test $H_0 : \mu_1 = \mu_2$ versus $H_a : \mu_1 < \mu_2$, with $\alpha = .05$. If $n_1 = 14$, $s_1^2 = 6$, $n_2 = 15$, $s_2^2 = 7$, and it is assumed that $\sigma_1^2 = \sigma_2^2 = \sigma^2$, what is the numerical value of the estimate of σ^2?

6.6. If each of 200 independent samples is used to construct a 95% confidence interval for $\mu_1 - \mu_2$, how many of the intervals would we expect to contain the actual (unknown) value of $\mu_1 - \mu_2$? Let $W = $ the number of times that the interval contains $\mu_1 - \mu_2$. Clearly, W is a random variable. Determine the variance of W. (*Hint*: Either the interval contains $\mu_1 - \mu_2$ or it doesn't, so there are two possible outcomes.) Let Y denote the width of the interval. Determine the variance of Y if σ_1^2 and σ_2^2 are assumed known.

6.7. There are various ways to construct a normal probability plot, depending on what is plotted. One approach is to plot the normal score (i.e., Z-score) on the vertical axis and the observations on the horizontal axis. Another approach, illustrated in Section 6.1.2, is to have the cumulative probability on the vertical axis. Explain why a line fit through the points on a plot using either approach could not have a negative slope.

6.8. Consider Example 6.3 in Section 6.2. Would it be fruitful to construct a normal probability plot of the differences to test for normality of the differences? Explain. Casting aside the issue of normality for a moment, perform a paired-t test on the data and draw a conclusion. How much faith would you place in your results? Explain.

6.9. A sample is obtained from each of two (approximately) normal populations and the following results are obtained: $n_1 = 8$, $\bar{x}_1 = 60$, $s_1^2 = 20$, $n_2 = 12$, $\bar{x}_2 = 75$, $s_2^2 = 15$.
   (a) What would be the numerical value of the estimate of $\mu_1 - \mu_2$?
   (b) Is it necessary to assume that the populations each have a normal distribution in order to obtain the estimate? Why or why not?

6.10. Assume that you are tutoring a student from your engineering statistics class who encounters the following statements in different sources. The first of these is: "The null hypothesis, denoted by H_0, is the claim about one or more population characteristics that is initially assumed to be true" (the "prior belief" claim). The second reads: "To ask and answer questions about engineering phenomena the engineer must pose two competing hypotheses, a working hypothesis (the hypothesis to be nullified, sometimes called the null hypothesis) and an alternative hypothesis. Generally the engineer intends to demonstrate that the data are supported by the alternative hypothesis."

Your fellow student is nonplused, especially since the first statement appears in a statistics text and practically contradicts the second statement(s), which seems to zero in on what "null" means in a generic sense. Are the statements in conflict? What would you say to the student?

6.11. Assume that two "treatments" (such as two blood pressure medications) are to be compared in a study with each of 20 people receiving each of the two medications. Explain why the appropriate t-test to use for this analysis is *not* the one that has $n_1 + n_2 - 2 = 38$ degrees of freedom.

6.12. For the paired t-test, how many degrees of freedom will the t-statistic have if there are 10 pairs?

6.13. What is the numerical value for the degrees of freedom of an independent sample t-test if $n_1 = 35$ and $n_2 = 30$?

6.14. The data in the file IMPEACHCLINTON.DAT can be used to illustrate some of the problems that are encountered when trying to use a two-sample t-test with actual data. The American Conservative Union assigns a "conservatism rating" to each U.S. senator. Assume that we decide to perform a t-test to test the hypothesis that the average conservatism rating is the same for Democrats and Republicans (acting as if we don't know what the outcome will be).

(a) We could perform a t-test if this is a sample from a population, but it wouldn't make any sense to do so if we had the entire population, as then we would know the answer, with certainty, since there would be no p-value, for example. With this in mind, would you suggest that a t-test be performed?

(b) Casting aside your answer to part (a) for a moment, proceed to perform the test, if possible, bearing in mind the assumptions for the test. Would you recommend using the exact test or the approximate test? Do the results differ? When the NORMPLOT command in MINITAB is used for each of the two groups of data, the conclusion is that both populations are non-normal. Considering the respective sample sizes, should this be a point of concern?

6.15. Gunst (*Quality Progress*, October 2002, vol. 35, pp. 107–108) gave "16 mutually independent resistivity measurements for wires from two supposedly identical processes" and showed that the p-value for testing that the two process means were equal was .010. The data are given below.

Process 1	Process 2
0.141	0.145
0.138	0.144
0.144	0.147
0.142	0.150
0.139	0.142
0.146	0.143
0.143	0.148
0.142	0.150

(a) Considering the small number of observations from each population, what is the first step you would take before you performed the hypothesis test?

(b) Take the action that you indicated in part (a) and then, if possible, perform the pooled-t test and state your conclusion. If it is not possible to perform that test, is there another test that could be used? If so, perform the test and state your conclusion.

6.16. Online grade information for the second semester of the 2001–2002 academic year at Purdue University showed for undergraduates that the average grade-point average (GPA) for all 17,098 male students was 2.720, and the average GPA for all 12,275 female students was 2.989.

(a) Could we perform some type of t-test for the null hypothesis of equality of the average GPA for male and female undergraduate students? If not, why not.

(b) If we view the results for that semester as a sample from all semesters for 1992–2002, could we perform a t-test with the information that is given? In particular, is dispersion information absolutely necessary in view of the differences in the GPAs? What would the conclusion be if equal variances were assumed and ultraconservative dispersion estimates based on ranges were used? Since we have moved beyond the 2001–2002 academic year, could a conclusion from a t-test be extended to the present and/or future? If so, under what conditions could this be done? If not, why not?

6.17. Using the data in Example 6.7 in Section 6.4, construct a 99% two-sided confidence interval for σ_1^2/σ_2^2 using the standard, F-statistic approach. Would it be possible to use the Brown–Forsythe approach with the information that is given? Why or why not?

6.18. You are given two samples of 100 observations each and asked to test the hypothesis that the population means are equal. Explain how you would proceed.

6.19. The U.S. Consumer Product Safety Commission stated in a November 5, 2001 memorandum (http://www.cpsc.gov/library/toydth00.pdf) that of the 191,000 estimated toy-related injuries in 2000, males were involved in 68% of them. Would it be possible to construct a 95% confidence interval for the difference in the proportion of males and females who were involved in toy-related injuries that year? If so, construct the interval. If not, explain why the interval cannot be constructed.

6.20. Nelson (*Journal of Quality Technology*, vol. 21, No. 4, pp. 232–241, 1989) compared three brands of ultrasonic humidifiers in terms of the rate at which they output moisture, this being what was measured. Since there was a missing observation for the third humidifier, we will consider only the first two. The maximum output rate (fl. oz/h) for a sample of each of the two brands is as follows:

Brand 1:	14.0	14.3	12.2	15.1
Brand 2:	12.1	13.6	11.9	11.2

You are asked by management to make a recommendation based on these numbers. Do so, if possible. If it is not possible, or at least practical, explain why not.

6.21. Assume that an experimenter wishes to test the equality of two means but fears that the populations may be highly nonnormal. Accordingly, he takes samples of $n = 25$ from each population and constructs a normal probability plot for each sample. The points on one of the plots practically form a straight line, but there is a strong S-shape on the other plot.

(a) What does an S-shape suggest regarding the tails of the distribution?

(b) Despite the signal from one of the plots, the experimenter decides to proceed with the test and will reject the null hypothesis if the absolute value of the t-statistic is greater than 2. Does the plot configuration make this more likely or less likely to occur than if normality existed for each population? Explain.

6.22. Respiration rates for each of nine experimental subjects in two breathing chambers (i.e., each subject spent time in each chamber) are given below. One chamber had a high concentration of carbon monoxide while the other did not. Our scientific interest is in establishing that exposure to carbon monoxide significantly increases respiration rate.

Subject	With CO	Without CO	
1	30	30	0
2	45	40	5
3	26	25	1
4	24	23	1
5	34	30	4
6	51	49	2
7	46	41	5
8	32	35	-3
9	30	28	2

(a) Based on the information given, is there a way to test for a significant increase in respiration rate? If so, perform the test. If not, explain why not.

(b) If the answer to part (a) is "no," is there any type of test that can be performed that would be of interest, provided that approximate normality can be assumed? If so, perform the test.

6.23. (Harder problem) Explain, in specific terms, why an inequality in an alternative hypothesis cannot be reversed before the value of the test statistic is computed if the data suggest that the direction of the inequality is incorrect.

6.24. The designer of a new sheet metal stamping machine claims that her new machine can produce a certain product faster than the machine that is presently being used. Twenty units of the product were randomly selected from a large group of units and were randomly assigned to the two machines, with the stamping then performed by each machine on ten of the units. The data obtained from the experiment are as follows:

	Average	Standard Deviation
Current machine	35.46 seconds	3.48 seconds
New machine	32.67 seconds	3.69 seconds

Recent calculations with other data showed that the time using the current machine appears to be approximately normally distributed. Using only this information, explain how you would proceed to analyze the data relative to the designer's claim, justify your approach, and then analyze the data and draw a conclusion.

6.25. Oxide layers on semiconductor wafers are etched in a mixture of gases to achieve the desired thickness. Low variability of the thickness of the layers is highly desirable, with variability of the thickness being a critical characteristic of the layers. Two different mixtures of gases are under consideration and there is interest in determining if one is superior to the other in terms of variability. Samples of 25 layers are etched with each of the two mixtures, with a standard deviation of 1.85 angstroms observed for the first mixture and 2.02 angstroms for the second mixture. Test the appropriate hypothesis under the assumption that approximate normality is a plausible assumption for each population.

6.26. A pollution control inspector suspected that a riverside community was releasing semi-treated sewage into a river and consequently this was changing the level of dissolved oxygen in the river. He would like to perform a statistical test to see if the level of dissolved oxygen downstream differs from the level of dissolved oxygen upstream. He is not quite sure how to collect the data.

    (a) Should he draw the water specimens at the same location upstream and the same location downstream, or should the samples be drawn from multiple locations upstream and downstream? What considerations should he make in reaching a decision about this?

    (b) He also isn't sure how many samples to take, but he knows that the samples will be relatively inexpensive to obtain. What would you suggest?

    (c) What problems, if any, would be presented by taking samples of size 8?

6.27. If the lower limit of a (two-sided) 95% confidence interval for $\mu_1 - \mu_2$ using t is 23.4, will the limit of a lower one-sided 95% confidence interval for $\mu_1 - \mu_2$ (using the same data) be less than 23.4 or greater than 23.4? Explain.

6.28. Explain why a confidence interval for $p_1 - p_2$ will never have the desired degree of confidence.

6.29. Explain why the numerator of the test statistic of the paired-t test is equivalent to the numerator of the independent sample t-test, but the denominators are not the same.

6.30. Generate 1000 pairs of samples of $n = 50$ from a $N(\mu = 10, \sigma^2 = 25)$ distribution. Compute the variance for each sample and form a fraction for each sample by placing the larger of the two variances in the numerator of the fraction and the smaller variance in the denominator. Since the critical value of F for a two-sided test of the equality of two population variances when $\alpha = .05$ is $F_{.975,49,49} = 1.7622$, we might be tempted to conclude that approximately 25 of the ratios should exceed this number.

(a) Did this happen in your case?

(b) Why should we *not* expect this to happen, but rather expect more than 25?

6.31. Explain why we do not have to assume that $\sigma_1^2 = \sigma_2^2$ whenever we use t^* (as given in Section 6.1.2) in constructing a confidence interval for $\mu_1 - \mu_2$.

6.32. An experimenter wishes to test the equality of two population means (corresponding to two processes), but erroneously sets $\mu_1 \neq \mu_2$ as the null hypothesis. Explain why the value of the test statistic cannot be calculated for this form of the null hypothesis.

6.33. The Dean of Student Affairs at a large midwestern university decided to conduct a study to determine if there was a difference in academic performance between freshmen male students who do and do not have cars on campus. A sample of 100 male students from each of the two populations is selected with the null hypothesis being that the mean GPA is the same for both populations, and the alternative hypothesis is that car owners have a lower GPA. The sample without cars had a GPA of 2.7 with a variance of 0.36, as compared with the car owners, whose average GPA was 2.54 with a variance of 0.40. The alternative hypothesis is the one that the dean suggested. Perform the appropriate test and reply to the dean. Given your result, if you have constructed a confidence interval or confidence bound on the mean GPA for car owners minus the mean GPA for students who do not have a car on campus, what can you say about how the interval would appear, relative to zero? (Don't construct the interval or bound.)

6.34. In 2003 the University of Georgia passed a rule (later rescinded) requiring entering freshmen to live on campus, including local residents (with few exceptions). One reason that was cited in support of the rule is that for the Fall 2002 semester, freshmen who lived on campus had an average GPA of 3.15, while the average for freshmen who lived off campus was 3.0. Do you believe that the results of a pooled t-test, if the variances were available, would necessarily be adequate support for the rule? Explain.

6.35. Consider Example 6.5 in Section 6.3 and construct a 99% confidence interval for the difference of the two proportions. We know, however, what the relationship will be between the two endpoints of the interval and zero. What will be that relationship?

6.36. An assembly operation in a manufacturing plant requires three months of training for a new employee to reach maximum efficiency. This length of time has been judged excessive by top management, so a new training program was developed that requires only six weeks of training. The new program will be adopted if it is judged to be more effective than the old method. To determine if this is true, one group of 12 employees was trained using the old method and another group of 12 employees was trained under the new method. The length of time that each trainee took to assemble a particular device was recorded. For the old program the average time was 34.28 minutes with a standard deviation of 4.86 minutes. For the new program the numbers were 31.36 and 4.63, respectively. You have been asked by management to analyze these data, for which you are to assume approximate normality for each method. Analyze the data and report your conclusion.

6.37. The management of a new company is trying to decide which brand of tires to use for its company cars when the tires currently on the cars have to be replaced. One of the executives stumbles onto a research report that provided the results of an accelerated testing study. (Accelerated testing is discussed in Section 14.3.) The study involved 100 tires of each of two types and showed that Brand A had an average useful life of 32,400 miles with a variance of 1,001,034 miles, whereas Brand B had an average useful life of 29,988 miles with a variance of 888,698 miles. Construct a 99% confidence interval for the mean of Brand A minus the mean of Brand B and draw a conclusion regarding which brand the company should select.

6.38. Verify the lower limit and upper limit of the 95% confidence interval on σ_1^2/σ_2^2 that was given in Section 6.4.

6.39. (MINITAB required) One of the sample datasets that comes with MINITAB is VOICE.MTW, which is in the STUDENT1 subdirectory. A professor in a theater arts department was interested in seeing if a particular voice training class improved a performer's voice quality. Ten students took the class and six judges rated each student's voice on a scale of 1–6, before and after the class. One method of analysis would be to obtain an average score for each student, before and after the class. What would be lost, if anything in using such an approach? Perform the appropriate test with these data, after first stating the assumptions for the test and whether or not those assumptions can be tested for this dataset. What do you conclude?

6.40. The following is a two-sample problem given by an instructor of an engineering statistics class, and which is available on the Web.

> Oxide layers on semiconductor wafers are etched in a mixture of gases to achieve the proper thickness. The variability in the thickness of these layers is a critical characteristic of the wafer and low variability is desirable for subsequent processing steps. Two different mixtures of gases are being studied to determine whether one is superior in reducing the variability of the oxide thickness. Sixteen wafers are etched in the first mixture and 15 in the second. The sample standard deviations of oxide thickness are $s_1 = 1.96$ angstroms and $s_2 = 2.13$ angstroms. Is there evidence to indicate that either gas is preferable? Use $\alpha = .02$.

(a) The answer given by the instructor was "$f = .85$; do not reject H_0." Comment on this.

   (b) How would you analyze the data based on the wording of the problem? Perform an appropriate analysis, if possible. If this is not possible, explain why not.

6.41. A new employee at a company performs a paired-t test on some company data with 15 pairs and finds that all of the differences have a positive sign, with an average of 4.9 and a standard deviation of 1.8.

   (a) What decision should be reached regarding the test if the alternative hypothesis is $H_a : \mu_d < 0$?

   (b) A person who is familiar with the data is surprised at the conclusion drawn by the new employee and contends that the differences should have been negative. What is the likely cause of this discrepancy?

6.42. A person erroneously performs a paired-t test when the samples are independent and obtains a p-value of .06. The raw data were discarded and only the differences were retained. If the population variances were approximately equal, would the p-value for the independent sample test likely have been greater than .06 or less than .06? Explain.

6.43. A paired-t two-sided test is performed and the p-value is .04, with the value of the test statistic being positive. What would be the relationship between the endpoints of a 95% confidence interval for μ_d and zero if the interval had been constructed using the same data?

6.44. Consider the camshaft dataset that was analyzed in Section 6.1.2. The respective degrees of freedom for the two methods of analysis that were discussed were 120 and 198, respectively. Since these numbers differ greatly and the variances differ considerably, does this suggest that one of the two methods should be preferred? Explain.

CHAPTER 7

Tolerance Intervals and Prediction Intervals

As far as statistical intervals are concerned, a statistics student is generally taught only a few confidence intervals and a prediction interval for a future observation in the context of regression analysis. The student of engineering needs to know much more than this, however, and in the past ten years or so there has been interest in meeting these needs. Indeed, Vardeman (1992) asked "What about the other intervals?"

In particular, there are federal regulations regarding products that specify the extent to which a product can deviate from what it is purported to be, in terms of both composition and weight. For example, every 8-ounce bag of potato chips is not 8 ounces, as some are more than 8 ounces and many are less. If all were less, then a company would have a major problem, whereas, say, 2% being less might be allowable, depending on the extent of deviation from 8 ounces. Thus, a production engineer would want to know what percentage of the units of production exceed a specified lower bound: that is, a *lower tolerance bound* would be needed.

Note how this differs from a confidence interval, which is an interval for an unknown parameter. In this chapter we consider tolerance intervals and prediction intervals. A consumer who plans to purchase a particular product would be interested in knowing the length of useful life *for the particular unit* that he/she intends to purchase. Thus, the consumer would be interested in a prediction interval. The manufacturer of course is interested in the population of units that the company produces, rather than the performance of any particular unit, and so would not be interested in a prediction interval. Instead, the company would be interested in a tolerance interval or a one-sided tolerance bound and could also be interested in a confidence interval for the mean, such as a confidence interval for the average miles per gallon of a certain make of car.

As a point of clarification, we should note that a tolerance interval can be constructed without any reference to engineering specifications, which are often called tolerances. For example, the specifications for the manufacture of 100-ohm resistors might state the acceptable range for the manufactured resistors as being from 95 to 105 ohms (i.e., the tolerances), but a tolerance interval constructed from a sample will almost certainly be at least slightly different. If the endpoints of the tolerance interval differed more than slightly from 95 and 105, this could signify that the process is not centered at a target value, and/or the process variability is too great, or perhaps the tolerances are simply unrealistic.

Various types of tolerance and prediction intervals are discussed in the following sections, some of which are based on the assumption of a normal distribution, while some are not based on any distributional assumption. Regardless of the type of interval used, it is important to keep in mind that whatever process is involved, the process should be in a state of statistical control (control charts are presented in Chapter 11). This means that the parameters of the distribution are stable, which implies that the distribution itself is stable. Even when *nonparametric* (distribution-free) tolerance intervals are used, one must still assume that the distribution, whatever it may be, is stable. (Nonparametric methods are discussed in Chapter 16.)

7.1 TOLERANCE INTERVALS: NORMALITY ASSUMED

Since tolerance intervals are for individual observations, they are very sensitive to any particular distributional assumption. A normal distribution is typically assumed and there is also the tacit assumption that the population of interest is either infinite or at least very large relative to the sample size. If the latter were not true, then the finite population correction factor given in Section 4.2.2 would have to be used if the sample size was much more than 5% of the population size. The correction factor is rarely necessary, however.

If, for example, we knew the mean and standard deviation of a particular normal distribution, then we would know that 95% of the observations fall in the interval $\mu \pm 1.96\sigma$. That is, we are 100% sure of this. Of course, we almost always do not know μ or σ, however, and it should be apparent that we could *not* say that we are 100% certain that 95% of the observations would be contained in the interval $\bar{x} \pm 1.96s$ because to make such a statement would be to ignore sampling variability. That is, since parameters are being estimated, we cannot be sure that whatever interval we construct will contain a specified percentage of the population unless we state, trivially, that our 100% tolerance interval is from minus infinity to plus infinity. In this unrealistic scenario we would be 100% confident that our interval includes 100% of the population. Therefore, in applications of tolerance intervals, both percentages will be less than 100.

Some reflection should suggest why this is the case. As stated, the interval with the estimated parameters is not fixed, but rather has sampling variability. There is an analogy here to Figure 5.1, which showed an arbitrary parameter being fixed and the constructed intervals being random. We have the same situation regarding tolerance intervals. The true proportion between two numbers is fixed, just as a parameter is fixed when a confidence interval is constructed. Before a confidence interval is constructed, there is a specified probability that the interval will contain the unknown parameter. So can we say, for example, that we will construct a tolerance interval such that we are 90% confident that our tolerance interval contains exactly 95% of the population values? No, because that is practically the same as saying that we are 90% confident that $\bar{x} = \mu$ and $s = \sigma$. Clearly, those probabilities are zero, for the same reason that the probability that a normal random variable equals a specified number is zero.

Therefore, we cannot have "exactness" on the proportion of a population contained by a tolerance interval. To see this, consider the standard normal distribution. We know that 95% of the observations are between -1.96 and 1.96. If we constructed 100 tolerance intervals, how many of them would we expect to have endpoints of exactly those two values? The answer is zero because the probability is zero that the combination of the values of $\bar{x}$ and s will produce those endpoints exactly. Therefore, we may say, somewhat trivially, that if we tried to construct a tolerance interval that contains exactly 95% (or some other percentage) of a population, we will always be 0% confident that we have done so. (The reader is asked to show this in Exercise 7.14.)

7.1.1 Two-Sided Interval

Thus, there is not an exact analogy with confidence intervals. Instead, a tolerance interval is constructed such that there is a specified degree of confidence that the interval contains *at least* a certain percentage of the population values.

■ **EXAMPLE 7.1**

Objective

Assume that we wish to determine a tolerance interval for a certain type of condenser measurement and we want our interval to contain at least 99% of the population (assumed to have a normal distribution) with 99% confidence.

Interval Computation

We would compute the interval as $\bar{x} \pm 3.74s$ if the sample size is 30, obtaining the value from Table E. Thus, if $\bar{x} = 6.78$ and $s = 1.12$, the interval is (2.59, 10.97). The theory behind the values in Table E is beyond the scope of this book. Readers seeking explanatory material are referred to Odeh and Owen (1980), which is unfortunately out of print, and Howe (1969). ■

Of course, it is preferable to use software, which would allow $\bar{x}$, s, and the interval to be computed simultaneously. Furthermore, the tables for normality-based tolerance intervals in Hahn and Meeker (1991) contain only a few values for n and do not go beyond 60.

As a simple example, solely for illustrative purposes, assume that our sample consists of 100 random numbers generated from a $N(50, 25)$ distribution and that we want a tolerance interval such that we are 95% confident that it will cover at least 90% of the population. The interval can be obtained in MINITAB, for example, by downloading the TOLINT macro from the Minitab, Inc. website. Specifically, if the data are in the first column, the tolerance interval would be produced using the command TOLINT C1. The output is as follows:

```
   Tolerance (Confidence) Level: 95%
 Proportion of Population Covered: 90%

    N         Mean        StDev       Tolerance Interval
    100     49.6790     5.17357       (39.9846, 59.3734)
```

The calculations for the above interval assumes that the data follow a normal distribution.

Note that although the output gives "proportion of population covered," this should not be interpreted literally, in accordance with the discussion in Section 7.1.

Note also that the interval is wider than $\mu \pm 1.645\sigma$, which is (41.775, 58.225). This is partly because the appropriate multiplier is larger than 1.645 since the parameters are estimated, and also because σ is overestimated by more than a small amount. If we use a very large sample, we would expect the interval to be very close to (41.775, 58.225), and when 5000 observations are generated from this distribution we obtain

```
  N      Mean     StDev    Tolerance Interval
5000   50.0162   5.03307   (41.5980, 58.4344)
```

so the interval is, as expected, very close to the theoretical interval.

The default settings for the TOLINT macro are 95% confidence and 90% proportion. If some other combination were desired, either one or two subcommands would have to be used. For example, if 99% confidence were desired with 95% proportion, the interval would be generated using

```
MTB>%TOLINT C1;
SUBC>TOLER 99;
SUBC>PROP 95.
```

Tolerance intervals can of course be produced using other popular statistical software and are obtained in JMP using the "Distribution" feature.

7.1.1.1 Approximations

Since the necessary tables for constructing tolerance intervals are given in very few statistics books and tolerance interval capability is not available in all statistical software, an approximation to the multiplier s of will sometimes be useful.

Two such approximations are in use. Howe (1969) gave one approximation as

$$k_1 = \sqrt{\frac{(N-1)(1+1/N)z^2_{(1-p/2)}}{\chi^2_{\alpha,n-1}}}$$

for a tolerance interval that is to cover at least $100(1-p)\%$ of the population with $100(1-\alpha)\%$ confidence. For Example 7.1, $k_1 = 3.735$, which does not differ very much from the tabular value of 3.742. Approximations such as obtained by using k_1 can be quite useful since, in particular, tables do not include most values of n, as mentioned previously, especially odd numbers. Another approximate multiplier has also been proposed, but it requires slightly more work to obtain than is required to compute k_1.

The multiplier of s decreases as n increases, approaching 2.58 for 99% confidence and 99% proportion as n approaches infinity. This of course is because the estimation error approaches zero as n approaches infinity, so the tolerance interval approaches a probability interval, as indicated previously. That is, it approaches $\mu \pm 2.58\sigma$ as n approaches infinity. Using 2.58 as the multiplier, as an approximation, will generally not be justifiable, however, because this number can be considerably different from the exact value even when n is large. For example, when both the proportion and the degree of confidence are 99%, the exact value is 3.043 when $n = 120$, so obviously there can be a considerable difference between the exact value and 2.58, even when n is large.

7.1.2 Two-Sided Interval, Possibly Unequal Tails

A symmetric two-sided interval is appropriate if the costs or consequences of being outside the interval are approximately equal. This won't always be the case, however. Therefore, it will often be desirable to construct the interval in such a way that, with a given degree of confidence, there is a desired maximum proportion of the population in the upper tail and a different desired maximum proportion in the lower tail. This requires the use of a different table.

To illustrate, consider the condenser example in Section 7.1.1 and assume that we want to construct a tolerance interval with 99% confidence that the interval will produce a maximum proportion of 0.5% in the upper tail and a maximum proportion of 2.5% in the lower tail. That is, we are more concerned about the upper tail than about the lower tail.

Using, for example, Table A.11b from Hahn and Meeker (1991), we obtain the limits as: lower limit $= \bar{x} - 3.67s$ and upper limit $= \bar{x} + 3.930s$.

Note that we would use 3.930 as the multiplier for both the upper and lower limits if we wanted to have a maximum 0.5% in each tail, and that this is different from the constant (3.742) that was used in Example 7.1 in which a 99% tolerance interval was constructed with a confidence coefficient of 99%—the same value as is used in this example. Since a maximum of 0.5% in each tail means that the interval will contain at least 99% of the population, it may seem counterintuitive that the multipliers differ since a symmetric distribution is assumed. The value of $\bar{x}$ will almost certainly not be equal to μ, however, so the interval will not be symmetric relative to the tails of the population distribution.

For example, if $\bar{x}$ were much greater than μ, the percentage of observations below the lower limit could be much larger than the percentage of observations above the upper limit. Therefore, it logically follows that in order to ensure that the specified maximum proportion of the population in each tail is less than the desired proportion, the tabular value must be larger than the tabular value that is used when the individual tails are not considered, as in the first example.

7.1.3 One-Sided Bound

Often a one-sided tolerance bound will be what is needed, such as when a company is faced with regulatory requirements. As stated earlier, there are federal regulations relating to food products that have a stated net contents weight, with regulations specifying the maximum proportion of units that can be a certain amount below the nominal net contents, as well as regulations that specify the allowable fraction of units that can be under the stated weight. A one-sided tolerance bound could be used to provide the desired information. Similarly, in civil engineering and building construction, materials and structures should exceed the minimum acceptable stress.

Probably the most useful approach to the food products problem would be to obtain a confidence interval on the proportion of units that are below a regulatory figure, and then see if the interval contains the allowable proportion. If not, the company would not be in compliance.

■ **EXAMPLE 7.2**

Problem Statement

A one-sided tolerance bound could also provide useful information, however. To illustrate, assume that 1% of manufactured units can be less than 15.8 ounces for a product with a stated weight of 16 ounces. A sample of 50 units is obtained, with $\bar{x} = 15.9$ and $s = 0.18$.

Confidence Bound Construction and Decision

Assume we want to construct a bound such that we are 99% confident that at least 99% of the values in the population exceed it. Using Table A.12b of Hahn and Meeker (1991), we find that the bound is $15.9 - 3.125(0.18) = 15.34$. Thus, the company has a problem as these sample data do not suggest compliance.

Would taking a larger sample help? Not necessarily, although of course we would increase the precision of our estimates of μ and σ. Evidence of compliance could come from either a sufficiently larger value of $\bar{x}$ or a sufficiently smaller value of s, or some combination of the two.

If an upper bound were desired with these data, with the same coverage and confidence level, the bound would have been computed as $15.9 + 3.125(0.18) = 16.46$. ■

One-sided bounds can also be produced using the `TOLINT` macro in MINITAB. For example, an upper bound can be produced by using `UPPER` as the subcommand, and a lower bound can be obtained by using `LOWER` as the subcommand, perhaps in tandem with other subcommands as discussed in Section 7.1.1.

One-sided tolerance intervals for populations with a normal distribution are based on the noncentral-t distribution, which is beyond the scope of this text, as is the theory behind two-sided intervals. The appropriate theory is given by Odeh and Owen (1980), which is out of print, however, as stated previously.

7.2 TOLERANCE INTERVALS AND SIX SIGMA

We can relate tolerance intervals to the Six Sigma movement in the following way. Widely available tables for the construction of tolerance intervals do not provide for more than 99% coverage, and for a normal distribution this corresponds to 2.58 sigma. Since process capability is computed using engineering tolerances (specifications), we might compute a tolerance interval with 99% confidence that covers at least 99% of the population. Then, if this interval is far inside the engineering tolerances, we know that there is good process capability for whatever is being measured.

7.3 DISTRIBUTION-FREE TOLERANCE INTERVALS

Since normality does not exist in practice, what happens if the distribution is not close to being normal? The higher the degree of confidence and the larger the percentage of population values that the interval is to contain, the larger the multiplier of s. Thus, if we want a high degree of confidence and a high confidence level, we are going far out into the tail of *some* distribution. If the distribution has heavier tails than a normal distribution and we believe we are constructing a 99% confidence interval to contain at least 99% of the population, the percentage of the time that, say, 1000 randomly constructed intervals actually contain at least 99% of the population could be far less than 990 (i.e., 99%). Conversely, if the population distribution has much lighter tails than a normal distribution, the interval could be much wider than necessary.

For skewed distributions, if one tail is light, then the other one must be heavy, so the errors can be somewhat offsetting for a two-sided tolerance interval with equal tails. This is not the case for a one-sided tolerance bound, however, which could be seriously in error, relative to the assumed properties, for a highly skewed distribution; nor would it be the case for a two-sided interval constructed with unequal tails, as in Section 7.1.2.

Therefore, it is very important to check the normality assumption before constructing a tolerance interval. If a test for normality leads to rejection of normality, the user can either try to transform the data or use a distribution-free procedure.

When a distribution-free approach is used, the interval is determined from the order statistics (i.e., the sample observations ordered from smallest to largest). The tables that are used specify the number of ordered observations that are to be removed from each end, with the extreme values that remain forming the tolerance interval.

For example, if $n = 50$ and we want a 95% interval to include at least 99% of the population, we would, following Table A.16 of Hahn and Meeker (1991), delete either one observation from the low end and nothing from the top end, or delete one from the top end and nothing from the bottom end. It is worth noting that this is the same action that would be taken for any value of n up to 400 (going up in increments of 100) and would additionally apply for $n \leq 60$ whenever the coverage is at least 95% and the degree of confidence is 90%, 95%, or 99%.

We should bear in mind, however, that we will generally not be able to achieve the desired degree of confidence unless we have a large sample, and in many cases we will miss it by a huge margin. For example, if we wanted a 95% interval to contain at least 99% of the population for $n = 30$ and we delete one observation from either end, the degree of confidence would actually be only 26.03%. If $n = 300$, however, the degree of confidence would be 95.1%, which is almost exactly equal to the nominal value.

If we wanted to do so, we could turn it around and solve for n, rather than accepting whatever degree of confidence results for a fixed value of n. For example, if we wanted a 95% interval that contained at least 95% of the population, we see from Table A.17 of Hahn and Meeker (1991) that the smallest sample size that will accomplish this is $n = 93$.

Because we can't necessarily come even remotely close to the desired degree of confidence for a small value of n, it would clearly be preferable to solve for n from tables such as Table A.17 and then decide whether or not the required sample size is practical. Since we would probably seldom want to dip below either 95% for the degree of confidence or the percent coverage, for all practical purposes a practitioner should think about constructing tolerance intervals from a sample of 100 or so observations.

One thing to keep in mind about distribution-free tolerance intervals is that there is no mechanism for selecting a combination of the degree of confidence and the percent coverage for a fixed sample size. Therefore, the selection of this combination is arbitrary. For example, using Table A.17 of Hahn and Meeker (1991), if we used $n = 89$ and used the largest and smallest observations to form the interval, we could view it as either a 75% interval that covers at least 97% of the population, or a 99.9% interval that covers at least 90% of the population. Obviously, there is a huge difference in these two combinations.

If this arbitrariness is disturbing to the user, which would be understandable, the alternative would be to try to transform the data to approximate normality by using, for example, a Box–Cox (1964) transformation and then construct a parametric interval based on the assumption of (approximate) normality. [This transformation method is presented in many books, including Ryan (1997).] Because of the sensitivity of normality-based tolerance intervals to nonnormality, something less than strong evidence of approximate normality after a transformation should lead the user back to considering a distribution-free interval.

Another point worth noting is that tolerance intervals computed from small samples will have a large variance, so the actual percentages in one application may differ considerably from the nominal values that would be the average percentages with the averaging done over a large number of samples.

Unless n is quite small and/or the observation that would be removed is an outlier, the tolerance interval should not differ very much from the interval that would result from simply using the largest and smallest observations in the sample. Of course, software can easily be used to order the observations and delete one of them, but we might argue that such an approximation could be justified on the grounds that when only one observation is to be deleted, the decision to delete it from either the top end or the bottom end is of course arbitrary. For other cases, such as when the desired coverage is at least 90%, it would be best to use the appropriate table, such as Table A.16 in Hahn and Meeker (1991).

Of course, when a one-sided bound is used, there is no arbitrariness as the single observation would be removed from the low end if a lower bound were to be obtained, and from the high end if an upper bound were to be produced. In this case, it would also be appropriate to use the table rather than simply use the smallest number in the sample.

For more information on tolerance intervals, see Hahn and Meeker (1991) and Patel (1986); see Jilek (1981) and Jilek and Ackerman (1989) for tolerance interval bibliographies.

7.3.1 Determining Sample Size

A rather large sample will be required for a two-sided tolerance interval if the confidence level is to be at least 95% and the proportion of covered items is to be at least 95%. In fact, for the "95 and 95" combination, $n = 93$ is required, as stated previously. This is one reason why a one-sided tolerance bound should be used whenever possible, as $n = 59$ is required for the same 95-and-95 combination. The necessary tables are given in various sources, including Hahn and Meeker (1991).

7.4 PREDICTION INTERVALS

W. Edwards Deming (1900–1993) stated "The only useful function of a statistician is to make predictions, and thus to provide a basis for action" (Wallis, 1980). Deming spoke similarly in distinguishing enumerative studies from analytic studies and stressing the importance of the latter. His point was that statistics should be used for improvement, which of course means moving forward in time and trying to make predictions and forecasts so that corrective action can be taken if specific future conditions are predicted to be something other than what is desired.

Students in statistics classes generally encounter prediction intervals in regression analysis (only), but we can construct a prediction interval without using regression (or forecasting in time series).

All of us use prediction intervals in one form or another. Consider the person who states: "I expect to get at least two more years out of my car, but I'll probably have to junk it not long thereafter." That is a lower prediction bound with an upper bound that is unspecified, although obviously not much higher than the lower bound. As another example, a family starts out on a vacation trip and the father states: "I believe we can make Denver in 8 hours, maybe 7." This is a prediction interval (i.e., 7–8). As another example, this time of a prediction without a prediction interval, assume that a flight is slightly late departing the gate, and the captain later states something like "we do have a tailwind so it looks like we will arrive at the gate at about 7:30." Or how about the executive who calls his wife and states: "Honey, I'll be home around 7:15." Will we ever hear a captain state: "My 95% prediction interval for when we will be at the gate is 7:26 to 7:34"? Not likely, of course, but the need for predictions and prediction intervals are all around us.

Whenever we construct a prediction interval (and this also applies to confidence intervals and tolerance intervals and to statistical inference, in general), we assume that the relevant processes will not change over time and, in particular, will not go out of statistical control (see Chapter 11). For the first example, the person who expects "to get at least two more years out of my car" might have to amend that prediction if there were a subsequent move from, say, a non-mountainous area to an area with steep hills, or to Florida, where rust can be a serious problem. The parametric methods for analytic studies given in this chapter and in the book do not apply when processes change; the nonparametric (distribution-free) methods would also not apply because they are similarly based on the assumption of a stable distrubution.

In a manufacturing context, a prediction interval will be of particular interest to a consumer, who will want to know how long one particular unit will last; namely, the one that he/she bought! A manufacturer, on the other hand, will be interested in average performance and thus will favor confidence intervals on parameters that relate to average performance.

In civil engineering, it is obviously important to be able to predict concrete strength, and prediction intervals are used in this context. One relatively recent application is described by Qasrawi (2000), who gave charts of 95% prediction intervals when the rebound hammer and ultrasonic pulse velocity tests were used in the study.

7.4.1   Known Parameters

The examples of everyday predictions given in the preceding section undoubtedly have a basis in fact, and countless more examples could be given, with the quantitative basis for the statements undoubtedly being shaky—if there is any quantitative basis at all—in many instances.

In this section we consider prediction intervals for one or more future observations, using statistical methodology. Recall the statement made in Section 7.1 that, if we had a normal distribution μ with σ and known, then we know that 95% of the observations are contained in the interval $\mu \pm 1.96\sigma$. Similarly, we could state that we are 95% confident that the next observation we observe will be contained in this interval. Thus, in the parameters-known case, a 95% tolerance interval and a 95% prediction interval are the same, although their interpretation is of course quite different.

Because of the close connection between tolerance intervals and prediction intervals, they are connected (and potentially confused) in the literature, as what has been termed a β-expectation tolerance interval is really a $100\beta\%$ prediction interval for a single observation. [The term "β-expectation" has been used by Guttman (1970), in particular.]

The general interpretation and view of a prediction interval is essentially the same as for a confidence interval. That is, if we constructed one hundred 95% prediction intervals, our best guess as to the number that will contain the next observation is 95, although, as with a confidence interval, the number probably wouldn't equal 95.

7.4.2 Unknown Parameters with Normality Assumed (Single Observation)

Of course, the parameters generally won't be known, so we need an expression for a prediction interval for that case, and of course the prediction interval will be a function of a t-value. The appropriate expression can be obtained by first recalling from Section 3.4.4 that the t-distribution results from dividing a $N(0, 1)$ random variable by the square root of a chi-square random variable divided by its degrees of freedom, with the normal and chi-square random variables being independent.

A new observation, x, will be independent of $\bar{x}$ computed from a sample, so $Var(x - \bar{x}) = Var(x) + Var(\bar{x}) = \sigma^2 + \sigma^2/n = \sigma^2(1 + 1/n)$. If we assume that the individual observations have a normal distribution, then $(x - \bar{x})/\sigma\sqrt{(1 + 1/n)}$ is $N(0, 1)$. Since $(n - 1)s^2/\sigma^2$ is χ^2_{n-1}, we then have

$$t_{n-1} = \frac{\dfrac{(x - \bar{x})}{\sigma\sqrt{(1 + 1/n)}}}{\sqrt{\dfrac{(n - 1)s^2/\sigma^2}{n - 1}}}$$

$$= \frac{(x - \bar{x})}{s\sqrt{(1 + 1/n)}}$$

with the t-statistic having $(n - 1)$ degrees of freedom because the chi-square component of the expression before it is simplified has $(n - 1)$ degrees of freedom. It then follows that

$$P\left(-t_{\alpha/2,n-1} \leq \frac{(x - \bar{x})}{s\sqrt{(1 + 1/n)}} \leq t_{\alpha/2,n-1}\right) = 1 - \alpha$$

Recall that this corresponded to the starting point for confidence interval construction, for which the objective was to manipulate the double inequality so that the parameter would be in the middle. Here we want x to be in the middle and with the necessary algebra we obtain

$$P\left(\bar{x} - t_{\alpha/2,n-1}s\sqrt{(1 + 1/n)} \leq x \leq \bar{x} + t_{\alpha/2,n-1}s\sqrt{(1 + 1/n)}\right) = 1 - \alpha \qquad (7.1)$$

as the expression for the $100(1 - \alpha)\%$ prediction interval. A $100(1 - \alpha)\%$ upper bound would then be given by the expression $\bar{x} + t_{\alpha,n-1}s\sqrt{(1 + 1/n)}$ and the lower bound would be given by the expression $\bar{x} - t_{\alpha,n-1}s\sqrt{(1 + 1/n)}$.

7.4.2.1 Sensitivity to Nonnormality

Whenever we construct a prediction interval for a single observation, it is imperative that there not be more than a slight deviation from the assumed distribution. Even though the distribution of $\bar{X}$ approaches normality as $n \to \infty$, the prediction interval approaches $\mu \pm Z_{\alpha/2}\sigma$, which of course is the form of a tolerance interval when the parameters are known. Thus, if the distribution is asymmetric or has heavier or lighter tails than a normal distribution, $\pm Z_{\alpha/2}$ is then inappropriate, and could differ considerably from what should be used instead. Thus, even very large samples can be of no help in regard to nonnormality.

Accordingly, Eq. (7.1) should not be used unless the population distribution is apparently very close to a normal distribution, so a test for normality must be used. Of course, "very close" and "close enough" are subjective determinations in the practice of statistics, as there are no precise guidelines for tolerance intervals or prediction intervals, or for statistical methods in general. If, however, there is clear evidence, such as from a normal probability plot, that the assumption of normality is more than slightly violated, one possibility would be to try to transform the data to approximate normality by using, for example, the Box–Cox approach, as was discussed for tolerance intervals in Section 7.3. Alternatively, or if the Box–Cox approach is unsuccessful, a distribution-free approach might be used (Section 7.4.7).

■ EXAMPLE 7.3

Objective and Data

Hahn and Meeker (1991, Table 13.1, p. 256) gave an example of a confidence bound for the probability of a manufactured engine passing an emissions test, with the bound to be determined from 20 historical engine emissions measurements of a particular pollutant. Those measurements were, in parts per million: 73.2, 70.9, 67.1, 74.2, 67.8, 65.4, 69.2, 69.1, 68.5, 71.2, 66.5, 64.0, 73.8, 72.4, 72.9, 68.9, 69.3, 69.6, 75.4, and 70.2. Assume that our objective is to use these data to construct an upper prediction bound for the next manufactured engine to be tested.

First, the data should be appropriately checked for more-than-slight nonnormality. If the data pass the test, the bound can then be computed. The reader is asked to pursue this in Exercise 7.4.   ■

7.4.2.2 Width of the Interval

One thing to note about the width of the prediction interval given by Eq. (7.1) is that, unlike a confidence interval, the width cannot be made arbitrarily small. What is gained by using a large sample size is a better estimate of σ and $\sigma_{\bar{x}}$ is made small. But σ is independent of the sample size that is used so the lower bound on the width is $2t_{\alpha/2,n-1}\widehat{\sigma}$ with $\widehat{\sigma} = s$. If $\widehat{\sigma}_{\bar{x}}$ is small for a selected value of n, the benefit from using a larger value of n would be quite small. In trying to determine a sample size, it would thus be reasonable to consider the width of $2t_{\alpha/2,n-1}\widehat{\sigma}\sqrt{1+1/n}$ relative to the lower bound of $2Z_{\alpha/2}\widehat{\sigma}$, as has been suggested in the literature. The U.S. Environmental Protection Agency (1992) has specified that at least eight samples (observations) be taken.

7.4.3 Nonnormal Distributions: Single Observation

If a normal probability plot or other test for normality suggests that the population distribution is markedly nonnormal, the user has three options: (1) try to fit another distribution to the data and hope that an adequate fit can be obtained; (2) attempt to transform the data to approximate normality; or (3) use a distribution-free approach

Regarding the second option, common transformations such as log, reciprocal, and square root transformations are often effective, as are power transformations in general for which the data are raised to a power between 0 and 1. Transformations are not an exact science, however, and expertise in using them comes with experience.

Prediction intervals for specific nonnormal distributions are generally not given in textbooks, and indeed the development of such intervals can be complicated. Papers in which such methods have been given are described in Chapter 11 of Hahn and Meeker (1991). See also Patel (1989) and Patel and Samaranayake (1991). Methodology given by the latter is presented in the next section.

7.4.4 Nonnormal Distributions: Number of Failures

Nelson (2000) considered the problem of predicting the number of equipment failures in a future time period based on the number of failures observed in a previous inspection, and provided prediction limits under the assumption that the failure times were independent observations from a population with a Weibull distribution. Nordman and Meeker (2002) subsequently considered the same scenario and provided approximate prediction intervals.

Nelson's work was motivated by the following problem. Nuclear power plants have large heat exchangers that transfer energy from the reactor to steam turbines. The exchangers typically have 10,000–20,000 stainless steel tubes that conduct the flow of steam. The tubes develop cracks over time due to stress and corrosion and are removed from service. Plant management wanted a prediction of the additional number of tubes that would fail by a specified time, and of course a prediction interval would be superior to a point estimate since the interval would incorporate the standard error of the point estimator used to obtain the prediction.

7.4.5 Prediction Intervals for Multiple Future Observations

There is often a need to construct a prediction interval for more than one future observation, including the average of a group of future observations. The need for such an interval can arise when a manufacturer needs to predict for a shipment of m items the average value of some product characteristic (e.g., voltage, tensile strength). Equation (7.1) must be modified slightly to produce such an interval, with the radicand in that equation replaced by $1/m + 1/n$.

Unlike prediction intervals for a single observation, these intervals are not sensitive to the assumption of normality unless either the group of future observations is small or the sample used in constructing the interval is small. Tables are available (e.g., in Hahn and Meeker, 1991) that allow such intervals to be constructed easily by providing the multiplier of s, although the slightly more involved computations will of course be needed for values of n not covered by tables.

In addition to a prediction interval for a single future observation and for the average from a sample of future observations, it is also possible to construct a prediction interval to contain all or some of the observations in a future sample. See Hahn and Meeker (1991) for details.

7.4.6 One-Sided Prediction Bounds

A one-sided prediction bound is of particular interest to manufacturers of products that have a warranty, just as a one-sided tolerance bound is of interest. Such manufacturers naturally want their products to be trouble-free during the warranty period.

Analogous to a one-sided confidence bound, a one-sided prediction bound can be obtained through appropriate modification of the expression for a two-sided prediction interval. For example, a $100(1 - \alpha)\%$ lower prediction bound for a single future observation would be obtained as $\bar{x} - t_{\alpha,n-1}s\sqrt{(1 + 1/n)}$, with other types of one-sided prediction bounds similarly obtained from the appropriate expression for the two-sided interval.

Neither one-sided nor two-sided prediction intervals are available in MINITAB and in general are not widely available in statistical software.

7.4.6.1 One-Sided Prediction Bounds for Certain Discrete Distributions

Patel and Samaranayake (1991) gave conservative one-sided prediction bounds for the binomial, Poisson, hypergeometric, and negative binomial distributions. (The latter distribution was mentioned briefly in Section 3.3.5.) Their approach entails numerical work, however, rather than simple expressions of the type that are given in Section 7.4.2. Since simple methods are the ones that are used most often, it would be necessary to either write appropriate code or search for simple expressions that would be satisfactory substitutes for the numerical approach.

7.4.7 Distribution-Free Prediction Intervals

As with distribution-free procedures in general (as discussed in Chapter 16), it is usually better to use a distribution-free (nonparametric) procedure than to use the wrong parametric procedure, but any distribution-free procedure is inferior to the corresponding parametric (distribution-specified) procedure when the latter is appropriate. Distribution-free prediction intervals are discussed in this section.

As the reader might guess, distribution-free prediction intervals are wider than their parametric counterparts because of the additional uncertainty caused by not knowing the distribution. This extra uncertainty can cause the degree of confidence to be lower than desired for a stated sample size, or can require a very large sample in order to attain a desired level of confidence.

Since no probability distribution is used, an interval must be based solely on the sample data. Accordingly, there are two approaches to obtaining a distribution-free prediction interval: (1) construct the interval using the extreme values in a sample with the sample size selected so as to produce a desired degree of confidence; or (2) determine which ordered values in a sample of predetermined size should be used to construct a prediction interval that has an approximate desired degree of confidence.

The theory involved in the construction of distribution-free prediction intervals is based on the theory of order statistics (as in the case of tolerance intervals), which is beyond the intended level of this text. Consequently, examples and heuristic explanations will be used rather than the more rigorous presentation that was used for normality-based prediction intervals.

Consider the first stated option for obtaining the interval. If we wanted a 99% interval formed from the extreme values of the sample, how large a sample should be taken? Obviously, the sample cannot be small because a small sample does not have extreme sample percentiles. For example, there is clearly not a 99th percentile of a sample of size 10. We would expect the answer to be approximately 200 because $1/200 = .005$ and $2(.005) = .01 = 1 - .99$. In fact, the answer is 199, as given in Table A.20a of Hahn and Meeker (1991). This result can also be obtained by solving the equation $(n - 1)/(n + 1) = 1 - \alpha$ for n.

If we wanted a one-sided prediction bound, we would expect the necessary sample size to be half the sample size needed for the two-sided interval. The value of n is obtained by solving the equation $n/(n + 1) = 1 - \alpha$. Thus, the necessary sample size for a 99% bound is 99, as is also given in Table A.21a of Hahn and Meeker (1991).

For the second option, one solves for the order statistic(s) in the existing sample so as to have a two-sided interval or one-sided bound with an approximate desired degree of confidence. Although this approach may sometimes be necessary if freedom of sample size determination is considerably restricted, it could be difficult to obtain a confidence level close to the desired value if the sample is at most of moderate size. (Recall from Section 7.3 that this problem also exists for distribution-free tolerance intervals.)

7.5 CHOICE BETWEEN INTERVALS

The type of interval that is appropriate for a particular scenario depends on the objective. Hahn (1998, p. 5.11) gave a good illustration of this relative to a miles-per-gallon problem for a particular type of automobile under specified conditions. He stated that if interest centered on the average gasoline consumption, then a confidence interval for the mean gasoline consumption would be appropriate. If, however, a manufacturer must pay a penalty for each automobile whose gasoline consumption is below a specified value, a lower tolerance bound should be constructed. A consumer with the intention of purchasing a single automobile of this type will have only passing interest in these two intervals, however, and would be interested primarily in a prediction interval for the gasoline consumption of a future automobile and, in particular, the automobile that the consumer plans to purchase.

Unfortunately, just as these various tolerance and prediction intervals have been neglected by textbook authors, they have also been generally ignored by software companies. This of course makes it hard to use available methods and thus limits usage and hence the popularity of the methods. Users do have some options, however. In particular, Hahn and Meeker (1991) gave Fortran subroutines for the various distribution-free intervals in their book. Tolerance intervals can also be computed using DATAPLOT (freeware), as is illustrated in Section 7.2.6.3 of the *NIST/SEMATECH e-Handbook of Statistical Methods* (http://www.itl.nist.gov/div898/handbook/prc/section2/prc263.htm), which provides the capability for constructing tolerance intervals and prediction intervals for certain nonnormal distributions. In addition to the tables given in Hahn and Meeker (1991), tables for both parametric and distribution-free intervals can also be found in Natrella (1963)—the forerunner of the *e-Handbook of Statistical Methods*.

It should be borne in mind that sometimes the best choice will be "no interval," as will be the case when there is no clear evidence that processes are in a state of statistical control (see Chapter 11). Along these lines, Hahn and Meeker (1991, p. 286) stated "Despite our enthusiasm for statistical intervals, we feel that there are numerous situations where the practitioner is better served by not calculating such intervals and by emphasizing instead the limitations of the available information, perhaps suggesting how improved data can be obtained."

7.6 SUMMARY

In order to be motivated to search for something, one has to know that it exists and engineering students (and students in other disciplines) have not seen much material on tolerance intervals or prediction intervals in statistics texts. This is unfortunate because there are many potential applications of these intervals in engineering, and the examples given in this chapter provide some indication of the diversity of applications. If we agree with the Deming quote in Section 7.4 (and certainly many would disagree), we would place more emphasis on prediction intervals than one sees in statistics texts. Of course, Deming meant that statistics should be used to provide some insight into what we can expect in the future unless the current state of affairs is altered in possibly necessary ways, and thus to make appropriate changes, if needed. When we make predictions we generally need to include the uncertainty of our predictions, which leads to prediction intervals. Furthermore, the use of lower prediction bounds can be motivated by manufacturers' warranties.

Tolerance intervals can also be motivated by, in particular, federal regulations regarding deviations from the stated contents of food products and other items.

Parametric tolerance intervals and prediction intervals are preferable when one can assume at most a slight departure from an assumed model, with nonparametric (distribution-free) intervals being useful when the appropriate distribution is not at all clear. The fact that a specific combination of degree of confidence and percent coverage of the population cannot be made with a distribution-free interval should be kept in mind, so a normality-based interval may be preferable if a transformation to approximate normality can be achieved.

Recall Figure 1.1, which depicted the sequential nature of experimentation. The construction of statistical intervals (including confidence intervals) should also be viewed as an iterative process, especially when not all processes relevant to the construction of a particular type of interval are in a state of statistical control. Furthermore, even if control has been attained, it might still be possible to reduce variability, which in turn would reduce the width of a tolerance interval or a prediction interval (or a confidence interval).

There is much more to tolerance intervals and prediction intervals than has been presented in this survey, and readers are encouraged to consult Hahn and Meeker (1991) and other references such as Hahn (1970, 1998) and Hahn and Nelson (1973) for additional information, including appropriate tables and additional applications.

REFERENCES

Box, G. E. P. and D. R. Cox (1964). An analysis of transformations. *Journal of the Royal Statistical Society B*, **26**, 211–243 (discussion: pp. 244–252).

Environmental Protection Agency (1992). *Environmental Protection Agency, Statistical Training Course for Ground-Water Monitoring Data Analysis*, EPA/530-R-93-003, Washington, DC: Office of Solid Waste. (600-page document available on the Web with extremely long URL)

Guttman, I. (1970). *Statistical Tolerance Regions: Classical and Bayesian*. Griffin's Statistical Monographs and Courses, No. 26. London: Griffin.

Hahn, G. J. (1970). Statistical intervals for a normal population. Part I. Tables, examples, and applications. *Journal of Quality Technology*, **2**(3), 115–125.

Hahn, G. J. (1998). Statistical intervals. Chapter 5 in *Handbook of Statistical Methods for Engineers and Scientists* (H. M. Wadsworth, ed.). New York: McGraw-Hill.

Hahn, G. J. and W. Nelson (1973). A study of prediction intervals and their applications. *Journal of Quality Technology*, **5**(4), 178–188.

Hahn, G. J. and W. Q. Meeker (1991). *Statistical Intervals: A Guide for Practitioners*. New York: Wiley.

Howe, W. G. (1969). Two-sided tolerance limits for normal populations—some improvements. *Journal of the American Statistical Association*, **64**, 610–620.

Jilek, M. (1981). A bibliography of statistical tolerance regions. *Statistics*, **12**, 441–456.

Jilek, M. and H. Ackerman (1989). A bibliography of statistical tolerance regions, II. *Statistics*, **20**, 165–172.

Natrella, M. G. (1963). *Experimental Statistics, Handbook 91*. Washington, DC: National Bureau of Standards, U.S. Department of Commerce.

Nelson, W. (2000). Weibull prediction of a future number of failures. *Quality and Reliability Engineering International*, **16**, 23–26.

Nordman, D. J. and W. Q. Meeker (2002). Weibull prediction intervals for a future number of failures. *Technometrics*, **44**(1), 15–23.

Odeh, R. E. and D. B. Owen (1980). *Tables for Normal Tolerance Limits, Sampling Plans, and Screening*. New York: Marcel Dekker.

Patel, J. K. (1986). Tolerance intervals—a review. *Communications in Statistics—Theory and Methods*, **15**(9), 2719–2762.

Patel, J. K. (1989). Prediction intervals—a review. *Communications in Statistics—Theory and Methods*, **18**, 2393–2465.

Patel, J. K. and V. A. Samaranayake (1991). Prediction intervals for some discrete distributions. *Journal of Quality Technology*, **23**(4), 270–278.

Qasrawi, H. Y. (2000). Concrete strength by combined nondestructive methods—simply and reliably predicted. *Cement and Concrete Research*, **30**(5), 739–746.

Ryan, T. P. (1997). *Modern Regression Methods*. New York: Wiley.

Vardeman, S. B. (1992). What about the other intervals? *The American Statistician*, **46**, 193–197.

Wallis, W. A. (1980). The Statistical Research Group, 1942–45. *Journal of the American Statistical Association*, **75**(370), 321.

EXERCISES

7.1. Which of the intervals presented in this chapter, if any, are of the general form $\widehat{\theta} \pm t s_{\widehat{\theta}}$? What is an interval called that is of this form?

7.2. Assume that a certain type of ball bearing has a diameter that is $N(2.4, 0.36)$. Construct a 99% prediction interval for the diameter of the next ball bearing that is produced. Is this also a tolerance interval? Explain.

7.3. When will a prediction interval and a tolerance interval be the same and when will they differ?

7.4. Consider Example 7.3 in Section 7.4.2.1. Determine if the data could have come from a population with a distribution that is approximately normal. If so, construct a 95% upper bound for the next engine to be tested. Since the reading must be below 75, it is desirable to have the upper bound be under 75. Is that the case here if the upper bound could be constructed? Explain. If the degree of nonnormality appears to be more than just slight, what approach would you recommend? Proceed with that approach, if necessary.

7.5. Will a prediction interval for a single observation always be wider than a confidence interval for μ (using the same data)? Explain.

7.6. Consider Example 7.1 in Section 7.1.1. If a 99% prediction interval were constructed using the same data as were used to construct the tolerance interval, which interval would be wider?

7.7. Which of the following intervals would not be appropriate for compliance monitoring: (a) confidence interval for μ, (b) prediction interval, (c) tolerance interval?

7.8. Which of the following will *not* be sensitive to nonnormality when an extremely large sample is taken and normality is assumed: (a) tolerance interval, (b) confidence interval, (c) prediction interval? Explain.

7.9. A steel mill has been asked to produce high-strength low-alloy steel (H.S.L.A.) that has a minimum yield strength of 60,000 psi. As the steel is produced, each coil is tested for strength. The product is a thin sheet steel that is several feet wide and several thousand feet long. The strip has been rolled into coils to make them easier to handle. The production department is anxious to send out a large shipment to a particular customer. There isn't time to test every item but you are given the task of determining if the minimum yield strength requirement has been met for each item. Would you use a method given in this chapter for addressing this issue? If so, how would you proceed? If not, explain what you would do.

7.10. Assume that you are an admissions officer at a university and you are using statistical methods to help decide whether or not to admit each student (as many universities actually do). In terms of student grades, which would be the more useful, a prediction interval or a tolerance interval? If you used a prediction interval, why would you not want to use the prediction interval given by Eq. (7.1)?

7.11. Explain to someone unfamiliar with tolerance intervals the difference between a 99% tolerance interval that has a confidence coefficient of 95% and a 95% tolerance interval that has a confidence coefficient of 99%.

7.12. A sample of 100 observations is taken, with $\bar{x} = 25.8$ and $s = 5.7$. An experimenter wishes to construct a 99% tolerance interval that has a confidence coefficient of 95%. What is the first step that should be taken?

7.13. A toothpaste was formulated with a silica content of 3.5% by weight. A manufacturer randomly samples 40 tubes of toothpaste and finds that the average silica content is 3.6% with a standard deviation of 0.28%. You are an employee of this company and you are asked to construct a 95% prediction interval, using this information (only). Could you do so using methodology presented in this chapter? If so, compute the interval; if not, explain why not.

7.14. Write a program (using MINITAB or other software) to generate 100 random samples of size 100 from a population that has a standard normal distribution. Then use the 100 samples to construct one hundred 90% tolerance intervals to contain at least 95% of the population.

(a) Did any of the intervals contain exactly 95% of the population?

(b) If not, should we say, generally, that a tolerance interval designed to contain at least 95% of the population will actually contain more than 95% in the same way that $P(X \geq 1.96) = P(X > 1.96)$ when X has a standard normal distribution? Explain.

7.15. Very large amounts of data are now being routinely maintained in datafiles. Assume that a company has such a file for a manufactured part, with the distribution being approximately normal with a mean of 0.82 inch and a standard deviation of 0.06 inch.

   (a) What would be the endpoints of an approximate 99% tolerance interval?

   (b) Why would this interval not have a degree of confidence attached to it, considering the information given in the problem?

7.16. A company decides to construct a 95% distribution-free prediction interval for a future observation that will be constructed from the largest and smallest observations in a sample that will be obtained.

   (a) Determine the sample size that is necessary to accomplish this.

   (b) What would be the required sample size if a decision were made to use a one-sided bound instead of the two-sided interval?

7.17. Assume that a company constructed a 95% prediction interval for the number of future failures of a particular component by a specified time. Assume that the distribution of failure times is close to being symmetric and there is subsequently an increase in the mean number of failures after the interval has been constructed, due to a process control problem. What action should be taken regarding the interval that was constructed?

7.18. A 99% prediction interval is constructed under the assumption of normality, but a considerable amount of data on the random variable are later examined and the conclusion is that the actual distribution has heavier tails than a normal distribution. Will the actual degree of confidence of the constructed interval be greater than 99 or less than 99? Explain.

7.19. Assume that a 99% tolerance interval to contain at least 95% of a population with a normal distribution has been constructed and shortly thereafter quality improvement efforts were successful in reducing the standard deviation. How will this affect the percentage of the population covered by the interval and the degree of confidence? Or will only one of these be affected, or will neither be affected? Explain.

7.20. A practitioner constructs a distribution-free tolerance interval with $n = 65$ and with the intention of it being an interval with 95% confidence that covers at least 95% of the population. If we accept the latter, would the actual degree of confidence be (at least) 95% or not? Explain. (*Note*: A table isn't needed to answer this question.)

7.21. Is it possible to solve explicitly for the sample size in a prediction interval, given the desired width of the interval, the degree of confidence, and an estimate of sigma? Explain.

CHAPTER 8

Simple Linear Regression, Correlation, and Calibration

Regression analysis is one of the two most widely used statistical procedures; the other is analysis of variance, which is covered in Chapter 12 as well as in this chapter. Regression is an important engineering statistics tool and a regression approach is the standard way of analyzing data from designed experiments, such as those illustrated in Chapter 12.

Many practitioners would debate the importance of regression methods in engineering, however, contending that the model for the variables they are studying is often known, at least approximately, so that a data-analytic approach won't be needed in those situations. It is true that mechanistic models, as they are called, often do perform quite well. This issue is discussed further in Chapter 10, wherein the suggestion is made to combine an empirical approach with whatever knowledge may be available to suggest a particular mechanistic model.

8.1 INTRODUCTION

The word "regression" has a much different meaning outside the realm of statistics than it does within it. The literal definition of the word is to revert back (i.e., "regress") to a previous state or form. In the field of statistics, the word was coined by Sir Francis Galton (1822–1911), who observed that children's heights regressed toward the average height of the population rather than digressing from it. This is essentially unrelated to the present-day use of regression, however, as regression is used primarily to predict something in the future, not to revert back to the past.

In this chapter we present regression as a statistical tool that can be used for (1) description, (2) prediction, (3) estimation, and (4) control.

8.2 SIMPLE LINEAR REGRESSION

In (univariate) regression there is always a single "dependent" variable, and one or more "independent" variables. For example, we might think of the number of nonconforming

Modern Engineering Statistics   By Thomas P. Ryan
Copyright © 2007 John Wiley & Sons, Inc.

units produced within a particular company each month as being dependent on the amount of time devoted to training operators. In this case the amount of time (in minutes, say) would be the single independent variable. This should not be construed to mean, however, that there is a cause-and-effect relationship. There might be such a relationship, but that is not assumed in regression analysis. Furthermore, a cause-and-effect relationship can be detected only through the use of designed experiments. The word *simple* is used to denote the fact that a single independent variable is being used.

Linear does not have quite the meaning that one would expect. Specifically, it does not necessarily mean that the relationship between the dependent variable and the independent variable is a straight-line relationship. The equation

$$Y = \beta_0 + \beta_1 X + \epsilon \tag{8.1}$$

is simply one example of a linear regression model, although this is the standard model. Notice how this closely resembles the equation for a straight line in algebra, with β_1 denoting the slope and β_0 representing the Y-intercept. The symbol ϵ represents the error term. This does not mean that a mistake is being made; it is simply a symbol used to indicate the absence of an exact relationship between the independent variable X and the dependent variable Y, as exact relationships do not exist in regression analysis.

Other examples of linear regression models with a single X include

$$Y = \beta_0' + \beta_1' X + \beta_{11} X^2 + \epsilon$$

and

$$Y = \beta_1^* X + \epsilon \tag{8.2}$$

A regression model is linear if it is linear in the parameters (the betas), and both of these last equations satisfy that condition.

DEFINITION

A *linear regression model* is a model that is linear in the parameters, regardless of the form of the predictors in the model.

We will note at the outset that the model given in Eq. (8.2) is rarely used, even though in some applications that might seem counterintuitive. For example, assume that X represents an input variable such as grams of a product and Y is an output variable that is related directly to the input and depends entirely on the input. If the input is zero, then the output will also be zero, so the model given in Eq. (8.2) would seem to be appropriate. But we generally are not interested in using a regression model when X is approximately zero, however, and to do so would generally require extrapolating beyond the range of X-values in the sample. Therefore, the model in Eq. (8.1) can be expected to almost always provide a better fit to the data for the range of X that is of interest. (In later sections and in Chapter 9 we will refer to X by any of its various labels: independent variable, predictor, or regressor.)

Regression analysis is not used for variables that have an exact linear relationship, as there would be no need for it. For example, the equation

$$F = \frac{9}{5}C + 32 \tag{8.3}$$

expresses temperature in Fahrenheit as a function of temperature measured on the Celsius scale. Thus, F can be determined exactly for any given value of C. This is not the case for Y in Eq. (8.1), however, as β_0 and β_1 are generally unknown and therefore must be estimated. Even if they were known [as is the case with the constants $\frac{9}{5}$ and 32 in Eq. (8.3)], the presence of ϵ in Eq. (8.1) would still prevent Y from being determined exactly for a given value of X.

It is important to recognize that a model such as Eq. (8.1) is simply an approximation to reality. Certainly, the world is nonlinear, so there is the need to determine if the linear regression model is an adequate approximation to reality for a particular application. Diagnostic tools are available for making such an assessment; these are discussed later in the chapter.

8.2.1 Regression Equation

The parameters β_0 and β_1 must be estimated. Various methods have been proposed for doing so, and the most frequently used method for doing so is given in Section 8.2.2. Once the parameters have been estimated, the equation

$$\widehat{Y} = \widehat{\beta}_0 + \widehat{\beta}_1 X \tag{8.4}$$

is called the *regression equation*. When it is to be used to predict future values of Y, it could be called the *prediction equation*, with the values of $\widehat{Y}$ the *predicted values*. (When the regression equation is used to compute the values of $\widehat{Y}$ to perhaps compare with the observed values of Y in the dataset that is used to estimate the parameters, the values of $\widehat{Y}$ are termed the *fitted values*.)

8.2.2 Estimating β_0 and β_1

Point estimates of β_0 and β_1 are needed to obtain the prediction equation given in general form in Eq. (8.4). A crude approach would be to draw a line through the center of the points and then use the Y-intercept and slope of the line as the estimates of β_0 and β_1, respectively.

Before one could even attempt to do so, however, it would be necessary to define what is meant by "center." We could attempt to minimize the sum of the slant distances (with each distance measured from the point to the line), but since we will be using X to predict Y, it would make more sense to try to minimize the sum of the vertical distances. If we use the signs of those distances we have a problem, however, because then the line is not uniquely defined (as the reader is asked to show in Exercise 8.4).

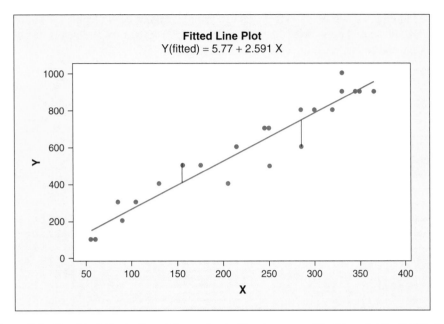

Figure 8.1 Graphical illustration of the method of least squares: fitted line plot for Y (fitted) $=$ 5.77 + 2.591 X.

The standard approach is to minimize the sum of the *squares* of the vertical distances, and this is accomplished by using the *method of least squares*. Figure 8.1 illustrates the general idea, with the line that is shown being the least squares fitted line, which minimizes the sum of the squared vertical distances of each point to the line.

For the purpose of illustration, we must assume that Eq. (8.1) is the correct model, although as stated previously, the correct model will generally be unknown.

The starting point is to write the model as

$$\epsilon = Y - (\beta_0 - \beta_1 X) \tag{8.5}$$

Since ϵ represents the vertical distance from the observed value of Y to the line represented by $Y = \beta_0 + \beta_1 X$ that we would have if β_0 and β_1 were known, we want to minimize $\sum \epsilon^2$. Recall that it was shown in Section 4.4.3 that this method minimizes the sum of the squares of the differences between the observations and their mean. As is discussed in Section 8.2.3, the mean of ϵ is assumed to be zero (i.e., the sum of the differences of the values on the right-hand side would be zero if the parameters were known). With this assumption, it follows that the mean of Y is $(\beta_0 + \beta_1 X)$, or properly stated as $\mu_{y|x}$, the conditional mean of Y given X since the mean depends on X. Thus, $\sum \epsilon^2 = \sum (Y - \mu_{y|x})^2$.

For convenience we define L as

$$L = \sum_{i=1}^{n} \epsilon_i^2 = \sum_{i=1}^{n} (Y_i - \beta_0 - \beta_1 X_i)^2 \tag{8.6}$$

with n denoting the number of data points in a sample that has been obtained. To minimize L, we take the partial derivative of L with respect to each of the two parameters that we are estimating and set the resulting expressions equal to zero. Thus,

$$\frac{\partial L}{\partial \beta_0} = 2 \sum_{i=1}^{n} (Y_i - \beta_0 - \beta_1 X_i)(-1) = 0 \tag{8.7a}$$

and

$$\frac{\partial L}{\partial \beta_1} = 2 \sum_{i=1}^{n} (Y_i - \beta_0 - \beta_1 X_i)(-X_i) = 0 \tag{8.7b}$$

Dropping the 2 and the -1 from Eqs. (8.7a) and (8.7b), the solutions for β_0 and β_1 would be obtained by solving the equations (which are generally called the *normal equations*)

$$\sum_{i=1}^{n} (Y_i - \beta_0 - \beta_1 X_i) = 0 \quad \text{and} \quad \sum_{i=1}^{n} (X_i Y_i - X_i \beta_0 - \beta_1 X_i^2) = 0$$

which become

$$n\beta_0 + \beta_1 \sum_{i=1}^{n} X_i = \sum_{i=1}^{n} Y_i \quad \text{and} \quad \beta_0 \sum_{i=1}^{n} X_i + \beta_1 \sum_{i=1}^{n} X_i^2 = \sum_{i=1}^{n} X_i Y_i$$

The solution of these two equations (for the estimators of β_0 and β_1) produces the least squares estimators

$$\widehat{\beta}_1 = \frac{\sum X_i Y_i - (\sum X_i)(\sum Y_i)/n}{\sum X_i^2 - (\sum X_i)^2/n} \tag{8.8}$$

and

$$\widehat{\beta}_0 = \overline{Y} - \widehat{\beta}_1 \overline{X} \tag{8.9}$$

For notational convenience we will denote the numerator of Eq. (8.8) by S_{xy} and the denominator by S_{xx}. Thus, $\widehat{\beta}_1 = S_{xy}/S_{xx}$. The fitted line that results from the use of these parameter estimates in Eq. (8.4), as depicted in Figure 8.1, minimizes the sum of the squared distances of the points to the line.

[The astute reader will recognize that L in Eq. (8.6) is not automatically minimized just by setting the first derivatives equal to zero and solving the resultant equations. It can be shown, however, that the determinant of the matrix of second-order partial derivatives of L is positive, thus ensuring that a minimum, rather than a maximum, has been attained.]

Consider the following simple example.

X	2	3	6	5	4
Y	6	4	5	3	1

Applying Eq. (8.8), which would have to be used first,

$$\widehat{\beta}_1 = \frac{83 - (20)(19)/5}{90 - (20)^2/5} = 0.7$$

Then, $\widehat{\beta}_0 = 4.8 - (0.7)4 = 2$. The fitted equation is then $\widehat{Y} = 0.7 + 2X$.

8.2.3 Assumptions

If we simply wanted to obtain the prediction equation and compute some measure of the worth of the model, we would need to assume only that the model is a good proxy for the true, unknown model. We almost always want to go beyond that point, however, and make various types of inferences (confidence intervals, etc.). In order to do so, we have to make a distributional assumption on the error term, ϵ, and the usual assumption is that ϵ has a normal distribution. Specifically, we assume that $\epsilon_i \sim NID(0, \sigma_\epsilon^2)$. That is, the errors are assumed to be normally distributed ("N"), independent ("ID"), with a mean of zero and a variance that is constant for each error term and also does not depend on any variable. It is important that the assumptions of normality, independence, and constant error variance be checked, as violation of these assumptions can cause problems, especially violation of the independence assumption.

8.2.4 Sequence of Steps

The first step in any regression analysis with a single predictor is to *plot the data*. This would be done to see if there is evidence of a linear relationship between X and Y as well as for other purposes such as checking for outlying observations (outliers). The importance of plotting the data cannot be overemphasized. Anscombe (1973) drove that point home with four sample sets he provided in his paper—all of which produced virtually the same regression equation but had completely different graphs. The reader is asked to examine and comment on the datasets in Exercise 8.1.

■ **EXAMPLE 8.1**

Objective

We will use a dataset given by T. Raz and M. Barad in their paper, "In-Process Control of Design Inspection Effectiveness" (*Quality and Reliability Engineering International*, **20**, 17–30, 2004) to emphasize the importance of first plotting the data. The objective was to model the defect escape probability (PROB) as part of a new methodology for the in-process control of design inspection. The probability was the dependent variable and size of the work product inspected was the independent variable. Clusters of the data were formed initially, with 2 of the 12 clusters discarded because the clusters contained fewer than 5 observations.

Data and Preliminary Analysis

In the first modeling phase, the researchers looked at both SIZE and LOG(SIZE) as the independent variable, and if we plot the data given below (see Figure 8.2), we can see why we should use LOG(SIZE). (The units for SIZE are thousands of lines of code.)

PROB	SIZE
0.82	1936.39
0.40	2766.31
0.80	4101.67
0.40	51.15
0.00	107.72
0.20	160.32
0.27	486.20
0.46	1039.34
0.33	295.68
0.60	657.47

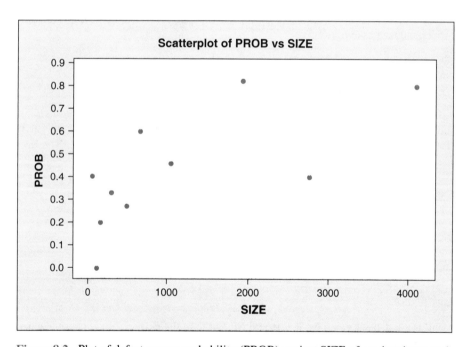

Figure 8.2 Plot of defect escape probability (PROB) against SIZE of product inspected.

Clearly, there is not a straight-line relationship; rather, there is evidence of a logarithmic relationship. Plotting PROB versus LOG(SIZE) in Figure 8.3 shows some evidence of a linear relationship, although a couple of points don't conform to that relationship.

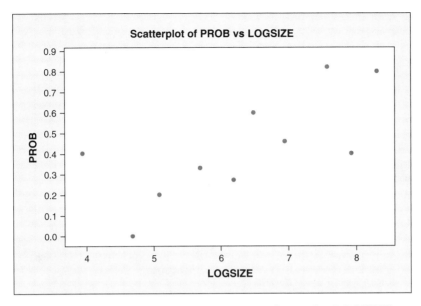

Figure 8.3 Plot of defect escape probability (PROB) against LOG(SIZE).

We continue this example in Section 8.2.7 and will then return to the example in Section 9.5 and examine and critique how the authors added predictors, thus creating a multiple regression model—the topic of Chapter 9. ■

8.2.5 Example with College Data

Data   We will begin with the data on the 49 schools in the top 50 (actually 52 because of a tie) national universities for 2002 that use the SAT test as the primary entrance test. Some of these data were used previously in Figure 1.4 to illustrate how a scatter plot can be enhanced by adding an additional variable to the two-dimensional display. The data needed for the example in this section are in file 75-25PERCENTILES2002.MTW. These data are good for illustrating how a regression analysis should be performed because they have some interesting hidden features that can be detected with regression analysis, and the subject is one to which every college student can relate, thus avoiding potential esotericism.

Plot the Data   We know there should be a strong relationship between the 75th percentile SAT score and the 25th percentile SAT score. The scatter plot for these variables is given in Figure 8.4.

As expected, the graph shows a strong linear relationship. There is some evidence of curvature in the upper right corner; that should not be surprising since the maximum possible score is being approached by the 75th percentile score. In other words, there is a horizontal asymptote at 1600, whereas there is no clear vertical asymptote, although obviously a 25th percentile score could not be particularly close to 1600.

This graph illustrates the fact that nonlinearity will be encountered for virtually any pair of continuous variables if the variables have a wide enough range, as natural boundaries will exist.

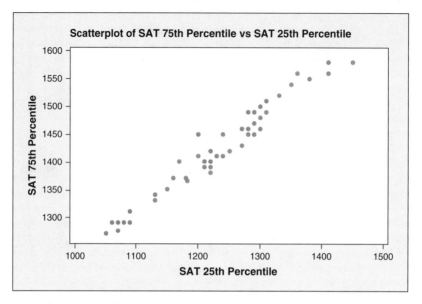

Figure 8.4   Scatter plot of SAT 75th percentile versus SAT 25th percentile for schools data.

8.2.5.1 Computer Output

Because of the strong signal of linearity, we will fit a simple linear regression model and perform the appropriate analyses. MINITAB has excellent regression capabilities and its output is given throughout the chapter. Given below is the output when SAT 75th percentile is regressed against SAT 25th percentile, with explanations of the components of the output given following the output.

```
Regression Analysis: SAT 75th percentile versus SAT 25th percentile

The regression equation is
SAT 75th percentile = 378 + 0.848 SAT 25th percentile

Predictor              Coef         SE Coef      T        P
Constant             378.20         33.44      11.31    0.000
SAT 25th               0.84761       0.02699   31.41    0.000

S = 18.08         R-Sq = 95.5%

Analysis of Variance

Source                 DF           SS           MS       F        P
Regression             1           322506       322506   986.57   0.000
Residual              47            15364          327
      Lack of Fit     26             8827          339     1.09    0.424
      Pure Error      21             6538          311
Total                 48           337870
```

```
Unusual Observations
Obs    SAT 25th   SAT 75th     Fit     SE Fit   Residual    St Resid
 4      1450       1580.00   1607.23    6.34     -27.23       -1.61X
20      1200       1450.00   1395.33    2.75      54.67        3.06R

R denotes an observation with a large standardized residual
X denotes an observation whose X value gives it large influence.
```

The printout shows the following. With $Y =$ SAT 75th percentile score and $X =$ SAT 25th percentile score, the regression equation is $\widehat{Y} = 378 + 0.848\ X$. Since this is in the general form of the equation of a straight line, the interpretation is similar. That is, 0.848 is the slope of the line; $\widehat{Y}$ increases by 0.848 for every unit increase in X. If the equation is a good fit to the data (which it is, as will be explained shortly), then the increase in Y per unit increase in X will not differ greatly from 0.848. When we order the X values from smallest to largest and compute the average increase in the corresponding Y values, we obtain 0.764, with the median change equal to 0.875.

The number 378 is where the regression line would cross the Y-axis if $X = 0$. But obviously X cannot be less than 400, so the 378 is not directly interpretable. This illustrates why β_0 is often viewed by statisticians as a nuisance parameter. In some applications, β_0 and its estimate *do* have a physical interpretation, but that is not the case here.

That doesn't mean we can do without the term, however. Even if we knew that $Y = 0$ when $X = 0$ (which clearly isn't the case here since neither can be zero), we will still generally obtain a better fit to the data when we use the intercept term, as was stated previously. Fitting a line through the origin just because $Y = 0$ when $X = 0$ would not be defensible if we don't intend to use the model when X is near zero.

Regarding the quality of the fit, we can see from Figure 8.5 that most of the points are very close to the line. Thus, our visual impression is that the line fits the data very well. Clearly, it would also be good to have a quantitative measure of the quality of the fit.

If we knew that $\beta_1 = 0$, then we wouldn't attempt to fit a line, as this would mean that there is no relationship between X and Y. It we fit the line anyway and obtained $\widehat{\beta}_1 = 0$ (which would be a sheer coincidence since $\widehat{\beta}_1$ is an estimator and thus has a variance), it should be apparent from Eq. (8.8) that we would have $\widehat{Y} = \widehat{\beta}_0 = \overline{Y}$. That is, every predicted value would be $\overline{Y}$, so the regression line would be just a horizontal line at the value of $\overline{Y}$. This of course is intuitive since if we knew there was no relationship between X and Y, we might as well just use the average of the Y-values in the sample to predict future Y-values.

A measure of the variability in Y is $\sum (Y - \overline{Y})^2$, which we recognize as the numerator of the sample variance of Y. It can be shown that

$$\sum (Y - \overline{Y})^2 = \sum (Y - \widehat{Y})^2 + \sum (\widehat{Y} - \overline{Y})^2 \tag{8.10}$$

The left-hand side of Eq. (8.10) can be viewed as the sum of the squared distances of the points from a horizontal line at $\overline{Y}$ (i.e., when X is of no value), and the first term on the right is the sum of the squared distances of the points to the line $\widehat{Y} = \widehat{\beta}_0 + \widehat{\beta}_1 X$. It follows then that the second term on the right-hand side is a measure of the incremental value in using the regression line to predict Y rather than using $\widehat{Y} = \overline{Y}$.

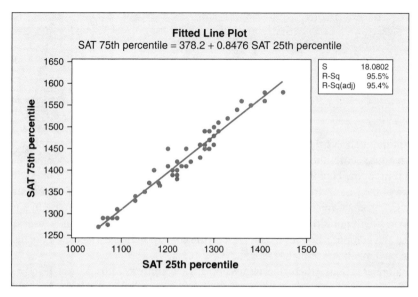

Figure 8.5 Regression line fit to the college data: SAT 75th percentile $= 378.2 + 0.8476$ SAT 25th percentile.

R^2

Therefore, it would be logical to use either $\sum (Y - \widehat{Y})^2$ or $\sum (\widehat{Y} - \overline{Y})^2$ as a measure of the worth of the prediction equation, and divide the one that is used by $\sum (Y - \overline{Y})^2$ so as to produce a unit-free number.

It is a question of whether we want the measure to be large or small. If we define

$$R^2 = \frac{\sum (\widehat{Y} - \overline{Y})^2}{\sum (Y - \overline{Y})^2} \tag{8.11}$$

then R^2 represents the percentage of the variability in Y [as represented by $\sum (Y - \overline{Y})^2$] that is explained by using X to predict Y, whereas if $\sum (Y - \widehat{Y})^2$ were used in the numerator, then the fraction would be the percentage of the variability in Y that was *not* explained by the regression equation. (R^2 has some well-known limitations, such as the fact that its value is affected by the slope of the regression line, but it is the most frequently used indicator of the worth of a regression model.)

Note that if $\widehat{\beta}_1 = 0$, then $\widehat{Y} = \overline{Y}$ and $R^2 = 0$. At the other extreme, R^2 would equal 1 if $Y = \widehat{Y}$ for each value of Y in the data set. Thus, $0 \leq R^2 \leq 1$ and we would want R^2 to be as close to 1 as possible.

For this dataset we have $R^2 = .955$, which is obviously close to 1. The value of R^2 appears in a section of the computer output (above) that allows the reader to see the worth of the regression equation in terms of other statistics. In particular, $\widehat{\beta}_1$ and its standard error (i.e., estimated standard deviation) are given, in addition to the value of $t = \widehat{\beta}_1 / s_{\widehat{\beta}_1}$. (The denominator of this fraction can be shown to equal $s / \sqrt{S_{xx}}$, as the reader is asked to show in Exercise 8.7.)

When R^2 is close to 1, this t-statistic will far exceed a nominal threshold value of 2. Since this statistic has a t-distribution with $n - 2$ degrees of freedom when the errors are assumed to have a normal distribution, it might be used to test the hypothesis H_0: $\beta_1 = 0$. Care must be exercised, however, as R^2 can be small even when the hypothesis is rejected, because it can be shown that $R^2 = t^2/(n - 2 + t^2)$. Thus, if $n = 15$ and $t = 2.165$, so that a two-sided test of H_0 would barely be rejected using $\alpha = .05$, $R^2 = 0.265$, a very low value. (Note that R^2 values will often be of approximately this magnitude when n is an order or two of magnitude greater than in this illustration, even when other statistics suggest that the model fits well.) Requiring the statistic to be more than double the tabular value is a more reasonable criterion for most sample sizes, although one might argue for a multiple slightly greater than two since here a multiplier of two would bring R^2 up to only .583.

Another problem with t-statistics is they can easily be misused when there is more than one X in the model, as is explained in Chapter 9.

Information is also provided in the output that would permit a hypothesis test on β_0, but such a test would generally not be used for deciding whether or not to use that term in the model. Such a decision would be made independent of any data, and as mentioned previously, we generally want an intercept term in the model.

The value of $s(= 18.08)$ is the estimate of the square root of the variance of the error term: that is, $s = \sqrt{\hat{\sigma}_\epsilon^2}$.

Analysis of Variance Section
The next section of the output might be called "analysis of variation" because the total variation of the Y-values, $\sum (Y - \overline{Y})^2$, is broken down into its component parts: the part that it is due to the regression and the part that is left ("residual"). The regression "sum of squares" (denoted as SS) and the total sum of squares are used in computing R^2, as $R^2 = 322{,}506/337{,}870 = .955$. The numerator and denominator are thus the numerical values of the expressions given in Eq. (8.11).

The degrees of freedom (DF) for regression is always equal to the number of independent variables (which in this chapter is of course always one), and the residual degrees of freedom is thus $n - 2$ since the total degrees of freedom must be $n - 1$.

The sums of squares are divided by their respective degrees of freedom to produce the "mean squares," and the ratio of the mean squares is given by F since that ratio has an F distribution, again provided that the errors have a normal distribution. In this case the F-value is so large that the p-value is given as zero to three decimal places. (Analysis of variance tables are discussed in greater detail in Chapter 12.)

When the X-values are repeated, as they are in this example, it is possible to break the residual sum of squares down into two components: lack of fit and pure error. The latter is vertical scatter on the scatter plot, which no model can accommodate. That is, we cannot fit a regression line that is a vertical line as the slope would be undefined. If we have regressor values that are repeated, then we will almost certainly have vertical scatter as the Y-values will not likely be the same. So the amount of pure error in the data is then a function of the total amount of vertical scatter. Mathematically,

$$SS_{\text{pure error}} = \sum_{j=1}^{n_i} \sum_{i=1}^{k} (Y_{ij} - \overline{Y}_i)^2 \tag{8.12}$$

with k denoting the number of X's that are repeated and n_i denotes the number of repeats of the ith repeated X-value. When this sum of squares is deducted from SS_{residual}, the result is $SS_{\text{lack of fit}}$, as is illustrated in the Analysis of Variance section of the output. The degrees of freedom for pure error is $\sum_{i=1}^{k} (n_i - 1)$, and the degrees of freedom for lack of fit can be obtained by subtraction since the pure error and lack of fit degrees of freedom add to the error degrees of freedom.

The mean squares for lack of fit and pure error are obtained in the same way that mean squares are always obtained, by dividing them by their respective degrees of freedom, and the ratio of the mean squares is the F-statistic shown in the output. In this case the lack of fit is not significant (as indicated by the p-value). When this occurs, the usual recommendation is to use the residual mean square as the estimate of the error variance. If the lack of fit were significant, we would want to use the pure error mean square as the estimate of the error variance, and the hypothesis test for the slope would then have this mean square in the denominator of the statistic.

Careful attention should be given to the last component of the output, which spotlights unusual observations. Unfortunately, information on unusual observations is generally not utilized by practitioners. This can have serious consequences, however, as certain observations can sometimes strongly influence the results.

Unusual Observations

This is unlike what occurs with other statistics as when we compute the average of ten numbers, for example, each number has the same influence on the average since each number has a weight of 1/10. This doesn't happen in regression, however, because we don't have linearity in the method of least *squares*. Therefore, since we will not have observations that will equally influence the results, there is the possibility that some observations could be highly influential.

In the MINITAB output we see one point that is labeled as influential. This designation requires some explanation. MINITAB spotlights an observation as being (potentially) influential when, in the case of a single independent variable, an observation has a value for that variable that is well-removed from the center of the other predictor values. This does not make the point automatically influential, however, as whether the point is influential depends on both the corresponding Y-value and, in the case of a model with a linear term only, the strength of the linear relationship between the other points.

MINITAB also spotlights points that have a standardized residual that is greater than two in absolute value. A standardized residual is a residual divided by its standard deviation. The standard deviation of a residual depends on the value of X, so the standard deviation of a residual is different for different values of X. Specifically,

$$e_i' = \frac{e_i}{s\sqrt{1 - h_i}} \tag{8.13}$$

with

$$h_i = \frac{1}{n} + \frac{(x_i - \bar{x})^2}{\sum (x_i - \bar{x})^2} \tag{8.14}$$

and s as previously defined will be called a *standardized residual* in this chapter and in the next chapter. Unfortunately, there is disagreement in the literature regarding the definition of the term. When we standardize a random variable, we subtract its mean and divide by its standard deviation, as was explained in Section 3.4.3. The expected value of a residual is zero, by assumption, although this is true strictly speaking only if the model that is used were the correct model, as the reader is asked to show in Exercise 8.5. Therefore, since the denominator in Eq. (8.13) is the standard error of e_i, it follows that e_i' given in Eq. (8.13) is a standardized residual.

In this example, observation #20 has a standardized residual of 3.06. Under the assumption of normality, the probability of observing a standardized residual that is at least this large in absolute value is .0022. Thus, this is an extreme value that suggests either (1) the distribution of the error term may have heavier tails than the tails of a normal distribution, or (2) the observation may be a bad data point, or at least be anomalous in some way.

The observations in this dataset are readily identifiable, and observation #20 is The University of California–Berkeley. The point has a large standardized residual because Berkeley has the largest difference between the 75th and 25th percentile scores of the 49 schools. Could that be a data recording error? Since it is a public university, we would not expect the 25th percentile score to be high, but since it is a prestigious public university we should not be surprised to see a high 75th percentile score. So the large difference can be explained, and we thus should not be surprised by the large standardized residual. Nevertheless, the data point is clearly unusual and points that have large standardized residuals (say, larger than 2.8 or so in absolute value) should be investigated.

8.2.6 Checking Assumptions

Many (most?) users of regression analysis stop after obtaining the output shown for this example. Unfortunately, doing so can cause serious problems. In particular, correlated errors can cause major problems. Perhaps the best example of this can be found in a paper by Coen, Gomme, and Kendall (1969). The latter thought they had shown that car sales seven quarters earlier could be used to predict stock prices, as $\hat{\beta}_1$ was 14 times its standard deviation. Unfortunately, they failed to examine the residuals, and a residuals plot would have provided strong evidence that the errors were correlated. After fitting an appropriate model, Box and Newbold (1971) showed that there was no significant relationship between the two variables.

So how do we detect correlated errors? Of course, we don't observe the errors because the only way we would know the errors would be if we knew the values of the parameters, and since we don't know those, we don't know the errors! Therefore, we must use the residuals as a substitute for the errors, but they are far from being a perfect substitute. In particular, the residuals are generally closer to being normally distributed than the errors when the errors are nonnormal. This means that a normal probability plot of the residuals could be misleading, but a full discussion and remedy of the problem are beyond the scope of this text. This is discussed further in Section 8.2.6.3.

8.2.6.1 Testing for Independent Errors
We can select from various available methods for testing the assumption of independent errors using the residuals. As stated in the preceding section, the residuals are not a perfect substitute for the errors, and this includes the independence assumption. Specifically, even if the errors are independent, the residuals won't be because they must sum to zero. This is due to Eq. (8.7a) and a second linear restriction on the residuals is represented by Eq. (8.7b). Therefore, it follows that only $n - 2$ of the residuals could be independent. To see this,

assume that we want to solve those two equations for the e_i. Since we have two equations to solve, we need two unknowns (residuals); the other $n - 2$ residuals are independent if the errors are independent. Thus, the residuals are not independent, but this isn't a major problem.

A simple method of checking the independence assumption on the errors is to construct a time sequence plot of the residuals or of the standardized residuals. If the residuals are strongly autocorrelated (i.e., "self" correlated), this will be apparent from the graph. When there is intermediate autocorrelation, this won't be so obvious from the graph, so a numerical measure of the degree of autocorrelation is then needed. In MINITAB, a graph of the autocorrelations for values that are k observations apart for $k \geq 1$ can be produced, in addition to providing decision lines that can be used to determine significant autocorrelations. The numerical values of the autocorrelations are automatically produced as output.

Various other statistics have been proposed for testing for autocorrelation; these are generally described in regression books such as Ryan (1997).

8.2.6.2 Testing for Nonconstant Error Variance

Figure 1.5 provided evidence of nonconstant variance of the variable on the vertical axis, although this wasn't discussed in Chapter 1. That figure is reproduced here, in a somewhat different form, as Figure 8.6. By assumption, the variance of Y is the same as the variance of ϵ when X is assumed fixed (i.e., not random). This is because $\beta_0 + \beta_1 X$ is then a constant (unknown, of course), and a constant added to a random variable (ϵ) produces a new random variable with the same variance, in accordance with Rule #4 in Section 4.2.2.

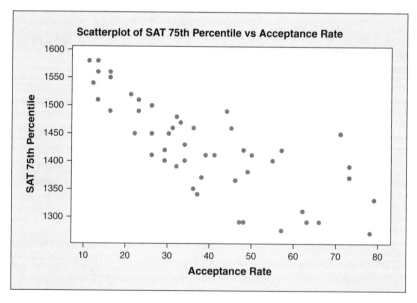

Figure 8.6 Example of nonconstant variance: scatter plot of SAT 75th percentile versus acceptance rate.

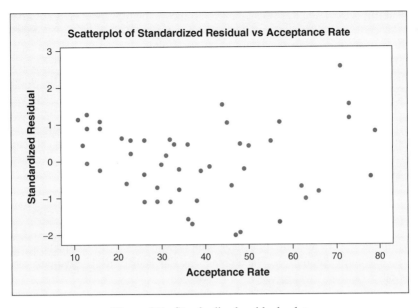

Figure 8.7 Standardized residuals plot.

The problem with a nonconstant variance is that the least squares estimators do not have the smallest variance that they could have. They are still unbiased, as can be shown, but they do not have minimum variance, as they do when the assumption is not violated. This problem should not be ignored when it exists, but methods for dealing with the problem are beyond the scope of this text. Briefly, the best approach is to try to model the variance as a function of $\hat{\mu}_{y|x}$. See Carroll and Ruppert (1988) or Ryan (1997, Chapter 2) for details.

Although with only a single predictor we can often see evidence of nonconstant error variance from an x-y scatter plot, we don't want to rely on such a plot for this purpose. Instead, we should plot the standardized residuals on the vertical axis. For this particular problem it is best to plot the predictor on the horizontal axis because the coefficient of the predictor is negative. This is because we want Figure 8.7 to be of the same general form as Figure 8.6, and to be revealing if the latter is not revealing. The other option is to plot $\hat{Y}$ on the horizontal axis, but when the predictor coefficient is negative, the plot will be the mirror image of the plot when X is on the horizontal axis, and that is generally not what we want.

Figure 8.7 gives us the same general message as Figure 8.6, as the spread of the standardized residuals increases considerably over increasing values of the acceptance rate.

When there is more than one predictor, a logical starting point would be to plot the standardized residuals against $\hat{Y}$. If there is evidence of a problem from this plot, the standardized residuals could then be plotted against each predictor.

8.2.6.3 *Checking for Nonnormality*
It was stated in Section 3.4.3 that there is no such thing as a normal distribution in practice. Therefore, it follows that an error term cannot have a normal distribution. Thus, what we realistically want to check for is a major departure from normality. The consequence of such a departure is that the method of least squares is then not necessarily the best method

to use, and if it is used, the statistics that are assumed to be (approximately) normal won't be, and this could greatly undermine the hypothesis tests and other inferences that may be made, all of which depend on the assumption of normality. Testing for nonnormal errors is a bit tricky because the residuals will be "less nonnormal" than the errors when the errors are nonnormal. [If this sounds confusing, the reader is advised to read the explanation for this phenomenon given by Weisberg (1980).] Thus, constructing a normal probability plot of the residuals—the usual approach for testing the normality assumption—is not entirely satisfactory. A complete discussion of this topic is beyond the scope of this text; see Ryan (1997, Section 2.1.2) for details, or any other source on simulation envelopes for residuals.

8.2.7 Defect Escape Probability Example (Continued)

Now that the necessary analytic tools have been presented, we can complete our analysis of the data given in Section 8.2.4. The output using LOG(SIZE) as the predictor variable is given below.

```
Regression Analysis: PROB versus LOGSIZE

The regression equation is
PROB = − 0.361 + 0.126 LOGSIZE

Predictor           Coef      SE Coef       T         P
Constant         −0.3613      0.2846     −1.27     0.240
LOGSIZE           0.12566     0.04426     2.84     0.022

S = 0.192307       R-Sq = 50.2%       R-Sq(adj) = 44.0%

Analysis of Variance

Source             DF         SS         MS        F        P
Regression          1      0.29810    0.29810    8.06     0.022
Residual Error      8      0.29586    0.03698
Total               9      0.59396
```

Although the regression model is statistically significant with a p-value of .022, clearly the fit isn't very good. One of the main problems is that point #4 doesn't fit with the others. If we delete that point, the value of R^2 increases from .50 to .70. Nevertheless, we can't delete valid data points in an effort to increase R^2, but the data that are used should all be from the same population, and of course errors of various types, including recording errors, do occur. Therefore, it would have been a good idea to check point #4.

   If all of the data points are used, it is apparent that more predictors will need to be employed in order to have a good model. This is what Raz and Barad (2004) did in their paper, so we will return to this example in Section 9.5 and discuss it further in the context of multiple regression.

8.2.8 After the Assumptions Have Been Checked

After the assumptions have been checked, the user can proceed to use regression for whatever purpose was intended. The assumptions say nothing about outliers and influential observations, however, so it is important to pay particular attention to the identification of residual outliers and potentially influential observations, as were identified in the SAT example in Section 8.2.4. Statistical software is geared toward detecting single outliers, as in that example. These methods will often be satisfactory, but the reader should be aware that they can fail when multiple outliers are present in the same general area, as a cluster of outliers could exert enough collective influence on the regression line so as to essentially camouflage themselves.

There are various influence statistics that have been developed for determining influence on the regression coefficients and on the predicted values. These are covered in most regression books.

8.2.9 Fixed Versus Random Regressors

A regressor might be "fixed," as when we fix the temperature in a lab experiment, or it may be random such as when a regression model is used to predict college grade point average and one of the regressors is high school GPA. Obviously, the latter is random, not fixed, and probably most of the applications of regression that we can think of would have one or more random regressors.

The random regressor case has been somewhat controversial over the years, with various methods of analysis proposed. The usual message to practitioners, however, is that in the random X case we can proceed the same as in the fixed X case provided that the conditional distribution of Y given X is (approximately) normal for each value of X. (Conditional distributions are not covered in this text but here the meaning of the term should be apparent.)

8.2.10 Transformations

Transformations can be very useful in regression. There are two general types of transformations: (1) transforming the model and (2) transforming Y and/or X.

8.2.10.1 Transforming the Model

Assume that we "know" that the true model is $Y = \beta_0 \beta_1^{X_1} \epsilon$. This is not a linear regression model because it is not linear in the parameters. If we take the natural logarithm of each side of the model, we then obtain a linear regression model: $Y' = \beta_0' + \beta_1' X_1 + \epsilon'$ with $Y' = \log(Y)$, $\beta_0' = \log(\beta_0)$, $\beta_1' = \log(\beta_1)$, and $\epsilon' = \log(\epsilon)$. Although this is straightforward mathematically, we need to know the error structure of the model and to know that it is as specified here in order for this approach to work. If the error term had been additive instead of multiplicative, then taking logs would not work. In that case we would have a nonlinear regression model. Such models are not covered in this book. The interested reader is referred to books on the subject such as Bates and Watts (1988). Oftentimes knowledge about the form of the error term will not be available or will vary from application to application even when the same nonlinear model is used.

8.2.10.2 Transforming Y and/or X

We never want to transform Y to try to improve the fit because if the assumptions were approximately met before the transformation, they won't be met after the transformation. Of course, we might want to transform Y in an effort to satisfy the assumptions. If we transform Y and want to see how good the fit is, we need to transform the values of $\widehat{Y}$ back to the original scale before computing R^2. If we have transformed Y in order to meet the model assumptions, the fit of the model may have seriously been degraded. In order to regain the quality of the fit, it is necessary to either transform the entire right-hand side of the model using the same transformation as was employed for Y, or transform X in that manner. That is, if Y^λ is the transformation, with $\log(Y)$ used for $\lambda = 0$ [this family of transformations is due to Box and Cox (1964)], the choices for the right-hand side of the model are $(\beta_0 + \beta_1 X)^\lambda + \epsilon$ or $\beta_0 + \beta_1 X^\lambda + \epsilon$. The former, which is called the transform-both-sides method, is preferable as that preserves the functional relationship. The other form has the advantage of simplicity, however, since it is a simple linear regression model (with a nonlinear term in X), whereas the other form is a nonlinear regression model. Nonlinear regression is not covered in this text, but as mentioned previously there are various nonlinear regression texts, including Bates and Watts (1988).

Frequently, a nonlinear term in X will be needed, either in place of the linear term or in addition to it. For example, the scatter plot may indicate the need for a quadratic term. Such a term would generally be used in addition to the linear term, rather than as a replacement for it. So if we are fitting polynomial terms, we usually want to include the lower-order terms, as justifying their exclusion could be difficult. If the scatter plot suggests a logarithmic or reciprocal relationship, however, such a term should be used without the linear term. A plot of the standardized residuals against each predictor is one of the plots used in multiple regression (i.e., more than one predictor) to check for the need to transform a predictor, but in simple linear regression all of the information needed to make that determination is contained in the scatter plot.

8.2.11 Prediction Intervals and Confidence Intervals

Prediction intervals were covered in Chapter 7, but not in the context of regression. Since one of the uses of regression is for prediction, it would be logical to construct an interval about a predicted value, and we naturally call the interval a *prediction interval*. A prediction interval in regression will be wider than the prediction interval given by Eq. (7.1) because in regression there is the added variability of $\widehat{\beta}_1$, which is of course not involved in constructing a prediction interval that isn't constructed using a regression model. (Of course, $\widehat{\beta}_0$ was also not involved in Chapter 7, but $\widehat{\beta}_0$ is a function of $\overline{Y}$ and $\widehat{\beta}_1$ and the variance of the former *was* involved in the prediction interval in Chapter 7.)

As was emphasized in Chapter 5, a confidence interval for a parameter θ is of the general form $\widehat{\theta} \pm t s_{\widehat{\theta}}$ whenever the use of t is appropriate. A prediction interval for Y is not an interval for a parameter, however, and will thus not be of that form. We will lead into the interval by first going through a somewhat similar interval, as that will make the transition easier.

As a simple example, assume that we wish to predict college grade point average using high school grade point average as the sole predictor. Thus, $Y = $ college GPA and $X = $ high school GPA. Would a confidence interval for the average GPA of all students who have entered a certain college with a particular value of X and subsequently graduated be of any value? Clearly, it is of no value if we are trying to reach a decision regarding the admittance of a particular student. What we need in that case is an interval for the student's GPA if the student were admitted and attended for four years, or at least until possibly encountering insurmountable academic problems.

The most intuitive way to obtain the expression for the prediction interval is to first obtain the expression for the confidence interval.

For the confidence interval, we want an interval for $\mu_{Y|X_0}$. In words, this is the mean of Y when $X = X_0$, which would also be called the conditional mean. Since this mean is our parameter θ, the next step is to determine $\widehat{\theta}$. Because the mean of ϵ is zero, it follows that the conditional mean is equal to $\beta_0 + \beta_1 X_0$. This of course would be estimated by $\widehat{Y}_0$, the predicted value of Y when $X = X_0$, as clearly $E(\widehat{Y}_0) = \mu_{Y|X_0}$ since the least squares estimators are unbiased. We then need $s_{\widehat{Y}_0}$. Obtaining the latter requires a moderate amount of algebra, which will not be given here. The result is

$$ s_{\widehat{Y}_0} = s\sqrt{\frac{1}{n} + \frac{(x_0 - \overline{x})^2}{S_{xx}}}, \quad \text{with} \quad s^2 = \widehat{\sigma}_\epsilon^2 = \sum \frac{(Y - \widehat{Y})^2}{(n-2)} $$

(This result is derived in regression books and the interested reader seeking details is referred to such books.) Thus, the $100(1 - \alpha)\%$ confidence interval is given by

$$ \widehat{Y}_0 \pm t_{\alpha/2, n-2} s\sqrt{\frac{1}{n} + \frac{(x_0 - \overline{x})^2}{S_{xx}}} \tag{8.15} $$

It is relatively straightforward to go from the confidence interval expression to the expression for the prediction interval. The latter, of course, is to be constructed about $\widehat{Y}_0$ since this is the predicted value, so this is common to both expressions. That should not be surprising because the only difference between the model expression for Y_0 ($=\beta_0 + \beta_1 X_0 + \epsilon$) and $\mu_{Y|X_0}$ is a random error term that has a mean of zero and thus is not a term that can contribute to either estimation or prediction.

Since we are trying to predict a random variable, Y_0, rather than estimate a parameter, $\mu_{Y|X_0}$, we must account for the variability of the random variable in constructing the interval. Since $Var(\epsilon) = Var(Y_0)$, it follows that we must add σ_ϵ^2 to $\sigma_{\widehat{Y}_0}^2$. That is, we want $Var(Y_0 - \widehat{Y}_0) = Var(Y_0) + Var(\widehat{Y}_0)$, with the variances being added because a new observation is independent of the data used in producing the regression equation, and thus independent of $\widehat{Y}_0$. The estimator of $Var(\widehat{Y}_0)$ is of course the same as is used for the confidence interval and the estimator of $Var(Y_0)$ is s^2. Therefore, using these estimators and taking the square root, we obtain the prediction interval:

$$ \widehat{Y}_0 \pm t_{\alpha/2, n-2} s\sqrt{1 + \frac{1}{n} + \frac{(x_0 - \overline{x})^2}{S_{xx}}} \tag{8.16} $$

Although the expressions are similar mathematically, conceptually they are quite different since a prediction interval is for the future and a confidence interval is for the present.

It is useful to know what the expression is for the prediction interval, but we don't want to obtain a prediction interval using hand calculation, at least not more than once. We can easily obtain a prediction interval in MINITAB, for example, and if we obtain a 95% prediction interval for the SAT example with $X_0 = 1200$, which must be for a school not included in the sample, the output is as follows. (Recall that X_0 denotes the 25th percentile SAT score.)

```
                      Predicted Value for New Observation

New Obs    Fit    SE Fit        95.0% CI            95.0% PI
    1    1395.33   2.75   (1389.79, 1400.87)  (1358.54, 1432.12)
```

The first interval is the confidence interval and the second is the prediction interval. Of course, the latter is always wider and we observe that here. Note that the observed value of 1450 (see Figure 8.4) for one of the sample data points lies outside the prediction interval. Why is that? This result should not be surprising because we picked a point that did not fit well with the other points. Therefore, the fitted value, in this case 1395.33, is not close to the actual value, so the amount that is added to (or subtracted from) the fitted value won't necessarily cover the actual value, especially when s is small. The prediction interval might still work for new data with $X_0 = 1200$, however, and of course a prediction interval is used for new X values.

This illustrates the importance of using *regression diagnostics*, as they are called, rather than just taking a perfunctory approach.

To give a "better," but less profound, example, observation #16 is (1300, 1480), so when we produce the prediction interval using this X-value, we obtain

```
                      Predicted Value for New Observation

New Obs    Fit    SE Fit        95.0% CI            95.0% PI
    1    1480.09   3.12   (1473.82, 1486.36)  (1443.18, 1517.00)
```

Thus, the actual value is virtually in the middle of the prediction interval because the predicted value is virtually the same as the observed value.

There are many areas of application in which prediction intervals are extremely important. Consider again the scenario with Y = college GPA and X = high school GPA and consider two possible prediction intervals given by $(1.8, 2.6)$ and $(2.1, 2.3)$. The value of $\widehat{Y}$ is 2.2 in each case (since $\widehat{Y}$ lies in the center of the interval), but an admissions officer would undoubtedly look more favorably upon the second interval than the first interval. In other areas of application, such as fisheries management, having a well-estimated prediction interval is thought to be probably as important as having good estimates of the regression parameters.

Another interval that is potentially useful is a confidence interval on β_1. As was emphasized in Chapters 5 and 6, the signal we receive from a confidence interval agrees with the signal that we receive from the corresponding hypothesis test. Therefore, for the example given in Section 8.2.4, we know that the confidence interval will not come close to containing zero because the p-value for the hypothesis test is zero to three decimal places. Since $\widehat{\beta}_1 = 0.84761$ and $s_{\widehat{\beta}_1} = 0.02699$, a 95% confidence interval for β_1 would be $\widehat{\beta}_1 \pm t_{\alpha/2, n-2} s_{\widehat{\beta}_1} = 0.84761 \pm 2.0117(0.02699)$, which gives us the interval (0.7933, 0.9019). Thus, the interval does not come close to containing zero, and as stated we knew that would happen before we constructed the interval.

There is an important point to be made before we leave this example. Although it is quite reasonable to construct a confidence interval for β_1 when there is only a single regressor, we should construct confidence intervals for the β_i in the case of more than one regressor only when the X's are fixed. This is discussed further in Section 9.1.

■ **EXAMPLE 8.2**

Objective and Data

There is a delay when data are read from a disk storage unit into a memory buffer. This is termed a latency delay and other latency delays can also occur. Both the number of such delays and the length of them are random variables. This means that the time T_i to read the ith unit of data, which is of length t_i, does not have a deterministic relationship and may not have a regression relationship to a significant degree. We will examine this for the following data sample, with T_i given in milliseconds and t_i given in kilobytes.

T_i	58.2	91.1	54.6	88.3	127.9	176.5	214.2	248.2	245.4	301.2	351.1	332.29
t_i	100	200	100	300	400	500	600	700	700	800	900	1000

T_i	158.2	291.1	59.6	88.3	102.9	206.5	244.2	288.2	345.4	321.2	331.1	362.29
t_i	500	800	100	300	300	400	500	600	900	800	900	900

If there is a regression relationship, it is desired to have a 95% prediction interval for T for $t = 300$ and $t = 500$.

Solution   The relevant MINITAB output is given below.

```
The regression equation is Time = 12.4 + 0.360 Length

Predictor          Coef    SE Coef       T       P
Constant          12.37      12.90    0.96   0.348
Length          0.35965    0.02075   17.33   0.000

S = 28.6005       R-Sq = 93.2%       R-Sq(adj) = 92.9%

Analysis of Variance

Source             DF        SS       MS       F       P
Regression          1    245710   245710  300.38   0.000
Residual Error     22     17996      818
Total              23    263706
```

This output shows that there is a strong relationship between Time and Length, so it is appropriate to obtain the prediction intervals:

```
Prediction Intervals

    New
Obs Value     Fit    SE Fit         95% PI
300        120.27    7.87    (58.75, 181.78)
500        192.20    5.95   (131.62, 252.78)
```

The reader is asked to address another aspect of this problem in Exercise 8.68.   ■

8.2.12   Model Validation

Bearing in mind the sequential, repetitive nature of experimentation, as expressed by Figure 1.1, a regression model should be "validated" before it is used. That is, its performance should be assessed on data other than the data used to develop the model and to estimate the parameters. So how is model validation accomplished? One approach that has been used is to split the data in half, making the two halves approximately equal in terms of a selected similarity measure, and use the second half for validation. An argument against such practice has effectively been made in the literature, however: the loss of efficiency caused by using only half the data for parameter estimation mitigates against the use of data splitting. The other alternative is to obtain new data, if possible and practical, and test model performance on the new data.

8.3   CORRELATION

A regression analysis can be performed when X is either fixed or random, but the *correlation* between two variables is defined only when they are both random. This convention is not strictly adhered to, however, as the term "correlation" is often used in conjunction with designed experiments in which the regressors are fixed. A correlation coefficient, such as r_{xy}, measures the extent to which the two random variables are related. The sample (Pearson) correlation coefficient is computed as

$$r_{xy} = \frac{S_{xy}}{\sqrt{S_{xx}S_{yy}}} \tag{8.17}$$

with S_{yy} defined analogous to the way that S_{xx} was defined in Section 8.2.2, and the other components of Eq. (8.17) are as previously defined. Formally, the population correlation coefficient, ρ_{XY}, is defined as the covariance between X and Y divided by the standard deviation of X times the standard deviation of Y. (Covariance was covered in Section 2.2.) When those parameters are estimated by the sample statistics, Eq. (8.17) results.

■   **EXAMPLE 8.3**

Data and Objective

The data listed below are part of a much larger dataset given by Hughes-Oliver, Lu, Davis, and Gyurcsik (1998). The full dataset consisted of measurements of polysilicon thickness at 13 sites on 22 wafers in semiconductor fabrication that were processed using rapid thermal chemical vapor deposition. The processing conditions were not constant for each wafer, as the thickness of the oxide applied to the wafer prior to deposition of polysilicon and the deposition time both varied. Given below is the deposition time (in seconds) for each of the 22 wafers, in addition to the corresponding polysilicon thickness (in angstroms) for the first location on the wafer. These are denoted by A and B, respectively. The objective is to investigate the relationship, if any, between deposition time and polysilicon thickness.

A	18	35	52	52	18	35	35	35	23	23	47	47
B	494	853	1090	1058	517	882	732	1143	608	590	940	920

A	47	23	23	23	23	47	47	47	35	35
B	917	581	738	732	750	1205	1194	1221	1209	708

A scatter plot, Figure 8.8, shows a linear relationship between time and thickness, although there is evidence that the relationship becomes nonlinear at the high thickness values.

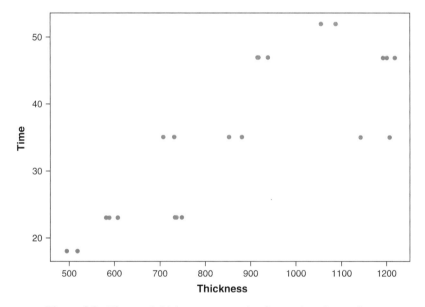

Figure 8.8   Time and thickness scatter plot for semiconductor data.

The value of the correlation coefficient, using Eq. (8.17), is .812, so the strength of the linear relationship is exemplified by this relatively high value.

The possible values of r_{XY} range from -1 to $+1$. The former represents perfect negative correlation, as would happen if all of the points fell on a simple linear regression line with a negative slope, and the latter represents perfect positive correlation. In regression there is nothing really "negative" about negative correlation, as a strong negative correlation is just as valuable as a strong positive correlation as far as estimation and prediction are concerned. For example, the correlation between the variables in Figure 8.6, both of which are obviously random variables, is $-.766$. This is a moderate negative correlation, with the magnitude of the negative correlation decreased by the vertical scatter in the lower right portion of the graph. A zero correlation between X and Y would signify that it would not be meaningful to construct a regression equation using that regressor, but when there is more than one regressor we would prefer that the regressors be "uncorrelated" (orthogonal) with each other. This is discussed and illustrated in Chapter 12.

A hypothesis test of $\rho_{XY} = 0$ might be performed, but if we decide to construct a simple linear regression model based on the X-Y scatter plot, the correlation would almost certainly *not* be zero. So testing the hypothesis will essentially be a waste of time in almost all applications. Furthermore, the test is equivalent to the test of H_0: $\beta_1 = 0$.

When there is only a single regressor, $r_{xy}^2 = R^2$. Thus, for the one-regressor, random-X case, R^2 is the square of the correlation between X and Y. When X is fixed, R^2 must then be viewed (and labeled) somewhat differently, and it is customary to refer to R^2 as the *coefficient of determination* (and the coefficient of multiple determination when there is more than one regressor).

Another connection between the square of a correlation coefficient and R^2 is $r_{Y\widehat{Y}}^2 = r_{xy}^2 = R^2$, as the reader is asked to show in Exercise 8.6. (Note that this last result holds whether X is random or not, since both Y and $\widehat{Y}$ are random variables.) Clearly, Y and $\widehat{Y}$ must be highly correlated for the regression model to have value.   ∎

8.3.1 Assumptions

Certain assumptions should be met, at least approximately, in order to use the correlation coefficient given in Eq. (8.17). Foremost is the fact that there must be a linear relationship between the two variables, which, as stated in Section 8.3, must both be random. The variables must also have continuous distributions and the joint distribution of the two random variables should be the bivariate normal distribution. Bivariate normality implies univariate normality for the distribution of each of the two random variables, but the converse is not true. Therefore, testing for normality of each of the two univariate distributions is not sufficient.

Of course, from a practical standpoint, if a univariate normal distribution doesn't exist in practice, then a bivariate normal distribution certainly does not exist either! As in the univariate case, a small departure from a bivariate normal distribution won't be a problem, but a large departure from bivariate normality will cause a serious problem: for example, see Lai, Rayner, and Hutchinson (1999) for details.

When the assumption of linearity or the assumption of bivariate normality is suspect, a nonparametric correlation coefficient might be used instead. See Section 16.3.2.

There is also the tacit assumption that the correlation that is computed must make sense. The term "spurious correlation" is often used to represent a nonsensical correlation, such as the correlation between college professors' salaries in a country like Great Britain and the total sales of alcoholic beverages in that country—an often-cited example. Such a correlation can be high simply because both variables are related to the state of the economy and to the gross domestic product.

8.4 MISCELLANEOUS USES OF REGRESSION

There are some important specialized uses of regression, one of which occurs extensively in engineering. Assume that we have two measuring instruments; one is accurate to the intended level of accuracy but is both expensive to use and slow, and the other is fast and less expensive to use but is inaccurate. If the measurements obtained from the two devices were highly correlated, the measurement that would have been made using the expensive measuring device could be predicted fairly well from the measurement that is actually obtained using the less expensive device.

8.4.1 Calibration

There are two approaches to calibration: *inverse regression* and the *classical theory of calibration*. We will discuss each of these in this section.

If the less expensive device has an almost constant bias and we define $X =$ measurement from the accurate instrument and $Y =$ measurement from the inaccurate device, and then regress X on Y to obtain $\widehat{X}^* = \widehat{\beta}_0^* + \widehat{\beta}_1^* Y$, we would expect $\widehat{\beta}_1^*$ to be close to 1.0 and $\widehat{\beta}_0^*$ to be approximately equal to the bias.

■ **EXAMPLE 8.4**

If $\widehat{\beta}_1^*$ was extremely close to 1.0 and $\widehat{\beta}_0^*$ was not significant, this would suggest that the difference in the two sets of measurements was essentially insignificant so that calibration would not be necessary. Such a dataset, with the corresponding analysis, can be found at `http://www.itl.nist.gov/div898/strd/lls/data/LINKS/DATA/Norris.dat`.

Calibration Hypothesis Test

The output at the above website (note that X and Y are reversed) shows that $\widehat{\beta}_1^*$ is significantly different from zero. The question to be asked, however, is whether it is significantly different from 1.0, using this as the hypothesized value and computing $t = (\widehat{\beta}_1^* - 1)/s_{\widehat{\beta}_1^*}$ instead of $\widehat{\beta}_1^*/s_{\widehat{\beta}_1^*}$ for the form of the t-statistic, as was used in Section 8.2.5.1. Performing the computation shows that $t = 4.93$. So it *is* significantly different from 1.0. This would seem to suggest that there is a problem with the customer's measuring instrument, which while perhaps not biased, does at least show evidence of being erratic.

Some thought must be given to this, however, before such a decision can be made. The standard error of $\widehat{\beta}_1^*$ is $s/\sqrt{S_{xx}}$, as given in Section 8.2.5.1, which will be small when S_{xx} is large, relative to the fit of the model as reflected by the value of s. Consider two datasets for which the values of S_{xx} differ considerably. If the data are such that the dataset with the (much) larger value of S_{xx} does not have a proportionately larger value of s, the standard errors and hence the t-statistics can differ greatly. The reader is asked to show this in Exercise 8.61.

Thus, there is the potential for declaring a result to be statistically significant that does not have practical significance—analogous to the discussion in Section 5.9 for hypothesis tests in general. For this problem the user would have to decide whether the difference between $\widehat{\beta}_1^* = 1.002$ and the desired value of $\beta_1 = 1.0$ is of any practical significance, aided by inspection of the absolute difference between the values of X and the values of Y.   ■

This calibration method is termed *inverse regression* because X is being regressed on Y instead of Y regressed on X.

Since X and Y might seem to be just arbitrary labels, why not reverse them so that we would then have just simple linear regression? Recall that Y must be a random variable, and classical regression theory holds that X is fixed. A measurement from an accurate device should, theoretically, have a zero variance, and the redefined X would be a random variable.

But we have exactly the same problem if we regress X on Y, with X and Y as originally defined. The fact that the independent variable Y is a random variable is not really a problem, as regression can still be used when the independent variable is random, as was discussed earlier in this chapter.

If the *dependent* variable is not truly a random variable, however, then classical regression is being "bent" considerably, and this is one reason why inverse regression has been somewhat controversial.

There is an alternative to inverse regression that avoids these problems, however, and that should frequently produce almost identical results.

In the "classical theory of calibration," Y is regressed against X and X is then solved for in terms of Y, for the purpose of predicting X for a given value of Y. Specifically, $\widehat{Y} = \widehat{\beta}_0 + \widehat{\beta}_1 X$, so that $X = (\widehat{Y} - \widehat{\beta}_0)/\widehat{\beta}_1$. For a given value of Y, say, Y_0, X is then predicted as $\widehat{X}_0 = (Y_0 - \widehat{\beta}_0)/\widehat{\beta}_1$. [A prediction interval for $\widehat{X}_0$, called a Fieller interval, could be constructed, but that will not be discussed here. The interested reader is referred to Iyer (2003).]

As in Section 8.3, let r_{xy} denote the correlation between X and Y. If $r_{xy} \approx 1$, then $\widehat{X}_0$ and $\widehat{X}_0^*$ (from inverse regression) will be almost identical. Recall that for inverse regression to be effective, the two measurements must be highly correlated and this also applies to the classical theory of calibration. Therefore, under conditions for which we would want to use either inverse regression or the classical method of calibration, the two approaches will give very similar results. Therefore, we need not be concerned about the controversy surrounding inverse regression.

Of course, it is implicit in the application of either of these approaches that an inaccurate measurement makes a significant difference. Obviously, the need for an accurate measurement is important if we are weighing gold, but probably not if we are weighing dirt.

■ EXAMPLE 8.5

Data

We can also encounter problems with nonconstant error variance in calibration work, just as in other applications. The Alaska pipeline data in Section 4.6.2.1 of the *NIST/SEMATECH e-Handbook of Statistical Methods* illustrates this type of problem. The data consisted of in-field ultrasonic measurements of the depths of defects in the Alaska pipeline. The depth of the measurements was then remeasured in the laboratory. The objective of course was to calibrate the bias of the field measurements relative to the laboratory measurements.

Graphical Analysis

The plot of the field defect size against the lab defect size is shown in Figure 8.9.

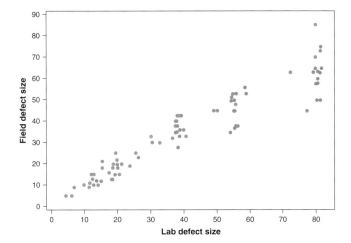

Figure 8.9 Calibration data—nonconstant variance.

Notice that the spread of the values of the dependent variable increases as the predictor variable increases. This will cause a plot of the standardized residuals against the predictor to have essentially the same shape, which will thus signal a violation of the constant error variance assumption.

The approach that was used to correct this problem for these data was the second approach that was discussed in Section 4.6.2.4 of the *NIST/SEMATECH e-Handbook*. ■

■ **EXAMPLE 8.6 (CASE STUDY)**

In using calibration, we of course expect to see a high R^2 value, but even a very high value does not mean that a better model cannot be found. This point was made in the case study of Ryan (2003). This was originally a load cell calibration study performed at the National Bureau of Standards (which was renamed the National Institute of Standards and Technology in 1988). The data are given at the text website and are also given below with Y being the Deflection and X the Load.

```
Row       Y            X

  1    0.11019      150000
  2    0.21956      300000
  3    0.32949      450000
  4    0.43899      600000
  5    0.54803      750000
  6    0.65694      900000
  7    0.76562     1050000
  8    0.87487     1200000
  9    0.98292     1350000
 10    1.09146     1500000
 11    1.20001     1650000
 12    1.30822     1800000
```

13	1.41599	1950000
14	1.52399	2100000
15	1.63194	2250000
16	1.73947	2400000
17	1.84646	2550000
18	1.95392	2700000
19	2.06128	2850000
20	2.16844	3000000
21	0.11052	150000
22	0.22018	300000
23	0.32939	450000
24	0.43886	600000
25	0.54798	750000
26	0.65739	900000
27	0.76596	1050000
28	0.87474	1200000
29	0.98300	1350000
30	1.09150	1500000
31	1.20004	1650000
32	1.30818	1800000
33	1.41613	1950000
34	1.52408	2100000
35	1.63159	2250000
36	1.73965	2400000
37	1.84696	2550000
38	1.95445	2700000
39	2.06177	2850000
40	2.16829	3000000

Because of the magnitude of the X-values, we will divide them by 10^4, as was done in Ryan (2003). The scatter plot in Figure 8.10 shows an extremely strong straight-line relationship, as is also indicated by the output that follows.

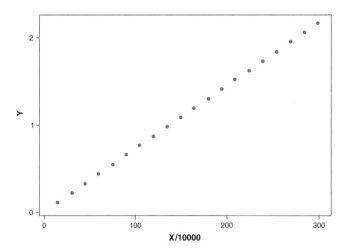

Figure 8.10 Scatter plot of load cell calibration data.

```
Regression Analysis: Y versus X/10000

The regression equation is
Y = 0.00615 + 0.00722(X/10000)

Predictor          Coef          SE Coef           T          P
Constant        0.0061497       0.0007132        8.62      0.000
X/10000         0.00722103      0.00000397     1819.29     0.000


S = 0.002171     R-Sq = 100.0%           R-Sq(adj) = 100.0%

Analysis of Variance

Source            DF      SS       MS          F           P
Regression         1   15.604   15.604    3.310E+06    0.000
Residual Error    38    0.000    0.000
Total             39   15.604
```

 With virtually any regression data we would not seek model improvement when the computer output shows an R^2 value of 100%. Calibration data are different, however, since we should expect a very high R^2 value, as stated previously. Interestingly, the plot of the standardized residuals against X shows pronounced curvature, as can be seen from Figure 8.11.

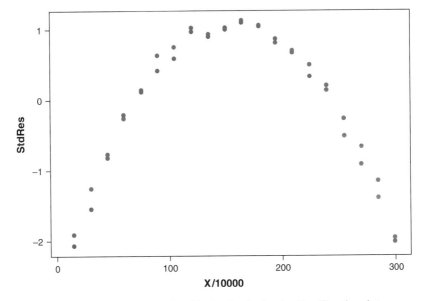

Figure 8.11 Standardized residuals plot for load cell calibration data.

 This graph suggests that a quadratic term should be added to the model, and when this is done the average absolute value of $(Y - \widehat{Y})$ improves from 0.00183 to 0.00016. Certain NIST scientists have indicated that this improvement is significant.

 The lesson to be learned from this case study is that the application of regression to calibration problems requires a different mindset regarding R^2 values. ∎

8.4.1.1 Calibration Intervals

As with other statistical procedures, we would like to go beyond obtaining either $\widehat{X}_0$ or $\widehat{X}_0^*$ and obtain a calibration interval for X. Since we obtain $\widehat{X}_0$ by solving for it in the regression equation for Y, it should not be surprising that we obtain the endpoints of the calibration interval from the expression for the endpoints of the prediction interval for Y_0. Of course, we don't have the latter, however, without X_0, so it is the expression that is used, not numerical values.

When we construct a prediction interval, we obtain the endpoint interval values for Y for a single observed value of X. Since we are, in essence, reversing things, we might expect that we would obtain the endpoint interval values for X from the single observed value of Y. That is, we set the expression for the prediction interval equal to the observed value of Y, say, Y_0, and solve for the interval endpoints for X_0 from Eq. (8.15). Obtaining the result requires considerable algebra, which will not be reproduced here.

Numerical results given by Schechtman and Spiegelman (2002) showed that their modification of a method given by Dahiya and McKeon (1991) outperformed the classical interval, however. Therefore, the best approach may be to use this modified interval. The expression for a $100(1 - \alpha)\%$ interval for x_0 using this approach is given by

$$\overline{x} + \widehat{\gamma}(Y_0 - \overline{Y}) \pm t_{\alpha/2, n-2} \sqrt{\frac{\widehat{\sigma}^2 \widehat{\gamma}^2 (1 + 1/n + (\widehat{x}_0 - \overline{x})^2 + \widehat{\sigma}^2 \widehat{\gamma}^2)}{S_{xx}(1 + \widehat{\sigma}^2 \widehat{\gamma}^2 / S_{xx})}} \tag{8.18}$$

with $\widehat{\gamma} = \widehat{\beta}_1 S_{xx} / (\widehat{\beta}_1^2 S_{xx} + \widehat{\sigma}^2)$ and the other components of the expression are as customarily defined in simple linear regression. We illustrate the use of this interval with the following example.

■ **EXAMPLE 8.7**

Data

We will use thermocouple calibration data described at http://www.itl.nist.gov/div898/handbook/pmd/section5/pmd521.htm with the response variable being pressure and the independent variable being temperature.

Analysis

The general idea is to estimate temperature from a pressure measurement, after first regressing pressure on temperature. The regression equation is

$$\widehat{Y} = 7.75 + 3.93X \tag{8.19}$$

with $R^2 = .994$. If we wanted to estimate temperature for a pressure measurement of, say, 178, we would solve for X in Eq. (8.19) after first equating 178 to $\widehat{Y}$, thus obtaining

$$X = \frac{178 - 7.75}{3.93}$$

so that $\widehat{x}_0 = 43.32$. The 95% calibration interval for x_0 using Eq. (8.18) is then

$$44.45 + 0.2544(178 - 182.43) \pm 2.02439W$$

where

$$W = \sqrt{\frac{18.48(0.2544)^2(1 + 1/40 + (43.32 - 44.45)^2 + (18.48)(0.2544)^2)}{7189.65\left(1 + \frac{(18.48)(0.2544)^2}{7189.65}\right)}}$$

so the interval is (43.27, 43.37).

For additional reading on calibration confidence intervals, see Section 4.5.2.1 of the *NIST/SEMATECH e-Handbook of Statistical Methods* (2002), the references cited in this section, Eberhardt and Mee (1994) who present constant-width calibration intervals for linear regression, and Osborne (1991), which is a review of calibration. ∎

8.4.2 Measurement Error

Measurement error was involved in the calibration example discussed in the preceding section, in which the objective was to work around the error using calibration and thus eliminate its effect. Measurement error is generally not incorporated into regression models, however, and only within the past ten years or so have statisticians and scientists given much attention to the effect of measurement errors. If measurement error in Y is random, normally distributed, and has a mean of zero, it adds to the model error (if the measurement error and model error are independent), leaves more variability in Y unexplained by the model, and thus reduces the value of R^2. If the measurement error is large relative to the random error and is markedly nonnormal, an alternative to least squares may have to be used.

The effect of measurement error in X will depend on the nature of the error and also on whether or not X is a random variable. In general, it is desirable to have the spread in X large relative to measurement error in X whenever the latter is anticipated.

8.4.3 Regression for Control

An important but infrequently discussed application of regression is to attempt to control Y at a desired level through the manipulation of X.

Doing so may be difficult, however, for a number of reasons. First, since a cause-and-effect relationship is being inferred, the prediction equation must have been produced from preselected X values, and there must not be any other independent variables that are related to Y. The latter is not likely to be true when only a single X is being used, however, and the consequences when it is not true may be great.

Box (1966) gave a good example of this in regard to a hypothetical chemical process in which Y = process yield, X_1 = pressure, and X_2 = an unsuspected impurity. The scenario is that undesirable frothing can be reduced by increasing pressure, but a high value of X_2 is what actually produced frothing and also lowers yield, with the latter unrelated to pressure.

Thus, although regression for control is potentially useful, there is also a high risk.

8.5 SUMMARY

Regression analysis is one of the most frequently used statistical tools, but it is often misused. It would be highly impractical to try to analyze regression data without good software, but software does not abrogate a user's responsibility to carefully check assumptions and check for unusual data points. Serious problems can ensue if regression data are not analyzed carefully and thoroughly. If one or more of the assumptions are not met, appropriate corrective action should be taken. Extreme points may be bad data points or they may be good data points that should be downweighted. See any regression book for details on actions that should be taken when assumptions are violated and/or extreme data points exist.

Since entire books have been written on regression analysis and books have also been written on subareas in regression, the material in this chapter and the next chapter constitute only an introduction to the subject. Engineering applications of regression often require sophisticated approaches and models. For example, it may be necessary to fit different regression lines (or, in general, different models) to different parts of a dataset. This is illustrated in Exercise 8.62 with some calibration data.

There have been many important applications of the use of regression with a single regressor, such as calibration, but there are far more applications of multiple regression (i.e., more than one regressor), which is discussed in Chapter 9. There are also more regression tools that must be used in multiple regression than are needed for simple regression. This is also discussed in Chapter 9.

In general, a good command of regression methods is important for more than just regression analysis, as certain regression methods are quite useful in the analysis of data from designed experiments. Accordingly, regression has been presented before design of experiments in this text, with the latter presented in Chapter 12.

REFERENCES

Anscombe, F. J. (1973). Graphs in statistical analysis. *The American Statistician*, **27**, 17–21.

Bates, D. M. and D.G. Watts (1988). *Nonlinear Regression*. New York: Wiley.

Box, G. E. P. (1966). Use and abuse of regression. *Technometrics*, **8**, 625–629.

Box, G. E. P. and D. R. Cox (1964). An analysis of transformations. *Journal of the Royal Statistical Society, Series B*, **26**, 211–243 (discussion: pp. 244–253).

Box, G. E. P. and P. Newbold (1971). Some comments on a paper by Coen, Gomme, and Kendall. *Journal of the Royal Statistical Society, Series A*, **134**(2), 229–240.

Carroll, R. J. and D. Ruppert (1988). *Transformation and Weighting in Regression*. New York: Chapman and Hall.

Coen, P. J., E. Gomme, and M. G. Kendall (1969). Lagged relationships in economic forecasting. *Journal of the Royal Statistical Society, Series A*, **132**, 133–152.

Dahiya, R. C. and J. J. McKeon (1991). Modified classical and inverse regression estimators in calibration. *Sankhya B*, **53**, 48–55.

Eberhardt, K. R. and R.W. Mee (1994). Constant-width calibration intervals for linear regression. *Journal of Quality Technology*, **26**(1), 21–29.

Hughes-Oliver, J. M., J. C. Lu, J. C. Davis, and R. S. Gyurcsik (1998). Achieving uniformity in a semiconductor fabrication process using spatial modeling. *Journal of the American Statistical Association*, **93**, 36–45.

Iyer, H. (2003). Statistical calibration and measurements. Chapter 20 in *Handbook of Statistics 22* (R. Khattree and C. R. Rao, eds.). Amsterdam: Elsevier Science B.V.

Lai, C. D., J. C. W. Rayner, and T. P. Hutchinson (1999). Robustness of the sample correlation—the bivariate lognormal case. *Journal of Applied Mathematics and Decision Sciences*, **3**, 7–19.

NIST/SEMATECH e-Handbook of Statistical Methods (2002). A joint effort between the National Institute of Standards and Technology and International SEMATECH. The handbook has the following URL: http://www.itl.nist.gov/div898/handbook.

Osborne, C. (1991). Statistical calibration: A review. *International Statistical Review*, **59**, 309–336.

Ryan, T. P. (1997). *Modern Regression Methods*. New York: Wiley.

Ryan, T. P. (2003). *Case Study—Pontius Data*. National Institute of Standards and Technology, Statistical Engineering Division. (see http://www.itl.nist.gov/div898/casestud/pontius.pdf).

Schechtman, E. and C. Spiegelman (2002). A nonlinear approach to linear calibration intervals. *Journal of Quality Technology*, **34**(1), 71–79.

Weisberg, S. (1980). Comment on "Some large-sample tests for nonnormality in the linear regression model," by H. White and G. M. MacDonald. *Journal of the American Statistical Association*, **75**, 28–31.

EXERCISES

8.1. Consider the four datasets given in the datafile ANSCOMBE.MTW, which is also one of the sample datasets that comes with MINITAB. Construct a scatter plot for each data set and then obtain the regression equation for each dataset, using MINITAB or other software. Comment. In particular, would we really want to use simple linear regression for each of these datasets? Which datasets, if any, would seem to be suitably fit by a simple linear regression equation?

8.2. Consider the information that is provided by the value of a correlation coefficient while considering the following statement: "It has been found that there is a positive correlation between the number of firefighters battling a blaze and the amount of damage that is done." That is, the more firefighters, the greater the dollar damage. Does this suggest that too many firefighters may be used at many fires with the result that more damage than expected may be resulting because the firefighters are getting in each other's way and not working efficiently?

8.3. Assume that $\widehat{\beta}_1 = 2.38$ in a simple linear regression equation. What does this number mean, *in words*, relative to X and Y?

8.4. Given the following data:

X	1.6	3.2	3.8	4.2	4.4	5.8	6.0	6.7	7.1	7.8
Y	5.6	7.9	8.0	8.2	8.1	9.2	9.5	9.4	9.6	9.9

compute $\widehat{\beta}_0$ and $\widehat{\beta}_1$ using either computer software or a hand calculator: Now make up five different values for $\widehat{\beta}_{1(\text{wrong})}$ that are of the same order of magnitude as $\widehat{\beta}_1$, and for each of these compute $\widehat{\beta}_{0(\text{wrong})} = \overline{Y} - \widehat{\beta}_{1(\text{wrong})}$.

Then compute $\sum (Y - \widehat{Y})$ and $\sum (Y - \widehat{Y})^2$ for each of these six solutions, using four or five decimal places in the calculations. Explain your results relative to what was discussed in Section 8.2.2.

8.5. Show that the expected value of a residual is zero when the simple linear regression model is the correct model. (*Hint*: Write the ith residual as $Y_i - \widehat{Y}_i$ and then obtain the expected value.)

8.6. For simple linear regression with the predictor being a random variable, show that both $r^2_{Y\widehat{Y}}$ and r^2_{YX} are equal to R^2. (*Hint*: What type of relationship exists between $\widehat{Y}$ and X?)

8.7. Show that $s_{\widehat{\beta}_1} = s/\sqrt{S_{xx}}$. [*Hint*: First show that $\widehat{\beta}_1$ can be denoted by $\sum k_i Y_i$ with $k_i = (X_i - \overline{X})/S_{xx}$.]

8.8. Explain the independence assumption for a simple linear regression model.

8.9. Critique the following statement: "The least squares estimates minimize the sum of the squares of the errors."

8.10. Assume that in simple linear regression a 95% prediction interval is to be constructed for Y_0, with $X_0 = \overline{x}$. If $n = 20$, $R^2 = .82$, and $S_{yy} = 40$, what number will be added to $\widehat{Y}_0$ to produce the upper limit?

8.11. The data in EX8-12ENGSTAT.MTW consist of two columns of 100 random numbers, each generated from the standard normal distribution. Thus, there is no relationship between the numbers, but when either column of numbers is used to represent the predictor variable and the other is used to represent the dependent variable, the p-value for testing H_0: $\beta_1 = 0$ is .026 (i.e., significant at the .05 level), but the value of R^2 is only .049 (i.e., 4.9%).

(a) What does this suggest about looking at p-values in regression, especially when n is large?

(b) Use the second column of numbers in the file to represent the dependent variable and construct a scatter plot. Does the plot suggest any relationship between the two "variables"?

(c) Can the numerical value of the t-statistic for testing H_0: $\beta_1 = 0$ be determined from what has been given in the problem statement (i.e., not using the data)? If so, what is the value? If not, explain why the value can't be computed.

(d) Can the absolute value of this t-statistic be determined? If so, what is the value?

8.12. Critique the following statement: "I am going to use a simple linear regression model without an intercept because I know that Y must be zero when X is zero."

8.13. What does $1 - R^2$ mean in words?

8.14. Consider the following set of predictor values: 4, 5, 6, 8, 2, and 3. Now compute values for the independent variable as $Y = 7 + 8X$. What will be the numerical value of the correlation coefficient? What would be the numerical value if $Y = 6 - 3X$?

8.15. Given the following numbers, fill in the blank:

Y	3	5	6	7	2	4	8
$\widehat{Y}$	2.53	—	5.30	6.68	3.22	5.30	8.06

Could the prediction equation be determined from what is given? Explain why or why not. If possible, give the equation.

8.16. Use MINITAB or other software to perform a lack-of-fit test for the following data:

X	3.2	2.1	2.6	3.2	3.0	2.1	3.2	2.9	3.0	2.6	2.4
Y	3.6	3.7	3.0	3.7	3.3	3.8	3.4	2.7	2.8	3.1	3.4

Now construct a scatter plot of the data. Does the plot support the result of the lack-of-fit test?

8.17. Consider the following data:

X	1	2	3	4	5	6	7
Y	5	6	7	8	7	6	5

First obtain the prediction equation and compute R^2. *Then*, graph the data. Looking at your results, what does this suggest about which step should come first?

8.18. The frequent misuse of regression was mentioned in Section 8.5. As an example, it has been folklore that expected life length is indicated by the length of the lifeline in one's hand. In their paper written in a somewhat whimsical manner, P. G. Newrick, E. Affie, and R. J. M. Corrall ("Relationship Between Longevity and Lifeline: A Manual Study of 100 Patients," *Journal of the Royal Society of Medicine*, **83**, 498–501, 1990) "proved" the relationship because the t-statistic for the slope in their simple linear regression model was sufficiently large. A scatter plot showed, however, that most of the points formed horizontal lines, except for several points in the lower left corner of the graph. Despite the fact that this may have been intended to be a joke, the wire services carried the story of this "discovery," and the story appeared in major newspapers. What would you say to these researchers?

8.19. For the data in the file `75-25PERCENTILES2002.MTW`, regress the 75th percentile SAT scores against acceptance rate and then plot the standardized residuals against $\widehat{Y}$. Compare your plot with Figure 8.4. Can you see from your graph what can be seen from Figure 8.4?

8.20. Critique the following statement (question): "How can we expect the least squares regression line to go through the center of the data, such that the residuals sum to zero, if we don't know the true model?"

8.21. It is possible to see the effect of individual observations dynamically by using one of the available Java applets. As of this writing there are several such applets available on the Internet. One of these is given at the following URL: `http://gsbwww.uchicago.edu/fac/robert.mcculloch/research /teachingApplets/Leverage/index.html`. Another one is available at `http://www.stat.sc.edu/~west/javahtml/Regression.html`. Use one of these two applets to show what happens when the rightmost point is moved up or down. Then show what happens when a point near the center, in terms of the x-coordinate, is moved up and down. Explain why the effects differ.

8.22. Use the data in Exercise 8.16 to fit a simple linear regression model *without* an intercept. (This can be accomplished in MINITAB by unchecking "Fit intercept" under "Options" when using menu mode or by using the `Noconstant` subcommand with command mode.) Compute the sum of the residuals and explain why the sum is not zero.

8.23. Consider a scatter plot of Y against X in simple linear regression. If all of the points practically form a line with a negative slope, what would be the approximate value of the correlation coefficient (i.e., r_{xy})?

8.24. Given the following values for Y and $\widehat{Y}$ from a regression analysis:

Y	7	5	4	2	4	6
$\widehat{Y}$	6.06	5.60	5.13	3.27	4.20	c

What is the numerical value of c?

8.25. Explain what the word "least" refers to in the "method of least squares."

8.26. Consider the following sample of (y, x) data : $(9, 1), (4, 2), (1, 3), (0, 4), (1, 5), (4, 6)$, and $(9, 7)$. Compute the value of the correlation coefficient. Does the value suggest that there is any relationship between X and Y? Then graph the data. What does this suggest about what should be done before computing the value of the correlation coefficient in Section 8.3?

8.27. Many, if not most, colleges and universities use regression analysis to predict what a student's four-year GPA would be and use this predicted value as one factor in reaching a decision as to whether or not to accept an applicant. Georgia Tech has been one such university, and in the mid-1980s they used a Predicted Grade Index (PGI) as an aid in reaching decisions. The regression equation, using their notation, was PGI $= -0.75 + 0.4$ GPA $+ 0.002$(SAT-M) $+ 0.001$(SAT-V). (*Source*: Student newspaper; of course, "GPA" in the equation represents high school GPA.) Bearing in mind that this was before the recentering of the SAT scores and also before grade inflation, answer the following:
   (a) What is the largest possible value of PGI?
   (b) Considering your answer to part (a) and any knowledge that you might have of the school, for what approximate range of actual college GPA values would you expect the fit to be the best?

(c) Similarly, for what range of values of the predictors (roughly) would you expect the equation to be useful, and could we determine the set of possible values of the predictors for which the equations *could* be used?

8.28. In checking the normality assumption in linear regression, explain why a normal probability plot should not be constructed using the residuals (i.e., the $e_i = Y_i - \widehat{Y}_i$ values).

8.29. Consider the following data for a simple linear regression problem. The sample size was $n = 20$ and when H_0: $\beta_1 = 0$ was tested against H_a: $\beta \neq 0$, it was found that the value of the test statistic was positive and was equal to the critical (tabular) value.

(a) What is the numerical value of the correlation coefficient, assuming that X is a random variable?

(b) What does your answer to part (a) suggest about relying on the outcome of the hypothesis test given in this exercise for determining whether or not the regression equation has predictive value?

8.30. It has been stated "Some times we throw out perfectly good data when we should throw out questionable models" (p. 527 of *Statistical Methods for Engineers* by G. G. Vining). Explain what you believe was meant by that statement.

8.31. Consider a simple linear regression prediction equation written in an alternative way. If $\widehat{\beta}_1 X$ is replaced by $\widehat{\beta}_1(X - \overline{X})$, what will be the numerical value of the intercept if $\sum X = 23$, $\sum Y = 43$, and $n = 20$? Can the numerical value of $\widehat{\beta}_1$ be determined from what is given here, combined with the value of $\widehat{\beta}_0$? Why or why not? If possible, what is the value?

8.32. Explain why no distributional assumption would be necessary if a person's objective were just to obtain a simple linear regression equation. Would you advise anyone to do the latter? Explain.

8.33. Hunter and Lamboy (*Technometrics*, 1981, Vol. 23, pp. 323–328; discussion: pp. 329–350) presented the following calibration data that were originally provided by G. E. P. Box:

Measured Amount of Molybdenum	Known Amount	Measured Amount of Molybdenum	Known Amount
1.8	1	6.8	6
1.6	1	6.9	6
3.1	2	8.2	7
2.6	2	7.3	7
3.6	3	8.8	8
3.4	3	8.5	8
4.9	4	9.5	9
4.2	4	9.5	9
6.0	5	10.6	10
5.9	5	10.6	10

(a) Does a sufficiently strong regression relationship exist between the known and measured amounts of molybdenum for calibration to be useful? Explain.

(b) If so, construct a 95% calibration confidence interval for the known amount when the measured amount is 5.8, using Expression (8.18).

(c) Regarding Expression (8.18), what would the width of the interval approach if $\widehat{\sigma}^2$ were virtually zero? Does the limit seem reasonable?

(d) Assume that two analysts work with this set of data. If the second analyst decides to use inverse regression instead of the classical theory of calibration, will the equation for $\widehat{X}^*$ (i.e., using inverse regression) be expected to differ very much from the expression for $\widehat{X}$ (classical theory of calibration)? Explain.

8.34. Sketch an X-Y scatter plot for which the correlation between X and Y (r_{XY}) is approximately zero.

8.35. The following calibration data were originally given by Kromer et al. (*Radiocarbon*, 1986, Vol. 28(2B), pp. 954–960).

Y	8199	8271	8212	8198	8141	8166	8249	8263	8161	8163
X	7207	7194	7178	7173	7166	7133	7129	7107	7098	7088

Y	8158	8152	8157	8081	8000	8150	8166	8083	8019	7913
X	7087	7085	7077	7074	7072	7064	7062	7060	7058	7035

Each Y-value is the radiocarbon age of an artifact, with the corresponding X-value representing the age of the artifact determined by methods that are closer to being accurate. The question to be addressed is whether calibration can be used to determine an estimate of what the X-value would be if it had been obtained.

(a) What is the first step in making this determination?

(b) If practical, determine the length of a 99% calibration confidence interval when Y is 8000. If this would not be practical, explain why.

8.36. Mee and Eberhardt (*Technometrics*, 1996, Vol. 38, pp. 221–229) gave data from a calibration of alpha track detectors that are used to measure indoor concentrations of radon, ^{222}Rn. The data are on radon exposure, the predictor variable, with the response variable being an optical count of the number of damaged tracks caused by radioactive decays over a specified area of the film. The data are given in file RADON2.MTW.

(a) Does there appear to be a strong enough relationship between radon exposure and the number of damaged tracks for a prediction interval and/or calibration confidence interval to be useful? Explain.

(b) Obtain a general expression, as a function of x, for a 99% prediction interval.

(c) Assume there was interest in obtaining a calibration confidence interval when the number of damaged tracks is 200. Obtain the 95% confidence interval.

8.37. Croarkin and Varner (*National Institute of Standards and Technology Technical Note*, 1982) illustrated the use of a linear calibration function to calibrate optical imaging systems. With Y = linewidth and X = NIST-certified linewidth, they obtained a regression equation of $\widehat{Y} = 0.282 + 0.977X$. With $n = 40$, $S_{xx} = 325.1$, and $S = 0.0683$, determine each of the following.

(a) What is the numerical value of the t-statistic for testing the hypothesis that $\beta_1 = 0$?

(b) Is the t-statistic large enough to suggest that the equation should be useful for calibration?

(c) What is the expression for $\widehat{X}_0$, assuming that the classical theory of calibration is used?

8.38. Assume, for the sake of the exercise, that the simple linear regression model is the correct model. Show that $\widehat{\beta}_1$ is an unbiased estimator of β_1 by writing $\widehat{\beta}_1 = \sum k_i Y_i$ with k_i appropriately defined. Does your result depend on the constant error variance assumption? Explain.

8.39. Construct an (x, y) scatter plot that shows very little linear relationship between X and Y when X varies from 10 to 20 but does show evidence of a linear relationship when the range is increased to 10–30. Would you fit a (common) simple linear regression line through the entire set of data or would you fit a line to each segment? (The latter is called *piecewise linear regression*.) Which approach would provide the best fit?

8.40. Construct an (x, y) scatter plot that has two outliers that greatly reduce the value of R^2, yet when they are deleted the regression line remains essentially unchanged.

8.41. In production flow-shop problems, performance is often evaluated by minimum makespan, this being the total elapsed time from starting the first job on the first machine until the last job is completed on the last machine. We might expect that minimum makespan would be linearly related, at least approximately, to the number of jobs. Consider the following data, with X denoting the number of jobs and Y denoting the minimum makespan in hours:

X	3	4	5	6	7	8	9	10	11	12	13
Y	6.50	7.25	8.00	8.50	9.50	10.25	11.50	12.25	13.00	13.75	14.50

(a) From the standpoint of engineering economics, what would a nonlinear relationship signify?

(b) What does a scatter plot of the data suggest about the relationship?

(c) If appropriate, fit a simple linear regression model to the data and estimate the increase in the minimum makespan for each additional job. If doing this would be inappropriate, explain why.

8.42. Construct an (x, y) scatter plot that has one point that does fit the (strong) regression relationship of the other points but the regression line changes very little when the point is deleted.

8.43. Assume a set of 75 data points such that $\widehat{Y} = -3.5 + 6.2X + 1.2X^2$. Would it make sense to report the value of r_{xy}? Why, or why not? If the answer is yes, can the value of r_{xy} be determined from the regression equation? Explain.

8.44. A simple linear regression model is fit to a set of 60 data points, for which $\bar{x} = 23.8$. If $\sum y = 612$, what are the coordinates of one point through which the regression line must pass?

8.45. What is a consequence, if any, of trying to fit a simple linear regression model to a dataset that has a considerable amount of pure error?

8.46. If possible, give an engineering application for which the use of regression for control could produce misleading results.

8.47. A student takes the data in `75-25PERCENTILES2002.MTW` that were used in Section 8.2.5 and also in Chapter 1. The student decides to use the 25th percentile score as the dependent variable, and the 75th percentile score as the independent variable, which is the reverse of what was done in Section 8.2.5. The student fits the model and computes R^2. Will it be the same value as was obtained in Section 8.2.5? Why or why not?

8.48. It is useful to be able to recognize outliers and influential data points in simple linear regression, in addition to roughly estimating the magnitude of the correlation coefficient and the position of the regression line. The Java applet available at `http://www.ruf.rice.edu/~lane/stat_sim/reg_by_eye/index.html` allows you to test your skill at making these identifications. Follow the directions and test your skill on ten data sets. How many of the correlations did you guess correctly? How did you do in placing the regression line?

8.49. A simple linear regression equation is obtained, with $\widehat{Y} = 2.8 - 5.3X$ and $R^2 = .76$. What is the numerical value of r_{xy}?

8.50. A prediction interval was given in Section 8.2.11, using MINITAB, for the data in `75-25PERCENTILES2002.MTW`. Could that interval have been obtained using the computer output in Section 8.2.5.1? Why or why not?

8.51. If you were asked to suggest a transformation of the predictor based on Figure 8.6, what would it be (if any)? Does Figure 8.7 suggest the same transformation that you are recommending (if any)? Explain. Do you believe that either of these two figures is misleading? Explain.

8.52. Respond to the following statement: "I don't see how splitting the data in half and then using the second half for model validation will accomplish anything since the split is performed in such a way as to make the halves extremely similar."

8.53. Consider the regression equation $\widehat{Y} = 13.4 + 6.8X + 2.3X^2$. Can we infer from the magnitude of the regression coefficients that the linear term is approximately three times as important in predicting Y as the quadratic term? Explain.

8.54. Can the regression equation be determined from the data given in Exercise 8.24 after the value of c is obtained? Why or why not? Can R^2 be determined after c is obtained?

8.55. Use any regression applet that is available on the Internet, such as one of the two given in Exercise 8.21, to see what happens when you have several points that plot as a very steep line with a positive slope and then you add a point that is far to the right and below these points. See how the line changes as you move that point down and away from the line.

8.56. Assume that a predictor is measured in feet and its coefficient in a simple linear regression equation is 6.72. What will be the value of the coefficient if the predictor were changed from feet to inches? Would the value of the intercept change? If so, what would be the change?

8.57. The correlation between SAT-Math and SAT-Verbal scores, using the average scores for each of the 50 states as one data point, is quite high (over .90). Is either one of these variables realistically the "dependent" variable and the other the "independent" variable in a regression model? If not, how can the high correlation be explained?

8.58. A homemaker asks the following question: "I prepare popcorn for my family every Friday night and I need to know how thick a layer of kernels to use in the cooker in order to produce the desired amount of popcorn. Of course, I know that some kernels won't pop and the number that won't pop is a direct function of how many are used. I found on the Internet the results of a group study performed in a statistics class, with a regression equation being given. I don't understand the use of the constant (i.e., intercept) in the equation, however, because obviously if I don't use any kernels, I won't have any popcorn. How can a regression equation with an intercept make any sense in my application?" Respond to the question.

8.59. Proceed as follows to demonstrate the difference between statistical significance in simple linear regression and practical significance, analogous to the discussion in Section 5.9. For one dataset, use the integers 20–30 for one set of X-values. Then generate 11 random errors from the standard normal distribution and add those to the X-values to create the Y-values. Then repeat using the integers 10–40 so that 31 random errors must be generated. Perform the regression analysis on each data set and compare the two values of the standard error of $\widehat{\beta}_1$. Compare the two values of s relative to the difference in the two standard errors of $\widehat{\beta}_1$ and relative to the two t-statistics for testing that $\beta_1 = 0$. What have you demonstrated?

8.60. In addition to the fact that all of the observations in a dataset should not necessarily be used in fitting a model, especially when there are outliers that may have to be discarded, different models may need to be fit to different intervals of the predictor variable. We will consider the data given in Section 4.6.1.1 of the *NIST/SEMATECH e-Handbook of Statistical Methods* (see http://www.itl.nist.gov/div898/handbook/pmd/section6/pmd611.htm) and also given in this chapter as Example 8.5. As stated in that section, this is load cell calibration data from the National Institute of Standards and Technology, with Y representing deflection and X being the load and the data were collected in two sets of increasing load. Some of the parts below ask you to reproduce the analysis in Example 8.5 and then go beyond that.

(a) Fit a simple linear regression model to the entire data set.

(b) Does the fit of the model suggest that model improvement should be attempted? Explain.

(c) Construct a plot of the standardized residuals against $\widehat{Y}$. Does this plot suggest that the model should be modified? Explain.

(d) Construct a normal probability plot of the standardized residuals. What does this plot suggest?

(e) In view of your answers to (c) and (d), what action, if any, do you believe should be taken?

(f) Fit the model suggested by the plot of the standardized residuals and then construct a plot of the standardized residuals and a normal probability plot for the new model. Did the model modification correct the problems?

(g) Since the first 20 observations were collected as one group and the next 20 observations as the second group, there is the possibility that a lurking variable could have affected the values in the second group, and this effect could easily go unnoticed. What type of scatter plot would you suggest be constructed to check on this possibility? Construct that plot. Does the plot suggest that separate models should be fit to each half, or should one model be fit to the entire dataset? If the evidence suggests that a model should be fit to each set, do so and compare the results.

8.61. Consider the following statement: "I know that $Var(\widehat{\beta}_1)$ is affected by the spread of the predictor values, so I am going to use a measurement unit of inches instead of feet for the model predictor so as to increase the spread." Will this work? Show what will happen algebraically when feet are used initially and then inches are used as the measurement unit. Comment.

8.62. A study was conducted at a large engineering firm to examine the relationship between the number of active projects (X) and the number of man-hours required per week for a graphics project (Y), using the firm's data for the preceding year. An approximately linear relationship was found and a simple linear regression model was fit to the data. The results obtained using $n = 75$ showed that $\widehat{\beta}_0 = 1.2$, $\widehat{\beta}_1 = 3.4$, $\overline{Y} = 9.6$, $S_{xx} = 74.2$, the residual mean square is 26.8, and a 95% prediction interval is desired for $X = 3$.

(a) Notice that the intercept is not zero. Should it be zero? Explain.

(b) Construct the prediction interval.

8.63. To see the connection between the least squares regression equation and the equation for a straight line from algebra, consider two points (x_1, y_1) and (x_2, y_2). Show that the least squares solution for $\widehat{\beta}_1$ is mathematically equivalent to $(y_2 - y_1)/(x_2 - x_1)$. What is the value of R^2 for this line?

8.64. Consider a (small) sample of three observations on X_1 and X_2. If the values on X_1 are $-1, 0$, and 1, respectively, give corresponding values for X_2 such that the correlation between X_1 and X_2 is zero.

8.65. J. A. Hoeting and A. R. Olsen gave a table in their article, "Are the Fish Safe to Eat? Assessing Mercury Levels in Fish in Maine Lakes" (in *Statistical Case Studies: A Collaboration Between Academe and Industry*, R. Peck, L. D. Haugh, and A. Goodman, eds., ASA-SIAM series) in which the *p*-value for a simple linear regression model is .0002 but the R^2 value is only .13. The messages are thus conflicting as to whether or not the model is an adequate fit to the data. Explain how this could happen.

8.66. (Harder problem) Prove Eq. (8.10). (*Hint*: There are various ways to do this. One way is to add and subtract $\widehat{Y}$ within the parentheses on the left hand side of the equation, square the resultant expression as a binomial, and then show that the middle term vanishes.)

8.67. The dataset SUGAR.MTW that comes with MINITAB was analyzed in the context of a paired-*t* test in Section 6.2. The objective required more than the use of a paired-*t* test, however, since there was a desire to see if there is a relationship between the expensive lab measurements and the less expensive field measurements so that what the lab measurement would be if made can be estimated from the field measurement. The results for the regression of field measurement on lab measurement are shown below.

```
Regression Analysis: Field versus Lab

The regression equation is
Field = 0.261 + 0.0536 Lab

Predictor         Coef      SE Coef     T       P
Constant       0.261046   0.002802   93.17   0.000
Lab            0.053554   0.005222   10.25   0.000

S = 0.00901890    R-Sq = 80.2%    R-Sq(adj) = 79.4%
```

This suggests a moderately strong linear relationship between the two sets of measurements, although the strength of the relationship might not be sufficient for the intended use of the equation. Now construct a scatter plot of field versus lab. What lesson have you learned?

8.68. Consider the output for Example 8.2 in Section 8.2.11. In particular, look at the *p*-value for the intercept term. Would you decide not to use the intercept in the model based on this *p*-value, or would you base your decision on other factors? Explain. Construct a scatter plot of the data. Does this plot support your conclusion, whatever that was? Explain.

CHAPTER 9

Multiple Regression

Most applications of regression analysis involve the use of more than one regressor, as we would realistically expect that there would be more than one regressor that would be correlated with Y. The model for *multiple linear regression* with linear terms is

$$Y = \beta_0 + \beta_1 X_1 + \beta_2 X_2 + \cdots + \beta_m X_m + \epsilon \tag{9.1}$$

with the corresponding prediction equation

$$\widehat{Y} = \widehat{\beta}_0 + \widehat{\beta}_1 X_1 + \widehat{\beta}_2 X_2 + \cdots + \widehat{\beta}_m X_m$$

As in Eq. (9.1), m will hereinafter denote the number of regressors.

Even though we may appropriately regard multiple linear regression as an extension of simple linear regression, with the same assumptions on ϵ, there are some questions that the user of multiple regression must address that are not encountered in simple regression. In particular,

- If data are available on, say, k variables that might seem to be related to the dependent variable, should all k variables be used? If not, which ones should be used?
- What is gained, if anything, by having $m < k$?
- Can we use scatter plots to determine which independent variables to include in the model?
- Can possible transformations of the regressors be determined simply by examining such scatter plots?
- How is the constant error variance assumption checked when there are multiple regressors?
- How can we "see" influential observations and outliers when there are more than two dimensions?
- Should alternatives to least squares be used under certain conditions? If so, under what conditions should they be used, and which alternatives should be considered?

Modern Engineering Statistics By Thomas P. Ryan
Copyright © 2007 John Wiley & Sons, Inc.

Specifically, should least squares still be used when there are high correlations among the regressors?

- Are regression coefficients interpretable when the predictors are correlated; and similarly, are confidence intervals for the corresponding parameters meaningful under such conditions?

In addition to these questions, there are other issues that are common to both simple regression and multiple regression. For example, if the regressors are to have fixed values, what configuration of values should be used?

These and other questions will be addressed in this chapter. Multiple linear regression is naturally more complex than simple linear regression because there is more to investigate and we literally cannot see things as well. With simple linear regression, all of the information that is needed to determine the relationship between Y and X is contained in the scatter plot. We lose our ability to see the data when we add regressors. For example, we obviously cannot see five-dimensional data. This forces us to use graphical methods that cannot be guaranteed to always emit the proper signal.

These disadvantages are more than offset, however, by the fact that by not being limited to a single regressor, we can do a much better job of fitting a model to the data.

9.1 HOW DO WE START?

Because of the dimensionality problem, the first matter to resolve is the determination of the first step(s) to be taken to help gain insight into which of the available variables seem important, and what their functional form should be in the model.

How can this insight be gained? In particular, can we simply construct a scatter plot of Y against each of the available regressors when the data have been obtained from an observational study, rather than from a designed experiment? If we did so, that would be equivalent to performing k simple linear regressions. In other words, each potential regressor would be viewed without regard for the presence or absence of other regressors in the model. Common sense should tell us that this could easily lead us astray.

There is nothing wrong with looking at these scatter plots initially, but the practitioner should be aware of their limitations. What is needed is a way of looking at the worth of a regressor when other regressors are in the model. The t-statistic for each corresponding parameter will give us this information, but the t-statistics when all of the available regressors are in the model cannot be used to tell us which regressors should be in the model, although unfortunately many regression users do use them for this purpose. For example, if two regressors were highly correlated Y, the two t-statistics could be small even if both regressors were highly correlated with Y. The t-statistics being small would simply tell us that we don't need both regressors, which should be intuitively apparent since the regressors are highly correlated.

In that instance the user would presumably realize that one of the regressors should be in the model, but when more than two regressors are in the model the problem of interpretation becomes harder and is not worth pursuing. The safest strategy is simply not to rely on t-statistics during the initial stages of trying to determine which variables to use. These statistics do have value, such as when the data have been obtained from a designed experiment, but they should not be used, or at least not relied on, during the initial stages of model building with data from an observational study.

If we had a very large number of candidate variables, an algorithmic approach would be needed to try to identify the most important variables, as is illustrated in the example in Section 9.4. With a relatively small number of candidate variables, we might simply put all of them in the model and look at the R^2 value for the set. If the R^2 value is small (with what is considered small dependent on the application), then one should seek to increase the size of the set of candidate variables. If R^2 is of reasonable size, then trial and error might be used to arrive at a reasonable subset. This would be strictly an empirical approach and would not be recommended in general. If prior information existed and suggested a particular model, then that model would logically be the starting point. This is discussed in detail in Chapter 10.

9.2 INTERPRETING REGRESSION COEFFICIENTS

The definition of β_i in multiple regression is that it is the expected change in Y per unit change in X_i when all of the other regressors in the model are held constant. The definition is applicable, however, only when regression is used to analyze data from a designed experiment in which the regressor values are literally fixed. When the regressors are random, however (i.e., random values of the regressors), the definition has no meaning. In particular, consider the case of two very highly correlated aptitude test scores, such that their relationship practically forms a straight line. If they were perfectly correlated, we could not change one of the two by a unit without simultaneously changing the other one, and two very highly correlated regressors will approach this situation.

Therefore, since we cannot apply the definition of β_i to the random regressor case, it should not be surprising that regression coefficients are not interpretable in that case, and thus confidence intervals should not be constructed for the corresponding parameters, as these would be constructed using the regression coefficients.

IMPORTANT POINT

Regression coefficients in multiple regression should *not* be interpreted as reflecting the relationship between Y and each predictor when the predictors are random variables.

Coefficients can even have the "wrong" sign in the sense that the sign of a regression coefficient can differ from the sign of the correlation between Y and the corresponding regressor. This has baffled many regression users but it is easily explainable, with the signs really not being wrong at all! Specifically, if two regressors are highly correlated with Y and also highly correlated with each other, the weaker of the two regressors could have a sign in the regression equation that is different from the sign of the correlation coefficient computed using that regressor and Y when the correlation coefficient is much less than the correlation between the other regressor and Y. These "wrong sign" problems can even occur when the three correlations are not large; the determining factor is the relationship between the correlations. This is illustrated and discussed in Ryan (1997, p. 131). Although what appears to be wrong signs can be bothersome, consider the following. If a regression equation had two regressors that were both highly correlated with Y and the coefficients were similar in magnitude and of the same sign, we would expect the predicted values to be practically double the observed values. Therefore, something has to give and what gives is the weaker of the two variables.

Although the regression coefficients are not directly interpretable in the random regressor case (which would also create a problem if regression were used for control, as in Section 8.4.3), this does not affect the usefulness of regression for prediction—the primary use of regression.

9.3 EXAMPLE WITH FIXED REGRESSORS

Regressors are generally fixed in an experimental design, and oftentimes one of the objectives in using such a design is to determine optimum combinations of the predictor values, which in a manufacturing setting would translate to optimum operating conditions for those predictor variables.

■ EXAMPLE 9.1

Czitrom, Sniegowski, and Haugh (1998) described an experiment to study the effect of three manufacturing process factors (bulk flow, CF_4 flow, and power) on three response variables in regard to the manufacture of semiconductor silicon wafers. We will consider one of the response variables, etch rate, which is the amount of oxide that is removed from a wafer per unit of time.

Objective

One of the objectives of the experiment was to see if the value of CF_4 could be set to at least 15 without having a deleterious effect on throughput, with a higher etch rate corresponding to higher manufacturing throughput.

Data

The data from the experiment are as follows:

Bulk gas flow (sccm)	CF_4 flow (sccm)	Power (watts)	Etch rate (Å/min)
60	5	550	2710
180	5	550	2903
60	15	550	3021
180	15	550	3029
60	5	700	3233
180	15	700	3679
60	5	700	3638
180	15	700	3814
120	10	625	3378
120	10	625	3295

Recall the definition of a regression parameter. That definition applies when the regressors are fixed and uncorrelated, as they will be most of the time when an experimental design is used. Therefore, the $\widehat{\beta}_i$ can be interpreted in the same way and in fact the coefficients are the same regardless of whether or not linear terms in all three variables are in the regression equation, or some subset of the three variables is used.

Analysis

When all three variables are in the model, the regression equation is

$$Etch = 6 + 1.71^*bulk\ flow + 24.4^*CF_4flow + 4.50^*power$$

and the output when the data are analyzed is as follows:

Predictor	Coef	SE Coef	T	P
Constant	6.5	324.9	0.02	0.985
Bulk gas flow	1.7146	0.6122	2.80	0.031
CF$_4$ flow	24.425	7.346	3.32	0.016
Power	4.5017	0.4898	9.19	0.000

S = 103.9   R-Sq = 94.5%   R-Sq(adj) = 91.8%

Analysis of Variance

Source	DF	SS	MS	F	P
Regression	3	1115907	371969	34.46	0.000
Residual Error	6	64763	10794		
Total	9	1180670			

Source	DF	Seq SS
Bulk gas flow	1	84666
CF$_4$	1	119316
Power	1	911925

Conclusions

We can see that the full model is certainly adequate as the R^2 value is large for a model with only three terms, and all of the predictor variables are significant at the .05 level of significance.

To see that the coefficient for bulk flow is physically interpretable, we can show, by pairing up points from the scatter plot such that the line connecting the two points has a positive slope, that the average rate of change in the etch rate per unit change in bulk gas flow is 1.48, which doesn't differ greatly from 1.71. (There is no reason why they should be the same, but they should not differ very much since a regression coefficient for data from an experiment such as this one essentially measures the "collective slope".) Similarly, if we regressed etch rate against only bulk gas flow we would see that the regression coefficient is (still) 1.71.

We could perform the same type of calculations for the other coefficients and we would see similar results. This is in stark contrast to the situation with random predictors, for which the coefficients are not interpretable.

An important result of the experiment is the fact that CF_4 is significant, and the coefficient is positive. This gives a signal that increasing the CF_4 value should increase throughput.

Since the regressors are fixed, confidence intervals for the regression parameters could be appropriately constructed, using the output given above. For example, a 95% confidence interval for the Power parameter would be obtained as $4.5017 \pm 2.447(0.4898) = 4.5017 \pm 1.1985 = (3.3032, 5.7002)$, with $2.447 = t_{.025, 6}$.

These data permit the fitting of additional terms in the model (specifically, cross-product terms) but we will defer that to Chapter 12 after interactions have been discussed. Large interactions (i.e., cross-product terms) can create problems in interpreting the significance of linear terms by causing them to appear to be of little significance, but that is not the case in this example. The deleterious effects of large interactions are illustrated in Section 12.5.1.

We should always proceed beyond the basic output and look for unusual observations and test the assumptions. Neither a normal probability plot of the standardized residuals nor a plot of the standardized residuals against the fitted values shows anything unusual for these data, however. Of course, there will not be any observations with extreme x-coordinate values when the predictor values are fixed, but y-coordinates could be extreme. This might cause one or more observations to be overly influential. One solution to this problem is to downweight influential observations. The reader is referred to regression books such as Ryan (1997) for further discussion of this issue. ■

9.4   EXAMPLE WITH RANDOM REGRESSORS

It is always of interest to see how regression methods perform on actual data, but of course one of the disadvantages of using actual data is that we don't know the "right answer." That is, we don't know the true model and may not have any insights that would provide us a starting point for approximating the true model. With simulated data we know the true model and can see if the methodology that is used tracks that model. The disadvantage of simulated data of course is that the data are not real data and do not correspond to an actual modeling situation.

Naturally, it would be desirable to combine the two: that is, we know the model and also have a practical problem. Unfortunately, this is rarely possible.

We can approximate this state of affairs, however, if we use the data on college and university rankings in the annual "Best Colleges" issue of *U.S. News and World Report*. (Some of these data from the 2002 issue were used for illustration in Chapters 1 and 8.) For simplicity, we will use only the data on the top 50 national universities and try to reproduce by using regression analysis the scores that each school received, with each school rated on each of 16 criteria.

Each criterion is weighted, with the weights adding to 100%. (See `http://www.usnews.com/USNEWS/edu/college/rankings/about/weight_brief.php` and `http://www.ericdigests.org/2003-3/rankings.htm` for discussion of the weights.) Listed in Table 9.1 are the criteria that are given a weight greater than 2%. An important feature of these data from an instructional standpoint is that these are real data, but yet we know the weights that were used, so we know the relative importance of all of the predictors. This is a rare scenario in which we have actual data and yet we know, generally, what the model should be.

TABLE 9.1 **Highest Weighted Criteria for University Rankings**

Criterion	Weight (%)
Academic reputation	25
Graduation rate	16
Financial resources	10
Faculty compensation	7
Percentage of classes under 20	6
SAT score	6
Percentage of students in top 10% HS class	5.25
Graduation rate performance	5
Alumni giving	5
Freshman retention	4
Percentage of faculty with terminal degree	3
Acceptance rate	2.25

The ratings of the leading schools have been controversial over the years, and with "academic reputation" receiving by far the highest weight, there is at least a moderate amount of subjectivity. The percentages in Table 9.1 add to 94.5%, with the six omitted factors accounting for the remaining 5.5%. One immediate problem is that the faculty compensation scores are not published—for obvious reasons. One way around this problem is to use the faculty resources rank, of which faculty compensation is a part. Faculty resources, with six component parts in addition to faculty compensation, has an assigned weight of 20%. Three of those components are represented in the table, so if those were replaced by faculty resources, the net effect would be to add 4%, which would bring the total to 98.5%.

We will use this as a data-driven example, with the data helping guide the analysis. If we had prior information to suggest that certain predictors could be important, we might look at some graphs involving those predictors. In this instance, our prior information is infallible since we know how the scores were obtained. We should not, however, expect to be able to reproduce the weights in a regression model because the regressors are correlated since these are observational data.

Nevertheless, we will treat this example as roughly corresponding to a prior information situation. Although an *x*-*y* scatter plot in simple regression contains all of the information in the data, scatter plots of Y against each predictor in multiple regression can be misleading. This is because they show us only the relationship between Y and X in a simple linear regression context, as was stated previously. It would obviously be much better to see that strength adjusted in some way for the presence of one or more other predictors in the model. There is no harm in looking at these individual scatter plots as long as we recognize their limitations.

9.4.1 Use of Scatterplot Matrix

It is convenient to use a scatterplot matrix for this purpose and in Figure 9.1 we see scatter plots of score against each of the six highest-weighted criteria, in addition to plots of those criteria against each other.

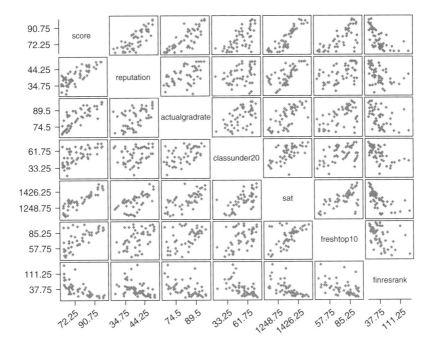

Figure 9.1   Scatterplot matrix of college rankings data.

We observe that a university's score is strongly related to the SAT score, which here is taken as the average of the published 25th and 75th percentile scores, with linear relationships also apparent with graduation rate and reputation. There is clear evidence of quadrature with financial resources and also with freshman in the top 10% of their high school class, so a nonlinear term may be needed in each of these variables. But we must keep in mind that we receive this message from a plot that ignores the other possible predictors, and we know that they are needed in the model. We will see later how we can test for the need for a nonlinear term without having to try specific terms.

9.4.2   Outliers and Unusual Observations: Model Specific

We should check for outliers and influential observations, but these can be defined only relative to a specific model, so we generally should begin the modeling process before we attempt to make such identifications, although data recording errors that are in error by an order of magnitude should of course be detectable before performing any type of formal analysis. Of course, we should also check the model assumptions after we have selected a model.

9.4.3   The Need for Variable Selection

It might seem as though we should simply use every variable in the model that would seem to be related to the dependent variable. We would indeed maximize R^2 that way, but we would also pay a heavy price. This is because $Var(\widehat{Y})$ is a nondecreasing function of the number of predictors, so if we kept adding predictors to the model, we mighty make $Var(\widehat{Y})$

unacceptably large. (Of course, this variance also depends on the values of the predictors.) Recall from Section 8.2.10 that the width of a prediction interval for Y depends in part on $Var(\widehat{Y})$, so inflating $Var(\widehat{Y})$ considerably could cause a prediction interval to be so wide as to be almost useless. For example, a prediction interval of $(0.45, 3.95)$ for the grade point average of a particular student after four years if the student were admitted is of no value at all.

Model building should generally be iterative and with a moderate-to-large number of candidate predictors we should not expect to arrive at the final model very quickly. We could start by fitting the model with all of the candidate predictors and see how good the fit is. If we have very many candidate predictors, however, and here we have 16, excluding graduation rank since that is a function of the actual graduation rate, we know that we will not use all of them, and perhaps not even half of them. Furthermore, the output will not tell us which ones we should use since this decision cannot be made on the basis of the magnitude of the t-statistics.

Therefore, a better approach is needed, and specifically some *screening* should be performed. Of course, in this case we know that all 16 predictors are part of the true model, but if a predictor contributes very little in terms of its contribution to R^2, the extent to which it inflates $Var(\widehat{Y})$ may more than offset that contribution.

9.4.4 Illustration of Stepwise Regression

One screening tool is *stepwise regression*. With this technique, which is discussed further in Section 9.6.3, variables are sequentially added to the model if they pass a threshold test. Variables that have been added might become redundant at a later stage, however, so variables are also tested for possible deletion at each step. These decisions are based on the outcome of partial-F tests. We won't be concerned with the mechanics, however, because it is not possible to perform stepwise regression by hand.

We want to perform stepwise regression with more than just linear terms in the 16 candidate predictors because there was evidence of curvature in two of the scatter plots. We could use a specific nonlinear term that might be suggested by a scatter plot, such as Figure 9.1, but that might not be the term that is needed when other predictors are in the model. A general procedure is to use $X_i \log(X_i)$ as a proxy for an arbitrary nonlinear term, and we will use this constructed variable for each of the two predictors that exhibited curvature, in addition to the 16 linear terms. (This specific form can be justified on the basis that it is part of a particular transformation approach: the Box–Tidwell transformation technique.)

An edited version of the output from the stepwise regression procedure in MINITAB is given as follows:

```
Stepwise Regression: score versus reputation, fresretention, ...

F-to-Enter: 4 F-to-Remove: 4

Response is score on 18 predictors, with N = 48
N(cases with missing observations) = 4 N(all cases) = 52
```

Step	1	2	3	4	5	6	7
Constant	−74.394	−62.838	−59.439	−31.979	−17.977	−4.247	11.671

sat	0.1148	0.0716	0.0503	0.0307	0.0221	0.0157	0.0043
T-Value	12.55	8.66	6.63	4.41	3.49	2.67	0.77
P-Value	0.000	0.000	0.000	0.000	0.001	0.011	0.448
reputation		1.122	0.963	1.063	1.017	0.923	0.983
T-Value		7.70	8.17	11.31	12.46	12.08	14.82
P-Value		0.000	0.000	0.000	0.000	0.000	0.000
graduation rate			0.381	0.345	0.354	0.377	0.288
T-Value			5.42	6.24	7.44	8.89	6.85
P-Value			0.000	0.000	0.000	0.000	0.000
faculty resources				−0.0615	−0.0550	−0.0505	−0.0543
T-Value				−5.38	−5.53	−5.70	−7.19
P-Value				0.000	0.000	0.000	0.000
finresrank				−0.046	−0.441	−0.429
T-Value				−4.03	−4.03	−4.63
P-Value				0.000	0.000	0.000
finresranklog(frr)						0.076	0.074
T-Value						3.62	4.14
P-Value						0.001	0.000
alumgrate							0.139
T-Value							4.13
P-Value							0.000
S	5.62	3.73	2.92	2.28	1.96	1.73	1.47
R-Sq	77.39	90.24	94.15	96.50	97.48	98.09	98.66

Step	8	9	10	11	12
Constant	28.93	25.56	24.18	22.12	41.09
sat	−0.0025				
T-Value	−0.47				
P-Value	0.641				
reputation	0.942	0.936	0.891	0.941	0.924
T-Value	15.67	16.09	15.39	17.46	17.86
P-Value	0.000	0.000	0.000	0.000	0.000
actualgr	0.233	0.235	0.228	0.214	0.221
T-Value	5.76	5.90	6.05	6.31	6.83
P-Value	0.000	0.000	0.000	0.000	0.000
facresourc	−0.0510	−0.0501	−0.0449	−0.0292	−0.0336
T-Value	−7.53	−7.81	−7.00	−3.92	−4.58
P-Value	0.000	0.000	0.000	0.000	0.000
finresrank	−0.460	−0.448	−0.459	−0.435	−0.395
T-Value	−5.55	−5.72	−6.20	−6.54	−6.02
P-Value	0.000	0.000	0.000	0.000	0.000
firerlog(frr)	0.079	0.077	0.080	0.075	0.067
T-Value	4.98	5.09	5.55	5.81	5.25
P-Value	0.000	0.000	0.000	0.000	0.000

alumgrate	0.163	0.155	0.174	0.166	0.171
T-Value	5.32	6.00	6.80	7.21	7.79
P-Value	0.000	0.000	0.000	0.000	0.000
acceptrate	−0.064	−0.060	−0.048	−0.039	−0.039
T-Value	−3.43	−3.54	−2.86	−2.55	−2.69
P-Value	0.001	0.001	0.007	0.015	0.011
freshtop10log(ftt)			0.0077	0.0135	0.2779
T-Value			2.44	4.06	2.39
P-Value			0.020	0.000	0.022
class50ormore				−0.138	−0.150
T-Value				−3.29	−3.74
P-Value				0.002	0.001
freshtop10					−1.40
T-Value					−2.28
P-Value					0.029
S	1.30	1.29	1.22	1.09	1.03
R-Sq	98.97	98.97	99.10	99.30	99.39

This example is most unusual in that the algorithm terminates after the 12th step, whereas with most problems the algorithm will terminate with about half this many steps. Of course, we know that there are several variables that have very small weights, so their existence as candidate variables is what causes the algorithm to run much longer than usual.

Notice that SAT is the first variable that enters the model, with $R^2 = 77.39(\%)$ when only that predictor is in the model. Thus, a high percentage of the variation in university scores is explained by this single predictor. This also means that SAT has the highest correlation with the university score among all of the candidate predictors, and that correlation is $\sqrt{.7739} = .88$. Of course, we would guess that SAT would enter first from the scatter plots in Figure 9.1. This might seem a bit odd since there are three variables that are given a higher weight than SAT in determining the university scores, but we should bear in mind that we don't have exactly the same data that were used in determining the scores since we don't have the faculty compensation values. Nevertheless, it does seem a bit odd that SAT would enter first.

The SAT variable would enter in the first place only if the value of its t-statistic exceeded 2.0 in absolute value since the default for F-to-enter is 4.0 and $t = \sqrt{F}$. It easily qualifies for entry since $t = 12.55$, as can be observed from the output.

The second variable to enter is reputation and that should not be surprising since reputation is assigned the highest weight. Notice that SAT is removed at the ninth step and is not included in the final model. This might seem odd, but its removal at that step means that it became redundant at that point with the presence of the other variables. It was removed after the rate of alumni giving was added to the model and that is not surprising since the correlation between the two variables is .78. Notice that R^2 does not decrease when SAT is dropped from the model, thus giving further evidence of the redundancy.

The model selected by the algorithm contains 10 predictors. That is stretching things a bit because we have only 48 (complete) observations since ACT was used instead of the SAT at three schools and information was missing on another variable for one school.

We haven't quite determined the model yet, however, because the two $X_i \log(X_i)$ terms were selected and we need to determine the specific forms of the nonlinear terms to use.

If we do a good job of identifying the needed terms, we might be able to get by with fewer terms in the model. Doing a "good job" can be a challenge, however. The scatter plot in Figure 9.1 strongly suggests a reciprocal term in "financial resources rank," but when we use a more sophisticated graphical approach for determining the nonlinear form, the reciprocal message is not as strong. That message is from a *partial residual plot*, which is a plot of $e + \widehat{\beta}_i X_i$ against X_i, with e representing the n raw residuals. This plot is generally superior to plotting the residuals or standardized residuals against a predictor, as the reader is asked to demonstrate in Exercise 9.44, although no graphical approach is guaranteed to emit the appropriate signal for a transformation when one is needed.

Even though the message isn't strong, when we use the reciprocal term as a candidate variable, it is selected by the stepwise algorithm in addition to the same six terms that were in the model previously at step #9, and the R^2 value is only slightly smaller (98.80 versus 98.97).

There are other methods for selecting a subset from a group of candidate variables, including examining all possible subsets implicitly, but there is little incentive to go beyond this analysis since the results are quite good.

We might view this as the end of "stage 1." That is, we have identified six important variables—reputation, graduation rate, faculty resources, financial resources rank, acceptance rate, and alumni giving rate—and have additionally used the reciprocal of financial resources rank. This model has a very acceptable R^2 value of 98.8 and this is about as many variables as we should attempt to use in the model with only 48 observations. Since there is hardly any room for model improvement, we might forego checks to see if any nonlinear terms are needed with the other variables. We must also forego a lack-of-fit test because there are no replicates. That is, all of the schools are distinct in terms of their combination of predictor values. Obviously, with random predictors, we are not likely to have any replicates, and this is especially true when there are more than a few predictors. Methods of constructing pseudoreplicates, such as the method of nearest neighbors, are discussed in regression books.

9.4.5 Unusual Observations

The unusual observations using this model are identified by MINITAB and are given as follows:

```
Unusual Observations

Obs    reputation     score      Fit    SE Fit    Residual    St Resid
 4          4.70    96.000    96.537     1.176      -0.537       -0.81X
 7          4.50    95.000    91.190     0.299       3.810        2.89R
34          3.60    70.000    73.022     0.317      -3.022       -2.30R

R denotes an observation with a large standardized residual
X denotes an observation whose X value gives it large influence.
```

We should be concerned by the fact that observation #7 has a standardized residual close to 3. We can quickly see why this occurred. This observation is the University of Pennsylvania, which has a reputation score of 4.5, compared to 4.9 for the two schools, Stanford and M.I.T., that have the same overall score. Since the regression coefficient for reputation is 9.95, this alone accounts for the fitted value being about four points below the actual value, and there are no other data values for the school that stand out as being unusual. This point is a borderline residual outlier, but we have identified the cause and there is nothing to suggest that a recording error was made. In general, outliers should be discarded (or corrected) only if an error has been made.

Observation #4 also stands out as being potentially influential and is labeled as such by MINITAB. We can easily see why because the school, Cal Tech, ranks only 25th in graduation rate, despite being first in financial resources and second in faculty resources, and of course is fourth on the overall score. These predictor value extremes cause the school to stand out as an unusual observation in the seven-dimensional predictor space, which is why it is flagged by MINITAB.

Why does an extreme position in the predictor space make Cal Tech a potentially influential data point? As indicated in Chapter 8, there are many Java applets for regression that are available on the Internet that allow a student to see the effect of adding an extreme point in simple linear regression. One such applet can be found at http://www.mste.uiuc.edu/activity/regression. These allow a student to see the effect of influential observations by dynamically creating such observations by using the computer mouse, as was discussed in Chapter 8. It is of course easy to see this in two dimensions, but it is certainly not possible to see it in seven dimensions. The concept, however, extends directly to higher dimensions.

Of course, a point that is potentially influential need not be influential, as whether or not it is influential will depend on its Y-value. This can also be seen very easily in the two-dimensional case. For example, consider the sequence of points: $(1, 1), (2, 2), (3, 3), (4, 4)$, and (a, b). If $a = 100$ and $b = 100$, whether this point is used or not will have no influence whatsoever on the position of the regression line. But what if the coordinates are $(1, 5)$? Now the Y-coordinate being much smaller than the X-coordinate will cause the point to be far below the line that connects the other four points and will cause a substantial tilt in the line if the point is used.

Ideally, we would want all of the points to have the same influence in determining the line or plane. Certainly, that is the case when we compute a simple average of five numbers, but it almost certainly won't happen in regression because we are using a nonlinear criterion to determine the line or plane. Points that are highly influential should be downweighted or perhaps eliminated. Downweighting is discussed in books and papers on robust regression.

One point to keep in mind is that the diagnostics for outliers and potentially influential data points are single-point diagnostics. That is, they attempt to identify anomalous data points one at a time. Such methods will generally be sufficient but can fail when there are multiple anomalous points in the same general area. Methods for detecting the latter are not yet widely available in statistical software, although they are becoming available and are available in MINITAB, at least as a macro. The reader is referred to books on regression analysis for influential data diagnostics.

9.4.6 Checking Model Assumptions

With the model that we have selected, the regression equation is

```
score = 16.7 + 9.95 reputation + 0.257 graduation_rate
        −0.0543 fac_resource_rank - 0.0448 financial_resource_rank
        + 6.39 1/financial_resource_rank + 0.163 alumni_giving_rate
        − 0.0500 accept_rate
```

Note that the six factors used had assigned weights that added to 78.25, and since we are also using a function of one of the six, we'll count that weight of 2.25 twice so that the sum is then 80.50. But our R^2 value of 98.8 is considerably higher than this. How can this be? The answer is that since there are high correlations among the variables, the variables that are present in the model at least partially represent those variables that are not included. Although it is quite difficult to disentangle correlated data and obtain independent effects when the predictors are not independent, Chevan and Sutherland (1991) did propose a method for doing so. The reader with much free time and considerable computing skill and capability might apply their method to these data and compare the results with the assigned weights.

How might this regression equation be used? A university that aspires to be ranked in the top 50 could see how changes in certain variables would affect its score, starting from its present score. Care would have to be exercised in doing so, however, as schools that are not ranked in the top 50 will have lower or higher values on these predictors than those in the top 50, including a lower overall score. Thus, there would be *extrapolation* beyond the range of the data that were used in obtaining the equation. If the extrapolation is severe, the results may be poor. Although it is easy to see when extrapolation is occurring when there are only one or two predictors, this becomes somewhat problematic in higher dimensions since there is no exact way of displaying the region when there is more than a few predictors. Only approximate methods are available.

We should note that all of the coefficients have the "right" sign. That is, variables that are positively correlated with score, such as reputation, have a plus sign and variables that are negatively correlated with score have a minus sign. In particular, note that of the four rank and rate variables, the only one that is positively correlated with score is the alumni giving rate, which is the percentage of alumni who contribute to the school. The reciprocal term should have a positive coefficient because a small number for the denominator is desirable and the smaller the denominator, the larger the fraction. So a large fraction should increase the score.

In general, there is no reason why the regression coefficients should have the same sign as the sign of the corresponding correlation coefficients, however, frequently the signs will differ. This is often a source of confusion for regression users but can be explained mathematically, as is done in certain regression books.

9.4.6.1 Normality
Before the equation can be used, the assumptions of course must be checked. If the assumptions appear to be at least approximately met, then we could proceed to use it. The standard approach for checking the normality assumption is to construct a normal probability plot of the residuals or of the standardized residuals. It is preferable to use the latter since the random variables that correspond to points in a probability plot should all have the same variance. This is not the case when the raw residuals are used.

The normal probability plot of the standardized residuals is given in Figure 9.2.

This plot shows no evidence of (pronounced) nonnormality, as the indicated p-value for the Anderson–Darling normality test is far in excess of .05. Such a plot can also help identify standardized residual outliers, and the extreme point in Figure 9.2 was previously identified as the University of Pennsylvania.

Although this is a standard method of checking for nonnormal errors, it is not the best approach because, as mentioned previously, the residuals are "less nonnormal" than the errors when the errors are nonnormal. A simulation envelope could be used instead of

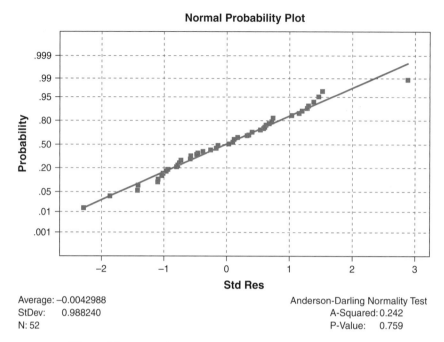

Figure 9.2 Normal probability plot of standardized residuals.

a normal plot, but a simulation envelope is beyond the scope of this book. Simulation envelopes are discussed in detail in Ryan (1997).

9.4.6.2 Constant Variance

We should check the constant error variance assumption by plotting the standardized residuals (not the raw residuals) against $\widehat{Y}$ and hopefully see a configuration of points that is practically a horizontal band of constant width. Although commonly used to check the constant error variance assumption, the raw residuals should not be used because their variance is not constant, even when the errors have a constant variance. The plots obtained using the raw residuals and the standardized residuals will generally be quite similar, so using the raw residuals usually won't lead the user astray, but there is no point in using a method that is theoretically improper when the proper plot can be constructed just as easily.

The configuration of points in Figure 9.3 is about as close to being ideal as we could realistically expect. Since the vertical spread of points is roughly constant over $\widehat{Y}$, we would expect to see roughly the same type of pattern when we plot the standardized residuals against the individual predictors. That in fact does happen so the plots are not shown here. Thus, the assumption of constant error variance appears to be at least approximately met.

9.4.6.3 Independent Errors

This assumption will often be violated when data are collected over time, such as occurs with economic data. In this example, the data were almost certainly not obtained all at once,

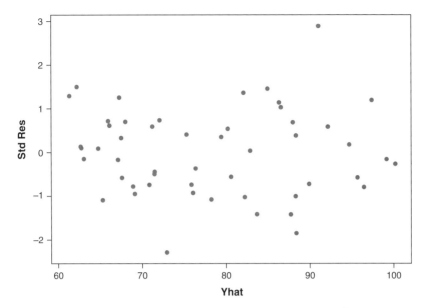

Figure 9.3 Plot of the standardized residuals against $\widehat{Y}$.

but we would have no reason to suspect that this assumption is violated because we are not measuring the schools at different points in time.

In general, we can check this assumption by constructing a time sequence plot of the residuals and/or by looking at the autocorrelation function of the residuals. These were discussed in Section 8.2.6.1.

For this example, there is no evidence of the errors being correlated, and there is no reason why they should be, even if the data were collected on different days, for example, since data on different schools are involved, not repeated measurements on the same schools.

9.4.7 Summary of Example

Analyzing regression data requires considerable effort, more so than just examining the output from standard statistical software. Analyzing the data when the predictors are random can be especially challenging and certain methods not presented herein may occasionally be necessary. The model assumptions should always be checked.

9.5 EXAMPLE OF SECTION 8.2.4 EXTENDED

In Section 8.2.4 we considered a modeling problem as discussed by T. Raz and M. Barad in their paper "In-process Control of Design Inspection Effectiveness" (*Quality and Reliability Engineering International*, **20**, 17–30, 2004). Recall that their objective was to model the defect escape probability (PROB) as part of a new methodology for the in-process control of design inspection.

In the first modeling phase, the authors decided that PROB regressed against LOG(SIZE) produced the best simple linear regression model, with SIZE designating the size of the product that was inspected, with the units being thousands of lines of code. After the preparation has been completed, the number of person-hours spent preparing for the inspection (PREP) is known, and that variable becomes a potential model predictor. The researchers decided to use a standardized version of PREP by dividing it by the measure of size, LOG(SIZE), that was used in the simple linear regression model. The data were then "reclustered" using this constructed variable and LOG(SIZE) as clustering variables. When backward elimination was used as the variable selection algorithm (i.e., starting with all of the candidate predictors in the model and then testing for deletion; see Section 9.6.2), LOG(SIZE) dropped out of the model, so the data were clustered for a third time, using only PREP/(LOG(SIZE)) as the clustering variable. This resulted in an R^2 value of .65 for that simple regression model.

After the inspection has been performed, the number of person-hours invested in the inspection meeting (INSP) is known and thus becomes a potential model predictor. When the data were clustered using all three variables and backward elimination was again performed, PREP/(LOG(SIZE)) dropped out of the model. This necessitated another reclustering, this time using the two remaining variables. The new data, which were used to develop the model, are as follows:

LOG(SIZE)	INSP/PREP	PROB
7.14	2.25	1.00
7.59	0.85	0.83
8.04	0.17	0.57
8.69	0.99	1.00
4.82	0.39	0.00
6.65	0.22	0.20
5.97	1.33	0.11
6.74	1.07	0.67

The regression output is shown below, with the R^2 value suggesting a good fit. This is potentially misleading, however, because only eight data points are being used to estimate three parameters, thus spreading the data rather thin. One suggested rule of thumb (Draper and Smith, (1998)) is there should be at least ten observations for each predictor. Here that requirement is not met, just as it won't generally be met in certain areas of application.

```
Regression Analysis: PROB versus LOG(SIZE), INSP/PREP

The regression equation is
PROB = - 1.48 + 0.257 LOG(SIZE) + 0.261 INSP/PREP
```

Predictor	Coef	SE Coef	T	P
Constant	−1.4756	0.3941	−3.74	0.013
LOG(SIZE)	0.25678	0.05538	4.64	0.006
INSP/PREP	0.26104	0.09779	2.67	0.044

```
S = 0.177228      R-Sq = 85.9%              R-Sq(adj) = 80.3%
```

```
Analysis of Variance

Source              DF          SS          MS          F          P
Regression          2      0.95970     0.47985      15.28      0.007
Residual Error      5      0.15705     0.03141
Total               7      1.11675

Source              DF      Seq SS
LOG(SIZE)           1      0.73587
INSP/PREP           1      0.22383
```

Of course, the model assumptions should be checked and when that is done there is no evidence of nonconstant error variance nor a strong departure from normality of the error term. The analysis also does not reveal any unusual observations.

9.6 SELECTING REGRESSION VARIABLES

Stepwise regression was introduced in Section 9.4 as one method that can be used for selecting regression variables. This technique is beneficial when we have a moderate-to-large number of available predictors and we want to identify the most important ones. Stepwise regression is especially useful when there are so many available predictors (say, 50 or more) that it would be impractical to consider all possible combinations of variables.

The motivation for selecting a subset of available predictors is that using all of the available predictors would unduly inflate $Var(\widehat{Y})$, as was stated in Section 9.4. This should be intuitively apparent, because $\widehat{Y}$ is a linear combination of the $\widehat{\beta}_i$, which of course are random variables that have their own variances. Thus, the more $\widehat{\beta}_i$ that are in the model, the greater will be $Var(\widehat{Y})$.

In the sections that follow, we present the methods that are available for selecting a subset of predictors, with no one method always being best. The reader should keep in mind, however, that variable selection is a double-edged sword. When available predictors are highly correlated, we generally don't want to use all of them, so the use of some variable selection approach certainly seems desirable. We should bear in mind, however, that variable selection is a bit risky under such conditions as small perturbations in the data can result in an entirely different subset of predictors being selected. Unfortunately, when the predictors are random variables (the usual case), we will almost certainly have some moderate-to-high correlations between available predictors if the number of such predictors is not small. Thus, if another sample were taken and a subset selected, the second subset could differ considerably from the first one. Although variable selection will often be necessary, this is something that should be kept in mind.

It is generally stated that variable selection is not a good strategy unless the correlations between the available predictors are small. In such cases it will be relatively easy to select a good subset.

It is possible to use all or almost all of the available predictors if one is willing to use *ridge regression* or some other alternative to least squares. These methods are discussed in books on regression.

9.6.1   Forward Selection

One simple approach would be to start adding variables, one at a time, to form a model, and then stop when some threshold value is not met. This approach has been termed *forward selection* by Draper and Smith (1998) and others.

Although we always use software for forward selection, it is helpful to know how the subset is obtained algorithmically. Partial-F (or t-) tests are performed, generally with a threshold value of 4 for an F-value and 2 for a t-value. That is, the first variable selected enters the model only if its t-statistic exceeds 2, which is equivalent to the F-statistic exceeding 4. The predictor that has the largest such t-statistic or F-statistic is the first predictor to enter the model if the statistic exceeds its threshold value.

Although forward selection is frequently used, it does have a few shortcomings. First, if there are k available predictors, there are only (at most) k possible models that are examined with this approach, whereas there are $2^k - 1$ possible subsets, in general. (Each variable can be either in or out of the model, and we exclude the case when they are all out, thus producing $2^k - 1$ possible subsets.)

Accordingly, when k is large, only a small percentage of possible subset models are examined, and there is no guarantee that the subset used will satisfy certain optimality criteria (such as maximum R^2 for a given subset size). See also the discussion in the chapter summary.

This is not a sufficient reason for eschewing forward selection, but it is something that should be kept in mind. The significance levels are also unknown because the best variable at each stage is tested for model inclusion, whereas one of the remaining candidate variables would have to be picked arbitrarily in order to have a true significance level for any one particular test. This uncertainty should not preclude the use of hypothesis tests in subset selection procedures, however; it simply means that some thought should be given to the selection of the significance level for each test. There has been much discussion of this problem in the literature regarding subset selection (and not restricted to just forward selection), but definitive guidelines are lacking, especially for the case of correlated regressors.

Papers in which this problem has been discussed include those by Pope and Webster (1972), Freedman (1983), Butler (1984), Pinault (1988), and Freedman and Pee (1989). The message that emerges is that there can be a considerable difference between the actual significance levels and the nominal levels when the ratio of the sample size to the number of regressors is small. Therefore, it is desirable to try to keep this ratio reasonably large, if possible. [As stated previously, a rough rule of thumb given by Draper and Smith (1998) is that the ratio should be at least 10:1, whenever possible. In certain applications, however, such as chemometrics, this won't be possible as the number of variables can even exceed the number of observations.]

The difference between the actual and nominal levels will also increase as the number of variables that have entered the model increases, as discussed by Butler (1984). Because of this result, Butler (1984) suggests that the nominal significance level be increased as more variables enter the model, as the nominal significance level is closer to the true level when the number of variables in the model is small, with $m = k/2$ given somewhat as a line of demarcation. (As before, m denotes the number of predictors in the model.)

Butler (1984) supports the use of forward selection, but the procedure has been found to be inferior to other selection procedures [e.g., see Draper and Smith (1998) or Seber (1977, p. 217)]. One condition under which forward selection will fail is when variables are jointly significant but not individually significant. In particular, an example could be constructed with neither of the two predictors being highly correlated with Y, but R^2 is virtually 1.0 when both predictors are used in the model. If forward selection is applied to that data set, the better of the two variables, X_2, barely enters the model, using $t = 2$ as a rough cutoff, as the value of the t-statistic is only 2.142. (If we had arbitrarily selected X_2, we would then have a valid hypothesis test with the nominal level the same as the true level, and the calculated value of 2.142 could then be compared directly to the .05 significance level t-value of 2.101.)

With a larger number of candidate variables, we might also encounter subsets for which "the whole is greater than the sum of the parts" (in terms of R^2), which could similarly have deleterious effects on the performance of the forward selection approach.

9.6.2 Backward Elimination

A related procedure is *backward elimination*, in which the starting point is the model with all of the available variables. The partial-F tests are then used to determine if the worst variable at each testing stage can be deleted, with "worst" determined by $\min_i (r_{YX_i.X})$, with X representing the other predictors that are in the model, and the parenthetical expression denoting the partial correlation between Y and X_i, adjusted for their relationships with the predictors that are already in the model. Although many data analysts prefer backward elimination to forward selection, the former has the same general shortcomings as the latter in that not all subset models are examined and the hypothesis tests are correlated.

9.6.3 Stepwise Regression

Stepwise regression, as illustrated for the example in Section 9.4.4, is a combination of these two procedures. The procedure starts in "forward mode," but unlike forward selection, testing is performed after each regressor has been entered to see if any of the other regressors in the model can be deleted. The procedure terminates when no regressor can be added or deleted. Although this procedure also has the aforementioned shortcomings of correlated hypothesis tests and the failure to examine all possible models, stepwise regression can be used when there are so many variables (say, more than 50) that it is impractical to implicitly or explicitly consider all possible subsets.

The results produced can be misleading, however, since all possible subsets are not considered. Of course, this also applies to forward selection and backward elimination. Another problem, which also applies to all three, is that there can be considerable variation in results when the variables are highly correlated: that is, small perturbations in the data can lead to different subsets being selected.

9.6.3.1 Significance Levels

One determination that must be made when stepwise regression is used is the choice of (nominal) significance levels for the hypothesis tests. As in forward selection, the true significance levels are unknown since the tests are not independent. The actual significance levels are determined by the specification of F_{IN} and F_{OUT} when statistical software is used for the computations. The general idea is to choose these two F-values in such a way that the nominal significance levels will translate into desired actual significance levels, while recognizing that this cannot be done exactly.

Some statistical software packages have default values for which $F_{IN} = F_{OUT}$, such as MINITAB, which uses 4.0 for each, but it isn't mandatory that they be equal. It is necessary, however, to have $F_{OUT} \leq F_{IN}$; otherwise, a variable that was added in a particular step could then be removed in the next step.

9.6.4 All Possible Regressions

The consideration of all possible subsets is generally referred to as *all possible regressions*. When the number of available variables is small, it is practical to compute all possible regressions explicitly. When k is larger than, say, five or six, however, it is desirable to have all possible regressions computed *implicitly*.

To illustrate the latter, let's assume that we wish to determine the subset of each size that has the largest R^2 value, and that we have ten available variables. Of the 1,023 subset models, we would expect that many of these would have low R^2 values simply because not all of the ten regressors are apt to be highly correlated with Y. If these inferior subsets can be identified, then the computations need not be performed for them.

For example, if the model with only X_1 has a higher R^2 value than the model with both X_4 and X_5, then we know that the latter subset will be inferior, in terms of R^2, to all subsets (X_1, X_i), $i = 2, 3, \ldots, 10$. This is the general idea behind the *branch and bound algorithm* of Furnival and Wilson (1974), which can be used to efficiently (and implicitly) perform all possible regressions so as to identify the best subset(s) of each size (and/or overall) according to some criterion.

Obviously, all possible regressions provides the user with more information than is available when forward selection, backward elimination, or stepwise regression is used. Another disadvantage of these last three is that they select a single subset, whereas it is unlikely that one subset will stand out as being clearly superior to all other subsets. It is preferable to provide the user with a list of good subsets and then allow the user to select a subset, possibly using nonstatistical considerations (e.g., costs, ease of measurement) in reaching a decision.

It should also be noted that our objective should be to obtain a good model for the *population* (univariate or multivariate depending on whether the regressors are fixed or random), not just for the sample data that we have. The all possible regressions approach utilizes only sample statistics, whereas the other three methods involve hypothesis tests relative to a population. As pointed out by Berk (1978), it is important that the four procedures be compared relative to known population values, as a comparison based solely on the sample data will show an inflated advantage for all possible regressions. Berk's comparison study showed the latter to have a very slight advantage over forward selection and backward elimination (stepwise regression was not included in the study).

All possible regressions can be performed in MINITAB as well as with major statistical software in general.

9.6.4.1 Criteria

We may use one or more criteria in conjunction with all possible regressions. We might choose to look at the best subset(s) of each size in terms of R^2, or perhaps in terms of adjusted R^2. [The latter is a modification of R^2 that is defined as $R^2_{adjusted} = 1 - (1 - R^2)(n - 1)/(n - p - 1)$. $R^2_{adjusted}$ can decrease as the number of regressors increases.] With the former we would have to determine the "point of diminishing returns," which might suggest (roughly) the number of regressors that should be used, whereas the latter would determine that point directly.

Mallows' C_p

Alternatively, one might use Mallows' C_p statistic. The statistic was developed by Colin Mallows (and named for another prominent statistician, *Cuthbert Daniel*) as a vehicle for determining a reasonable range for the number of regressors to use. It was not intended to be used to select a particular subset, although it is often used for that purpose. Originally presented at a regional statistics conference in 1964, the C_p statistic is described extensively in Mallows (1973) and Seber (1977, pp. 364–369).

The statistic is defined here as

$$C_p = \frac{SSE_p}{\widehat{\sigma}^2_{\text{FULL}}} - (n - 2p) \tag{9.2}$$

with p denoting the number of parameters in the model, $\widehat{\sigma}^2_{\text{FULL}}$ is the estimator of σ^2 obtained when all of the available variables are used, SSE_p is the residual sum of squares for a particular subset model with $p - 1$ regressor terms, and n is the number of observations. [*Note*: We are tacitly assuming that the full model is the correct model so that $\widehat{\sigma}^2_{\text{FULL}}$ is an unbiased estimator of σ^2. Prior information on σ^2 might also be used in obtaining an estimate, but we will assume in this section that σ^2 is to be estimated using all of the available regressors. Mallows (1973) did not define C_p with σ^2 estimated in any specific manner, but the possible use of $\widehat{\sigma}^2_{\text{FULL}}$ was implied, and in fact this is the customary approach.]

The reason that the C_p statistic can be used for determining the approximate number of regressors to use can be explained as follows. If we use a subset model and assume that the full model is the correct model, then $E(Y_s) \neq \xi$, with $E(Y) = \xi$ for the full (true) model, and Y_s denoting the left side of a subset model. Thus, there is bias in the subset model: the least squares estimators are biased so that $E(\widehat{Y}_s) \neq \xi$, in addition to $E(Y_s) \neq \xi$. This bias might be more than offset, however, by a reduction in the variance of the predicted response resulting from use of the subset model.

When two estimators are to be compared and at least one of the two estimators is biased, it is reasonable to use *mean squared error* (variance plus squared bias) to compare the two estimators, as was explained in Section 4.2.4.

Here we are comparing two models over n data points rather than comparing estimators, however, so a slightly different approach is required. The *total squared error* is defined as $E(\sum_{i=1}^{n}(\widehat{Y}_{s,i} - \xi_i)^2)$, and we may easily show that

$$E\left(\sum_{i=1}^{n}(\widehat{Y}_{s,i} - \xi_i)^2\right) = E\left(\sum_{i=1}^{n}(\widehat{Y}_{s,i} - \eta_{s,i} + \eta_{s,i} - \xi_i)^2\right)$$

$$= E\left(\sum_{i=1}^{n}(\widehat{Y}_{s,i} - \eta_{s,i})^2\right) + \sum_{i=1}^{n}(\eta_{s,i} - \xi_i)^2$$

since $\eta_{s,i}$ here denotes $E(\widehat{Y}_{s,i})$, so the middle term vanishes; $E(\eta_{s,i} - \xi_i)^2 = (\eta_{s,i} - \xi_i)^2$; and we will use SSB to denote $\sum_{i=1}^{n}(\eta_{s,i} - \xi_i)^2$. By definition,

$$E\left(\sum_{i=1}^{n}(\widehat{Y}_{s,i} - \eta_{s,i})^2\right) = \sum_{i=1}^{n} Var(\widehat{Y}_{s,i})$$

It is desirable to divide by σ^2 so as to have a unit-free quantity, so putting these simplified components together and then dividing by σ^2 produces the *total standardized squared error*:

$$T = \frac{SSB + \sum_{i=1}^{n} Var(\widehat{Y}_{s,i})}{\sigma^2}$$

$$= \frac{SSB}{\sigma^2} + p$$

since it can be shown that $\sum_{i=1}^{n} Var(\widehat{Y}_{s,i}) = p\sigma^2$.

When the model is correct, $E(\widehat{\sigma}^2) = E(\sum(Y - \widehat{Y})^2/(n - p)) = \sigma^2$, so that $E(\sum(Y - \widehat{Y})^2) = E(SSE) = (n - p)\sigma^2$. When a subset model is used, however, $\widehat{\sigma}^2$, will be a biased estimator of σ^2, just as the least squares estimators are biased. Specifically, $E(\widehat{\sigma}^2) = SSB/(n - p) + \sigma^2$, as can be shown.

Thus, $SSB = E(SSE) - (n - p)\sigma^2$, so SSB could be estimated as $\widehat{SSB} = SSE - (n - p)\widehat{\sigma}^2$. It follows that we would estimate T as

$$\widehat{T} = \frac{SSE - (n - p)\widehat{\sigma}^2}{\widehat{\sigma}^2} + p$$

$$= \frac{SSE}{\widehat{\sigma}^2} - (n - p) + p$$

$$= \frac{SSE}{\widehat{\sigma}^2} - (n - 2p)$$

and, as indicated previously, one approach would be to estimate σ^2 using the full model, so that $\widehat{\sigma}^2 = \widehat{\sigma}^2_{FULL}$.

With such an estimator, we then have Eq. (9.2), so C_p is an estimator of the total standardized squared error.

If a subset model were the true model (and $\widehat{\sigma}^2 = \sigma^2$), we would have

$$E(C_p) = \frac{E(SSE_p)}{\sigma^2} - (n - 2p)$$

$$= \frac{(n - p)\sigma^2}{\sigma^2} - (n - 2p)$$

$$= p$$

Thus, a reasonable starting point would be to examine subsets for which C_p is close to p and, in particular, less than p.

One approach would be to plot C_p against p and draw the line $C_p = p$ on the chart, with subsets that have small C_p values then plotted on the chart. Here the emphasis is on looking at those subsets that have C_p close to p. Should we look at subsets that have $C_p \approx p$, or should we look at those subsets that have the smallest C_p values? Draper and Smith (1998) suggest that this is largely a matter of preference, whereas Mallows (1973) argues against automatically selecting the subset with the smallest C_p value.

If, however, we have two subsets of different sizes and $C_p \approx p$ for each subset, it will frequently be desirable to use the smaller of the two subsets. The logic behind such a decision is as follows. If we select a (larger) subset with a larger C_p value (with C_p having increased by more than the increase in p), we have in essence proceeded past the point of diminishing returns in terms of R^2. As mentioned briefly in Section 9.6.4.1, R^2_{adjusted} can be used to determine this point and Kennard (1971) has noted the relationship between R^2_{adjusted} and C_p. [See also Seber (1977, p. 368).] In particular, it can be shown (as the reader is asked to do in Exercise 9.26) that if the relationship $C_p = p + a$ holds for two values of p (with a denoting a constant), the smaller value of R^2_{adjusted} will occur at the larger of the two values of p.

When we use all of the available variables, we have $C_p = p$ (as can easily be shown). This will often be the smallest value of C_p, but that doesn't necessarily mean that we should use all of the variables. In particular, this can happen when the full model has a very high R^2 value (such as $R^2 > .99$), and σ^2 is estimated from the full model.

Although it would seem to be highly desirable to use the full model when it has a very high R^2 value, such an R^2 value will cause $\widehat{\sigma}^2$ to be quite small, which will in turn magnify the difference between subsets in terms of SSE. This follows because SSE is divided by $\widehat{\sigma}^2$ in the computation of C_p.

Specifically, C_{p+1} will be smaller than C_p if $(SSE_p - SSE_{p+1})/\widehat{\sigma}^2 > 2$, which could happen solely because of a very small value of $\widehat{\sigma}^2$. This underscores the fact that C_p should be used only as a guide in selecting the best subset size.

The difference between the two SSE values divided by $\widehat{\sigma}^2$ is the usual F-statistic for testing the addition of a new regressor, with the minor difference that σ^2 is typically estimated using the larger of the two models, whereas here it is estimated from the model using all of the available variables. Thus, we may say that seeking the minimum value of C_p is roughly equivalent to using an F_{IN} value of 2.

9.7 TRANSFORMATIONS

Variable transformations in regression are used to improve the quality of the fit of the model and to help satisfy the assumptions. As stated in Section 8.2.10.2, transformation of Y should generally not be done to improve the fit because if the assumptions were approximately met before the transformation, that won't be the case after the transformation. Instead, transforming Y should be done to help meet the assumptions of normality and constant error variance. The Box–Cox transformation approach is frequently used to identify a transformation that will produce approximate normality. (This method is a likelihood-based approach that is explained and illustrated in regression books.) Sometimes a transformation that produces approximate normality will also tend to equalize the variances. An alternative to trying to equalize the variances would be to use weighted least squares, which adjusts for the unequal variances.

Transforming Y to provide approximate normality and/or to roughly equalize the variances will generally cause the quality of the fit to be lost. That is, if R^2 were .90 before the transformation of Y, it will likely be less after the transformation. (*Note:* In computing R^2 after the transformation of Y, it is necessary to transform the fitted values on the transformed scale back to the original scale before computing R^2. Not doing so would cause the R^2 values to not be comparable because of the different scales.) The loss in the quality of the fit can be avoided by using the "transform-both-sides" approach. That is, the transformation applied to Y is also applied to the right-hand side of the equation, excluding the error term. The transformed model $Y^{0.4} = (\beta_0 + \beta_1 X_1 + \beta_2 X_2)^{0.4} + \epsilon$ is an example. Note that this is not a linear regression model, however, so nonlinear regression would have to be applied (see Section 9.10). Alternatively, one could use X_1 and X_2 each raised to the 0.4 power and use linear regression, but this will usually be a less satisfactory approach.

9.8   INDICATOR VARIABLES

It will sometimes be desirable to see if a common regression model applies to each level of one or more categorical variables, which might represent different physical plants, ball bearings of different diameters, and so on. This determination can be made through the use of *indicator variables*.

When there are only two categories, a single indicator variable would be used, with the values of the variable being zero and one. This results in an additional model parameter, β_i, and a test of $H_0: \beta_i = 0$ would determine whether or not there is a difference in the two categories. If there is a difference, β_i is added to β_0 [since $\beta_i(1) = \beta_i$], so that the difference appears in the intercept, which in turn affects the expected value of the response variable for a given set of predictor values.

When there are three categories, two indicator variables would be used. (You might ask why we can't just use a single variable with three levels. The reason is that an indicator variable is used to represent presence or absence, and a "2" would correspond to neither. Furthermore, there would also be a problem quantitatively since, for example, plant #2 is probably not double plant #1 in any sense.)

To illustrate the use of two indicator variables, assume that there are three categories of a qualitative variable so that two indicator variables must be used. The values of the two indicator variables (say, X_3 and X_4) used to represent the three levels (categories) would be (1, 0), (0, 1), and (0, 0), respectively. The value of X_3 would indicate whether or not the first level was used, with "1" indicating that it was used and "0" indicating that it was not used; and the same for X_4. The combination (0, 0) would thus mean that neither the first level nor the second level was used, which of course means that the third level was used.

Operationally, if MINITAB were used, for example, two columns of zeros and ones would be appended to the other columns, with the zeros and ones positioned so as to correspond to the appropriate data.

A hypothesis test of $H_0: \beta_i = 0$ would be performed for each of the two additional parameters that are created. Failing to reject both hypotheses would mean that two of the categories do not differ from the third, and thus there is no difference between any of the categories. Rejecting both hypotheses would mean that two of the categories differ from the third (but not necessarily from each other), whereas rejecting only one of the two hypotheses would mean that the corresponding category differs from the other two.

There are other uses of indicator variables, such as for interaction effects and in Analysis of Variance (see Chapter 12). Indicator variables are covered in many regression books.

9.9   REGRESSION GRAPHICS

The methods described in Section 9.5 are strictly numerical. There are graphical methods that have value for determining a useful model, in addition to the methods that have been presented to this point. In particular, a *partial residual plot*, mentioned in Section 9.4.4, will generally work better than a standardized residuals plot in suggesting a useful transformation of a candidate variable or a transformation of a variable that is already in the model. As stated previously, a partial residuals plot for X_i is a plot of $e + \widehat{\beta}_i X_i$ against X_i. If the configuration of plotted points has a definite shape, such as quadrature, then a term suggested by the plot may improve the fit of the model, although there is no guarantee that this plot will emit the correct signal.

There are many other useful graphical methods for regression. The interested reader is referred to Cook (1998) and Cook and Weisberg (1994, 1999).

9.10 LOGISTIC REGRESSION AND NONLINEAR REGRESSION MODELS

The use of least squares has been assumed to this point in the chapter and also in the preceding chapter. Least squares won't always be appropriate, however, because the error distribution won't always be normal, or even symmetric. For example, if Y is binary (such as 0 or 1), the error term accordingly can also assume only two values and thus cannot have a normal distribution.

Logistic regression is typically used for regression modeling when the response variable is binary, provided that the logistic function seems appropriate. The model is given by

$$\pi(X_1, X_2, \ldots, X_m) = \frac{\exp(\beta_0 + \beta_1 X_1 + \cdots + \beta_m X_m)}{1 + \exp(\beta_0 + \beta_1 X_1 + \cdots + \beta_m X_m)} + \epsilon \tag{9.4}$$

with $\pi(X_1, X_2, \ldots, X_m,)$ representing $P(Y{=}1 \mid X_1, X_2, \ldots, X_m)$. This is obviously not in the form of a linear model, but that can be accomplished by using the equivalent form

$$\log\left(\frac{\pi(X_1, X_2 \ldots, X_m)}{1 - \pi(X_1, X_2 \ldots, X_m)}\right) = \beta_0 + \beta_1 X_1 + \cdots + \beta_m X_m + \epsilon \tag{9.5}$$

The left-hand side of Eq. (9.5) is quite different from the left-hand side of a linear regression model, however, and it is not possible to obtain a closed-form expression for the parameter estimates, so the estimates must be obtained iteratively. The method of maximum likelihood is almost always used, but there is evidence (see King and Ryan, 2002) that maximum likelihood can perform poorly under certain conditions.

In general, logistic regression is more complicated than linear regression because there are certain pitfalls in logistic regression that one does not encounter in linear regression, and graphical methods, in particular, are not very useful in logistic regression. Although logistic regression is used primarily in health and medical applications, it is also applicable in engineering. In particular, it could be used to model nonconforming units and *polytomous logistic regression* (more than two possible outcomes) might be used to model nonconformities. (See Chapter 11 for a discussion of nonconformities and nonconforming units.) See Hosmer and Lemeshow (2000) for details on logistic regression.

A nonlinear regression model is often needed when the relationship between the response variable and one or more predictor variables is nonlinear. That is, the model is nonlinear in the parameters. An example would be

$$Y = \beta_0 X^{\beta_1} + \epsilon$$

This model is nonlinear in the parameters and is thus a nonlinear model. If this were actually a mechanistic model (see Chapter 10) and the scientists believed that ϵ was quite small, the logarithm of each side of the model might be taken (acting as if ϵ were zero). This would produce a linear model (without an error term) that could be used either as a substitute for the nonlinear model or to obtain starting values for the parameter estimates for the nonlinear model since the estimates are obtained iteratively. See Bates and Watts (1988) for additional reading on nonlinear regression models.

During the past ten years, nonparametric regression has become increasingly popular, but the field is still underdeveloped compared to linear regression. With nonparametric regression there is no model and the user essentially lets the data "model themselves." Probably the most popular method is *loess* (also "LOWESS"), which stands for "locally weighted regression." Since there is no model, inference does become a problem, however.

9.11   REGRESSION WITH MATRIX ALGEBRA

Regression books contain expressions that utilize matrix algebra. For example, the values of the predictors can be used to form a matrix, with a column of ones being the first column in the matrix (which permits estimation of the intercept). Such expressions are useful for understanding literature articles, but not for computation since it would be impractical to use regression analysis without appropriate software, especially when there are more than two predictors. Readers interested in regression analysis with matrix algebra are referred to Ryan (1997) and other books on the subject.

9.12   SUMMARY

Regression is almost certainly the most widely used statistical technique. It has great applicability and has been used in applications as wide ranging as modeling acid rain data, predicting spatial patterns of cattle behavior, estimating percent body fat, and predicting college grade point averages. A very general and important application is the use of regression tools in analyzing data from designed experiments. This is discussed in Chapter 12.

Regression is also the most misused statistical technique and this is because of the many ways in which it can be misused. A common misuse is to apply it to data collected over time without checking to see if this has induced at least moderate correlation in the regression residuals, over and above the small correlation that is due to the fact that the residuals sum to zero. Another misuse is to try to interpret regression coefficients for correlated predictors as if there weren't any other predictors in the model, failing to recognize that coefficients are not interpretable when the predictors are correlated. Thus, there is no such thing as a wrong sign for a regression coefficient computed using correlated regressors, a point that is frequently overlooked in the literature and by practitioners. Such interpretation can be made only when regression is applied to data from a designed experiment with the predictors uncorrelated. In general, it is important to check the assumptions and also to check for anomalous data points. Bad data points should of course be removed, with unusual data points perhaps downweighted. Various methods are available for doing so.

In any given year, the number of applications of logistic regression far exceeds the number of applications of linear regression, and this is because logistic regression is used so extensively in the medical and health areas. It should be noted that logistic regression also has many uses in engineering. Because mechanistic models are generally nonlinear models, nonlinear regression has considerable relevance to engineering, and nonlinear regression tools might be used to refine mechanistic models. This is discussed further in Chapter 10.

In general, there is a large collection of regression tools that are available, ranging from the very simple to the rather complex. Entire books have been written on regression (many such books), and books have also been written on certain regression topics, so this chapter and the preceding chapter should be viewed as just an introduction to a very extensive subject.

This introduction to multiple regression and other regression methods was almost completely devoid of computing formulas and multiple regression expressions since multiple regression is not performed by hand calculation or hand calculator and an understanding of multiple regression can be gained without becoming mired in the many formulas that are inherent in multiple regression. The reader is referred to the references that have been cited for additional reading.

REFERENCES

Bates, D. M. and D. G. Watts (1988). *Nonlinear Regression and its Applications*. New York: Wiley.

Berk, K. N. (1978). Comparing subset regression procedures. *Technometrics*, **20**, 1–6.

Butler, R. W. (1984). The significance attained by the best-fitting regressor variable. *Journal of the American Statistical Association*, **79**, 341–348 (June).

Chevan, A. and M. Sutherland (1991). Hierarchical partitioning. *The American Statistician*, **45**, 90–96.

Cook, R. D. (1998). *Regression Graphics: Ideas for Studying Regressions Through Graphics*. New York: Wiley.

Cook, R. D. and S. Weisberg (1999). *Applied Regression Including Computing and Graphics*. New York: Wiley.

Cook, R. D. and S. Weisberg (1994). *An Introduction to Regression Graphics*. New York: Wiley.

Czitrom, V., J. Sniegowski, and L. D. Haugh (1998). Improving integrated circuit manufacture using a designed experiment. Chapter 9 in *Statistical Case Studies: A Collaboration Between Academe and Industry* (R. Peck, L. D. Haugh, and A. Goodman, eds.). Philadelphia: Society of Industrial and Applied Mathematics (jointly published with the American Statistical Association, Alexandria, VA).

Draper, N. R. and H. Smith (1998). *Applied Regression Analysis*, 3rd ed. New York: Wiley.

Freedman, D. A. (1983). A note on screening regression equations. *The American Statistician*, **37**(2), 152–155.

Freedman, L. S. and D. Pee (1989). Return to a note on screening regression equations. *The American Statistician*, **43**(4), 279–282.

Furnival, G. M. and R. W. Wilson (1974). Regression by leaps and bounds. *Technometrics*, **16**, 499–511.

Hosmer, D. W., Jr. and S. Lemeshow (2001). *Applied Logistic Regression*. New York: Wiley.

Kennard, R. W. (1971). A note on the C_p statistic. *Technometrics*, **13**, 899–900.

King, E. N. and T. P. Ryan (2002). A preliminary investigation of maximum likelihood logistic regression versus exact logistic regression. *The American Statistician*, **56**, 163–170.

Mallows, C. L. (1973). Some comments on C_p. *Technometrics*, **15**, 661–675.

Pinault, S. C. (1988). An analysis of subset regression for orthogonal designs. *The American Statistician*, **42**(4), 275–277.

Pope, P. T. and J. T. Webster (1972). The use of an F-statistic in stepwise regression procedures. *Technometrics*, **14**, 327–340.

Ryan, T. P. (1997). *Modern Regression Methods*. New York: Wiley.

Seber, G. A. F. (1977). *Linear Regression Analysis*. New York: Wiley.

EXERCISES

9.1. It was stated in Section 9.3 that for the example in that section a normal probability plot of the residuals and a plot of the standardized residuals against the predicted values did not reveal any problems with the assumptions, nor did those plots spotlight any unusual observations. Construct those two plots. Do you agree with that assessment?

9.2. Lawson (*Journal of Quality Technology*, 1982, Vol. 14, pp. 19–33) provided data from a caros acid optimization study and the data are given in LAWSONCH12.MTW. Do you believe that all three of the independent variables are important (i.e, should be used in the study)? Explain.

9.3. Notice that the plot of score against reputation in Figure 9.1 exhibits curvature. Does this suggest that a nonlinear term in reputation should be included in the model? Fit the model (the data are in RANKINGS2002.MTW) with linear terms in all predictors shown in Figure 9.1 and construct a partial residuals plot for reputation and compare the message that plot gives compared with the message of the scatter plot. Comment.

9.4. Explain why a regression user will generally not construct a regression model using all of the independent variables that may seem to be correlated with Y.

9.5. The following output was obtained using the "trees" data that comes with the MINITAB software, with the objective being to predict the volume of a tree from knowledge of its diameter and height (in feet), with the latter being easier to determine than the volume.

```
Regression Analysis: Volume versus Diameter, Height

The regression equation is
Volume = - 58.0 + 4.71 Diameter + 0.339 Height

Predictor          Coef        SE Coef          T           P
Constant        -57.988          8.638       -6.71       0.000
Diameter         4.7082          0.2643       17.82       0.000
Height           0.3393         _____        2.61       0.346

S = 3.882           R-Sq =     _____     R-Sq(adj) = 94.4%

Analysis of Variance

Source             DF        SS        MS          F          P
Regression          2     7684.2    3842.1      _____     0.000
Residual Error     28      421.9    _____
Total              30     8106.1
```

(a) Fill in the four blanks.

(b) When regression is used for prediction, it is generally for predicting what will happen in the future. If, however, in this application we know the diameter and height of a particular tree, are we trying to predict the future height of that tree, or another tree, or neither tree? Explain.

9.6. A particular multiple regression equation contains three independent variables. The equation was determined using observational data, with the correlations between the independent variables ranging from .6 to .8. Explain why confidence intervals on the $\beta_i (i = 1, 2, 3)$ would not be very meaningful.

9.7. If $H_0: \beta_1 = \beta_2 = \cdots \beta_k = 0$ is rejected, does this mean that all k regressors should be used in the regression equation? Why or why not?

9.8. Consider the following regression output:

```
                    Regression Analysis

The regression equation is
Y = - 5.13 - 0.025 X1 + 0.212 X2 + 0.993 X3
```

Predictor	Coef	StDev	T	P
Constant	-5.131	4.036	-1.27	0.216
X1	-0.0255	0.1821	-0.14	0.890
X2	0.2125	0.1317	1.61	0.120
X3	0.9933	0.1102	9.01	0.000

```
S = 1.504    R-Sq = 79.1%              R-Sq(adj) = 76.4%

Analysis of Variance
```

Source	DF	SS	MS	F	P
Regression	3	204.908	68.303	30.19	0.000
Residual Error	24	54.293	2.262		
Total	27	259.201			

Source	DF	Seq SS
X1	1	0.063
X2	1	21.083
X3	1	183.762

```
Unusual Observations
```

Obs	X1	Y	Fit	StDev	Residual	St Resid
15	0.00	13.300	10.284	0.529	3.016	2.14R
16	1.00	10.400	12.380	1.211	-1.980	-2.22RX

(a) Would you recommend that all three independent variables be used in the model? Why or why not?

(b) What percent of the variability in Y is explained by the three independent variables that were used in the regression equation?

(c) Since model building is a sequential procedure, what would you recommend to the practitioner?

9.9. Assume that two predictors are used in a regression calculation and $R^2 = .95$. Since R^2 is so high, could we simply assume that both predictors should be used in the model? Why or why not?

9.10. Assume that there are nine candidate independent variables for a regression project. All possible subsets of these variables are considered, either implicitly or explicitly. Explain why there will *not* be $2^9 = 512$ possible regression models.

9.11. Assume that a multiple linear regression equation is determined using two independent variables. Sketch an example of how we would expect a plot of the standardized residuals against X_i to look if the model is a reasonable proxy for the true model and the regression assumptions are met.

9.12. Explain why it would be hard not to extrapolate in using a regression equation with two independent variables if the correlation between the two predictors in the dataset used to obtain the regression coefficients was .93.

9.13. A regression equation contains two regressors, both of which are positively correlated with Y. Explain how the coefficient of one of the regressors could be negative.

9.14. Give three examples from your field of engineering or science in which multiple regression could be useful.

9.15. Explain why a transformation of one or more of the predictors is preferable to transforming Y when trying to improve the fit of a regression model.

9.16. Assume that we have four predictors from which it has been decided to select two in order to form a regression model. If the predictors are all pairwise uncorrelated, $R^2 = .80$ when all four are in the model, $r_{YX_2} > r_{YX_1} = \sqrt{.20}$, and $r_{YX_4} < r_{YX_2} < r_{YX_3}$.

(a) What is the worst individual predictor?
(b) For which subset of size 2 is R^2 maximized?
(c) What are the lower and upper bounds on R^2 for this subset?
(d) Would it be practical to use this subset model for prediction if the actual value of R^2 was halfway between the upper bound and the lower bound? Why or why not?

9.17. D. A. Haith ["Land Use and Water Quality in New York Rivers," *Journal of the Environmental Engineering Division*, American Society of Civil Engineers, **102**(EE1), Paper 11902, February 1–15, 1976] investigated the relationship between water quality and land usage. He sought to determine if relationships existed between land use and water quality, and if so, to determine which land uses most significantly impact water quality. If such relationships do exist, a model would be sought that would be useful "for both prediction and water quality management in northeastern river basins with mixtures of land uses comparable to the basins used in the study." Haith stated:

> In the present case, a suitable regression would be one which explains a significant portion of water quality variation with a small number of land-use variables. In addition, these

land use variables would be uncorrelated (independent). Such a regression would indicate the extent to which water quality is related to land uses and the relative importance of the various land uses in determining water quality.

In working this problem, you may ignore the second sentence in this quote since methods for attempting to obtain an estimate of the independent effects of correlated variables are not well known and are beyond the scope of this text. You are welcome to address the issue, however, if desired.

Haith used correlation and regression analyses and cited papers that described similar analyses of water quality data. He gave nitrogen content as one measure of water quality, with one set of land-use variables given as: (1) percentage of land area currently in agricultural use; (2) percentage of land use in forest, forest brushland, and plantations; (3) percentage of land in residential use; and (4) percentage of land area in commercial and manufacturing use. Haith also broke land-use variables into various subcategories, but we will be concerned only with the dataset that has the four land-use variables.

The land-use information for each river basin was obtained from New York's Land Use and Natural Resource Laboratory. Additional background information can be found in Haith (1976). You are encouraged to read the latter. Discussions of the Haith paper were given by Chiang in the October issue of the same journal and by Shapiro and Küchner in the December issue.

The data are given below.

River Basin	Total Nitrogen[a]	Area[b]	Agriculture (%)	Forest (%)	Residential (%)	Commercial—Industrial (%)
Olean	1.10	530	26	63	1.2	0.29
Cassadaga	1.01	390	29	57	0.7	0.09
Oatka	1.90	500	54	26	1.8	0.58
Neversink	1.00	810	2	84	1.9	1.98
Hackensack	1.99	120	3	27	29.4	3.11
Wappinger	1.42	460	19	61	3.4	0.56
Fishkill	2.04	490	16	60	5.6	1.11
Honeoye	1.65	670	40	43	1.3	0.24
Susquehanna	1.01	2,590	28	62	1.1	0.15
Chenango	1.21	1,830	26	60	0.9	0.23
Tioughnioga	1.33	1,990	26	53	0.9	0.18
West Canada	0.75	1,440	15	75	0.7	0.16
East Canada	0.73	750	6	84	0.5	0.12
Saranac	0.80	1,600	3	81	0.8	0.35
Ausable	0.76	1,330	2	89	0.7	0.35
Black	0.87	2,410	6	82	0.5	0.15
Schoharie	0.80	2,380	22	70	0.9	0.22
Raquette	0.87	3,280	4	75	0.4	0.18
Oswegatchie	0.66	3,560	21	56	0.5	0.13
Cohocton	1.25	1,350	40	49	1.1	0.13

[a]Total nitrogen is given as mean concentration in milligrams per liter.
[b]Area is given in square kilometers.

What model would you recommend to accomplish Haith's objective, and would you recommend using all of the data points?

9.18. When a partial-F test is performed for a dataset with two predictors, the test for the coefficient of X_2 is usually written as testing for $X_2 \mid X_1$ or as $\beta_2 \mid \beta_1$. This is generally not stated for the corresponding t-test, even though we know that the tests are equivalent. Explain why it isn't necessary to use such notation for the t-test whenever all of the other regressors (just one in this case) are in the model.

9.19. Consider the following dataset for two predictors:

Y	5	3	8	7	5	4	3	7	5	4
X_1	5	3	5	5	4	3	2	6	4	7
X_2	2	1	3	2	4	3	2	3	2	5

(a) Construct the regression equation with only X_2 in the model and note the sign of the slope coefficient.

(b) Then put both predictors in the model and compare the sign of the coefficient of X_2 with the sign when only X_2 is in the model. Explain why the sign changes.

(c) Would you recommend that the model contain either or both of these predictors, or should a nonlinear form of either or both of them be used?

9.20. Discuss the conditions under which scatter plots of Y against the available predictors could be useful.

9.21. Choudhury and Mitra (*Quality Engineering*, **12**(3), 439–445, 2000) considered a problem in manufacturing tuners that required attention. Specifically, the design of the printed circuit boards (PCBs) had changed and the shop floor workers felt that the quality of soldering was poor. Since rework was expensive, there was a need to study the process and determine optimum values for the process variables for the new PCB design. The resultant experiment used nine controllable factors and two uncontrollable (noise) factors. (The latter are, in general, factors that are either very difficult to control during manufacture, or else cannot be controlled at all during normal manufacturing conditions but can be controlled in an experimental study such as a pilot plant study.)

Of the two uncontrollable factors, the engineers knew that the dry solder defect was the more important of the two types of defects (shorting was the other defect).

An analysis was performed for each of the two types of defects. Since the two response variables are each count variables, possible transformation of each variable should be considered since counts do not follow a normal distribution. The study was performed by taking, for each response variable, the square root of the count under each of the two levels of the uncontrollable factor and combining those two numbers into a single performance measure: $Z(y_1, y_2) = -\log_{10} \sum_{i=1}^{2} y_i^2 / n$. Converting the counts into a single performance measure is in accordance with approaches advocated by G. Taguchi (see Section 12.17), but it should be noted that the use of such performance measures has been criticized considerably during the past fifteen years.

Rather than use that approach, however, we will use a more customary approach and sum the defects under the two conditions of the uncontrollable factor and then take the square root of the sum. Since it is very unlikely that all nine factors under study are significant, we will view this as a screening experiment. The data are in CH9ENGSTATEXER20.MTW.

(a) Use all nine of the available predictors in the model and identify those variables for which the p-value is less than .05. Which variables are selected using this approach?

(b) Then use stepwise regression to identify the significant effects (use $F_{IN} = F_{OUT} = 4.0$). What variables are selected using this method?

(c) Compute R^2 for each of the two models and check the assumptions. Is either model or both models acceptable in terms of R^2? Which model do you prefer?

9.22. Explain why the sign of the t-statistic for testing H_0: $\beta_i = 0$ is the same as the sign of $\widehat{\beta}_i$.

9.23. Apply stepwise regression (with $F_{IN} = F_{OUT} = 4.0$) to the etch rate data that were analyzed in Section 9.3. Does the stepwise algorithm select the same model that results from selecting variables for which the p-values are less than .05 when all of the variables are in the model? Explain why the t-statistics and p-values that are displayed when the stepwise algorithm is used differ from the corresponding statistics when all of the predictors are used in the model.

9.24. Give an example of a nonlinear regression model; specifically, one that could not be transformed into a linear regression model.

9.25. A company that has a very large number of employees is interested in determining the body fat percentage as simply and subtly as possible, without having to resort to underwater weighing or the use of body calipers. The reason for doing so is to identify employees who might be expected to benefit greatly from an exercise program, and who would be encouraged to participate in such a program. The company executives read that bodyfat percentage can be well-approximated as $[(\text{height})^2 \times (\text{waist})^2]/(\text{bodyweight} \times c)$, with $c = 970$ for women and 2,304 for men. The company would like to use a linear regression equation and would also like to have some idea of the model worth. Accordingly, the company intends to select a small number of employees to use to validate the model, using underwater weighing. If the model checks out, employees' waist measurements could be obtained along with their height and weight during their annual physical exam. Is it possible to transform the stated model and obtain a multiple linear regression model? If so, how would you proceed to obtain the model? If it is not possible, explain why it isn't possible.

9.26. A case study focused on determining if physical strength, as measured in two tests, relates to performance in certain jobs for which physical strength is necessary. A study of 149 individuals, including electricians, construction and maintenance workers, auto mechanics, and linemen, was performed to see if job performance ratings by supervisors and/or performance on simulated tasks correlates with arm strength and grip strength as measured by a testing device. Data on 147 of the study participants (two observations are missing) are given in RICETESTREG-DATA.MTW. (This study was once available on the Internet but is no longer available there.)

(a) Can a useful linear regression model be developed using supervisors' ratings as the response variable? Why or why not?

(b) Do the two predictor variables correlate more strongly with simulated tasks, and can a useful model be developed for that response variable? Explain.

(c) For each dependent variable, roughly how many observations would we expect to have standardized residuals greater than 2 in absolute value if the errors have a normal distribution?

9.27. Show that if the relationship $C_p = p + a$ holds for two consecutive values of p (where a is a constant), the smaller value of R^2_{adjusted} will occur at the larger of the two values of p.

9.28. Show that $C_p = p$ when linear terms (only) in all of the available predictors are used in the model.

9.29. Consider the following regression equation: $\widehat{Y} = 13.4 + 6.0X_1 + 1.9X_2 - 2.6X_3 + 6.1X_4$. Looking at the equation coefficients, state the conditions, if any, under which we could conclude that X_1 is more than three times as important as X_2 in the prediction of Y, X_3 is negatively correlated with Y, and the predictive value of X_4 is approximately equal to the predictive value of X_1.

9.30. Consider the output given in Exercise 9.8. Is it possible to determine what the value of R^2 would have been if only X_3 had been used in the model? If so, what is the value? If it isn't possible, explain why.

9.31. For the example given in Section 9.3, what would be the width of a 90% confidence interval for the bulk gas flow parameter?

9.32. Using appropriate software, such as MINITAB, construct a 99% prediction interval for Y for the example in Section 9.3 when bulk flow $= 60$, CF_4 flow $= 5$, and power $= 550$.

9.33. Consider a problem with two regressors.

(a) If there are four selected values for each regressor and a scatter plot of all combinations of the two sets of values plots as a rectangle, would it be practical to construct a 95% prediction interval for any combination of regressor values such that each value is within its extremes? Explain.

(b) Now assume that the regressors are random and the regressors have a correlation of .91. Would it still be practical to construct the prediction interval for any combination of regressor values within their respective extremes? Explain.

9.34. For the example given in Section 9.3, can the value of R^2 be determined from the output for a subset model that contains only linear terms in Power and Bulk Gas Flow in the model? If so, what is the value? If not, explain why not.

9.35. Consider the data in Exercise 9.5. Could the value of R^2 when *only* Diameter is used in the model be determined from the output once the blanks are filled in? If so, what is the value? If not, explain why not.

9.36. Respond to the following statement: "I ran a regression analysis with four predictors, and with the values of each predictor occurring at random. The t-statistics showed that only one predictor is significant, but I am fairly certain that two particular predictors, and possibly more, should be in the model. Consequently, I don't have any real confidence in these results."

9.37. Consider Exercise 9.19, for which the sign of X_2 changes, depending on whether X_1 is in or out of the regression equation. Because of this, would it be better to construct a confidence interval for β_2 when only X_2 is used in the equation, or should the confidence interval be constructed at all, based on what is given in the problem?

9.38. Respond to the following statement: "The regression equation in Exercise 9.5 doesn't make any sense because the intercept is a large negative number. Consequently a small tree will have a negative predicted volume, which would be nonsensical."

9.39. Explain why there are no extreme predictor values in the example in Section 9.3. Does the fact that there are no extreme predictor values preclude the possibility of having data points that could be overly influential? Explain.

9.40. What should be suspected, and checked out, whenever regression is applied to data collected over time?

9.41. (Harder problem) Assume that there are two predictors in a regression equation and the unit of measurement for one of them is feet. If the unit is changed to inches, what would you expect to be the effect on the regression coefficient, and would you anticipate any change in the coefficient for the other predictor?

9.42. Consider the output in Exercise 9.8. Does the number of standardized residuals that exceed 2 in absolute value seem excessive in regard to the number of observations? What would be the expected number of model *errors* outside the interval $(-2, 2)$ if the errors had a normal distribution?

9.43. Assume that you have constructed a normal probability plot of a set of standardized residuals in a regression analysis and the plot exhibits a borderline result. What action would you take?

9.44. The least squares estimators of the β_i are all unbiased in regression only when the fitted model is the true model. What does this result imply is likely to occur in a given application?

9.45. Consider the following data, which are also in `EX944ENGSTAT.MTW`:

Row	Y	X_1	X_2
1	21.31	1.35	16.10
2	22.06	2.34	10.17
3	20.25	2.85	8.81
4	20.36	1.55	14.77
5	17.51	2.50	10.32
6	20.67	1.48	15.22
7	17.15	1.79	14.03
8	14.52	1.84	14.49
9	13.47	3.50	7.93
10	11.20	1.67	18.62
11	28.37	1.99	10.46
12	18.08	2.25	11.24
13	25.60	3.04	7.78
14	30.00	2.59	8.38
15	24.56	3.56	6.88
16	23.40	3.87	6.47
17	24.35	2.05	10.92
18	15.27	1.68	15.51
19	26.67	1.45	13.54
20	14.48	2.12	12.69

(a) Fit the model $Y = \beta_0 + \beta_1 X_1 + \beta_2 X_2$ and plot the standardized residuals against X_2; then plot the raw residuals against X_2; and finally construct the partial residual plot for X_2. Notice that the last plot differs considerably from the other two plots.

(b) What transformation is suggested by that plot, if any? The true model contains a reciprocal term in X_2, not a linear term. Which plot gave the appropriate signal?

(c) Fit the true model and compare the R^2 value for this model with the R^2 value for the original model. Comment.

(d) What have you just demonstrated?

9.46. It might seem as though the most intuitive way to construct indicator variables would be to use a variable for each level of a categorical variable, with a one in a column indicating the values obtained when the variable is that level, and a zero in all other places in the column. Take any data set and then add two columns constructed in this manner, so as to represent a categorical variable with two levels. Use MINITAB or other statistical software. Can you explain why this didn't work?

9.47. An operations engineer is interested in modeling the length of time per month (in hours) that a machine will be shut down for repairs as a function of machine type (1 or 2) and the age of the machine (in years). What type of variable will machine type be and how should the values of the variable be assigned?

9.48. A multiple regression model is fit to a set of data and all five of the predictors are judged to be necessary because the p-values are all less than .05. The value of R^2, however, is only .31. Explain how this can happen.

9.49. Assume that there are three predictors under consideration for use in a regression model and the correlations between the predictors are reported to be the following: $r_{12} = .88$, $r_{13} = .92$, and $r_{23} = .02$.

(a) Would you use all three of the predictors in a regression equation, or would you need more information to make that determination? Explain.

(b) Is there any reason to believe that at least one of the reported correlations may be incorrect? Explain.

9.50. The BEARS . MTW dataset comes with MINITAB, with the data having been provided by Gary Alt. The objective is to develop a way for hunters to estimate the weight of a bear since a bear cannot be weighed in the wilderness, but other measurements could be taken if the bear was dead or at least incapacitated. Determine if a useful regression model can be developed for this purpose, paying particular attention to the following.

(a) You will note that there are many missing observations on the variable AGE. When data are missing, it is important to determine whether data are missing at random or missing systematically. Of course, this determination generally cannot be easily made without imputing missing data. In this instance, would you be inclined to use a variable that is highly correlated with age, if there is one, instead of age, so as to be able to use more observations in building the model?

(b) You will notice that the weight of the 143 bears varies greatly. Does your model apply equally well across the wide range of weights or would you suggest that different models be used for different weight intervals? If so, how would you proceed?

9.51. (Harder problem) Assume that there are four predictors in a regression model and a partial residuals plot has been constructed for X_3. If a least squares line is fit to the points in the plot, what will be the slope and intercept of the line?

CHAPTER 10

Mechanistic Models

For years statisticians and engineers have been somewhat at odds over the importance of statistics in engineering work. This debate has essentially centered on the need for statistically derived models versus the use of physical (mechanistic) models. Engineers have often maintained that they know the appropriate model, so they don't need to use statistical techniques to arrive at a model. Indeed, in the "Letters" section of the September 2002 issue of *RSS News* (a publication of the Royal Statistical Society), D. C. Lloyd stated: "I found that my students had a very deterministic view of the universe and found probability and uncertainty difficult concepts to grasp.... I fear that young engineers approach the world with an attitude which leaves little room for the uncertainty and doubt most of us encounter."

Statisticians, on the other hand, while recognizing the existence of uncertainty and subsequently quantifying it, have often failed to recognize that subject matter specialists may indeed have some theoretical models that are excellent but can be improved. Consequently, statisticians have focused primarily on data-driven models. Such an approach has sparked somewhat derisive remarks, such as the following, which pertains to predicting natural disasters (once stated at the website of the Department of Earth Science and Engineering, Imperial College of Science, Technology, and Medicine—http://www.ese.ic.ac.uk):

> Some people think prediction is just a matter of number crunching historic data sets. If we are to get a handle on short-term life-saving predictions, then there is no substitute for understanding the mechanisms causing the disasters.

A compromise from both camps seems desirable. As G. E. P. Box has been quoted countless times (including at least by a CEO, see http://seattle.bizjournals.com/seattle/stories/2002/09/16/focus11.html): "All models are wrong, but some are useful" (e.g., see Box, 1976). Therefore, it should be possible to improve upon theoretical models, but if a theoretical model exists in a particular application, then that model should usually be the starting point. Thus, the best approach will often be to combine an empirical (data-driven) approach with a mechanistic approach and thus use an empirical–mechanistic modeling approach. Such an approach is discussed in Section 10.2.

10.1   MECHANISTIC MODELS

Theoretical models are almost always nonlinear models, whereas empirical modeling generally starts with a linear model because there is no obvious starting point for a nonlinear model in the absence of subject matter knowledge.

Theoretical models are also termed *mechanistic models* as the intent is to model the mechanism under study. A check of the Institute for Scientific Information's Web of Science shows that over 500 papers have been published since 1945 for which the term "mechanistic models" is in either the title or abstract of a paper.

■   EXAMPLE 10.1

Background

Consider the following scenario, which occurred some years ago. Data were collected on a chemical/mechanical planarization (CMP) process at a particular semiconductor processing step. During the process, wafers were polished using a combination of chemicals in a polishing slurry using polishing pads. A number of wafers were polished for differing periods of time in order to calculate material removal rates.

Nonlinear Model

The removal rate is known to change with time. At first, the removal rate is high but declines as the wafer becomes more planar. Accordingly, the removal rate was modeled as

$$Y = \beta_0 + \beta_1 \exp(\beta_2 X) \tag{10.1}$$

with Y denoting removal rate and X denoting time.

Linear regression was presented in the two preceding chapters but it should be immediately apparent to the reader that Eq. (10.1) is not a linear regression model because the model is not linear in β_1 and β_2. This is a nonlinear model, with an unspecified form for the error term. The model was fit to the data using a nonlinear regression routine. Such programs perform iterative calculations and require initial estimates of the parameters. Such estimates can often be obtained by "linearizing" an intrinsically linear model. An intrinsically linear model is one that can be made linear with a transformation. For example, if the model, with a specified error term, were $Y = \beta_0 \beta_1 \exp(\beta_2 X) \epsilon$, taking the natural logarithm of each side of the equation produces a simple linear regression model, with $\ln(\beta_0)$ and $\ln(\beta_1)$ combined to form a single constant term.

The model in Eq. (10.1) cannot be transformed to a linear regression model, regardless of the error structure that is assumed. Consequently, good initial starting values based on physical considerations would be needed.   ■

10.1.1   Mechanistic Models in Accelerated Life Testing

One area in which mechanistic models are frequently used is accelerated life testing, which is covered in Section 14.3. In the next section we present one model that has been used extensively in accelerated life testing and mention how the model assumptions can be checked. This and other models used in reliability analysis are discussed in more detail in Chapter 14.

The reader might wonder what makes this different from selecting a model from Chapter 3 and checking the fit of the model. The difference is that the models discussed in this section are not covered in Chapter 3, are not generally presented in statistics books, and are not well known outside the application areas in which they are used.

10.1.1.1 *Arrhenius Model*

This model, which is discussed in some detail in Section 14.3.1, is given by

$$\mu(S) = \log(R_0) + \frac{E_a}{C(S + 273)} \tag{10.2}$$

with R_0 denoting the rate constant, E_a is the activation energy, C is Boltzmann's constant, S is the stress (temperature in degrees Celsius), and $\mu(S)$ is the mean at stress S. [This model is given in Chapter 14 as Eq. (14.4).] The model assumptions are that life has a lognormal distribution (see Section 3.4.8) at each temperature and each distribution has the same value of σ. These assumptions are checked using lognormal plots—see Nelson (1982, p. 114).

10.2 EMPIRICAL–MECHANISTIC MODELS

Although the literature on mechanistic models is voluminous, the term "empirical–mechanistic models" is rarely used in the literature. This is not to suggest that data are not being used to verify mechanistic models, as this is described in various papers and some people would contend that this is how science proceeds generally. For example, Stewart, Shon, and Box (1998) described a goodness-of-fit approach for discriminating among multiple candidate multiresponse mechanistic models and Keane and Murzin (2001) looked at the deviations of observed from predicted (i.e., residuals) as a means of selecting from a set of possible mechanistic models. Porte and Bartelink (2002) suggested that mechanistic approaches be added to empirical approaches in modeling mixed forest growth in forest management. Prozzi, Gossain, and Manuel (2005) illustrated the use of empirical–mechanistic models in regard to the reliability of pavement structures, with regression methods employed. Davies, Trantner, Wiggington, Eshleman, Peters, Van Sickle, De Walle, and Murdoch (1999) described the use of mechanistic modeling in predicting episodic acidification, with regression techniques heavily employed.

Although the examination of raw residuals can be helpful in spotlighting model inadequacies, and indeed is a step in the right direction, it is well known from the regression literature that raw residuals are the least effective of the different forms of residuals. Furthermore, nonlinear modeling requires the use of different types of residuals from those that are used in linear modeling (e.g., see Chapter 13 of Ryan, 1997). Therefore, one contribution that statisticians can make to improving the current state of affairs is in working with scientists to make them aware of various residual analysis methods and when each one should be used.

Although one does find the words "empirical" and "mechanistic" used together in the title or abstract of a very small number of papers, the absence of the term "empirical–mechanistic models" from the literature suggests that there is no formalized approach or even an attempt at creating one. The only publications of any sort that I am aware of that contain an attempt at formalizing such an approach are the technical report "Empirical–Mechanistic Modeling of Air Pollution" by Phadke, Box, and Tiao (1976), and the paper "Empirical–Mechanistic Model to Estimate Roughness" by Proffi and Madanat (2003).

Tan, Wang, and Marshall (1996) stated that a number of empirical models have been proposed for substrate inhibition of microbial growth, but these models were developed for enzyme kinetics and there is no theoretical basis to support their derivation for microorganisms. They proposed a general equation, supported by theory, with the parameters estimated using a standard nonlinear regression approach. Adequacy of model fit is then judged graphically using three datasets and an overall F-test is also performed. Although the fit is extremely good for each dataset, there is a very small number of points in each dataset.

Although there was no iterative model building—which might have seemed unnecessary anyway to the experimenters—at least data and theory were combined to produce the equation, so this was somewhat of an empirical–mechanistic approach, devoid of model criticism.

■ EXAMPLE 10.2 (CASE STUDY)

Background

Caroni (2002) presented a case study that was described as being "a useful example of multiple linear regression applied to an engineering problem." Although not labeled an empirical–mechanistic modeling example by the author, the application is indeed such an example, or at least should be such an example. That is, the starting point was a mechanistic model that was based on a "theoretically derived relationship" (Caroni, 2002). The theoretical model was the starting point for the development of a multiple linear regression model that was examined critically through the use of model criticism methods, such as were given in Chapters 8 and 9.

Objective

Specifically, the objective, of Lieblein and Zelen (1956) who performed the original analysis in a study for the American Standards Association and the National Bureau of Standards, was to model a percentile of the lifetime distribution for ball bearings, starting from the model for bearings in general that is given by ISO Standard 281, with ISO representing International Standards Organization. That model is

$$L_{10} = (C/P)^p \tag{10.3}$$

with L_{10} denoting the 10th percentile of the distribution of the lifetime of an arbitrary type of bearing, C denotes the constant load that would give a rating life of one million revolutions, and P represents the load of the bearing in operation, with $p = 3$ for ball bearings.

Model Acceptance

Lieblein and Zelen (1956) stated that the general form had been accepted by ball-bearing manufacturers, based on many years of experience, but there was a lack of agreement on whether $p = 3$ should be used.

The obvious approach should then be to obtain data to be used for model criticism, which in this case could be used to help settle the argument regarding the choice for p and which might even lead to an improved form of the model. That is, a mechanistic model is critiqued using empirical, data-driven methods.

Lieblein and Zelen (1956) stated that the form of the numerator of Eq. (10.3) that is used for ball bearings is written generally as $f Z^a D^b$, with Z denoting the number of balls in the bearing and D denoting the diameter of the ball, with f, a, and b representing constants.

If we consider Eq. (10.3) with $C = f Z^a D^b$ and with f, a, b, and p constants to be estimated, it is immediately apparent that this is not a linear model. It should also be apparent, however, that taking the logarithm of each side of the equation will produce a linear regression model if the model error is assumed to be multiplicative. Alternatively, the linear regression model could be considered a good approximation if the error was not multiplicative but was quite small. An additive error that is not very small would dictate the use of a nonlinear regression model, however, with the parameter estimates from the linear regression analysis serving as starting values in a nonlinear regression estimation procedure.

It is worth noting that the values of the dependent variable, the 10th and 50th percentile values of lifetime, were not observed but rather were estimated by fitting a separate Weibull distribution to the failure times of each of the three batches of data, one batch from each of the three companies that were performing the endurance tests. There were also three bearing types for each company.

Starting from the model

$$L = \left(\frac{f Z^a D^b}{P} \right)^p \tag{10.4}$$

as given by Caroni (2002), if we linearize the model by taking (natural) logarithms, we obtain

$$\ln(L) = p \, \ln(f) + ap \, \ln(Z) + bp \, \ln(D) - p \, \ln(P) \tag{10.5}$$

Since both p and f are constants, their product is a constant, as are the products ap and bp. If we write this in the usual form of a multiple linear regression model [Eq. (9.1) in Chapter 9], we obtain

$$\ln(L) = a + \beta_1 \, \ln(Z) + \beta_2 \, \ln(D) + \beta_3 \, \ln(P) + \epsilon \tag{10.6}$$

[Note that the minus sign before the last term in Eq. (10.5) is converted to a plus sign in Eq. (10.6) since minus signs are not used in regression models. Minus signs will often appear in fitted equations, however, as was illustrated in Section 8.2.7.]

Strictly speaking, the additive error in Eq. (10.6) is appropriate only if there is a multiplicative error term in Eq. (10.4). (Of course, this is not written that way here since the error structure is unknown.)

A major problem in trying to use all of the data is that the number of data values for each company are highly unequal, being 50, 148, and 12 for the three companies. This creates a problem if the data are not homogeneous across the companies. Noting that the number of bearings per test varied considerably, from 8 to 94, Caroni (2002) assumed that the variance of the error term in Eq. (10.6) was inversely proportional to the number of bearings per test and consequently used weighted least squares. The graph in Figure 10.1 does not support such an assumption, however, as there is no clear evidence to refute the assumption of equal variances. It is worth noting that such graphs should be viewed after covering any few plotted points that might result in a visual distortion, and here there are two points that should be covered. When this is done, there is no evidence of an unequal variances problem.

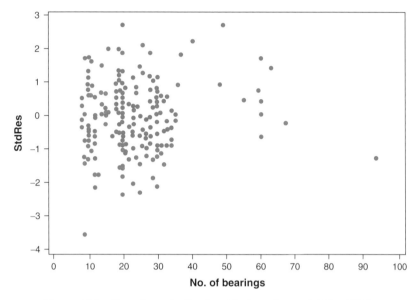

Figure 10.1 Plot of standardized residuals under model Eq. (10.6).

Therefore, we will proceed to use the linearized model and initially use all of the data. The output is as follows:

```
Regression Analysis: lnL10 versus lnno.balls, lnDiameter, lnLoad

The regression equation is
lnL10 = 19.6 + 1.45 lnno.balls + 3.82 lnDiamet - 2.25 lnLoad

Predictor              Coef         SE Coef              T              P
Constant             19.561           1.464          13.36          0.000
lnno.balls           1.4490          0.3749           3.87          0.000
lnDiamet             3.8236          0.3916           9.77          0.000
lnLoad              -2.2471          0.1864         -12.05          0.000

S = 0.6900          R-Sq = 45.5%          R-Sq(adj) = 44.7%
```

The R^2 value is only small-to-moderate, but this is not uncommon for good models when the sample size is in the hundreds. Caroni (2002) concluded that the parameter values were not the same for each company, but that conclusion was based on the results of an apparently unnecessary weighted least squares analysis. Furthermore, the value of the test statistic was not particularly large, so the apparent significance may be due primarily to the large sample size. There is no pattern of badly fit data when the data from all three companies are used. The only badly fit data point is observation #81, which is from the second company. When only the data from the second company are used, the R^2 value is even lower, 36.1, so merging the data does not seem inappropriate.

Therefore, we will obtain the starting values from the parameter estimates in the linearized model. We have a slight problem, however, as the parameters in Eq. (10.6) represent the products in Eq. (10.5), whereas we need estimates of the components of the products. We obtain 2.2471 from the regression as the estimate of p, however, and we can use this estimate to obtain estimates for a, b, and f. That is, setting $p \ln(f) = 2.2471 \ln(f) = 19.6$, we obtain $\ln(f) = 8.72235$, so $f = \exp(8.711) = 6138.61$. Similarly, $2.2471a = 1.45$, so $a = 0.645276$ and $2.2471b = 3.82$, so $b = 1.69997$.

Given below is the JMP output, with b0, b1, b2, and b3 representing f, a, b, and p, respectively.


```
Nonlinear Fit

Control Panel

Report
Converged in Objective Function

Criterion        Current              Stop Limit
Iteration        20                   60
Shortening       0                    15
Obj Change       4.1311918e−8         0.0000001
Prm Change       0.0006327919         0.0000001
Gradient         0.0066119318         0.000001


Parameter        Current Value
b0               628.46810064
b1                 1.4730840967
b2                 1.6874752187
b3                 3.3372449281


  SSE 121114.91533 N 210
  Edit Alpha

  0.050Convergence Criterion
  0.00001Goal SSE for CL

Solution

SSE                          DFE        MSE         RMSE
121114.91533                 206    587.93648    24.247402

Parameter            Estimate        ApproxStdErr
   b0              628.46810064        131.788382
   b1                1.4730840967        0.10742749
   b2                1.6874752187        0.07939645
   b3                3.3372449281        0.16774519
```

The value of b0 (i.e., f) changed considerably from the starting value, but that is not surprising since this corresponds to the intercept in the linearization of the nonlinear model.

If we use the form of R^2 recommended by Kvålseth (1985)—namely, $R^2 = 1 - \sum (Y - \widehat{Y})^2 / \sum (Y - \overline{Y})^2$—we obtain $R^2 = .656$. This is not a particularly good value, so we need to determine why the fit is not very good. Of course, we should bear in mind that we do not have the actual values of L_{10}, so using estimated values contributes to the difficulty in trying to obtain a good-fitting model.

We should of course go beyond the calculations given to this point and look at some plots. Figures 10.2–10.4 are plots of L_{10} against each of the three predictors.

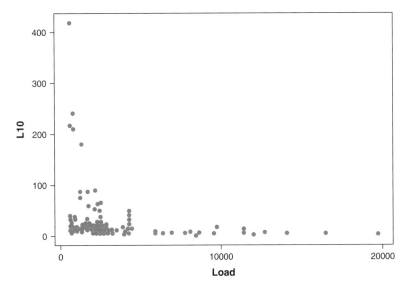

Figure 10.2 Plot of the dependent variable against load.

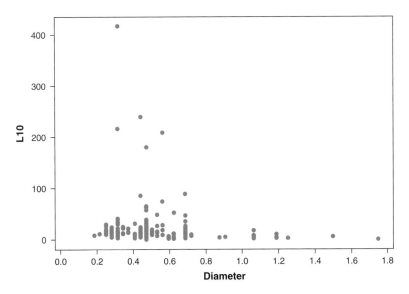

Figure 10.3 Plot of the dependent variable against diameter.

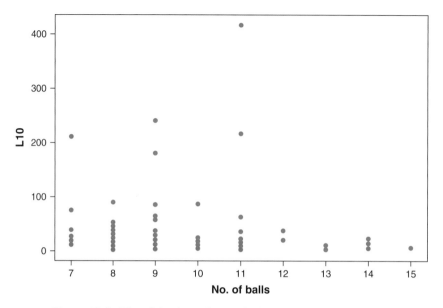

Figure 10.4 Plot of the dependent variable against number of balls.

Of course, we are fitting a nonlinear model that is not a simple function of these predictors, so scatter plots won't necessarily be very informative. Nevertheless, none of the plots suggests a relationship between L_{10} and any of the predictors.

A slightly different picture emerges if we plot L_{10} against the suggested function, without the exponents (Figure 10.5).

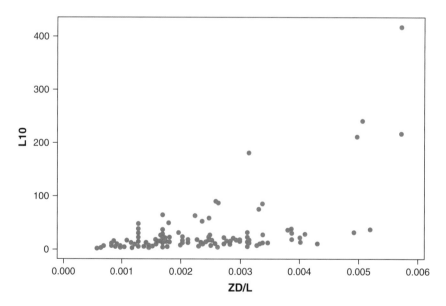

Figure 10.5 Plot of the dependent variable against ZD/L.

Now we see a hint of a relationship, although a rather weak one. A plot of the observed against the predicted values, Figure 10.6, is obscured by the large values that create a scaling problem.

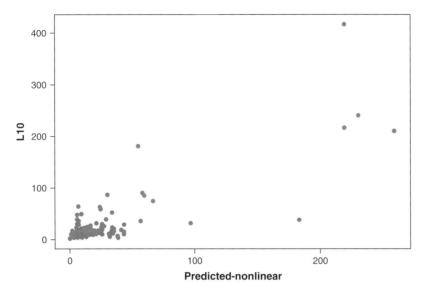

Figure 10.6 Plot of the observed versus predicted (fitted) values.

If we delete the five largest predicted values (which clearly are separate from the others) and compute R^2 using the remaining 205 values, we obtain $R^2 = .15$. Thus, there is a huge drop from .656 to .15 when those five points are excluded from the computation. Of course, this does not mean that the drop would be the same if those five points were excluded from the original computations, but it does suggest that the drop will be approximately of this size. When those points are excluded and the coefficients reestimated, they change greatly to 2896.77559856, 1.030429581, 1.8730882795, and 2.42254911, respectively, and R^2 drops from .656 to .327. These are significant changes, especially the change in the last coefficient since it was assumed to be 3, but there was disagreement about that. Now the difference between the coefficient and 3 is greater than the difference before the points were excluded. Just as importantly, R^2 is now less than half of the previous value, a much larger than normal drop when only 5 of 210 points are deleted.

```
Linear Regression of L10 vs. predicted values from nonlinear fit

Summary of Fit

RSquare    0.662102
RSquare Adj   0.660478
Root Mean Square Error    23.92157
Mean of Response    21.36082
Observations (or Sum Wgts)    210
```

```
Analysis of Variance

Source    DF    Sum of Squares    Mean Square    F Ratio      Prob > F
Model     1     233228.84              233229    407.5706      <.0001
Error     208   119026.26                 572
Total     209   352255.10


Lack Of Fit

Source         DF    Sum of Squares    Mean Square    F Ratio    Prob>F
Lack Of Fit    103        90428.07        877.942     3.2234     <.0001
Pure Error     105        28598.19        272.364
Total Error    208       119026.26


Max RSq        0.9188


Parameter Estimates

Term                      Estimate      Std Error    t Ratio    Prob > |t|
Intercept                3.5747673      1.871131       1.91       0.0574
Fitted Column            0.9573442      0.047421      20.19       <.0001


Effect Tests

Source    Nparm    DF    Sum of Squares     F Ratio    Prob > F
Fitted    Column    1    1233228.84        407.5706    <.0001
```

10.3 ADDITIONAL EXAMPLES

The fitting of a nonlinear regression function can be viewed as a form of mechanistic model building because the process cannot begin without a postulated model based on scientific/engineering knowledge. A model suggested by such knowledge might be applied to a set of data and then perhaps modified through the use of residual analysis. An example of this was given in Section 4.6.3.2 of the *NIST/SEMATECH e-Handbook of Statistical Methods*, with the data being ultrasonic reference block data. The response variable was ultrasonic response and the predictor was metal distance. The scientist chose to fit the following theoretical model:

$$Y = \frac{\exp(\beta_1 X)}{\beta_2 + \beta_3 X} \tag{10.7}$$

The reader will observe that this model does not have an error term, and indeed it can often be difficult to determine the appropriate error structure for a nonlinear model. For example, Ruppert, Cressie, and Carroll (1989) reported that there does not appear to be a typical error structure for the Michaelis–Menten model, which is given by $f(x; \theta) = \theta_1 X / (\theta_2 + X)$.

In the application using the model in Eq. (10.7), a shotgun approach was used to arrive at a starting value of 0.1 for each of the three parameters. Nonlinear least squares was then used to arrive at the final parameter estimates, which were more than a small percentage difference away from the initial estimates. (See http://www.itl.nist.gov/div898/handbook/pmd/section6/pmd632.htm for additional details.)

This is a typical nonlinear regression/mechanistic model example. In the case of a single predictor, in the absence of mechanistic modeling input it would be necessary to graph the data and then try to match a plot of the data against curves of various functions, such as those given in Ratkowsky (1990).

Another approach, which has been used to some extent by industrial personnel, is to use software that will sift through hundreds if not thousands of models and identify the one that provides the best fit (e.g., see the description of IGOR at `http://www.wavemetrics.com`). Such software should be used cautiously, if at all, because almost all data sets have errors and a strictly numerical approach with no scientific input could fail to detect the bad data and/or could falsely identify good data as bad data. The use of automated procedures could be especially hazardous for small data sets, as one data point could be driving the model selection. People who have criticized such automated shotgun approaches have pointed out that there may not be any scientific basis for the model that is selected. Some industrial personnel have used such software to identify a group of useful models. The software has utility for this purpose if the group so identified can be supported with scientific knowledge.

As a final and very practical example, Taguchi (2005, p. 1415) discussed the problem of trying to preheat or precool an office so that it is at the desired temperature when workers arrive for work at the start of the day. Heat transfer theory provides a mechanistic model that is a function of various factors, including of course the outside temperature. The objective is to determine the heating time, T_0, that is necessary to produce a target room temperature of θ_0, with the function for T_0 given by Taguchi (2005) as

$$T_0 = \frac{1}{\alpha \ln\{1 - [\theta_0 - \theta_n/k(1 + \beta\Delta\theta_A)]\}}$$

with θ_n denoting the temperature at which the heating began and $\Delta\theta_A$ denotes the difference between the outside temperature and the target temperature. The parameters α, β, and k must be estimated, and these will be different for different building structures and other conditions. Thus, even though a mechanistic model based on heat transfer theory provides a starting point for the work, an empirical–mechanistic modeling approach must obviously be used to estimate the parameters and to critique the resultant model.

The problem is trying to develop a single model, using indicator (0-1) variables to cover a variety of conditions with perhaps something less than a large amount of data. There simply may not be enough data in a sample to estimate the parameters for certain conditions. The interested reader is referred to Taguchi's (2005, pp. 1415–1419) discussion and illustration of the parameter estimation for this problem.

In general, scientific input can be expected to provide a better starting point for the iterative model-building process depicted in Figure 1.1.

Regarding the development of mechanistic models in different engineering fields, the construction of chemical engineering mechanistic models is described at `http://eweb.chemeng.ed.ac.uk/courses/control/restricted/course/advanced/module5.html`.

10.4 SOFTWARE

Mechanistic modeling generally requires the use of software with nonlinear regression capability. JMP was used for the nonlinear regression modeling in this chapter. Although such capability is not available directly in MINITAB, the macro `NONLIN.MAC` can be downloaded from the website at www.minitab.com and used for any nonlinear regression model.

10.5 SUMMARY

This chapter was relatively short simply because the use of mechanistic modeling combined with the type of diagnostic methods that are used in linear and nonlinear regression (to produce empirical–mechanistic modeling) has been practically nonexistent. This state of affairs should clearly change as scientists would benefit from the use of model criticism methods, even when they strongly believe that they have a good handle on the model, or at least an appropriate model. Similarly, statisticians need to recognize that mechanistic models are often very effective and can thus frequently serve as an excellent starting point for the iterative modeling process that should generally be used. Tan et al. (1996) combined theory with data in arriving at a model, but it isn't clear how they would have sought model improvement if the model fit had not been acceptable.

REFERENCES

Box, G. E. P. (1976). Science and statistics. *Journal of the American Statistical Association*, **71**, 791–799.

Box, G. E. P., J. S. Hunter, and W. G. Hunter (1978). *Statistics for Experimenters*. New York: Wiley.

Caroni, C. (2002). Modeling the reliability of ball bearings. *Journal of Statistics Education*, **10**(3) (electronic journal).

Daniel, C. and F. S. Wood (1980). *Fitting Equations to Data*, 2nd ed. New York: Wiley.

Davies, T. D., M. Trantner, P. J. Wiggington, Jr., K. N. Eshleman, N. E. Peters, J. Van Sickle, D. P. De Walle, and P. S. Murdoch (1999). Prediction of episodic acidification in northeastern USA: An empirical/mechanistic approach. *Hydrological Processes*, **13**, 1181–1195.

Keane, M. A. and D. Y. Murzin (2001). A kinetic treatment of the gas phase hydrodechlorination of chlorobenzene over nickel: Beyond conventional kinetics. *Chemical Engineering Science*, **56**(10), 3185–3195.

Kvålseth, T. O. (1985), Cautionary note about R^2. *The American Statistician*, **39**, 279–285.

Lieblein, J. and M. Zelen (1956). Statistical investigation of the fatigue life of deep-groove ball bearings. *Journal of Research of the National Bureau of Standards*, **57**, 273–316.

Nelson, W. (1982). *Applied Life Data Analysis*. New York: Wiley.

Phadke, M. S., G. E. P. Box, and G. C. Tiao (1976). Empirical–mechanistic modeling of air pollution. University of Wisconsin Department of Statistics Technical Report #456, May.

Porte, A. and H. H. Bartelink (2002). Modelling mixed forest growth: A review of models for forest management. *Ecological Modeling*, **150**(1–2), 141–188.

Proffi, J. A. and S. M. Madanat (2003). Empirical–mechanistic model to estimate roughness. Paper presented at a Transportation Research Board conference, January 14, 2003.

Prozzi, J. A., V. Gossain, and L. Manuel (2005). Reliability of pavement structures using empirical mechanistic models. Manuscript (http://www.ce.utexas.edu/prof/Manuel/Papers/ProzziGossainManuel_05--1794.pdf).

Ratkowsky, D. A. (1990). *Handbook of Nonlinear Regression Models*. New York: Marcel Dekker.

Ruppert, D., N. Cressie, and R. J. Carroll (1989). A transformation/weighting model for estimating Michaelis–Menten parameters. *Biometrics*, **45**, 637–656.

Ryan, T. P. (1997). *Modern Regression Methods*. New York: Wiley.

Stewart, W. E., Y. Shon, and G. E. P. Box (1998). Discrimination and goodness of fit for multiresponse mechanistic models. *AIChE Journal*, **44**(6), 1404–1412.

Taguchi, G. (2005). Applications of linear and nonlinear regression equations for engineering. Case 94 in *Taguchi Quality Engineering Handbook* (G. Taguchi, S. Chowdhury, and Y. Wu, eds.) Hoboken, NJ: Wiley.

Tan, Y., Z.-X. Wang, and K. C. Marshall (1996). Modeling substrate inhibition of microbial growth. *Biotechnology and Bioengineering*, **52**, 602–608.

EXERCISES

10.1. Assume you expect that a one-predictor model will be sufficient for predicting a particular response variable. No one with subject matter expertise is available to suggest a particular mechanistic model, but it is believed that a nonlinear relationship exists between these two variables. You have a sample of 50 (x, y) observations and are given a week to develop some type of model that gives good results. How would you proceed?

10.2. Consider Example 10.2 in Section 10.2. A mechanistic model, namely, Eq. (10.4), was used as the starting point for developing a multiple regression model. The data were originally given by Lieblein and Zelen (1956) but are also available on the Web at `http://www.amstat.org/publications/jse/datasets/ballbearings.dat`, with a description of the dataset available at `http://www.amstat.org/publications/jse/datasets/ballbearings.txt` and in `CH10ENGSTATEXER3.MTW`. Using the dataset, proceed as follows:

(a) Write the model that Eq. (10.4) transforms into when the logarithm of each side of the equation is taken, acting as if the error is multiplicative.

(b) Will the constants f, a, b, and p be estimated directly with the model that you have written? If not, how would p, in particular, be estimated with your model?

(c) Analyze the data and determine if $p = 3$ seems to be a reasonable choice.

(d) In the process of reaching your decision, determine whether or not the error term would seem to have a constant variance. Caroni (2002) claimed that there is no evidence that the constant variance assumption was violated, which would render unnecessary the use of weighted least squares, as was used by Lieblein and Zelen (1956). Do you agree with Caroni?

(e) The author concluded that $p = 3$ falls within a 95% confidence interval and is therefore a plausible value. Do you agree with this conclusion? More broadly, would you depend on the result from such a confidence interval when there are nonzero correlations between the transformed values, as there are in this case? Explain.

10.3. (Harder problem) Taguchi (2005) stated that the volume of a tree, Y, can be estimated from the following function of chest-height diameter, D, and height, H:

$$Y = a_1 D^{a_2} H^{a_3}$$

If a_1, a_2, and a_3 are unknown, as Taguchi assumed and which seems quite likely, they would have to be estimated. Thus, a nonlinear model is being postulated, which we observe could be transformed into a linear regression model if the error is multiplicative. If the latter is not true and the error is additive, the transformation would still give good results if the error is small.

The idea of estimating the volume of trees in a forest is an important practical problem and various models have been presented in the literature for the MINITAB trees data, which is one of the sample datasets that comes with the MINITAB software. That dataset contains volume, diameter, and height data for 31 trees in the Allegheny National Forest. Volume was measured in cubic feet, the diameter was in inches (measured at 4.5 feet above the ground), and the height is in feet.

Taguchi (2005) gave the following model:

$$Y = 0.0001096 \, D^{1.839} H^{0.8398}$$

with the coefficients obtained by applying nonlinear regression to a dataset of 52 trees that were cut down. Volume was stated as being measured in square meters, but that is obviously a translation error as Volume must be a cubic measurement and in fact is indicated as being in cubic meters in subsequent tables. Diameter was stated to be the "chest-high diameter" and was measured in centimeters, and height was measured in meters.

Because of the pervasive influence of Taguchi, let's assume that the foresters at Allegheny National Forest decided to use an empirical–mechanistic modeling approach. That is, they would start with Taguchi's model and see how it worked with their data rather than trying to applying their meager (we shall assume for this example) knowledge of statistics.

(a) What adjustment(s), if any, would have to be made to the model given by Taguchi (2005) before it can be used to see how well it applies to the MINITAB trees data?

(b) Make any necessary adjustment(s) and compare the predicted values with the values for volume in the TREES dataset. Also compute the correlation between these two sets of values.

(c) Note the very high correlation between the two sets of values, but also the relatively large differences between the two sets of values. To what might the high correlation and the large differences be attributed? In particular, could the difference between the "chest-high diameter" used by Taguchi and the diameter taken at 4.5 feet above the ground in the MINITAB data be a factor? Explain.

(d) What would be your recommendation to the foresters at Allegheny National Forest? In particular, would you suggest that the foresters fit a simpler multiple linear regression model obtained after taking the logarithm of volume, diameter, and height? Is this what would result from taking the logarithm of each side of the appropriate model? Explain.

10.4. In his online article "Using a Mechanistic Eutrophication Model for Water Quality Management of the Neuse River," civil engineering professor J. D. Bowen explains the use of mechanistic models in water quality management and addresses some direct questions regarding the use of such models. Read that article, which is available at http://www.coe.uncc.edu/~jdbowen/jdb.101.pdf, and comment on it. In particular, comment on the statement: "Mechanistic models need not be deterministic, if the modeling effort includes an analysis of prediction uncertainty." Would you expect a mechanistic model applied to water quality management to ever be deterministic (i.e., without error)? What does this suggest about the type of models that should be used in water quality management?

10.5. Wolfram (*A New Kind of Science*, Wolfram Media, 2002) makes some provocative statements regarding mechanistic models in the section "Ultimate Models for the Universe" in Chapter 9, Fundamental Physics, of his book. In particular, he states that "since at least the 1960s mechanistic models have carried the stigma of uninformed amateur science." In continuing to discuss physics, Wolfram states: "And instead I believe that what must happen relies on the phenomena discovered in this book—and involves the emergence of complex properties without any obvious underlying mechanistic physical set up." Since the author is one of the world's leading scientists and is the developer of the very popular *Mathematica* software, such statements should not be taken lightly. Do you adopt a counterviewpoint in regard to the use or possible use of mechanistic models in your field? Explain.

CHAPTER 11

Control Charts and Quality Improvement

Controlling and improving manufacturing and nonmanufacturing processes are important objectives to which engineering statistics in general and control charts in particular can make meaningful contributions. Control charts have also been used for a variety of other reasons, some of which are described in the next section. Control charts and other methods for quality improvement should be of particular interest to manufacturing engineers, industrial engineers, and process engineers, although control charts can also be used in various other engineering areas, even biomedical engineering.

11.1 BASIC CONTROL CHART PRINCIPLES

A control chart, presented in Section 1.4.2, is a plot of data (or of a statistic computed from data) over time—a desirable way to plot any set of data. The general idea is to identify common causes (of variation) and separate them from assignable causes, which are also called special causes, although the distinction between the two types of causes is often difficult to make.

An example may help. All processes have inherent variation, and I have used one particular example for many years in introducing the concept of a control chart to college students. Assume that I decide to chart my driving time to work for 50 consecutive school days. Assume further that my past experience indicates that my driving time *should* be about 30 minutes, so I will use this number as the centerline on the chart. I believe that the standard deviation of my driving time is about 2 minutes. I would guess that a normal distribution is a reasonable approximation to the unknown distribution of my driving times, so almost all of my driving times should be in the range $30 \pm 3(2) = (24, 36)$. Everything goes smoothly for the first two weeks, but then on Monday of the third week I arrive at a certain point on an expressway where there has been an accident 30 minutes earlier, 200 yards in front of me. Traffic is at a standstill and consequently my driving time to work that day is 58 minutes, which is above the "upper control limit" of 36 (i.e., the approximate longest time that it should take me to drive to work under normal conditions). Thus, the

Modern Engineering Statistics   By Thomas P. Ryan
Copyright © 2007 John Wiley & Sons, Inc.

process of me driving to work was "out of control" that day because of forces beyond my control. That is, there was an assignable cause—the accident.

If I arrived late for a class as a consequence, I could explain what happened but a piece of machinery obviously cannot explain why it is out of control in regard to a certain characteristic. Thus, some type of technique or chart is necessary to identify times at which one or more assignable causes are likely to be present.

The driving example can also be used to illustrate the blur that sometimes exists between common causes and assignable causes. Assume that the (24, 36) interval was determined based on driving in the autumn, but the chart limits are also applied to winter driving. Snow of any amount on the roads would certainly be viewed as an (unusual) assignable cause in early autumn in, say, Wisconsin, but would not be an assignable cause during the winter. So control limits for autumn driving would be inappropriate for winter driving.

Similarly, manufacturing processes can change, so the control limits should correspondingly be changed when necessary: that is, when "the system" has changed.

Although control charts are primarily used in engineering and industrial applications, there have been some novel nonindustrial applications. In particular, Charnes and Gitlow (1995) used control charts in reverse (i.e., going backward in time) to detect bribery in jai alai and to help gain a conviction. Another novel application, described at `http://www.minitab.com/resources/stories/ExhibitAPg3.aspx`, occurred after a house had been sold and the buyer sued the previous owners, alleging that the basement became flooded during heavy rains and that such information was not disclosed to the buyer. Since the previous owners had not encountered any such problems, they were baffled by the allegations. Consequently, they obtained historical precipitation data from the National Oceanic and Atmospheric Administration (NOAA) and constructed a control chart of precipitation for a six-year period. The chart showed that the rainfall during the period in question was above the upper control limit, which was apparently constructed using the data up through the time that they lived in the house. This swayed the judge and the ruling was in favor of the defendants.

These last two examples simply serve to illustrate that there are processes all around us, and these processes can easily go "out of control" in one way or another. Therefore, there are almost endless opportunities for successful control chart applications.

11.2 STAGES OF CONTROL CHART USAGE

In general, control charts can be used to determine if a process (e.g., a manufacturing process) has been in a state of statistical control by examining past data. This is frequently referred to as a retrospective data analysis. We shall also refer to this stage as Stage 1. More importantly, recent data can be used to determine control limits that would apply to future data obtained from a process, the objective being to determine if the process is being maintained in a state of statistical control. Real-time monitoring is Stage 2. (*Note*: Most writers have referred to these two stages as Phase I and Phase II, respectively, and we will follow custom and henceforth use the same terminology.) The meaning of "real time" differs among the various charts. For some charts, data might be acquired every 15 or 30 minutes, with points plotted on a control chart at such intervals. For other charts, such as a p-chart presented in Section 11.9.1, data might be acquired and plotted at the end of a day.

GENERAL PROCEDURE

Control limits are first constructed using historical data and the limits are used to test the historical data for control (Phase I). After perhaps some historical data corresponding to out-of-control states are discarded and the control limits are recomputed, the limits are used for real-time process monitoring (Phase II). The parameter estimators that are used for each stage shouldn't necessarily be the same.

If control charts are being used for the first time, it will be necessary to determine *trial control limits*. To do so, it is desirable to obtain at least 20 subgroups (of size 5, say) or at least 100 individual observations (depending on whether subgroups or individual observations are to be used) from either past data, if available, or current data. If collected from past data, they should be relatively recent data so that they adequately represent the current process.

Consider Figure 11.1. The points in Figure 11.1 might represent such data, with UCL and LCL denoting the upper control limit and the lower control limit, respectively, and the midline is typically the average value of the statistic that is plotted—the average of the historical data, which would be the average of the plotted points for the Phase I chart. Because there are two points outside the control limits, these points should be investigated since they were included in the calculation of the trial control limits. Whether or not these trial limits should be revised depends on the outcome of the investigation. If each of the points can be traced to a "special cause" (e.g., a machine out of adjustment), the limits should be recomputed *only* if the cause can be removed. If the cause cannot be removed, it should be regarded (unfortunately) as a permanent part of the process, so the trial limits

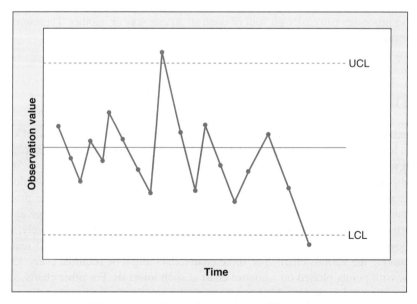

Figure 11.1 General control chart illustration.

should not be recomputed. Thus, the trial limits should be recomputed only if the cause of points lying outside the limits can be both detected *and* removed. Occasionally, points will fall outside the limits because of the natural variation of the process. This is much more likely to happen when a *group* of points is plotted rather than the sequential plotting of individual points. Specifically, if a is the probability that a single point is outside the control limits when points are plotted individually, the probability that at least k plotted points will be outside the limits is approximately ka.

The question arises as to what should be done when one or more points are outside the new limits that are obtained from deleting points that were outside the old limits. That could happen with the points in Figure 11.1, since deleting points outside the limits will cause the new limits to be closer together. Thus, the points that are just barely inside the old limits could be just barely outside the new limits. Technically, points that are outside the new limits should also be deleted and the limits recomputed if and only if an assignable cause can be both detected and removed, and this cycle should be continued until no further action can be taken.

After a process has been brought into a state of statistical control, a process capability study can be initiated to determine the capability of the process in regard to meeting the specifications. (Process capability indices are discussed briefly in Section 11.10.) It would be illogical to undertake such a study if the process were not in control, since the objective should be to study the capability of the process after all problematic causes have been eliminated, if possible. (Many companies do use *process performance indices*, however, which show process performance when a process is not in a state of statistical control. These indices are not covered in this chapter, however.)

Control charts alone cannot produce statistical control; that is the job of the people who are responsible for the process. Control charts can indicate whether or not statistical control is being maintained and can provide users with other signals from the data. They can also be used in studying process capability.

IMPORTANT POINT

There are many different control charts to select from, and also various ways of computing control limits. Regarding the latter, it is very important that the control limits be computed using a method that ensures the existence of a LCL for charts for which points falling below the LCL is a sign of improvement.

With the progression from Phase I to Phase II, there should be a progression to using more observations in computing the control limits. While 100 observations, either individual or in subgroups, is adequate for Phase I, the control limits to be used in Phase II should be computed from at least 300 observations. The latter number has been recommended in the literature (e.g., see Quesenberry, 1993) as the general idea is to have the control limits computed from enough observations so that the limits have approximately the same properties as they would if the parameter values were known. Clearly, it would be undesirable for the limits to have considerable sampling variability, but this will happen when only a small-to-moderate number of observations is used for computing the limits. This can be explained as follows. Control limits are extreme values (see Section 11.3), so consider small movements in the upper tail of the standard normal distribution. For example, $P(Z > 3.0) = .00135$ and $P(Z > 3.1) = .00087$. These probabilities differ very slightly, but control chart properties are determined from the reciprocals of tail probabilities and $1/.00135 = 740.74$ and $1/.00087 = 1149.43$ differ greatly.

Thus, if 100 observations were used in Phase I, data when the process is apparently in control in Phase II should be added to its counterparts in Phase I, with the (semipermanent) Phase II control limits computed from the requisite number of observations. Of course, these control limits should eventually be recomputed, especially when there is evidence of a process change. There are no rules as to when limits should be recomputed in the absence of discernible events, so this must be a matter of judgment. Having control limits computed from recent data is always a good idea, so it is a matter of *when* it should be done rather than *if* it should be done.

11.3   ASSUMPTIONS AND METHODS OF DETERMINING CONTROL LIMITS

The standard way to compute control limits is to use 3-sigma limits. That is, the limits are computed as $\widehat{\theta} \pm 3\sigma_{\widehat{\theta}}$, with $\widehat{\theta}$ denoting the statistic or function of the statistic (or individual observation) that is plotted on the chart. Although this method works satisfactorily in many applications, it is a poor method for certain types of charts and in general should not be used automatically.

The use of "$\pm$ something" means that a symmetric distribution is being assumed, and the user of 3-sigma limits is assuming that it is extremely unlikely for points to fall outside the limits when the process is in control. For example, *if* normality existed and the parameters were known, the probability of a point plotting below the LCL when the process was in control would be .00135, which would also be the probability of a point plotting above the UCL. Thus, the probability of a false signal is quite small, and obviously that is desirable. Of course, it is also desirable to detect a process change as quickly as possible, and the smaller the false alarm probability is, the longer it will take to detect a process change. Thus, there must be a compromise between the goal of quick detection and the desire for a small false alarm probability. The use of 3-sigma limits constitutes a reasonable compromise, provided that the distribution of the statistic whose values are plotted on a chart is reasonably symmetric and doesn't differ greatly from a normal distribution, especially in terms of the size of the tails of the distribution. If the data are asymmetric, and not few in number, this suggests that the population is also probably asymmetric. A control chart user might try to transform the data to normality, or fit a nonnormal distribution to the data and determine the control limits from the percentiles of the fitted distribution, or use a nonparametric control chart.

The plotted points must also be *independent*: that is, knowing the numerical value of any one of the plotted points is of no help in guessing/predicting the values of any of the other points. This assumption should always be checked, as a different approach would have to be used if the assumption were violated. A simple time series plot of the data can be used for this purpose, and statistics such as the autocorrelation function can also be used. The latter is discussed in Example 11.2 in Section 11.6.

There are various statistics used in process monitoring work whose distributions are quite asymmetric. For these statistics (such as the proportion nonconforming), the use of 3-sigma limits can cause the computed LCL to be negative, whereas hardly any (uncoded) measurement could ever be negative. There are very few exceptions; temperature in degrees Fahrenheit is one exception. When what is being measured cannot have a negative value, there will be no LCL because by definition a control limit is a value that could be exceeded (i.e., above the UCL or below the LCL) by a plotted point. A negative computed LCL will happen very frequently when 3-sigma limits are used for charts for nonconforming units and for nonconformities. This will also be a problem when the range of observations from a sample is plotted because the range also has a highly skewed distribution.

11.4 CONTROL CHART PROPERTIES

If the parameters were known, the expected length of time before a point plots outside the control limits could be obtained as the reciprocal of the probability of a single point falling outside the limits when each point is plotted individually. This expected value is called the average run length (ARL). It is desirable for the in-control ARL to be reasonably large, so that false alarms will rarely occur. With 3-sigma limits, the in-control ARL is $1/.0027 = 370.37$ under the assumption of normality and known parameter values. The in-control ARL being equal to the reciprocal of the sum of the tail areas results from the fact that the geometric distribution is the appropriate distribution, and the mean of a geometric random variable was given in Section 3.3.5 as $1/p$.

When a parameter change of an amount that is considered to be consequential occurs, we would like to detect the change as quickly as possible. Accordingly, the *parameter-change* ARL should be small. Unfortunately, Shewhart-type charts do not have good parameter-change ARL properties. For example, assume that a $1\sigma_{\bar{x}}$ increase in the mean occurs when an $\bar{X}$ chart is used. Again assuming that the parameters are known, we may determine the probability of a point falling outside the control limits by first computing two Z-values.

Specifically, we obtain $Z = [\mu + 3\sigma_{\bar{x}} - (\mu + \sigma_{\bar{x}})]/\sigma_{\bar{x}} = 2$ for the UCL, and $Z = [\mu - 3\sigma_{\bar{x}} - (\mu + \sigma_{\bar{x}})]/\sigma_{\bar{x}} = -4$. The ARL is then obtained as $1/[P(Z > 2) + P(Z < -4)] = 43.89$. ARLs for other mean changes as a multiple of $\sigma_{\bar{x}}$ could be computed by using $k\sigma_{\bar{x}}$ in place of $\sigma_{\bar{x}}$.

The reader should realize that parameters will generally have to be estimated, however, and doing so will inflate both the in-control ARL and the parameter-change ARLs. This can be explained as follows. For simplicity, assume that σ is known so that only μ must be estimated. With the assumption of normality, the UCL is as likely to be overestimated by a given amount as it is to be underestimated by the same amount. If the UCL is overestimated by $0.1\sigma_{\bar{x}}$, the reciprocal of the tail area of the normal distribution above the UCL is 1033.41, whereas for underestimation by the same amount the reciprocal is 535.94. The average of these two numbers is 784.67, which is noticeably greater than the nominal value of 740.74. Estimation errors of larger amounts would produce a larger discrepancy.

The determination of ARLs when parameters are estimated has been addressed by several authors. Quesenberry (1993) used simulation to obtain the ARLs and concluded that it is desirable to have the trial control limits based on at least 100 observations, with "permanent" limits based on at least 300 observations, as indicated previously. Thus, if a subgroup size of 5 is being used, the historical data that are being used to compute the trial limits should be comprised of at least 20 subgroups. See also Ghosh, Reynolds, and Hui (1981), who determined the ARLs using an analytical approach.

11.5   TYPES OF CHARTS

The most commonly used control charts are those that are referred to as "Shewhart charts." Walter A. Shewhart (1891–1967) sketched out the idea of a control chart in a May 1924 memorandum. The first chart that he proposed was a p-chart, a chart for nonconforming units. Shewhart charts are also used for controlling a process mean, process variability, number of nonconformities, and so on.

These charts are such that the control limits are in the form of $\widehat{\theta} \pm 3\sigma_{\widehat{\theta}}$ that was given in Section 11.3. As indicated in that section, however, there are strong reasons for not automatically computing the limits in this manner. Similarly, there are charts that have better properties than Shewhart charts. The latter do not accumulate information, so when a process goes out of control, the charts do not utilize data after the process change that would obviously be helpful in detecting the change. Furthermore, in discussing statistical process control (SPC), Montgomery and Woodall (1997) stated: "In many cases the processes to which SPC is now applied differ drastically from those which motivated Shewhart's methods." Alternatives to Shewhart charts are often needed, with many of them described in the chapter of the same name in Ryan (2000) and some presented in this text in Section 11.9.

Since Shewhart charts are so frequently used, they will be presented first, but caveats will also be given in appropriate places.

11.6   SHEWHART CHARTS FOR CONTROLLING A PROCESS MEAN AND VARIABILITY (WITHOUT SUBGROUPING)

The simplest control chart is the X-chart, which is the chart for individual observations. This has also been called an I-chart or an individual observations chart, and this terminology will be used often in this section since MINITAB uses this terminology and the I-charts that are used for illustration herein were produced using MINITAB. Data are often in the form of individual observations such that it would be impractical to try to group them in some way for the purpose of using charts for subgrouping. For example, it will generally be impractical to try to group temperature and pressure measurements since, for example, five temperature measurements made in quick order would likely be virtually the same. Clerical and accounting data would also have to be charted using individual numbers rather then subgroups.

If the process mean and standard deviation were known, the control limits would be set at $\mu \pm 3\sigma$, provided that it was reasonable to assume a normal distribution as an appropriate model. As with statistical methods in general, assumptions should be checked, and the normality assumption is critical for an X-chart.

Since parameter values are generally unknown, they must be estimated. The usual estimator of μ is $\overline{X}$, which would be computed from a set of recent historical data. Although there is a clear-cut choice for the estimator of μ, this is not the case when deciding how to estimate σ. The best approach is to use two estimators: one for analyzing the set of historical data and the other for process monitoring (i.e., Phase I and Phase II). The reason for this recommendation is that one estimator is preferable for Phase I, whereas the other is superior for Phase II.

These two estimators are based on the moving range and standard deviation, respectively. The following example illustrates how the moving range (of size 2) is computed.

X	Moving Range
14	
	4
18	
	2
16	
	1
17	
	4
21	

That is, the absolute value of the difference between consecutive values is taken as the moving range. It can be shown that the average moving range $(\overline{MR})$ divided by an appropriate constant (d_2) is an unbiased estimator of σ, with the value of d_2 obtained from Table F for $n = 2$. That is, $\widehat{\sigma} = \overline{MR}/d_2 = 2.75/1.128 = 2.44$. The estimator that should be used in Phase II is s/c_4, with c_4 the constant, from Table F, that makes s/c_4 an unbiased estimator of σ. For this simple example, $\widehat{\sigma} = s/c_4 = 2.59/0.9400 = 2.76$, with the value of c_4 based on the number of observations.

This leads to the following definition.

DEFINITION

The control limits for an X-chart are obtained as $\overline{X} \pm 3\widehat{\sigma}$, with, preferably, $\widehat{\sigma} = \overline{MR}/d_2$ in Phase I and $\widehat{\sigma} = s/c_4$ for Phase II.

We start with Phase I if a control chart has not been used before in a particular application. Without the use of any process control procedures, a process is probably out of control when control chart usage is initiated. Accordingly, the task is to separate the data that have come from the in-control distribution (if any) from the data that have come from the out-of-control distribution(s). That is, the process may have been in control for part of the historical data but be out of control for the rest of it.

Studies have shown that the moving range estimator is a better estimator of σ than s/c_4 in the presence of bad data. We would hope that we can identify all of the bad data in the first phase and then just use the good data in the second phase. If we have done this, and technically we should not proceed to Phase II until we have done so, we should use s/c_4 for that phase because $Var(s/c_4) < Var(\overline{MR}/d_2)$. That is, the control limits for Phase II will have less sampling variability if $\widehat{\sigma} = s/c_4$ is used for that stage. Stated differently, we would need more observations using the moving range estimator in order to have the same variability as results from use of the other estimator. This is important because unnecessary sampling variability is not only undesirable because we naturally prefer estimators with small variability, but also because it affects location (i.e., mean) properties.

Although using the moving range to estimate σ is desirable in Phase I, a moving range chart, for controlling the variability, with moving ranges of size 2 plotted in Phase II, adds very little, as shown by Roes, Does, and Schurink (1993). The problem is that an individual observation does not contain any information about variability, so when we are plotting individual observations in real time, we might expect that we are not likely to gain very much from using a contrived measure of variability, such as one based on a moving range. Thus, it would be reasonable to use an X-chart to control both the mean and the standard deviation, although there is no harm in additionally using a moving range chart.

The following example of an actual problem illustrates the construction and use of an X-chart.

■ **EXAMPLE 11.1**

A manufacturer of plate glass by the method of vertical drawing from a molten interface maintains daily records on the overall yield of plate glass panels produced relative to the total glass drawn.

Problem Statement and Data

Day-by-day yield fluctuations and monthly levels of yield were unsatisfactory, as Figure 11.2 shows. The 31 data points are as follows and are also at the textbook website (given at the start of the Chapter 1 exercises).

Yield

79 75 73 76 75 67 75 76 74 73 59 64 76 68 72 79
77 81 73 75 75 66 68 65 70 57 70 66 72 70 68

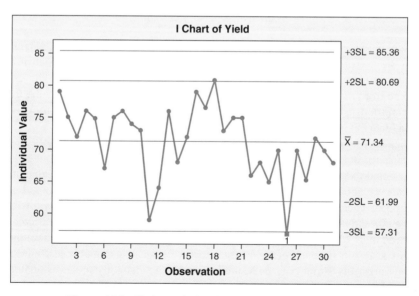

Figure 11.2 X-chart of yield in glass sheet manufacture.

The chart shows that the average daily yield was 71.3%. The maximum possible yield is 84% since 16% of glass drawn is expected to be reprocessed as cullet. Thus, the average daily yield was considered to be quite unsatisfactory and the day-to-day fluctuations were also considered to be unsatisfactory. It is important to note, however, that the latter conclusion was based on the fact that 2-sigma limits were used, not 3-sigma limits. (The chart in Figure 11.2 shows both.) Using the former instead of the latter increases the false alarm rate, which is offset somewhat by having better capability to detect out-of-control conditions. Here, however, the wrong signal may have been emitted as there are three points outside the 2-sigma limits, but only one point outside the 3-sigma limits, and that point is almost on the line.

The company sought to improve the average yield to bring it closer to the 84% figure. Separate charts were prepared for each of eight machines, three shifts, and five thicknesses ranging from 3 to 7 mm. The company also formed quality teams to critically analyze nonconforming units, categorizing them by severity of defects, and to determine which defects were the most prevalent. (Charts for nonconforming units are covered in Section 11.9.) ■

We will use the following example to illustrate some mistakes that are commonly made and to discuss how to avoid such mistakes.

■ **EXAMPLE 11.2**

Data

One of the sample files that comes with the MINITAB software is TILES.MTW. The file contains data on the amount of warping of floor tiles, with 10 tiles measured for each of 10 consecutive days. This provides 100 observations, which as previously stated is essentially the minimum number that should be used for computing the initial set of control limits.

The data are as follows and can be found at the textbook website.

Warping

1.60103	0.84326	3.00679	1.29923	2.24237	2.63579	0.34093
6.96534	3.46645	1.41079	2.31426	2.55635	4.72347	1.75362
1.62502	5.63857	4.64351	3.95409	4.38904	3.24065	0.52829
1.01497	1.12573	2.56891	4.23217	1.34943	2.84684	0.76492
2.78092	0.63771	2.89530	2.86853	2.18607	1.05339	1.25560
1.97268	0.84401	3.32894	4.15431	2.57873	0.44426	2.48648
3.91413	2.28159	0.96705	4.98517	5.79428	2.52868	3.08283
3.82585	5.31230	1.92282	1.22586	0.76149	2.39930	4.96089
1.96775	1.35006	4.79076	2.20538	1.22095	6.32858	3.80076
4.22622	4.33233	0.42845	1.20410	3.44007	2.51274	8.09064
4.24464	3.21267	3.48115	6.66919	2.44223	3.51246	8.03245
1.13819	4.27913	2.05914	1.12465	0.78193	4.14333	5.30071
3.79701	3.24770	5.04867	3.06800	2.45252	4.69474	0.28186
0.57069	0.70532	2.84843	6.25825	3.37523	3.23538	6.08121
1.66735	2.12262					

It is important to check for nonnormality when control charts are constructed and normality is assumed. We can see the importance of doing so with this data set.

Figure 11.3 shows the control chart constructed for the data in this file, with σ estimated using the moving range.

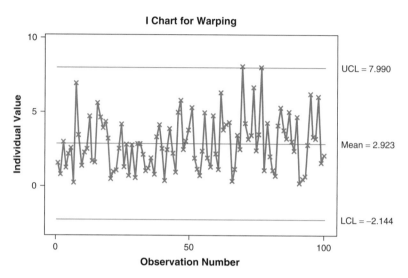

Figure 11.3 *I*-chart for Minitab TILES data.

Impossible Computed LCL

From a glance we can see that constructing the chart with 3-sigma limits was not the proper thing to do. In particular, the chart (produced using MINITAB) shows the LCL to be negative, whereas "amount of warping" cannot be negative! Of course, the software cannot be blamed for not knowing that a negative value is impossible, but a user can be blamed for unthinking construction of a control chart!

There is no LCL when the computed LCL is an impossible value, which means that by constructing the chart without first examining the data, we have in essence "forfeited" a LCL. It was stated in the "Important Point" of Section 11.2 that it is very important for control limits to be computed in such a way that there will be a LCL for charts for which points falling below the LCL signals improvement.

Clearly, that would be the case here as a company would certainly like to keep the amount of warping to a minimum. If that minimum were not being achieved, then we would want to see observations plotting below the LCL over time, as that would be a sign of progress.

Both a histogram and a normal probability plot show that the data are highly skewed, so it would be foolish to construct control limits that are symmetric about the centerline. So what corrective action should be taken? One possibility would be to fit a distribution to the data and determine the control limits from the percentiles of the fitted distribution. Some control chart software packages have a distribution-fitting capability.

Although MINITAB doesn't have this capability for control charts, it does have the capability of fitting life distributions (for survival analysis, not covered in this text, and reliability analysis, which is covered in Chapter 14). For example, a two-parameter Weibull distribution has considerable flexibility for fitting skewed distributions (since the distribution has two parameters rather than one), so we might consider trying to fit the data with that distribution. The commands to do so and to examine the fit relative to the data are given at the textbook website.

The reader who fits this distribution using some software and compares the percentiles of the data with the percentiles for the fitted Weibull distribution will observe that the percentiles match up reasonably well. For example, the first percentile of the fitted distribution is 0.217, whereas the smallest observation is 0.282, the 50th percentile of the fitted distribution is 2.64 which differs only slightly from the 50th ordered observation of 2.58, and the 99th percentile is 8.08, whereas the 99th observation is 8.03 and the 100th (largest) observation is 8.09.

Of course, another approach would be to use the sample percentiles and not attempt to fit a distribution. That would be a *nonparametric approach* (nonparametric methods, in general, are covered in Chapter 16). In that case, our sole interest would be in estimating the (extreme) percentiles that would be used for the control limits, but it generally takes more than a thousand observations to be able to do so with any degree of precision. When we fit a distribution, we are indirectly estimating the percentiles by fitting the distribution, and by using all of the data we have a more powerful procedure since we are not looking at just a couple of data points.

Here we have $\widehat{\alpha} = 3.2781$ and $\widehat{\beta} = 1.6937$ as the estimates of the Weibull parameters for the distribution written in the second form in Section 3.4.6. Since the distribution fits the data reasonably well, we might determine the control limits from the 0.135 and 99.865 percentiles of this distribution. Using MINITAB or other software, we obtain LCL = 0.0663 and UCL = 9.9953. All 100 warping measurements are within these limits, whereas two of the observations were (slightly) above the 3-sigma UCL.

Although the Weibull LCL is greater than zero, it is still very small. This is going to happen, however, whenever a distribution is skewed (and invariably data are right skewed), so that there is no "tail" on the left side.

One solution to this problem would be to obtain more observations, hoping that a tail might then occur. The other alternative would be to transform the data to approximate normality, if possible. MINITAB code to perform a Box–Cox transformation is given at the textbook website.

When data are right skewed, we need a transformation such that if we use a power transformation, the power (i.e., exponent) should be less than 1.0. Here the power that results from the iterative process is 0.337, and when a normal probability plot is applied to the transformed data, the data pass the normality check. The data should also pass the independence test, and one way to do this in MINITAB is to use the ACF command. This gives the autocorrelations between adjacent observations, between observations two units apart, three units apart, and so on. If the data are independent, the autocorrelations should all be close to zero, as they are for the transformed and original data. Confidence limits are also provided so that the user can decide whether or not an autocorrelation is significant.

When an X-chart is constructed for the transformed data, the LCL does exist and all of the points are well within the control limits. Although transformations generally make the data look "nicer," they do have the distinct disadvantage of being on a scale that generally will not have any physical meaning. For example, if data are measured in inches, $12.50^{0.337}$ has no physical interpretation.

Two approaches to charting these data have been given. Either would give reasonably good statistical results in this application, although many users would object to using transformed data. The important point is that one cannot simply assume that using 3-sigma limits will be satisfactory. This point cannot be overemphasized, so a second example will be given, albeit with less discussion than in this example. ■

■ **EXAMPLE 11.3**

Objective

Chou, Halverson, and Mandraccia (1998) presented a case study in which the objective was to determine whether or not a process that involved a new piece of equipment was in control.

Data

A sample of 116 observations was obtained, a sufficient number for computing initial control limits. The measured variable was equipment-generated particle counts on wafers. The data are as follows and are also given in EX113ENGSTAT.MTW.

```
27 16 16 34 14 13 10 43 23 13  8 22 14  8 15 20 79  9 13  6 18 13 34
16 15 11  4  9 12  9 38  7 15  7  4 31  9  7 35  7  8 15 13  5  4  4
13  7 39 61 27 11 10 18 14  3 15 14  8 12  9 13 35 11 23 11  9 11 42
12  4  4 15  9  8 10 11 25 10 19  8 19 11 13 12 37 44 12  9 11 74 12
27 43  4  6  5 15  3 24 22 10 23 16  5 40 27 16  5 12 23  5 30 19  8
 9
```

Incorrect Check on Assumptions

Again to illustrate why assumptions must be checked before a control chart is constructed (or immediately thereafter), we present the *X*-chart (*I*-chart) for these data, which is given in Figure 11.4, with "1" denoting a point outside the control limits.

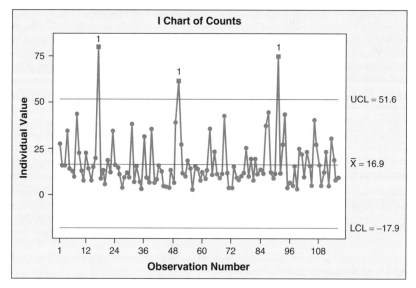

Figure 11.4   *I*-chart for new equipment data.

We again see the folly of constructing a standard control chart for these data as the LCL is far below zero, whereas all of the data are of course positive. We also observe three points that are above the UCL, and two of those points far exceed the UCL. These points would be outliers to be investigated under a normal distribution model, but the computed LCL being an impossible value tells us that a normal model doesn't make any sense.

Some experts contend that it is okay to construct a control chart without first checking assumptions, as the chart will indicate when the assumptions may be violated. That is true for this example, but will not necessarily be true for less extreme examples. It is better to check the assumptions first. A dotplot of the data is given in Figure 11.5.

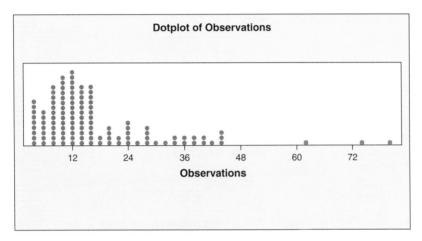

Figure 11.5 Dotplot of new equipment data.

The skewness in the data is apparent. The three largest values might all be bad data points (this would have to be investigated), but the LCL is still negative even when they are deleted.

As in the previous example, the user could choose between fitting a distribution and transforming the data. Chou, Halverson, and Mandraccia (1998) tried three transformations and selected the one (a Johnson transformation) that produced the closest transformation to normality. The second best transformation was a logarithmic transformation. This implies that a lognormal distribution would provide a reasonable fit to the untransformed data, and this can easily be verified. ∎

We should be somewhat concerned by the latter, however, as Ott (1975, p. 4) pointed out that a lognormal distribution is often incorrectly assumed when data are unknowingly obtained from a mixture of two normal distributions with different means, with the proportion of data coming from each being considerably different. If one of the distributions is the in-control distribution, failure to identify this by transforming the data or fitting a lognormal distribution to the original data could have serious consequences for Phase II, as the limits for that stage could have been computed from a sizable percentage of bad data.

We will not pursue separating a mixture distribution into component parts as that is beyond the scope of this book. The objective here is not to alarm the reader but rather to illustrate and emphasize that the use of control charts is not as straightforward as it might appear to be. Care must be exercised, as when any statistical method is applied.

Before concluding our discussion of control charts for individual observations, it should be mentioned, at least in passing, that moving average control charts are also used, with such usage undoubtedly stemming from the popularity of moving averages, in general. (Moving averages were discussed in Section 2.1.) A moving average control chart has some obvious shortcomings, however. In particular, when control charts are used, the user looks for patterns, but the moving average chart will have a nonrandom configuration even when the plotted points represent random variables with a common mean and variance. This is because the plotted points will be correlated since consecutive moving averages, for example, are computed using some of the same observations, and this property is not limited just to consecutive observations.

11.7 SHEWHART CHARTS FOR CONTROLLING A PROCESS MEAN AND VARIABILITY (WITH SUBGROUPING)

Historically, the $\overline{X}$-chart has been the most frequently used chart, and it seems reasonable to assume that it is still used more often than any other chart. It is used for the same purpose that an X-chart is used: for controlling a process mean.

11.7.1 $\overline{X}$-Chart

With an $\overline{X}$-chart, however, subgroups of data are formed, unlike an X-chart. For example, five ball bearings might be selected from an assembly line every 30 minutes, and the diameter of each one is measured. Why select five instead of one? With five we have more power to detect a change than with individual observations, but the comments in Section 5.9 on confidence intervals and hypothesis testing also apply here. That is, the larger the sample size, the more likely we are to detect a change that does not correspond to practical significance. With more sophisticated procedures than a Shewhart chart, such as a CUSUM procedure presented in Section 11.10.1.1, we can set the parameters of a control chart scheme in such a way that we focus on a shift of at least a certain magnitude.

Since subgroups are formed, the question arises as to how this should be done. The objective is generally to construct *rational subgroups*. The general idea is to have data in the subgroups that are from the same population (i.e., not from different machines, operators, etc.). Observations obtained close together in time won't necessarily meet this requirement. For example, assume that units of production are coming off an assembly line from three machines that have slightly different heads. If subgroups of units are formed a bit downstream without regard for the machine through which the unit came, the subgroups might differ considerably in terms of machine composition, with the fact that the units are close together in time not necessarily having any bearing on the subgroup composition.

11.7.1.1 Distributional Considerations
Since 3-sigma limits are typically used with an $\overline{X}$-chart, the distribution of $\overline{X}$ should be reasonably symmetric and close to a normal distribution. Some writers have stated that we shouldn't be concerned with nonnormality when an $\overline{X}$-chart is used because of the Central Limit Theorem (Section 4.3.1). But that is a limiting theorem that has nothing to say about what happens when we have a small sample. Although the distribution of $\overline{X}$ will be closer

to normal than will the distribution of X when the latter is nonnormal, the distribution could still be considerably asymmetric.

The minimum number of observations for the computation of trial control limits that was given for an X-chart also applies to an $\overline{X}$-chart. That is, if we have subgroups of size 5, we need at least 20 of them so as to provide at least 100 observations. It would be difficult to check the distribution of $\overline{X}$ if we have only a small number of subgroups, so we might simply check the distribution of the raw data. If the latter doesn't deviate greatly from a normal distribution, we should be okay in using 3-sigma limits on an $\overline{X}$-chart.

DEFINITION

The control limits for an $\overline{X}$-chart are computed as $\overline{\overline{X}} \pm 3\widehat{\sigma}_{\overline{X}}$, with $\overline{\overline{X}}$ denoting the average of the subgroup averages and $\widehat{\sigma}_{\overline{X}}$ denoting some estimator of the standard deviation of the subgroup averages. That estimator is typically obtained as either $(\overline{s}/c_4)/\sqrt{n}$, with $\overline{s}$ denoting the average of the subgroup standard deviations and c_4 the tabular value for a given subgroup size, or $(\overline{R}/d_2)/\sqrt{n}$, with $\overline{R}$ the average of the subgroup ranges and d_2 the tabular value for a given subgroup size.

■ **EXAMPLE 11.4**

A study was performed on the mass of polyvinyl chloride (PVC) bottles produced on a blow molding machine. The study was motivated by records that showed insufficient production of bottle units for PVC consumption. The study was performed for one month and the data that were recorded were on PVC consumption, actual production, and expected production based on a desired nominal mass. This study led to significant savings and motivated a study of the blow molding machine. The variation in the bottle mass was of particular interest.

An initial study was performed that consisted of samples of five consecutive bottles drawn from each of the two molds every 15 minutes. This resulted in 14 subgroups of size 5, which unfortunately is short of the minimum number of 100 observations that are preferred to be used to compute the initial control limits, as discussed in Section 11.2. It would have been better if more observations had been used so that the control limits could have been estimated with greater precision, especially since there was considerable variation in the subgroup averages when an $\overline{X}$-chart was constructed, which here is shown in Figure 11.6. (The data, given below, are used with the permission of H. M. Wadsworth.)

Bottle mass

33.6	33.6	33.6	33.8	33.7	33.6	33.7	33.6	33.6	33.7	33.4
33.3	33.4	33.3	33.6	33.4	33.4	33.5	33.4	33.3	33.5	33.5
33.4	33.4	33.5	33.6	33.5	33.5	33.5	33.5	33.4	33.4	33.4
33.3	33.3	33.3	33.4	33.4	33.4	33.3	33.5	33.5	33.4	33.4
33.4	33.5	33.5	33.4	33.5	33.5	33.4	33.4	33.3	33.3	33.5
33.2	33.4	33.2	33.4	33.4	33.7	33.9	33.9	33.9	33.8	33.9
33.9	33.8	33.8	33.8							

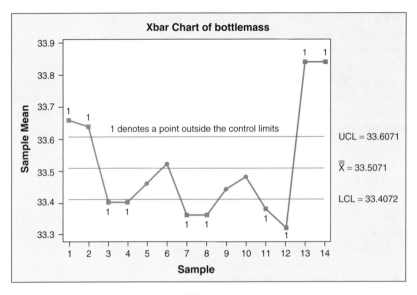

Figure 11.6 $\overline{X}$-chart of bottle mass.

Clearly, the chart shows a lack of control, with most points outside the control limits instead of inside them. There are actually four points that don't fit with the others, however—the two sets of measurements made at the start of the study and the two sets at the end of the study. This was blamed largely on operator overadjustment. The R-chart (see Section 11.7.2), which the reader is asked to construct in Exercise 11.66, showed that the variation was in a state of statistical control. There is another aspect to this dataset that the reader is asked to consider in Exercise 11.67.

Another conclusion that was drawn from the study was that the measurements from the two molds were sufficiently similar that future studies could be conducted using only one mold.

Based on the results of this initial study, a second study was performed in which ten samples (subgroups) of three consecutively produced bottles each were drawn directly from the machine at 5-minute intervals using mold #4 (only). Operators were instructed not to make any machine adjustments while the samples were obtained, as the initial study showed that they had a tendency to overadjust the machine. $\overline{X}$- and R-charts were constructed for both the bottle mass and the width of the bottom piece and both $\overline{X}$-charts exhibited a lack of statistical control, although the $\overline{X}$-chart limits are naturally somewhat suspect since they are based on only 30 observations.

The negative correlation between bottle mass and width of the bottom piece was evident from a comparison of the charts, as the charts were almost mirror images of each other, with the negative correlation becoming more apparent when a scatter plot was constructed. Because of the high negative correlation (the correlation coefficient was −0.95), it was necessary to control the width of the bottom piece at approximately 24.7 mm in order to be able to control the bottle mass at an average of 33 grams.

Although only process control methods for a single variable are covered in this chapter, this is not an atypical example. Accordingly, the more complex multivariate control chart methods, which are beyond the scope of this chapter, will often be needed. ■

11.7.1.2 Parameter Estimation

The logical way to estimate the process mean is to compute the average of the subgroup means, which is denoted as $\overline{\overline{X}}$. A choice must be made when σ is estimated, however, as there are essentially three options: (1) $\overline{R}/d_2$ with $\overline{R}$ denoting the average of the subgroup ranges (the largest observation minus the smallest in each subgroup), and d_2 (from Table F) based on the subgroup size; (2) $\overline{s}/c_4$ with $\overline{s}$ denoting the average of the subgroup standard deviations and c_4 also obtained from Table F for the appropriate subgroup size; and (3) $\sqrt{ave(s^2)}$ with $ave(s^2)$ denoting the average of the subgroup variances. The latter is the best estimator in the sense that it has the smallest variance, but it is only slightly superior to the second estimator. Thus, either of the last two estimators should be used.

11.7.2 s-Chart or R-Chart?

For measurement data in subgroups, a chart for controlling the mean is used in conjunction with a chart for controlling variability. The latter should be used first in a new application because the variability should generally be under control before one tries to control the mean. The reason for this recommendation is that unless the variability of the process is in a state of statistical control, we do not have a stable distribution of measurements with a single fixed mean. Stated differently, instead of having a single distribution, there might be different distributions over different time periods. A state of statistical control exists if the mean and the variability are stable over time, even if the mean is stabilized at an undesired value. Statistical control does not exist if the variability is unstable.

The two most frequently used charts for controlling the variability of data in subgroups are an R-chart and an s-chart. The former is a *range chart* and the latter is a *standard deviation chart*. Although the former has frequently been used, it is inefficient compared to a standard deviation chart in the same way that a range is inefficient to a standard deviation as a measure of variability. This is because a range is computed from only the two most extreme observations in a subgroup, whereas all of the observations are of course used in computing a standard deviation.

Whichever chart is used, it is important that 3-sigma limits *not* be used. As stated previously, it is important to have a LCL on any chart for which points falling below the LCL is a sign of probable improvement. A chart for controlling process variability is such a chart because we want to see variability reduced and be as small as possible for virtually all processes, and points falling below the LCL signal a likely reduction in variability. Unfortunately, there will be no LCL on an R-chart with 3-sigma limits when the subgroup size is less than 7 (as is usually the case), and there will not be a LCL on an s-chart with 3-sigma limits when the subgroup size is less than 6.

Thus, probability limits should be used instead of 3-sigma limits. The limits for the S-chart can be obtained by using the chi-square distribution (see Section 3.4.5.1) in conjunction with the following well-known theorem:

$$\text{If } X \sim N(\mu, \sigma^2), \quad \text{then} \quad \frac{(n-1)S^2}{\sigma^2} \sim \chi^2_{n-1}$$

with $(n-1)$ denoting the degrees of freedom. It follows from this result that

$$P\left(\chi^2_{.001} < \frac{(n-1)S^2}{\sigma^2} < \chi^2_{.999}\right) = .998$$

and so

$$P\left(\frac{\sigma^2}{n-1}\chi^2_{.001} < S^2 < \frac{\sigma^2}{n-1}\chi^2_{.999}\right) = .998$$

By taking square roots we obtain

$$P\left(\sigma\sqrt{\frac{\chi^2_{.001}}{n-1}} < S < \sigma\sqrt{\frac{\chi^2_{.999}}{n-1}}\right) = .998$$

Thus, if the process variability is in control at σ, 99.8% of the time the subgroup standard deviation, s, will fall between the endpoints of the interval. [The .998 is roughly equal to the (assumed) area between the 3-sigma limits on an X-chart or $\overline{X}$-chart under the assumption of normality, which is .9973.]

If an estimate, $\widehat{\sigma}$, of σ is available from past experience or past (but relatively recent) data, that estimate could be used to obtain the control limits as

$$\text{LCL} = \widehat{\sigma}\sqrt{\frac{\chi^2_{.001}}{n-1}}$$

$$\text{UCL} = \widehat{\sigma}\sqrt{\frac{\chi^2_{.999}}{n-1}}$$

If σ is to be estimated from data collected in subgroups (as in the present example), an unbiased estimator of σ is $\overline{s}/c_4$, with c_4 given in Table F. The control limits would then be

$$\text{LCL} = \overline{s}/c_4\sqrt{\frac{\chi^2_{.001}}{(n-1)}}$$

$$\text{UCL} = \overline{s}/c_4\sqrt{\frac{\chi^2_{.999}}{(n-1)}}$$

Note that we technically do not have limits with a specified probability after we estimate σ, and this is true in general whenever what we are trying to control is involved in the control limit expressions and must be estimated. This should be only a minor problem, however, as control limits should generally be constructed using a large amount of data. Furthermore, any slight deviation from the desired probability is more than offset by the fact that there is a LCL.

An R-chart is a chart on which subgroup ranges are plotted. A 3-sigma R-chart will not have a LCL unless the subgroup size is at least 7, as stated previously, and the larger the subgroup size, the more inefficient an R-chart will be relative to an s-chart. Thus, it is virtually impossible to justify the use of an R-chart with 3-sigma limits, although the general usage of the chart will undoubtedly continue at a high level. In general, the chart should be properly viewed as a popular chart in the days before modern computing devices were developed, which now of course enable a control chart user to easily compute a standard deviation.

Thus, a range chart is inadvisable, just as it is undesirable to use subgroup ranges in estimating sigma with an $\overline{X}$-chart.

11.8   IMPORTANT USE OF CONTROL CHARTS FOR MEASUREMENT DATA

A secondary, but important use of control charts is to ensure that processes are under control before an experiment is performed. Consider a manufacturing process for which a modification has been proposed, and an experiment is to be performed to compare the standard process with the proposed process. For example, the standard process might be used for two weeks and then the modified process used for the next two weeks. Process yield would be measured each day for the four weeks and a *t*-test could be performed (Section 6.1.2). One or more control charts should be used to monitor any factor that might go out of control during the experiment and affect the results of the experiment. For example, humidity could change near the time of the shift from the standard process to the new process and materially affect the results, possibly leading to an erroneous conclusion. Whether a variable can be controlled or not, it is important to monitor variables that could affect the results of experiments.

11.9   SHEWHART CONTROL CHARTS FOR NONCONFORMITIES AND NONCONFORMING UNITS

A unit, such as a unit of production, could be either conforming (to specifications) or nonconforming. Similarly, a unit could have one or more nonconformities, with the unit not necessarily being declared nonconforming, as the nonconformities could be minor (such as a small scratch on a radio or a small cut in a wood product).

Before studying and implementing charts for the number or proportion of nonconforming units, or the number of nonconformities per sample or per inspection unit, it is important to remember that, ideally, nonconforming units and nonconformities should not be produced. Consequently, attempting to control the number of nonconforming units or nonconformities at a particular level would generally be counterproductive. The objective should be, of course, to continually reduce the number of such units. When used for that purpose, such charts can be of value in indicating the extent to which the objective is being achieved over time.

Regarding the time element, another important point is that there is no attempt to have tight control over a process when these charts, which are called *attributes charts*, are used. Unlike an $\overline{X}$-chart, on which points might be plotted as frequently as every 15 minutes or so, points on a chart for nonconforming units might be plotted at the end of each day. So the word "control" in "control charts" applies to attributes charts only in a very loose sense.

Thus, such charts might be used in *conjunction* with measurement charts, such as $\overline{X}$-charts, rather than simply in place of them. If either type of chart could be used in a particular situation, it would be wasteful of information to use an attributes chart by itself. Accordingly, such charts should be used alone only when there is no other choice.

11.9.1 *p*-Chart and *np*-Chart

There are charts for both nonconformities and nonconforming units, and the first control chart developed by control chart inventor Walter Shewhart was a *p*-chart, which is for the proportion of nonconforming units. (An *np*-chart is for the number of nonconforming units per sample of size *n*.)

As emphasized previously in this chapter, it is important to have a LCL on charts for which points plotting below the LCL signal probable improvement, and a *p*-chart is such a chart, as are charts for nonconformities (*c*-chart and *u*-chart). Therefore, 3-sigma limits, which are given by

$$\bar{p} \pm 3\sqrt{\frac{\bar{p}(1 - \bar{p})}{n}}$$

with $\bar{p}$ denoting the percentage of nonconforming units in the recent historical data, should generally not be used since the computed LCL will be negative whenever $\bar{p} < 9/(9 + n)$. This could occur if $\bar{p}$ is small and/or *n* is large. Of course, we want $\bar{p}$ to be small in engineering applications, but, paradoxically, a small value would cause the LCL to not exist and thus make it harder to see improvement.

The corresponding *np*-chart limits would be obtained by multiplying the *p*-chart limits by *n*, but the same argument can be made against their use since a scale factor doesn't change a chart's properties. Thus, 3-sigma control limits should not be used for either type of chart, although this is what is commonly done.

Various alternative methods have been given for determining the control limits for these two charts. Ryan and Schwertman (1997) used a regression approach to produce control limits such that the probability of a point plotting outside the control limits when the process is in control will be approximately what it is under normal theory, provided there is not a small sample size. This approach is presented in the next section.

11.9.1.1 *Regression-Based Limits*
Ryan and Schwertman (1997) gave regression-based limits for an *np*-chart as

$$\text{UCL} = 0.6195 + 1.00523np + 2.983\sqrt{np} \tag{11.1a}$$

$$\text{LCL} = 2.9529 + 1.01956np - 3.2729\sqrt{np} \tag{11.1b}$$

The corresponding *p*-chart limits would be obtained by dividing each of the limits produced by Eqs. (11.1a) and (11.1b) by *n*. These limits will closely approximate the normal theory tail areas of .00135 when $p \approx .01$. When *p* is larger than, say, .03, the limits could be obtained using the program given in Schwertman and Ryan (1997). When *p* is smaller than .01, a *p*-chart or *np*-chart could be a poor choice. This is because there will be a concentration of probability at $X = 0$ when *p* is very small.

The reader will observe that these limits differ from the corresponding 3-sigma limit expressions in that the latter do not contain a constant, and the regression-based limits do not contain a factor of $1 - p$ in the third term. The latter is due to the fact that the regression-based limits were first constructed for a *c*-chart (see Section 11.9.2), with the limits for the *p*-chart and *np*-chart obtained as a by-product.

■   **EXAMPLE 11.5**

Problem Statement

A printed circuit board manufacturer is having considerable difficulty producing high-quality boards. In particular, an average of 4 nonconforming boards per 100 were found over the past three months, with 200 boards sampled each day. The sample size is thus 200 and $\overline{p} = .04$. The Ryan–Schwertman limits will be used because the proportion is so small that $\overline{p}$ is less than $9/(9 + n)$, so there is no 3-sigma LCL.

Control Limits

Using $\overline{p}$ in place of p in Eqs. (11.1a) and (11.1b) and then dividing by 200 so as to produce the p-chart limits, we obtain LCL $= .009$ and UCL $= .085$. These limits would then be applied to Phase II, provided that there were no points over the past three months that would plot outside the limits. If that were to occur, then a search for an assignable cause would have to be initiated.   ■

   The limits given by Eqs. (11.1a) and (11.1b) should generally be used whenever p is small, especially when p is close to .01. If p is, say, close to .50, as can occur in medical applications, the 3-sigma limits given at the beginning of Section 11.9.1 should be suitable. In Exercise 11.62 the reader is asked to compare the Ryan–Schwertman limits with the 3-sigma limits for an industrial example and to determine if the 3-sigma limits will suffice for that application.

11.9.1.2   Overdispersion
As was discussed in Example 3.4 in Section 3.3.2, it is worth noting that the binomial distribution isn't automatically the appropriate distribution whenever "successes" and "failures" are involved. Items coming off an assembly line will often be correlated in terms of whether or not they are in conformance with specifications. If so, then the binomial variance is not the correct variance for the process, with the likely result that using the binomial variance will cause the actual variance to be underestimated. This would cause the control limits to be too narrow and many false signals to be received. This issue is discussed in further detail in Section 6.1.7 of Ryan (2000).

11.9.2   c-Chart

A c-chart can be used to control the number of nonconformities per inspection unit, with the latter perhaps comprised of one or more physical units. For example, the inspection unit may consist of a single bolt or a container of bolts. The chart can be used for controlling a single type of nonconformity or for controlling all types of nonconformities without distinguishing between types.
   As mentioned in Section 11.8, a physical unit could have one or more (generally minor) nonconformities without being labeled (and discarded) as a nonconforming unit. Examples would include a minor blemish on a tire and wrapping on a food item that is not fastened properly. Ideally, such nonconformities should not occur, but elimination of all types of imperfections might be too expensive to be practical.

The standard approach is to use 3-sigma limits. The 3-sigma limits for a c-chart, with c representing the number of nonconformities, are obtained from

$$\bar{c} \pm 3\sqrt{\bar{c}} \tag{11.2}$$

with adequacy of the normal approximation to the Poisson distribution assumed, in addition to the appropriateness of the Poisson distribution itself (c is assumed to have a Poisson distribution). Specifically, the probability of observing a nonconformity in the inspection unit should be small, but a large number of nonconformities should be theoretically possible. The size of the inspection unit should also be constant over time.

The mean and variance of the Poisson distribution are the same, and $\bar{c}$ is the estimator of each. The "±3" implies the use of a normal approximation, as was the case with the p-chart and np-chart. With those two charts, the normal approximation to the binomial distribution was assumed to be adequate, although this generally won't be true when p is small. The normal approximation to the Poisson distribution has been assumed to be adequate when the mean of the Poisson distribution is at least 5, but that really applies to estimating Poisson probabilities rather than to the related problem of determining control limits.

When applied to the c-chart, this requires that $\bar{c}$ should be at least 5. Although this requirement will often be met in practice, control limits obtained from Eq. (11.2) should not automatically be used when it is met. In particular, when $5 \leq \bar{c} < 9$, there will be no LCL since $\bar{c} - 3\sqrt{\bar{c}}$ would be negative. Therefore, improvement could be hard to detect, as there would not be points plotting below a lower limit, which would signify likely improvement. (Recall that this is the same type of problem that can exist with a p-chart and an np-chart.) The tail areas beyond the control limits of the c-chart can also be off considerably from the assumed (nominal) areas of (approximately) .00135. For example, when $\bar{c} = 10$, the LCL is approximately 0.5, but $P(X = 0) = .000045$ when the Poisson mean is 10, which is one-thirtieth of the assumed area. If the objective were to come as close as possible to .00135, the LCL would be set at 2, and the tail area would be .00050.

11.9.2.1 Regression-Based Limits

This type of problem can be avoided by using the regression-based limits for a c-chart given by Ryan and Schwertman (1997). Those limits are obtained as

$$\text{UCL} = 0.6195 + 1.0052\bar{c} + 2.983\sqrt{\bar{c}} \tag{11.3a}$$

$$\text{LCL} = 2.9529 + 1.01956\bar{c} - 3.2729\sqrt{\bar{c}} \tag{11.3b}$$

The latter approach is not generally available in statistical software. A c-chart with 3-sigma limits (or, in general, k-sigma limits) can be produced in MINITAB. There is no option for the Ryan–Schwertman limits given by Eqs. (11.3a) and (11.3b), however, so they would have to be calculated manually and entered in a somewhat roundabout way as user-specified limits are easily entered unless they are outside the k-sigma limits. The LCL with the Ryan–Schwertman method will be tighter than the 3-sigma LCL, but the relationship is reversed for the UCL. Consequently, the numerical value of the UCL in Eq. (11.3a) would have to be equated to $\bar{c} \pm k\sqrt{\bar{c}}$ and then k solved for from that equation. That value of k would be used for the UCL and then the LCL would be entered directly as the value computed from the LCL expression in Eq. (11.3b).

■ **EXAMPLE 11.6**

Data

One of the MINITAB sample datasets is PLATING.MTW. Thirty samples, each containing 50 assembled parts, were obtained and the number of parts that had a plating nonconformity (defect) was recorded. The data are as follows:

$$1\ 6\ 5\ 5\quad4\quad3\ 2\ 2\ 4\ 6\ 2\ 1\ 3\ 1\ 4$$
$$5\ 4\ 1\ 6\ 15\ 12\ 6\ 3\ 4\ 3\ 3\ 2\ 5\ 7\ 4$$

Control Limits

The average number of defects per sample is 4.3, so there will be no LCL on a 3-sigma chart since the average is less than 9. Furthermore, since the average is less than 5, the normal approximation would not be expected to fare very well. Therefore, the control limits should be regression-based limits.

$$\text{UCL} = 0.6195 + 1.0052(4.3) + 2.983\sqrt{4.3} = \ 11.1275$$
$$\text{LCL} = 2.9529 + 1.01956(4.3) - 3.2729\sqrt{4.3} = 0.5502$$

Control Chart Interpretation

Figure 11.7 shows that two consecutive points are above the UCL, which is the same signal that is received from the 3-sigma c-chart. The UCL in Figure 11.7 is noticeably higher than the UCL of 10.52 for the 3-sigma c-chart, and unlike the latter, there is a LCL of

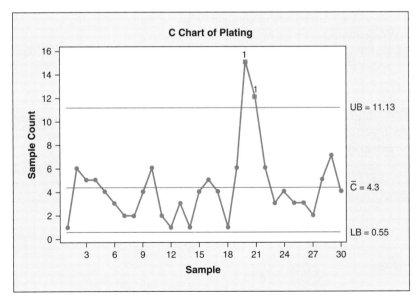

Figure 11.7 Regression-based limits for plating data.

0.55, whereas there is no LCL with the 3-sigma chart. Since every sample had at least one nonconformity, in this case the existence of the LCL of 0.55 has no effect; in many other applications it will have an effect.   ∎

11.9.2.2   Robustness Considerations

The 3-sigma approach to a c-chart, although greatly flawed, is based on the assumption of a Poisson distribution only through the assumption that the mean and variance are the same and both are estimated by $\bar{c}$. That is, the "3" is used in the same way that it is used for every other Shewhart chart, without regard to the actual distribution of the plotted statistic. The Ryan–Schwertman approach, however, is based on the assumption of a Poisson distribution and is not robust to departures from that distribution, nor was it intended to be. Although there has been some work on developing a robust c-chart, such a chart with good statistical properties has not yet been developed.

11.9.3 u-Chart

When the area of opportunity for the occurrence of nonconformities does not remain constant, a u-chart should be used rather than a c-chart, with $u = c/n$ and n is the number of inspection units from which c is obtained. When n is constant, either a u-chart or a c-chart can be used, but when n varies, a u-chart must be used. (Note that n is not involved in the construction of c-chart limits.)

The control limits for a u-chart are related to the limits for a c-chart. Specifically, $Var(c/n) = (1/n^2)Var(c)$, which is estimated by $(1/n^2)\bar{c}$. The latter is obviously equal to $(\bar{c}/n)/n$, which equals $\bar{u}/n$. Thus, the control limits are obtained from

$$\bar{u} \pm 3\sqrt{\frac{\bar{u}}{n}} \qquad (11.4)$$

with $\bar{u} = \bar{c}/n$ and n is assumed to be constant. When the latter is true, the u-chart limits are simply the c-chart limits divided by n.

When the sample size varies, n_i would replace n in Eq. (11.4). The control limits will thus vary as n_i varies, but the centerline will still be constant.

A u-chart can thus be used under each of the following conditions:

1. As a substitute for a c-chart when the (constant) sample size contains more than one inspection unit and there is a desire to chart the number of nonconformities per inspection unit.

2. When the sample size varies so that a c-chart could not be used. The control limits would then be (a) variable limits using individual sample sizes, or possibly (b) constant limits using the average sample size when the sample sizes differ only slightly.

It should be noted that a u-chart could be produced when a transformation is used for c. For example, if $y = \sqrt{c} + \sqrt{c + 1}$ and n is constant, the control limits would then be obtained from

$$\frac{\bar{y}}{n} \pm \frac{3}{n}$$

with $\bar{u} = \bar{y}/n$ denoting the average number of "transformed nonconformities" per inspection unit. Similarly, when the sample size varies, the exact variable limits would be obtained from

$$\frac{\Sigma y}{\Sigma n_i} \pm \frac{3}{n_i}$$

The motivation for using transformations in conjunction with a u-chart is the same as the motivation for using them with a c-chart. Specifically, if the normal approximation is not adequate for nonconformity data, it is not going to be adequate for nonconformities per unit either.

11.9.3.1 Regression-Based Limits
Regression-based limits can also be produced for a u-chart. The control limits for a u-chart that utilize the improved regression-based c-chart limits would have to be obtained as follows. Let the UCL for a c-chart using this approximation be represented by $\bar{c} + k_1\sqrt{\bar{c}}$ and the LCL be represented by $\bar{c} - k_2\sqrt{\bar{c}}$. The values of k_1 and k_2 can easily be solved for after the limits are computed using the approximations. Then, for variable n_i, the control limits for the u-chart would be given by UCL $= \bar{u} + k_1\sqrt{\bar{u}/n_i}$ and LCL $= \bar{u} - k_2\sqrt{\bar{u}/n_i}$.

When the n_i are equal, this has the effect of multiplying each of the c-chart limits by $1/n$, as stated previously. Accordingly, the accuracy of the u-chart limits will depend on $1/n$ and the accuracy of the corresponding c-chart limits. This suggests that n be chosen to be as large as is practical, which in turn will make λ large. As shown in Ryan and Schwertman (1997), the approximation improves as λ increases.

The command UCHART will produce a u-chart in MINITAB, although the specification of the Ryan–Schwertman limits on the chart will require the same type of calculations that are needed in order to specify the limits on a c-chart. A u-chart can also be produced with the PCAPA command in MINITAB, which provides a process capability analysis with a u-chart as one of the components.

11.9.3.2 Overdispersion
Just because nonconformities are to be charted does not necessarily mean that the Poisson distribution should be used in producing the control limits. Friedman (1993) discussed the use of u-charts in integrated circuit fabrication and showed that the in-control ARL for a u-chart (with 3-sigma limits) can be less than 10 due to clustering. Consequently, it is important that the assumption of a Poisson process be checked (see Friedman, 1993, for additional details).

As discussed in Section 3.3.4.1, Snedecor and Cochran (1974, p. 198) gave a simple but somewhat crude method of comparing the observed variance with the Poisson variance. The test is based on the assumed adequacy of the normal approximation to the Poisson distribution. That approximation is generally assumed to be adequate when the Poisson mean is at least 5, as stated previously in Section 11.9.2, although we have seen that for control chart purposes this rule of thumb is inadequate. Other tests for Poisson overdispersion have been proposed, such as those described by Bohning (1994) and Lee, Park, and Kim (1995). Evidence suggests that overdispersion is a problem in many applications, so a test for overdispersion should be used.

11.10   ALTERNATIVES TO SHEWHART CHARTS

Shewhart and Shewhart-type charts are the most commonly used control charts. Some have even argued that Shewhart charts are superior to other types of control charts/procedures and should be used exclusively. In particular, Deming (1993, p. 180) states: "The Shewhart charts do a good job under a wide range of conditions. No one has yet wrought improvement."

Times change, however, and charts with superior properties have been developed. This is to be expected. How many people would want to drive a car that was made in 1924, the year that Shewhart sketched out the idea for a control chart? This doesn't necessarily mean that modern procedures should necessarily permanently supplant Shewhart charts, but clearly modern methods should be studied and given serious consideration for use in applications for which they are well suited.

Assume that an $\overline{X}$-chart has been used in a particular application but when the process goes out of control, it does so by a small amount, not by a large amount. *If* normality existed and the parameters were known, the ARL for detecting a mean shift equal to $\sigma_{\overline{x}}$ is 43.9 subgroups, as the reader is asked to show in Exercise 11.2. If such a shift is considered to be consequential, we would want to detect the shift much sooner than this.

In order to do so, however, we need to use a procedure that "accumulates" data, starting directly after the process change, since such data should be revealing. For example, consider Figure 11.8.

The first 50 observations were generated from the standard normal distribution, with the last 50 generated from a $N(1, 1)$ distribution. Since the parameter values for the in-control standard normal distribution are known, they were used in computing the control limits. Notice for the first 50 observations the values fluctuate randomly about the midline, whereas there is clearly a mean shift after observation #50. The change is a 1-sigma shift, which will be consequential in many applications, yet a point does not plot above the UCL until observation #87. Thus, 37 additional points must be plotted before the shift is detected, with the theoretical average value being 44.

Notice that before observation #87, almost all of the plotted points starting with observation #51 are above the midline. If this information could be utilized in some way, then the mean shift could be detected much faster than with a Shewhart chart, which uses only data

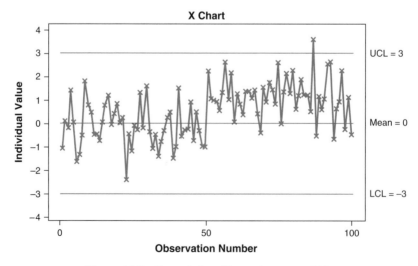

Figure 11.8   Control chart illustrating mean shift.

at each point in time, not data over time. That is, the cumulative evidence of many points a moderate distance above the midline should suggest that a mean shift has occurred.

11.10.1 CUSUM and EWMA Procedures

Therefore, what is needed is one or more methods that will accumulate deviations from the midline. A user can select from two methods that accomplish this: cumulative sum (CUSUM) methods and exponentially weighted moving average (EWMA) methods.

11.10.1.1 CUSUM Procedures

Different types of CUSUM procedures have been proposed, and one decision to be made is whether or not the deviations from the midline are to be standardized. Using standardized values has some obvious advantages, as this facilitates comparison across different process characteristics and across different processes. For individual observations, one computes

$$Z = \frac{x - \bar{x}}{\widehat{\sigma}_x}$$

with $\widehat{\sigma}_x = s_x$ if the CUSUM procedure is to be applied in Phase II, and $\widehat{\sigma}_x = \overline{MR}/d_2$ if the procedure is being used in Phase I. For data in subgroups,

$$Z = \frac{\bar{x} - \bar{\bar{x}}}{\widehat{\sigma}_{\bar{x}}}$$

would be used, with $\widehat{\sigma}_{\bar{x}}$ computed in the same way in which it would be computed if an $\bar{X}$-chart were used. A Z-value would be computed for each observation or subgroup. Then with a basic CUSUM procedure the following pair of cumulative sums would be computed:

$$S_{Hi} = \max[0, (Z_i - k) + S_{Hi-1}] \qquad S_{Li} = \min[0, (Z_i + k) + S_{Li-1}]$$

The first sum is for detecting a mean increase and the latter is for detecting a mean decrease. The sums would initially be set to zero, with the computations starting at time $i = 1$. The value of k is selected to be one-half of the minimum mean shift, in the appropriate standard deviation units, that one wishes to detect. Thus, if it were desired to detect a one $\sigma_{\bar{x}}$ mean change (increase or decrease), k would be set equal to 0.5 (and similarly if there were a desire to detect a one σ_x mean change using individual observations).

Note that there would be no change in either cumulative sum whenever the standardized observation or subgroup mean does not exceed 0.5 in absolute value. Both sums would be plotted on the same chart, with a threshold value of h for S_{Hi} and $-h$ for S_{Li}. These serve as decision limits in the same way that control limits on a Shewhart chart serve as decision limits. A reasonable choice would be $h = 5$ as this balances the need for a large in-control ARL with a small ARL for mean shifts that are considered to be consequential. This choice provides an in-control ARL of 465 and an ARL of 10.4 for detecting a mean shift equal to σ_x if μ and σ were known. These numbers serve as large-sample estimates of the actual ARLs, which depend on the number of observations that are used in estimating the parameters, as indicated previously.

Note that 465 and 10.4 are much better than their counterparts when a Shewhart chart is used, as those corresponding numbers are 370 and 44, respectively. This illustrates the importance of utilizing *all* of the data after a mean shift, not just data at individual points.

It is often said that a CUSUM procedure is good for detecting small changes, such as mean shifts, with a Shewhart chart being better at detecting large changes. It isn't necessary to choose between the two, however, as a combined Shewhart–CUSUM procedure could be used. Such a procedure involves the use of Shewhart limits in addition to the CUSUM decision rule. Since the latter reduces the in-control ARL, it is desirable to use 3.5-sigma limits rather than 3-sigma limits, however, so as not to make the in-control ARL smaller than might be considered acceptable.

Another CUSUM procedure that is frequently used is a fast initial response (FIR) CUSUM. Since when repair action on a process is initiated, it is possible that the repair work might not have restored the process to an in-control state, possibly because there might have been an undetected cause or two, it would be desirable to protect against this possibility. This can be done by resetting the CUSUM values not to zero after repair action has been performed, but to $\pm h/2$. If the process is still out of control, a signal will be received faster than if the sums were reset to zero, and fortunately these headstart values increase the in-control ARL value very little.

11.10.1.2 EWMA Procedures

An exponentially weighted moving average (EWMA) chart also utilizes past information and for subgroup averages is of the form

$$w_t = \lambda \overline{X} + (1 - \lambda)w_{t-1}$$

with w_t denoting the value of the exponentially weighted moving average at time t and λ is chosen to try to balance the in-control ARL and mean shift ARL, just as one tries to do with a CUSUM procedure. Whereas a CUSUM procedure can be used with or without a chart, this EWMA procedure is strictly a control chart procedure. For subgroup averages the limits are

$$\overline{\overline{X}} \pm 3 \frac{\hat{\sigma}}{\sqrt{n}} \sqrt{\frac{\lambda}{2 - \lambda}}$$

for use with $t \geq 5$. (Limits that are a function of t should be used for $t < 5$, although many users use the constant limits for all values of t.) A reasonable choice is $\lambda = 0.25$, which would provide an in-control ARL of 492.95 and an ARL of 10.95 for detecting a mean shift of $\sigma_{\overline{x}}$ *if* the parameters were known.

As with CUSUM procedures, there are EWMA charts for various purposes, including a Shewhart–EWMA chart. See Chapter 8 of Ryan (2000) for details.

One important point to be made regarding these charts is that an $\overline{X}$-chart with "runs rules" is *not* a satisfactory substitute for a CUSUM procedure or an EWMA chart. Runs rules have been used in industry, however, and these are rules that specify when a signal is to be received, analogous to a point plotting outside the limits of a control chart. For example, if eight consecutive points plotted above the midline on an $\overline{X}$-chart, this would suggest that the mean has probably increased since the probability of this occurring if the mean were unchanged is quite small. Eight consecutive points on one side of the centerline is in fact one of the rules, but the problem with the entire set of rules is that they drastically reduce the in-control ARL, bringing it down below 100 (see Champ and Woodall, 1987).

11.10.1.3 CUSUM and EWMA Charts with MINITAB

CUSUM and EWMA charts can be constructed with various software, including MINITAB. The command `CUSUM` is used to produce CUSUM charts for data either in subgroups or as individual observations. The command has a long list of subcommands/options.

The default is to accumulate deviations from a target of zero, so the mean of the historical data would have to be entered as the target value, using the `TARGET` subcommand. Another difference from what was presented in Section 11.9.1.1 is that Z-scores are not plotted: that is, the deviations from the mean value are not divided by the estimate of the standard deviation. Similarly, the UCL and LCL values that are shown on the chart are $\pm h\widehat{\sigma}_x$ for individual observations and $\pm h\widehat{\sigma}_{\bar{x}}$ for data in subgroups. (Of course, this is based on TARGET $= 0$.) The default values for h and k are 4 and 0.5, respectively, so if there was a preference for 5 and 0.5, for example, the subcommand `PLAN` would have to be used as PLAN 5 0.5.

EWMA charts can also be produced in MINITAB. The value of λ in Section 11.9.1.2 must be specified, or the default value of 0.2 can be used.

11.11 FINDING ASSIGNABLE CAUSES

Once a control chart signal has been received, the process engineer or control chart user in general searches for an assignable cause. There is hardly anything in the control chart literature about such searches (although a few companies have formal search procedures), and whether statistical methods can be of any value in the search. Indeed, there seems to be almost a tacit assumption that statistical methods have no role to play in the search because they generally aren't mentioned.

■ **EXAMPLE 11.7**

Problem Statement

Bisgaard and Kulahci (2000) illustrated, however, how graphical methods might be used for identifying assignable causes. In an actual industrial situation, a quality engineer who worked for a company that made cosmetic products was asked to solve a problem with off-center labels on a particular type of cosmetic bottle. Management was concerned that the excessive variability in the position of the labels degraded the look of the high-priced product, and company officers suspected that the variability had an adverse effect on the company's market share.

A drum that was used in the label-application process had six positions and the engineer decided to ascertain whether there were any differences between the positions. So the engineer and a maintenance mechanic ran off 60 bottles during a lunch break, noting the position on the drum from which each label came and measuring the vertical position of each label on the bottle.

Graphical Analysis

Two graphs that were subsequently constructed showed there was indeed a drum effect, as a time sequence plot of the 60 measurements exhibited an obvious cyclical effect, and a scatter plot of the difference between the height and the nominal height against the label drum position showed a clear position effect.

The cyclical effect should be apparent from Figure 11.9.

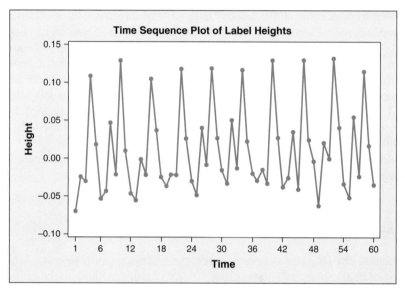

Figure 11.9 Time sequence plot of label heights minus the nominal value.

The cyclicality should be apparent but an autocorrelation plot (a plot of the correlation between observations two units apart, three units apart, etc.) shows the cyclicality very clearly. The plot is shown in Figure 11.10.

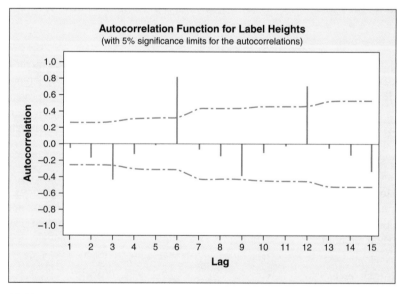

Figure 11.10 Autocorrelation plot for label heights data (with 5% significance limits for the autocorrelations).

The correlation between observations six units apart is apparent from the spike at lag 6. There is also a significant (at the .05 level of significance) negative autocorrelation at lag 3. Bisgaard and Kulahci (2000) did not construct an autocorrelation plot, so they did not detect this relationship. The top of the label applicator drum is position #1; the bottom is position #4. Should there be a significant negative correlation between these two positions, as there appears to be from Figure 11.9? This is something that should be investigated.

Thus, the assignable cause was found without doing any statistical analysis. However, if the message from the scatter plot had not been as clear as it was in this instance, an analysis of variance could have been performed, treating the different positions on the drum as six different levels of a factor. (See Chapter 12 for the analysis of this type of data.)

All of this would have been missed if an $\overline{X}$-chart and R-chart had been constructed from the data, as neither showed any evidence of lack of control. The $\overline{X}$-chart is shown in Figure 11.11.

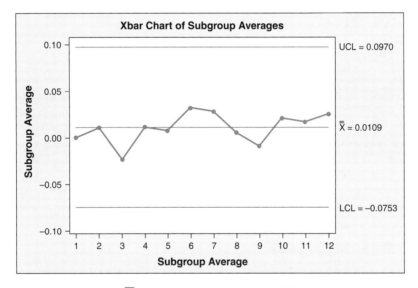

Figure 11.11 $\overline{X}$-chart of subgroup averages of label height data.

As Bisgaard and Kulahci (2000) pointed out, however, this chart looks suspicious because all of the points are near the centerline, whereas at least some points should be near the control limits, even when the process is in control. That does not happen here because the assumption of independent observations was violated; the failure to recognize this resulted in the control limits being too wide. ■

It is much easier to detect assignable causes when out-of-control conditions are not marked by intermittent spikes and then an apparent return to normal. Those cases present special problems. If it is a temporary electrical problem (the residential counterpart being lights flickering occasionally), then possible causes of that might be considered, perhaps by constructing a cause-and-effect diagram (covered in quality control books). Low electrical power to a residential area can cause intermittent power problems, so the cause of those problems could be questioned. The point is that when there is a sustained problem that has caused a mean shift, detecting the cause might not be difficult, but problems that exist for only a very brief period of time will be much harder to detect because we can't search for an assignable cause until we receive a control chart signal, and by the time we receive the signal, the problem may be gone (or going).

11.12 MULTIVARIATE CHARTS

The charts that have been discussed in preceding sections were all "univariate" charts: that is, they were for a single variable (e.g., a single type of measurement such as diameter). Most practical process control problems involve more than one variable, however, but multivariate control charts are beyond the scope of this book. Interested readers are referred to Ryan (2000, Chapter 9) and Section 6.5.4.3 of the *NIST/SEMATECH e-Handbook of Statistical Methods* (Croarkin and Tobias, 2002).

11.13 CASE STUDY

11.13.1 Objective and Data

Lynch and Markle (1997) described a study performed to develop an understanding of nonuniformities in the oxide etch process in a wafer fabrication. Data were collected from 27 runs of evaluating process stability and capability and identifying opportunities for process improvement. Although 18 wafers were involved in each run, only 6 were measured so as to reduce the amount of work. There were 9 sites that were sampled on each wafer. The dataset contains 1,458 datapoints so it will not be listed here, but is given at the textbook website.

Before the data were collected, it was believed that the process was out of control and within-wafer nonuniformity was believed to offer the greatest opportunity for improvement.

11.13.2 Test for Nonnormality

Lynch and Markle (1997) collapsed the data over wafers and positions within each wafer to obtain control charts with 27 points, representing the 27 runs. They constructed an $\overline{X}$-chart and an s-chart but we should perform a normality check before performing either. A normal probability plot test of the 1,458 observations results in rejection of the hypothesis of normality but that is due to some small values coupled with the fact that the test is highly sensitive to small departures from normality because of the very large sample size. The plot does not suggest a major departure from normality, nor does a histogram. Therefore, it would be appropriate to construct an $\overline{X}$-chart and an s-chart in the usual manner.

11.13.3 Control Charts

The $\overline{X}$-chart is given in Figure 11.12.

We can essentially see, as Lynch and Markle (1997) did, three groups of data: runs 1–9, 10–17, and 18–27. In fact, the equipment was worked on after run #9 and "there was a recalibration of the CHF3 Mass Flow Controller" after run #18. Clearly, there was a dramatic reduction in the variability of the averages after run #18, so we would certainly guess that there must have been some change. (Note that the data are averaged over both the positions and the wafers to produce the averages that are plotted.)

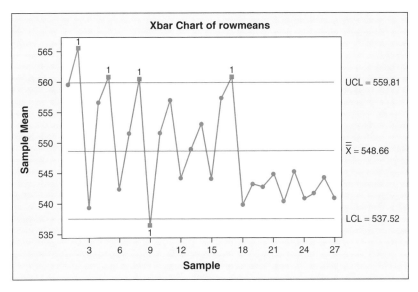

Figure 11.12 $\overline{X}$-chart of etch data, averages per run.

A reduction in the variability can also be seen from the *s*-chart of the averages in Figure 11.13. (Note that both this chart and the $\overline{X}$-chart are constructed as if there were a single number representing the measurement for each wafer, not the average of the measurements over the 9 positions.)

This is the type of chart that would be appropriate if interest is centered on wafer-to-wafer variability. If the within-wafer variability were of primary interest, then the averaging

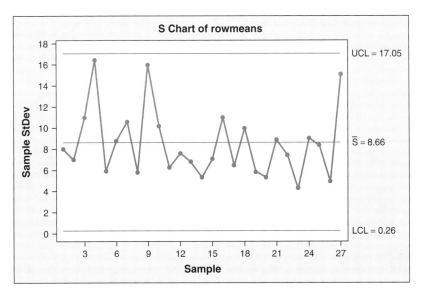

Figure 11.13 *s*-Chart of averages per run, etch data.

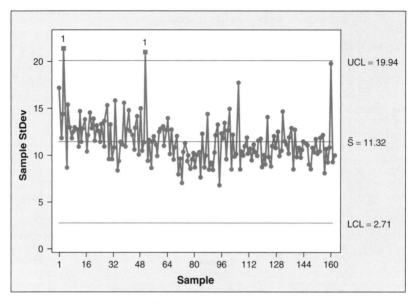

Figure 11.14   *s*-Chart showing within-wafer variability.

would be performed over the wafers instead of over the positions within each wafer. Lynch and Markle (1997) performed a variance components analysis and determined that the greatest opportunity for improvement was in reducing variability across the wafers. The variability is shown in Figure 11.14. This is an *s*-chart with a standard deviation computed for each of the 162 wafers that was used.

The chart shows a reduction in variability over time as almost all of the points near the end are below the midline. The variability of the standard deviations seems to be roughly in control as two points outside the limits out of 162 is not surprising. The practitioners would have to look at the magnitude of the standard deviations over time to determine if that magnitude was acceptable or unacceptable.

11.14   ENGINEERING PROCESS CONTROL

There are various ways to pursue process control—with and without statistical methods. Process control is often regarded as being the domain of the control engineer and is then called *engineering process control* (EPC) and is implemented using automatic equipment. That is, feedback and feedforward methods are employed, with the objective of keeping a process on target. We can contrast this with statistical process control (SPC), which is used to determine if a process is off-target using statistical methods, and in particular, not assuming that a process is off-target simply because the values of sample statistics don't fall on the midline.

The obvious question then is: When should processes be adjusted? Adjustment is often quite costly and it is well-known that unnecessary adjustment inflates the variance of whatever is being monitored. For example, the variance of what is being monitored is doubled if the size of the adjustment that is made at time t is equal to the absolute value of the difference between the observation at that time and the target value.

If processes stayed in a state of statistical control, adjustments would be improper, as deviations from the target value would simply be random fluctuations and would not be indicative of a process change. Processes do not always stay in statistical control, however, even reasonably well-behaved processes, and with some processes a state of statistical control is extremely difficult to achieve, if not impossible. For such processes, EPC methods can be quite useful, but it is of course preferable to first determine if statistical control can be achieved by using SPC methods.

This suggests that SPC and EPC should be used together, and Box and Luceño (1997) stated (p. 3), "Serious inefficiency can occur when these tools are not used together and appropriately coordinated." The message is obviously being heeded, at least in some quarters, as M. S. Daskin states in the Research Reports section of the November 2002 issue of *IIE Solutions*. "The integration of SPC and EPC has attracted much attention from academia and industry recently." This attention comes as no surprise to me because in my review of the Box and Luceño book I wrote (Ryan, 1998): "...I believe that the book will stimulate research, ... and also on optimal ways of combining EPC and SPC."

Recent evidence of the combining of the two approaches in the industrial engineering literature is the paper by Tsung and Apley (2002), who proposed an SPC scheme that automatically incorporates information related to the dynamics and autocorrelation present in feedback-controlled processes. Of course, engineers have long used feedback and feedforward techniques.

11.15 PROCESS CAPABILITY

Control charts are used for monitoring processes, but the commonly used charts do not provide information as to how well a process is performing relative to its specifications, since "control" is relative to the chart midline or a target value, not relative to engineering tolerances.

Therefore, various process capability indices have been proposed. Some writers have recommended that process capability indices not be used, however, and other writers have pointed out problems with the use of these indices.

Probably the most frequently used index is C_{pk}, which is defined as

$$C_{pk} = \frac{1}{3}\min\left[\frac{\text{USL} - \mu}{\sigma}, \frac{\mu - \text{LSL}}{\sigma}\right]$$

In words, this is one-third of the standardized distance that μ is from the closest engineering tolerance. One might argue that it would be better to drop the 1/3, as the definition of C_{pk} would then be in accordance with "Six Sigma" in that the statistic would then measure the number of standard deviations that the mean is from the closest tolerance, with six being the goal. (See Section 11.16 for a discussion of Six Sigma.) With the current definition, 2.0 would be the target. In the automobile industry, a minimum acceptable value of 1.33 has been cited.

Another process capability measure that has been used extensively is C_p, which is defined as

$$C_p = \frac{\text{USL} - \text{LSL}}{6\sigma}$$

A major shortcoming of this index is that it is not a function of μ, so it does not measure the relationship between the specifications and the center of the process distribution, as any good process capability index should do.

Regardless of which index is used, μ and/or σ must be estimated, and the estimates should be obtained when the process is in a state of statistical control. Therefore, an X-chart or an $\overline{X}$-chart would be needed to check for control. We will let $\widehat{C}_{pk}$ denote the statistic with the estimated parameters, with the parameter estimates perhaps coming from the same data that were used in demonstrating a state of statistical control. When process capability indices came into vogue about twenty years ago, $\widehat{C}_{pk}$ was generally used as if it had no sampling variance. Thus, for example, a value of 1.16 was used as if it were the value of C_{pk}.

A moderate amount of research has been devoted to developing confidence intervals for process capability indices. It is very difficult to obtain exact confidence intervals for most indices, however, and for some indices it is impossible. For example, the only way to obtain an exact confidence interval for C_{pk} is to use a computer program to solve a particular equation; a simple expression does not exist (see Ryan, 2000, p. 197). Nevertheless, an approximate confidence interval that will often be adequate is

$$\widehat{C}_{pk} \pm Z_{\alpha/2} \sqrt{\frac{1}{9n} + \frac{\widehat{C}_{pk}}{2n-2}}$$

when individual observations are used in computing $\widehat{C}_{pk}$.

Another problem with the use of process capability indices is that there were also instances of companies trying to finesse the numbers in efforts to achieve certain values, rather than using the $\widehat{C}_{pk}$ values as evidence of whether quality improvement was being achieved.

A further problem is that the most frequently used indices are based on the assumption of a normal distribution. Since engineering tolerances are individual values, nonnormality can seriously undermine the results. Another problem is the estimation of σ by s, as the distribution of the latter is not robust to nonnormality. Although extensive use of process capability indices is likely to continue, they should be used with caution. See Chapter 7 of Ryan (2000) for a detailed discussion of process capability indices and also see Kotz and Johnson (2002) and the discussions of that paper.

11.16   IMPROVING QUALITY WITH DESIGNED EXPERIMENTS

Control charts and process capability indices are "listening devices" that essentially describe the current state of affairs. As G. E. P. Box has emphasized, to see what happens when a process in interfered with, we must interfere with it (in a controlled, systematic way). That is, it is necessary to see what happens to product and process quality when the values of process variables are changed. This can be done either off-line or online, as there are methods that can be used for each case. Most experimentation is performed off-line; for example, see Section 12.17 for discussion of improving quality through the use of designed experiments.

11.17 SIX SIGMA

No discussion of quality improvement methods would be complete without at least mentioning what has become known as "Six Sigma." This is (1) a goal that specifies the minimum acceptable distance (six) from the engineering specifications to the center of a probability distribution, (2) a collection of statistical and nonstatistical tools for improving quality, and (3) a mindset.

Simply stated, the objective is to reduce variation. We may think of engineering tolerances as being a distance of $k\sigma$ from the center of a probability distribution (here making the simplifying assumption that the tolerances are equidistant from the center). Since the tolerances are fixed, if σ is reduced, then k must increase. This in a nutshell is what Six Sigma is all about—working toward having $k = 6$.

Unfortunately, however, the definition of Six Sigma in terms of $k\sigma$ has become garbled because of some misunderstanding by one of its proponents. Specifically, articles in which "six sigma" is mentioned state that this corresponds to 3.4 nonconforming units per million. This is true only under some rather unrealistic conditions. First, this is based on the assumption of normality, which is only an approximation to reality, and it is also based on the assumption that the process really has drifted *exactly* 1.5 sigma and stays there, with the specification limits assumed to be six sigma away from the process mean if there were no process drift. So this is actually 4.5-sigma capability. For true six-sigma capability (i.e., the process mean was constant), there would be 2 nonconforming units per *billion*, under the assumption of normality.

This may be impossible to achieve, whereas five-sigma capability with a fixed mean would result in 5.7 nonconforming units per 10 million. Virtually every company would consider this to be acceptable, and also perhaps achievable.

Clearly, however, what should be considered acceptable depends on what is being measured. In trying to give a picture of six sigma in concrete terms, Boyett, Schwartz, Osterwise, and Bauer (1993) indicated that deaths from airplane disasters occur about 4.3 times per 10 million flights, which translates to 4.92-sigma quality if we assumed normality, and using just one side of the distribution. (Of course, this is just for illustration, as a normal distribution probably would not be a suitable approximation to the actual distribution. The intent here is just to translate the "percent nonconforming" into a value of k in k-sigma.) For this application we would probably want to see more than six-sigma capability!

Thus, a variation of the six-sigma theme seems desirable. It also seems desirable to not seek six-sigma capability for noncritical processes, as doing so may not be time well spent.

Another question is: Is it practical to expect that six-sigma capability could be verified, even if it existed? This is analogous to estimating *very extreme* percentiles of a distribution. Clearly, it is impractical to think that one could take a large enough sample to estimate far-tail percentiles with any degree of precision. If actual six-sigma capability existed, a sample size of about 100 billion would probably be required to estimate the percentage of nonconforming units. One alternative would be to assume normality and estimate the percentage from the estimates of μ and σ and their relation to the specification limits, but that could produce a poor estimate.

Despite these slightly bothersome nuances, the Six Sigma paradigm is a sound one. Many books have been written on the subject, so readers seeking additional information will find a wealth of available reading material.

11.18   ACCEPTANCE SAMPLING

Books on quality control and improvement typically contain material on acceptance sampling. The latter encompasses sampling plans with decision criteria that lead to the acceptance or rejection of incoming or outgoing shipments. Several decades ago, quality control was equated with inspection, and acceptance sampling was viewed as a compromise between no inspection and 100% inspection. Although the latter is sometimes necessary, control charts and other process control procedures have bridged the gap between no inspection and 100% inspection. Furthermore, as the prominent industrial statistician Harold F. Dodge (1893–1976) once stated: "You cannot inspect quality into a product." This is because the quality of an item has been determined by the time it reaches the final inspection stage. Therefore, 100% final inspection will not ensure good quality.

Acceptance sampling is obviously no substitute for process control and should not be used after processes have attained a state of statistical control. This is the position taken in the acknowledged best reference on acceptance sampling, Schilling (1982), as well as in other sources. Even if acceptance sampling plans are used only temporarily, there are problems even in doing that because the plans are not in accordance with modern thinking in regard to acceptable quality in terms of the percentage of nonconforming units, for example. The plans are especially in conflict with Six Sigma goals.

11.19   MEASUREMENT ERROR

Statistical process control (SPC) and improvement can be very difficult when large measurement error is present, and Dechert, Case, and Kautiainen (2000) discuss the fact that the application of SPC in the clinical diagnostics industry is particularly challenging because of measurement error. Because measurement error can be the effect of one or more factors (such as technician, instrument, and/or environment), it is highly desirable to identify the components of measurement error, especially when the error is large. The methods described in Chapter 13 can be used to make such determinations.

11.20   SUMMARY

Process and quality improvement can be effected through the use of various statistical and nonstatistical tools, including control charts. The emphasis should be on improvement, however, not control. For this reason, it is important to use methods of determining control limits that produce a lower control limit on types of charts for which plotted points below the LCL signify improvement.

There are two stages of control chart use: analysis of historical data and real-time process control, and methods used to estimate the parameters need not be the same in each stage.

It is also important to remember that control chart assumptions should be checked. This is especially important for an X-chart, for which nonnormality could cause the chart to have properties that are much different from what is assumed. It is also important for other charts, including an $\overline{X}$-chart. If the normality assumption appears to be violated, the user might try to transform the data to approximate normality or fit a nonnormal distribution to the data and determine the control limits from the percentiles of the fitted distribution, or use a nonparametric control chart. The assumption of independent observations is also important and should be checked.

Process capability indices have been controversial and the use of these indices has not been without criticism. Process capability must be addressed and measured, however, so it is likely that usage of the most popular indices will continue. Care should be exercised in their use, however, and assumptions should be carefully checked, as with any statistical procedure.

Although the emphasis in this chapter has been on basic methods, it will frequently be necessary and desirable to monitor more than just means, standard deviations, and proportions. Functional relationships could also be monitored, such as linear regression relationships. Systems change, so no model can be assumed to be a good approximation to reality for an indefinite period of time. Methods for monitoring a simple linear regression relationship are given by Kim, Mahmoud, and Woodall (2003) and Mahmoud and Woodall (2004).

REFERENCES

Bisgaard, S. and M. Kulahci (2000). Finding assignable causes. *Quality Engineering*, **12**(4), 633–640.

Bohning, D. (1994). A note on a test for Poisson overdispersion. *Biometrika*, **81**(2), 418–419.

Box, G. E. P. and A. Luceño (1997). *Statistical Control by Monitoring and Feedback Adjustment*. New York: Wiley.

Boyett, J. H., S. Schwartz, L. Osterwise, and R. Bauer (1993). *The Quality Journey*. Alpharetta, GA: Boyett and Associates.

Champ, C. W. and W. H. Woodall (1987). Exact results for Shewhart control charts with supplementary runs rules. *Technometrics*, **29**(4), 393–399.

Charnes, J. and H. Gitlow (1995). Using control charts to corroborate bribery in jai alai. *The American Statistician*, **49**(4), 386–389.

Chou, Y.-L., G. D. Halverson, and S. T. Mandraccia (1998). Control charts for quality characteristics under nonnormal distributions. In *Statistical Case Studies: A Collaboration Between Academe and Industry* (R. Peck, L. D. Haugh, and A. Goodman, eds.) Philadelphia: Society for Industrial and Applied Mathematics, and Alexandria, VA: American Statistical Association.

Croarkin, C. and P. Tobias, eds. (2002). *e-Handbook of Statistical Methods*. A collaboration of the National Institute of Standards and Technology and International SEMATECH (http://www.itl.nist.gov/div898/handbook).

Dechert, J., K. E. Case, and T. L. Kautiainen (2000). Statistical process control in the presence of large measurement variation. *Quality Engineering*, **12**(3), 417–423.

Deming, W. E. (1993). *The New Economics for Industry, Government, and Education*. Cambridge, MA: Center for Advanced Engineering Study, Massachusetts Institute of Technology.

Friedman, D. J. (1993). Some considerations in the use of quality control techniques in integrated circuit fabrication. *International Statistical Review*, **61**, 97–107.

Ghosh, B. K., M. R. Reynolds, Jr., and Y. V. Hui (1981). Shewhart $\overline{X}$-charts with estimated process variance. *Communications in Statistics—Theory and Methods*, **10**(18), 1797–1822.

Kim, K. P., M. A. Mahmoud, and W. H. Woodall (2003). On the monitoring of linear profiles. *Journal of Quality Technology*, **35**(3), 317–328.

Kotz, S. and N. L. Johnson (2002). Process capability indices—a review, 1992–2000. *Journal of Quality Technology*, **34**(1), 2–19, discussion, pp. 20–53.

Lee, S., C. Park, and B. S. Kim (1995). Tests for detecting overdispersion in Poisson models. *Communications in Statistics—Theory and Methods*, **24**(9), 2405–2420.

Lynch, R. O. and R. J. Markle (1997). Understanding the nature of variability in a dry etch process. Chapter 7 in *Statistical Case Studies: A Collaboration Between Academe and Industry* (R. Peck, L. D. Haugh, and A. Goodman, eds.). Philadelphia: Society for Industrial and Applied Mathematics, and Alexandria, VA: American Statistical Association.

Mahmoud, M. A. and W. H. Woodall (2004). Phase I analysis of linear profiles with calibration applications. *Technometrics*, **46**(4), 380–391.

Montgomery, D. C. and W. H. Woodall (1997). Concluding remarks. *Journal of Quality Technology*, **29**(2), 157.

Ott, E. R. (1975). *Process Quality Control*. New York: McGraw-Hill.

Quesenberry, C. P. (1993). The effect of sample size on estimated limits for $\overline{X}$ and X control charts. *Journal of Quality Technology*, **25**(4), 237–247.

Roes, K. C. B., R. J. M. M. Does, and Y. Schurink (1993). Shewhart-type control charts for individual observations. *Journal of Quality Technology*, **26**, 274–287.

Ryan, T. P. (1998). Review of *Statistical Control by Monitoring and Feedback* by G. E. P. Box and A. Luceño. *Journal of Quality Technology*, **30**(4), 409–413.

Ryan, T. P. (2000). *Statistical Methods for Quality Improvement*, 2nd ed. New York: Wiley.

Ryan, T. P. and N. C. Schwertman (1997). Optimal limits for attribute control charts. *Journal of Quality Technology*, **29**(1), 86–98.

Schilling, E. G. (1982). *Acceptance Sampling in Quality Control*. New York: Marcel Dekker.

Schwertman, N. C. and T. P. Ryan (1997). Implementing optimal attributes control charts. *Journal of Quality Technology*, **29**(1), 99–104.

Snedecor, G. W. and W. G. Cochran (1974). *Statistical Methods*, 7th ed. Ames, IA: Iowa State University Press.

Tsung, F. and D. W. Apley (2002). The dynamic T^2 chart for monitoring feedback-controlled processes. *IIE Transactions*, **34**(12), 1043–1054.

EXERCISES

11.1. Assume that a practitioner decides to use a control chart with 2.5-sigma limits. If normality can be assumed (as well as known parameter values), what is the numerical value of the in-control ARL?

11.2. Assuming normality and known parameter values, show that the ARL for detecting a mean shift of one $\sigma_{\overline{X}}$ on an $\overline{X}$-chart with 3-sigma limits is 43.9.

11.3. Assume that you have 200 observations on customer waiting time and you want to construct a control chart to see if there is any evidence that the process of waiting on customers was out of control for the time period covered by the data. What is the first thing that you should do?

11.4. Assume that you have 100 individual observations that plot approximately as a sine curve. What assumption, if any, would be violated if an I-chart were constructed using these data?

11.5. An R-chart is constructed with subgroups of size 5 being used. Assuming normality of the individual observations, what will be the lower control limit if 3-sigma limits are used, if indeed there will be a lower limit? What would you recommend to the person who constructed this chart?

11.6. Explain why an s-chart would be strongly preferable to an R-chart for subgroups of size 10 but not for subgroups of size 2.

11.7. In a study that involved the salaries of chemical engineers, the following information is available, through July 1, 2000 (*Source*: Webcam):

Number of Firms Responding	Number of Employees	Mean Average Minimum Salary	Mean Average Salary (June 30, 2000)	Mean Average Maximum Salary
159	1,725	$46,732	$61, 261	$75,843

This is how the data were reported.

(a) Would you suggest any changes? In particular, what is meant by "mean average salary"? Is that the same as "average salary"? Explain.

(b) Can you estimate σ from these data using the range approach even though ranges are not explicitly given? Explain.

(c) Assume that you wanted to use a Stage 1 control chart approach to analyze these data so as to try to identify any companies that are unusual in terms of the variability of their salaries, as well as their average salaries. Could any of the control charts presented in this chapter be used for this purpose? Why or why not? Could this be accomplished if you had the entire data set and you sampled from that set? If so, how would you perform the sampling?

11.8. In order to demonstrate the superiority of an $\overline{X}$-chart over an X-chart for detecting mean shifts, we may determine the ARL for a one σ_x mean shift. We know that the ARL for the X-chart is 43.89. Determine the ARL for an $\overline{X}$-chart with a subgroup size of: (a) 4, (b) 5, and (c) 6. Comment.

11.9. Explain why it is difficult to control variability when individual observations are obtained rather than subgroups.

11.10. Explain why CUSUM procedures have better ARL properties than Shewhart charts.

11.11. If quality improvement is the goal, as it should be, should all attributes control charts have a LCL? Explain.

11.12. For what values of $\widehat{p}$ will the LCL of a p-chart with 3-sigma limits not exist? Will the LCL of an np-chart not exist for the same range of values of $\widehat{p}$? If so, explain why this occurs. If not, explain why the ranges differ. What can and should be done to correct the problem of not having a LCL on these charts?

11.13. Assume that a basic CUSUM procedure is used for individual observations and the first three Z-values are 0.8, 0.4, and 1.3, respectively. Assuming that $k = 5$ is used and both sums had been set to zero, what are the three values of S_H and S_L that correspond to these Z-values? What do these Z-values and your CUSUM values suggest about the process?

11.14. You are given the following numbers of nonconforming units for daily data that were collected over a period of one month, with $n = 500$ each day: 4, 8, 3, 2, 3, 9, 7, 11, 2, 6, 7, 5, 8, 12, 3, 4, 2, 8, 9, 7, 6, 3, 4, and 4. Compute the regression-based control limits for an np-chart given by Eqs. (11.1a) and (11.1b) and then compute the 3-sigma control limits. Do they differ very much? Which set of limits seems preferable, or can that be determined?

11.15. Compute the value of the process capability index C_{pk} when the engineering tolerances are given by LSL = 20 and USL = 40 and assume that $\mu = 30$ and $\sigma = 10/3$.

11.16. Indicate which chart you would prefer to use, an R-chart or an s-chart, when the subgroup size is (a) 2, (b) 5, and (c) 8. Explain.

11.17. A department manager wishes to monitor, by month, the proportion of invoices that are filled out correctly. What control chart could be used for this purpose?

11.18. Assume that an np-chart has been constructed and the control limits are 106.629 and 163.371. If $n = 400$, what would be the corresponding limits on the p-chart?

11.19. Assume that a control chart user fails to test for normality and independence before using an $\bar{X}$-chart. Assume further that the data are independent but have come from a nonnormal distribution. If the control limits that are constructed turn out to be at the 0.01 and 98.2 percentiles of the (nonnormal) distribution of $\bar{X}$, what is the in-control ARL?

11.20. Assume that 50 individual observations are plotted on an X-chart for the purpose of determining trial control limits. Assume normality and that all of the data have come from the same population, with μ and σ assumed known. What is the probability that all of the points plot *inside* the 3-sigma limits?

11.21. The UCL on an np-chart with 3-sigma limits is 39.6202 and the midline is 25. What would be the LCL on the corresponding p-chart if $n = 500$?

11.22. Assume that an FIR CUSUM approach is used with a headstart value of $h/2 = 2.5$ and $k = 0.5$. The data are in subgroups and the process that is being monitored is stopped at subgroup #62 and the sums are reset. Complete the following table.

Subgroup Number	Z	S_H	S_L
63	2.64	—	—
64	—	2.95	—
65	—	—	0.64

11.23. Determine the 3-sigma control limits for an $\overline{X}$-chart from the following information: subgroup size $= 4$, number of subgroups $= 25$, $\overline{\overline{X}} = 56.8$, and $\overline{s} = 3.42$.

11.24. Read the Charnes and Gitlow (1995) article that is listed in the references and write a report on the novel way that control charts were used as an aid in detecting criminal activity. Can you think of other applications for their approach?

11.25. Assume that an $\overline{X}$-chart is constructed using 3-sigma limits and parameter values that are assumed to be known, and the assumption of normality is met. If the process mean remains constant, what is the probability that the next plotted point will be outside the control limits?

11.26. Assume that a company wishes to have "six-sigma capability." For the sake of simplicity, assume that there is only a lower specification limit (LSL), with LSL $=$ 25. If $\mu = 35$ and $\sigma = 2$, to what value must σ be reduced in order to have six-sigma capability?

11.27. Consider the baseball data in the file ERAMC.MTW that was used in Exercise 1.1.

  (a) Construct the normal probability plot and note that the hypothesis of normality is rejected if we use $\alpha = .05$ since the p-value for the test that MINITAB uses is .018. Would you suggest that an X-chart be constructed for these data? Why or why not?

  (b) Even though doing so might be highly questionable in view of the normal probability plot, construct an X-chart for American League earned run averages (ERAs), using the data from 1901 to 1972. (Note that "averages" are being treated as individual observations since the averages are computed over teams and there is no interest in the individual teams.) Use the moving range approach to estimate sigma.

  (c) Would you have constructed the control limits using a different approach? Explain. If you would have constructed the control limits differently, do so. Are any points outside the control limits now?

  (d) Now plot the earned run averages starting in 1973, one at a time in simulating real time. Construct an X-chart with 3-sigma limits, and a CUSUM chart using $k = 0.5$ and $h = 5$. Is there a sufficient number of observations to enable the parameters to be estimated for the X-chart and have the chart perform in a reliable manner? Similarly, would you expect the CUSUM procedure to be reliable (i.e., have a small performance variance)?

  (e) Do the two chart methods give different signals after 1972? Since the designated hitter rule was started in 1973, which method gives the correct signal?

11.28. Would you expect a CUSUM or EWMA procedure to be effective in the analysis of Phase I data if the process was out of control at more than one point in time? Explain.

11.29. Chiang et al. [*Quality Engineering*, **14**(2), 313–318, 2001–2002] gave a case study for which the specifications for density (R1, g/cm^3) are 1.14–1.15, with $C_p = 0.67$ and $C_{pk} = 0.61$.

   (a) Determine the value of σ (here assumed known since the capability indices are not estimated).

   (b) Using this value of σ, can μ be determined? Why or why not? If not, give two possible values of μ. Are these the only possible values? Explain.

11.30. Consider the data in `DELTA.DAT` that were used in Exercise 1.38. This datafile contains the number of minutes that each Delta flight from Atlanta on June 30, 2002 was late in departing. (A negative number means that the flight departed early.)

   (a) Construct an individual observations chart without first checking for approximate normality (since some people assume that normality isn't important).

   (b) How many points are outside the control limits?

   (c) How many points would you expect to be outside the limits if all of the data came from the same normal distribution?

   (d) Do you believe that the "extra" number of points outside the control limits should be investigated? Stated differently, what do you believe should be the general shape of the in-control distribution?

   (e) Now test the assumption of a normal distribution by constructing a normal probability plot.

   (f) What does this suggest about the steps that should be followed when constructing a control chart for individual observations?

11.31. If the LCL for an *np*-chart is 0.83, what must be the LCL for the corresponding *p*-chart if $n = 100$?

11.32. Why is it important *not* to use 3-sigma limits on a *p*-chart, *np*-chart, *c*-chart, *u*-chart, or *R*-chart? When, if ever, should 3-sigma limits not be used for an *X*-chart or an $\overline{X}$-chart?

11.33. Twenty samples of size 200 are obtained and $\overline{p} = .035$. What was the total number of nonconforming units for the 20 samples?

11.34. Assume that the control limits for an *X*-chart are UCL = 82.5 and LCL = 68.1, with the chart indicating control so that process capability can be estimated from the data that were used in computing the control limits. What is the numerical value of $\widehat{C}_{pk}$ if USL = 90 and LSL = 65?

11.35. A quality engineer computes $\widehat{C}_{pk}$ and finds that $\widehat{C}_{pk} = 1.02$. If the process characteristic were normally distributed and the specification limits were equidistant from $\overline{\overline{X}}$, the average of the subgroup averages, what percentage of units would be nonconforming (i.e., outside the specification limits)?

11.36. Assume that a practitioner decides to use an $\overline{X}$-chart with 2.5-sigma limits so as to increase the chances of detecting a parameter change quickly. Since this also increases the chances of a false alarm, the process engineer decides to look for an assignable cause only if two points in a row plot outside the control limits.

(a) Assume normality, known parameter values, and that the plotted points are independent and compute the ARL for a one $\sigma_{\bar{x}}$ increase in μ and also compute the in-control ARL.

(b) Compare these numbers to the corresponding in-control and parameter-change ARLs when 3-sigma limits are used. Which of these two procedures would you recommend? Explain your choice.

11.37. Quality improvement results when variation is reduced. Assume a normal distribution with a known standard deviation, with 34% of the observations contained in the interval $\mu \pm a$. If the standard deviation is reduced by 50% and the mean is unchanged, what percentage of observations will lie in the interval $\mu \pm a$?

11.38. A process that is operating in the presence of assignable causes is said to be _____ (more than one word).

11.39. Critique the following statement: "Process capability must be good if control charts indicate that a company's processes are in a state of statistical control."

11.40. The British use 3.09-sigma limits instead of 3-sigma limits.

(a) Compute the in-control ARL for an X-chart under the assumption of normality and explain the choice of the number 3.09.

(b) Compute the ARL for a 1-sigma shift and compare this and the in-control ARL with the corresponding numbers for 3-sigma limits. Which would you recommend? Explain.

11.41. What are the two assumptions for an X-chart?

11.42. Consider two approaches for determining the limits for an X-chart for Phase II, which have the following forms: $\overline{X} \pm 3(\overline{MR}/d_2)$ and $\overline{X} \pm 3(s/c_4)$. Which limits will have the greater sampling variability? Which limits would you recommend?

11.43. An $\overline{X}$-chart is to be constructed for Phase I. Given that number of subgroups $= 24$, subgroup size $= 5$, $\overline{\overline{X}} = 36.1$, and $\bar{s} = 6.2$, compute the 3-sigma control limits. If you had been given the data, what would you have done first before computing the limits?

11.44. In certain fields of application, such as medical applications, a high false alarm rate is not a problem. Assume that 2-sigma limits are used for a c-chart, and the Poisson mean increases by 20% (from $\bar{c}$).

(a) Can the exact ARL for detecting this shift be computed if $\bar{c} = 9$? If so, what is it?

(b) Can the exact in-control ARL be determined? Explain.

11.45. We know that mean shifts (and shifts in general) are detected much faster with a CUSUM procedure than with a Shewhart chart. Explain why that happens.

11.46. Assume that a control chart user wishes to construct an $\overline{X}$-chart for Phase I, with L-sigma limits, such that the probability of receiving at least one false signal is .01. Assume that all of the 50 points are from the same normal population and the parameters are known. What value of L should be used to accomplish this objective?

11.47. Assume normality and that there has been a $1.6\sigma_x$ increase in the process mean. If X-chart limits were computed using known parameter values:

    (a) How many points would we expect to have to plot on the X-chart before observing one that is above the UCL?

    (b) If an $\overline{X}$-chart with subgroups of size $n = 4$ were used instead of an X-chart, how many points would we expect to observe before seeing a point above the UCL on the $\overline{X}$-chart?

    (c) Compare your answers and explain why your answer to part (b) is smaller than your answer to part (a).

    (d) Are your answers to (a) and (b) parameter-change ARLs or not? Explain.

11.48. Assume that 30 subgroups of size 4 are used to compute the control limits for an $\overline{X}$-chart with 3-sigma limits and one of the subgroup ranges plots above the UCL. What is the smallest possible difference between the original UCL and the recomputed UCL if the subgroup with the average above the UCL is discarded after it is discovered that the data were collected at a time when the process was temporarily out of control?

11.49. Explain why control chart parameters should not necessarily be estimated using the same estimators for Phase I and Phase II.

11.50. Consider the basketball data in the file BASKETBALL.MTW. These are the highest men's team field goal percentages in NCAA Division 1 during 1972–2006. If a control chart for these data were constructed, the chart should be either a p-chart or an np-chart since these are percentages. However, since the total number of field goal attempts of course varies from year to year and the numbers are not given in the file, construct an X-chart of the data as an approximation. Notice the points that fall outside the control limits.

    (a) Although it is difficult to draw conclusions from only 35 data points, what does the configuration of points suggest? In particular, considering the number of points outside the control limits (on *both* sides) when only 35 points are plotted, would you suspect a process that is temporarily out of control at certain times or would you suspect that the data are from two populations? In particular, how would the configuration of plotted points appear if the data had come from two populations with values of p, the binomial parameter, that differ more than slightly?

(b) Control charts can spotlight times at which there may be a problem, but determining the time at which the problem began won't necessarily be easy. Even if this could be determined with certainty, there is then the matter of determining if a problem does indeed exist, and if so, what caused the problem. The latter is the job of an engineer or the chartist, in general. There is no commonly used approach for trying to determine when a change occurred, although some have been proposed. In this instance the population shift is known because the year of the rule change that caused the shift is known. Which year does the chart suggest as being the approximate year of the change?

11.51. Assume that a company is using an $\overline{X}$-chart with $n = 4$ to monitor a process and is simultaneously using an X-chart because it is worried about the possible occurrence of some extreme observations that might not be picked up by the $\overline{X}$-chart. Assume that the control limits for each chart were constructed using assumed values for both $\mu = 20$ and $\sigma = 4$. If a point plots above the UCL on the $\overline{X}$-chart, by how much does it have to exceed the UCL in order to be certain that at least one of the four subgroup observations must plot above the UCL on the X-chart?

11.52. Assume that subgroups of size 2 are obtained and a point plots right at the UCL of an R-chart with 3-sigma limits. (This type of chart was not recommended in Section 11.7.2 but is in common use.) Where would the point plot on an s-chart with 3-sigma limits relative to its UCL? Could this question have been answered if subgroups of size 3 had been obtained? Why or why not?

11.53. As discussed in Section 11.9.1, the use of runs rules is inadvisable. Explain why the in-control ARL of an $\overline{X}$-chart with even one run rule (such as 8 points above the midline) must be less than the in-control ARL without the addition of the run rule?

11.54. An $\overline{X}$-chart and an R-chart are used in Phase I and two points plot outside the limits of the $\overline{X}$-chart, with one point above the UCL and one below the LCL. It is subsequently discovered that the process was out of control at the time corresponding to each of those points. The user decides not to recompute the limits, however, contending that the new $\overline{X}$-chart limits would be about the same as the old limits since the discarded points were at opposite extremes. Do you agree? Explain. Could the person make the same argument relative to the R-chart? Explain.

11.55. A combined Shewhart–CUSUM scheme is used for averages with 3.5-sigma limits and $h = 5$ and $k = 0.5$ for the CUSUM scheme. At what point will a signal be received if five consecutive subgroup averages transform to Z-values of 3.4, 1.86, 1.25, 0.91, and 0.55? Explain. Although a signal was received, what does the pattern of $\overline{X}$ values suggest to you? In particular, what course of action would you take as the process engineer, if any?

11.56. An employee gives you the following control chart for individual observations, constructed from historical data, and is using the control limits shown on the chart for Phase II process monitoring. The employee has erred in multiple ways. What has the employee done wrong?

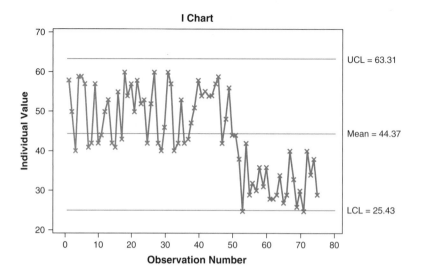

11.57. Consider the control chart given at http://www.minitab.com/resources/
stories/ExhibitAPg3.aspx that was described in Section 11.1. Can we
tell from the graph if the points in red were included or not when the control limits
were computed? Does the text offer any hint?

11.58. Consider the (*I*-chart, moving range chart) pairs given at http://www.
minitab.com/resources/stories/EatonSCI2002.aspx, which were
used by Eaton Corporation in one of the applications described therein.

   (a) The two moving range charts each show a LCL. Do you concur that the two
   charts each have a LCL? Explain.

   (b) Consider the configuration of points in the "left lead range-holes" chart. Does
   the configuration of points in that pair support the contention of Roes, Does,
   and Schurink (1993) that almost all of the information in the data points is
   contained in the *I*-chart? Explain.

   (c) Consider the configuration of points for the same chart as in part (b). Notice that
   there is no point above the UCL on the moving range chart. Does the *I*-chart
   suggest that there is a problem with variability? Explain. What would be your
   recommendation to Eaton Corporation?

11.59. Moving average charts were mentioned briefly, but not advocated, in Section 11.5.
Compute the correlation between consecutive moving averages of (a) size 3, (b)
size 5, and (c) size *n*. What would be one advantage of using a moving average
chart with averages computed from at least a moderate number of observations (say,
at least 5)? If a user insists on using a moving average chart, what would be your
recommendation?

11.60. The following statement has appeared on a Kodak web page: "Out-of-control
conditions due to temperature changes are difficult to solve. They can appear
and disappear rapidly because they are usually caused by intermittent electrical

problems". What actions, if any, would you take if you had responsibility for such a process and you saw points frequently plotting outside the control limits of an $\overline{X}$-chart but there was no evidence of a mean shift?

11.61. An application of a *p*-chart is given in the *NIST/SEMATECH e-Handbook of Statistical Methods* at `http://www.itl.nist.gov/div898/handbook/pmc/section3/pmc332.htm`. Specifically, chips on a wafer were investigated and a nonconforming unit was said to occur whenever there was a misregistration in terms of horizontal and/or vertical distances from the center. Thirty wafers were examined and there were 50 chips on each wafer, with the proportion of misregistrations over the 50 chips per wafer recorded for each wafer. The average of the proportions is .2313.

    (a) Would it be necessary to use the Ryan–Schwertman control limits in Section 11.9.1.1, or should 3-sigma limits suffice? Explain.

    (b) Compute both sets of limits and comment.

11.62. It is often claimed that engineering tolerances are somewhat arbitrarily set. Some tolerance figures are given in the paper "A New Chart for Analyzing Multiprocess Performance" by S. C. Singhal [*Quality Engineering*, **2**(4), 379–390, 1990], in addition to various statistics for illustrating a proposed control chart. The tolerances and standard deviations for 25 different in-process parameters related to diffusion are given below. Are any of these 25 tolerances and corresponding standard deviations suggestive of (a) a very poor process control situation, or (b) almost unbelievably good process control?

USL	LSL	Standard Deviation
35	15	2.253
35	15	2.299
430	300	9.684
430	300	15.137
1800	1600	36.182
1800	1600	50.000
250	150	9.010
680	520	17.310
240	180	6.220
255	165	14.630
5000	3500	95.632
7525	6475	140.015
35	15	4.490
1800	1600	108.614
1800	1600	26.069
250	150	6.730
680	520	21.846
220	180	19.154
255	165	40.000
5000	3500	68.775
7525	6475	134.100

USL	LSL	Standard Deviation
250	150	55.550
250	150	7.520
1800	1600	20.830
1800	1600	8.547

11.63. One of the sample datasets that comes with the MINITAB software is `PIPE.MTW`. A company that makes plastic pipes is concerned about the consistency of the diameters and collects the following diameter measurements over a 3-week period.

Week 1	Week 2	Week 3	Machine	Operator
5.19	5.57	8.73	1	A
5.53	5.11	5.01	2	B
4.78	5.76	7.59	1	A
5.44	5.65	4.73	2	B
4.47	4.99	4.93	1	A
4.78	5.25	5.19	2	A
4.26	7.00	6.77	1	B
5.70	5.20	5.66	2	B
4.40	5.30	6.48	1	A
5.64	4.91	5.20	2	B

Construct the appropriate control chart to determine if the diameter measurements are in control. Does the chart suggest control? If not, is there evidence of a lack of statistical control relative to either machine, or either operator, or a particular machine–operator combination?

11.64. The process of a customer replacing one provider of a service, such as a long-distance service, with another one is called a churn. In the paper "SPC with Applications to Churn Management" by M. Pettersson [*Quality and Reliability Engineering International*, August 2004, Vol. 20, pp. 397–406), the author explains that it is important for companies to monitor the churn rate of its customers. The author defines a ratio $X(t) = y(t)/N(t)$ with $y(t)$ denoting the number of churns at time t and $N(t)$ denoting the number of customers at risk. $E(X(t)) = p_0$ when the process is in control. Only the upper control limit was of interest in the paper, and the expression for it was given as

$$z \left[\sqrt{\frac{p_0(1 - p_0)}{N(t)}} + \sqrt{\hat{\sigma}_p^2} \right]$$

with z being the standard normal variate and $\hat{\sigma}_p^2$ denoting the estimated variance of p as p_0 is considered to vary substantially over time.

(a) Does the control limit expression seem reasonable? Would you have used a different expression? Explain.

(b) Considering what a p-chart should be used for, would you suggest that a company additionally use a lower limit? Explain.

11.65. The sample dataset PAPER.MTW that comes with the MINITAB software contains measurements on rolls of paper used by a large metropolitan newspaper. The engineer who inspects each roll of paper is especially concerned about small holes in the paper that are larger than 1 millimeter in diameter. The dataset contains 35 1.5×2 sections of paper, with each being sampled 6 times. There are thus 210 numbers, each of which represents the number of holes of at least 1 millimeter for the given section. Would it be reasonable to use the 210 counts to determine Phase I control limits for a c-chart? If so, what assumption(s) would you be making? Determine the control limits, using whatever method seems best in view of the small number of counts. Then test for control of the 210 measurements.

11.66. As indicated in Section 11.7.2, the R-chart will undoubtedly continued to be used extensively, despite its shortcomings, especially when 3-sigma limits are used. Therefore, construct the R-chart with 3-sigma limits for the data in Example 11.4, using appropriate software since the expressions used for computing the 3-sigma limits were not given.

11.67. Consider the $\overline{X}$-chart in Figure 11.6. Even though subgroup averages are plotted rather than individual observations, should a normal probability plot have been constructed before the $\overline{X}$-chart was constructed? Explain.

CHAPTER 12

Design and Analysis of Experiments

The preceding chapter contained what might be called "listening tools," that is, tools for listening to a process. But as George Box (1966) stated, "to find out what happens to a system when you interfere with it you have to interfere with it (not just passively observe it)." Evidence of the beneficial effect that could result from widespread use of experimental designs by engineers is reflected by the following quote from Box (1990).

> For there are hundreds of thousands of engineers in this country, and even if the 2^3 design was the only kind of design they ever used, and even if the only method of analysis that was employed was to eyeball the data, this alone could have an enormous impact on experimental efficiency, the rate of innovation, and the competitive position of this country.

As is intuitively obvious, the statistical design of experiments is most relevant to design engineers, but other types of engineers also need to understand the basic concepts since statistically designed experiments can be useful in solving a wide variety of engineering problems.

The design of an experiment is more important than the analysis, although clever analyses can often reveal certain facts that are by no means obvious. Much care and attention should be devoted to the design of an experiment. Morrow, Kuczek, and Abate (1998) gave two quotes that support this point of view and are relevant. The first, from an unknown source, is "A well-designed and executed experiment is usually more important than 'high-powered' statistical methods and will usually result in a trivial analysis with clear, unambiguous conclusions." The second quote, attributed to Gerry Hahn, a prominent statistician who is the author of several statistics books and is retired from General Electric, is "The world's best statistical analysis cannot rescue a poorly planned experimental program." It has often been said that data from a well-designed experiment will practically analyze themselves, although this is certainly an oversimplification.

The best way to design a particular experiment probably isn't going to be obvious, however. In fact, Box (1992) stated that Sir Ronald A. Fisher, who made great contributions to design of experiments, noted "the best time to design an experiment is after you have done it." Accordingly, designed experimentation should not be viewed as a one-shot affair, but rather should generally consist of at least two experiments, with experimentation beyond the second experiment performed over time as warranted.

Experimental design was alluded to in Section 9.3, in which regression was used to analyze data from designed experiments. When we construct an experimental design, our objective is to determine what variables affect the response variable, Y, just as we do in regression analysis. That is, modeling is common to both, as we are trying to seek a good proxy for the true, unknown model. The particular regression methods of analysis that are used with an experimental design depend on the type of design that is employed. Specifically, most designs are "orthogonal" in that the correlations (using the term loosely) between the variables are zero. Nonorthogonal designs, such as supersaturated designs mentioned in Section 12.11, are also useful and these require the use of stepwise regression (see Section 9.6.3) and similar methods.

We will also use some new terminology as a variable in Chapter 9 becomes a *factor* in this chapter.

12.1 PROCESSES MUST BE IN STATISTICAL CONTROL

One point that cannot be overemphasized is the need to have processes in a state of statistical control, if possible, when designed experiments are carried out. If control is not achieved, the influence of factors that might affect the mean of the response variable is confounded (i.e., confused) with changes in that variable caused by the process being out of control, as has been emphasized and illustrated in various places, including Ryan (2000, p. 362). Box, Bisgaard, and Fung (1990, p. 190) recommended that processes be brought into the best state of control possible and that experimentation be performed in such a way that blocks are formed to represent intervals of time when the mean appears to be constant. One potential impediment to the implementation of such an approach, however, is that it may be difficult to determine when the mean changes have occurred, unless the changes are large. Another obvious problem is being able to determine mean changes in real time as we certainly can't look at a graph and determine how to do the blocking since we can't go backward in time.

IMPORTANT POINT

It is important that processes be in a state of statistical control at the time that experiments are performed so that the results of the experiment will not be confounded with the possible effects of an out-of-control process.

Clark (1999) stated the importance of statistical control about as well as it can be stated in discussing the importance of the tooling process being in control for a resistance welding operation experiment that was undertaken: "Control of the tooling proves to be as important as parameter evaluation is in achieving optimum nugget size in resistance welding. Otherwise you are trying to make a prediction about an unpredictable event. Without stability, you have nothing."

As in previous chapters, we will rely on MINITAB for computations. In addition to regression, the methods of analysis will include Analysis of Variance (ANOVA), which was introduced in Chapter 8, and Analysis of Means (ANOM), a method that can be used in place of or in tandem with ANOVA. We will also utilize some Java applets that are available on the Internet so that the reader can learn in an interactive setting.

12.2   ONE-FACTOR EXPERIMENTS

The simplest type of experiment is one in which we have one variable (factor) and two levels. The method of analysis is then simply a t-test for two means, as was presented in Section 6.1.2. Although this is how the data will be analyzed, it is important to try to make sure that the analysis is not contaminated by "lurking" variables.

For example, assume that we are conducting a simple experiment to determine if a new process will improve process yield. We could measure yield for a certain number of days under the standard process and then under the new process. But assume that some process characteristic suddenly goes out of control near the time of the change from the standard process to the new process, and this causes process yield to decline. Assume further that if the process had not gone out of control, the yield from the new process would have been much higher than the yield from the standard process. Therefore, with the process out of control, we could easily reach the wrong conclusion and fail to conclude that the new process is superior.

This is a point that is rarely made in the literature. Morrow et al. (1998) do, however, stress the importance of maintaining a process in control and show control charts used with an experiment at Eastman Chemical Company for trying to determine how to produce plastic pellets with a response as close as possible to a specified target value.

Assume that we have taken the proper precautions by monitoring the process and the process characteristics stayed in control during the experiment so that the results are not contaminated. A decision will be made to switch to the new process only if the average yield for that process is at least 2 units greater than the yield for the standard process, with this difference to be detected with probability approximately .90.

Given below is MINITAB output with the commands that produced it given at the text website.

```
Power and Sample Size


2-Sample t Test


Testing mean 1 = mean 2 (versus >)
Calculating power for mean 1 = mean 2 + difference
Alpha = 0.05 Sigma = 1


              Sample   Target   Actual
Difference    Size     Power    Power
    2           6       0.9000   0.9420
```

Observe that we cannot come particularly close to the desired power of .90 for the specified difference in the means of 2 and $\hat{\sigma} = 1$. Of course, we don't have an estimate of σ until we take a sample, and if we don't estimate σ using data from the two samples, then the t-distribution would not be involved. So there is some "approximating" that is going on because, as mentioned earlier, we can't solve for the sample size directly.

So we should keep in mind the approximations that must be made to solve for n. Here that solution is $n_1 = n_2 = 6$. This solution is obtained by using the noncentral t-distribution. The distribution is not covered in Chapter 3 and is beyond this scope of this book.

An example of a one-factor experiment with two levels will not be given here since an example was given in Section 6.1.2.

One point to keep in mind, however, is that complete randomization should be employed, if practical, for one-factor experiments with two levels as well as for experiments with more factors and/or levels. That is, the levels of the single factor should be assigned to the experimental units (people, machines, etc.) at random, and the level should be reset after every run, even if it isn't to be changed. This may sound impractical if, for example, the factor being studied is temperature and the furnace must be continually heated up and cooled down, even when the same temperature is being used on the next experimental run. Indeed, it will often be both impractical and costly to do so, but the failure to do so creates a restricted randomization situation. This is discussed further in Section 12.14.

12.2.1 Three or More Levels

There is no extension of a t-test to three or more levels, so we need a new method of analysis for such a case. Analysis of Variance for regression was illustrated in Chapter 8, as it was used to isolate the three variation components. We similarly have three components for a one-factor experiment. Although Analysis of Variance (ANOVA) should be performed using statistical software, we will briefly discuss in Section 12.2.1.2 how some of the computer output is obtained.

Consider the following hypothetical example, with the data given below for each of three levels.

1	2	3
4.01	4.03	4.04
4.00	4.02	4.02
4.02	4.01	4.02
4.01	4.02	4.03
4.01	4.02	4.04
4.02	4.03	4.02
4.00	4.01	4.04

The averages for the three levels are 4.01, 4.02, and 4.03, respectively. Although these averages differ only slightly, when we test the hypothesis that the three population means are equal (using methodology to be given shortly), we easily reject that hypothesis because the p-value for the test is .002. If this had been actual data, the sample mean differences of .01 and .02 between the extremes would probably not be of practical significance. We obtain statistical significance, however, and the explanation for this will be given shortly.

At the other extreme, consider the following data:

1	2	3
9.06	6.91	6.87
5.13	11.04	10.01
6.12	7.95	8.76
7.59	5.86	9.98
7.10	8.24	9.38

Here the averages are 7.0, 8.0, and 9.0, respectively, but the hypothesis of equal means is not rejected, as the p-value is .184. Thus, we reject the null hypothesis when the means differ by only 0.01, but we fail to reject the null hypothesis when the means differ by 1.0. The explanation for this is that in the second example there is considerable within-level variability, which drowns out the between-level variability. The reverse occurs in the first example, as the within-level variability is so small that it does not offset the between-level variability, although the latter is obviously small.

IMPORTANT POINT

Analysis of Variance is a statistical technique for analyzing variation due to different sources, and with a single factor it is a matter of assessing the magnitude of the variation within groups.

12.2.1.1 Testing for Equality of Variances
Of course, before we test for equality of the population means, we should test for equality of the population variances. Recall from Section 6.1.2 that an approximate t-test can be used when there is evidence that the two population variances are not approximately equal. There is no direct counterpart when there are more than two levels, as in the two preceding examples, so if a test for equality of the three or more variances leads to rejection of the hypothesis of equality of the variances, an appropriate approach is to use a nonparametric test, specifically the Kruskal–Wallis method described in Section 16.4.1. [Heteroscedastic ANOVA methods have been proposed but are beyond the scope of this text. See Bishop and Dudewicz (1981).]

The test originally proposed by Levene (1960) and later modified by Brown and Forsythe (1974) was used in Section 6.1.2 to test for the equality of two variances. It can be used for any number of variances, however. We can tell by inspecting the first dataset in the preceding section that the sample variances do not differ greatly, so there would be no point in applying Levene's test when there is such evidence of approximate equality of the variances.

A somewhat different picture emerges for the second data set, however, so using Levene's test wouldn't be a bad idea. The p-value for the test of the equality of the variances is .815 even though the largest variance is more than twice the smallest one, so the assumption of equal variances would not be rejected. (The small sample sizes contribute considerably to the large p-value.)

Recall from Section 6.1.2 that Levene's test can be performed in MINITAB.

12.2.1.2 Example with Five Levels

Study Objective

In a study described by K. N. Anand ("Increasing Market Share Through Improved Product and Process Design: An Experimental Approach," *Quality Engineering*, 361–369, 1991), the effect that the brand of bearing has on motor vibration was investigated, with five different motor bearing brands examined by installing each type of bearing on different random samples of six motors. The amount of motor vibration (measure in microns) was recorded when each of the 30 motors was running, and the data are as follows:

```
Brand 1   13.1   15.0   14.0   14.4   14.0   11.6
Brand 2   16.3   15.7   17.4   14.9   14.4   17.2
Brand 3   13.7   13.9   12.3   13.8   14.9   13.3
Brand 4   15.7   13.7   14.5   16.0   12.3   14.7
Brand 5   13.5   13.4   13.1   12.7   13.4   12.3
```

Test for Equality of Variances

Each population should be approximately normally distributed when ANOVA (or ANOM) is used, but it is not practical to test that assumption with only six observations from each population. Fortunately, ANOVA is not strongly affected by small departures from normality. Therefore, the first step in the analysis is to test the assumption of equality of variances, and this would logically be performed using Levene's test or another test that is not sensitive to nonnormality. When Levene's test is applied to these data, the p-value for the test statistic is .549, so the hypothesis of equal variances is not rejected. (It is worth noting that Bartlett's test, which is a commonly used test that is unfortunately sensitive to nonnormality, has a p-value of .067 when applied to these data, which would cast some suspicion on the equal variances assumption.)

Test for Equality of Means

The MINITAB output when ANOVA is applied is as follows:

```
One-way ANOVA: Brand1, Brand2, Brand3, Brand4, Brand5

Analysis of Variance
Source       DF        SS        MS         F         P
Factor        4     30.67      7.67      6.75     0.001
Error        25     28.41      1.14
Total        29     59.08
                                    Individual 95% CIs For Mean
                                    Based on Pooled StDev
Level    N    Mean     StDev    ---------+---------+---------+-----
Brand1   6   13.683    1.194        (-----*-----)
Brand2   6   15.983    1.212                        (-----*-----)
Brand3   6   13.650    0.848        (-----*-----)
Brand4   6   14.483    1.357             (-----*-----)
Brand5   6   13.067    0.476    (-----*-----)
                                    ---------+---------+---------+-----
Pooled StDev = 1.066                   13.5      15.0      16.5
```

Conclusion

The hypothesis of equal means is rejected because of the small p-value and from the confidence intervals on each population mean it is clear that this is due to the mean of Brand 2 being noticeably larger than the other means. Since the objective is of course to minimize vibration, Brand 2 should not be considered further. The confidence intervals for the other four brands are overlapping, suggesting that the average amount of vibration for those brands may not differ to an extent that would be considered of practical significance.

Nonparametric Alternative

If the assumption of equal variances had been rejected, it would have been a good idea to use a nonparametric test for the means. The appropriate test would be the Kruskal–Wallis test, which is presented in Section 16.4.1.   ∎

12.2.1.3 ANOVA Analogy to t-Test

Recall from Section 6.1.2 that a *t*-test for one or two means performs in accordance with the explanation at the end of Section 12.2.1. That is, the numerator of the *t*-statistic is a measure of the variability "between" the levels relative to the variability *within* the levels.

Just as with ANOVA for regression, when ANOVA is applied to data from designed experiments, we are decomposing the total variability into its component parts. Hand computation will be greatly deemphasized in this chapter, so we will simply sketch out the way in which certain computations are performed.

The "total sum of squares" in an ANOVA table with Y as the dependent/response variable is always $\sum (y - \bar{y})^2$, which of course is the numerator of the sample variance of Y. By starting with the expression for the two-sample *t*-test with equal variances, we can work toward the expressions that we have in ANOVA in a way that should be intuitive.

Although we can have an unequal number of observations for each factor level in one-factor ANOVA without causing any trouble (as is also true for a two-sample *t*-test), we will assume an equal number of observations for illustration. With n observations in each of the two samples and using y rather than x, we can write the square of the two-sample *t*-statistic as

$$t^2 = \frac{n(\bar{y}_1 - \bar{y}_2)^2}{2s^2} \tag{12.1}$$

which the reader can easily verify with minimum algebra. (The reason for using t^2 instead of t will soon become apparent.)

We know from Section 3.4.7 that $t_{\nu_1}^2 = F_{1,\nu_1}$, so the left-hand side of Eq. (12.1) can be written with F in place of t^2. Let $\bar{y}_{all}$ denote the average of all of the data, which in the case of an equal number of observations per level (n), as in these examples, equals $(\bar{y}_1 + \bar{y}_2)/2$. With a small amount of algebra we may show, after substituting F for t^2, that Eq. (12.1) can be written equivalently

$$F = \frac{n \sum_{i=1}^{k} (\bar{y}_i - \bar{y}_{all})^2 / (k - 1)}{s^2} \tag{12.2}$$

with k denoting the number of levels. Of course, for a two-sample *t*-test $k = 2$, so the "number of levels minus one" is 1, but Eq. (12.2) is written in this form because this is the general form of the *F*-test for testing the equality of the population means. For that test we have

$$H_0: \mu_1 = \mu_2 = \cdots = \mu_k$$
$$H_a: \text{not all } \mu_i \text{ are equal}$$

There is no one-sided test as there can be with a *t*-test because, for one thing, F cannot be negative, as should be apparent from Eq. (12.2) because of the squared terms. A large value of F suggests that H_0 is likely false, whereas a small value of F would lead to nonrejection of H_0.

From a practical standpoint, we have the same type of situation here that we have with virtually all hypothesis tests: we know the null hypothesis is probably false before we collect any data. For example, do you believe that five population means could possibly be equal?

Thus, the real question is whether the sample data strongly contradict the null hypothesis. If so, then the population means probably differ more than slightly and the analysis would then lead to rejection of the null hypothesis.

The analysis should be performed using statistical software but the chapter appendix provides computing formulas for readers who also prefer to see how to perform the computations

Following is the (slightly edited) MINITAB output for the second example in Section 12.2.1:

```
One-way ANOVA:


Analysis of Variance
Source    DF          SS            MS        F        P
Between    2     0.0014000    0.0007000    9.00    0.002
Error     18     0.0014000    0.0000778
Total     20     0.0028000
                                  Individual 95% CIs For Mean
                                  Based on Pooled StDev
Level   N     Mean      StDev     --------+---------+---------+--------
-1      7   4.01000    0.00816    (------*------)
 0      7   4.02000    0.00816            (------*------)
 1      7   4.03000    0.01000                      (------*------)
                                  --------+---------+---------+--------
Pooled StDev = 0.00882              4.010     4.020     4.030
```

The last part of the output essentially tells the story, as when H_0 is rejected we conclude that there is some difference in the means, and the output shows that the confidence interval for the first mean does not overlap with the confidence interval for the third mean. Therefore, in this instance we don't need a more sophisticated *multiple comparison approach* to determine why the p-value for H_0 is only .002. In general, there are many such procedures and the choice of a particular procedure can be guided by how conservative the user wishes to be, as well as by whether the user wishes to specify what comparisons to test before or after looking at the data. Being conservative is guarding against having significant differences declared that are not really different. Since the selection of a multiple comparison procedure to use is somewhat arbitrary, using an ad hoc approach such as looking for nonoverlapping confidence intervals is somewhat defensible.

12.2.2 Assumptions

Since one-factor ANOVA with two levels is mathematically equivalent to an independent sample *t*-test, it follows that ANOVA must have the same assumptions that are made for the *t*-test: equal variances and normal populations, in addition to independence of observations within samples. Clearly, it is very difficult to test normality with only a small number of observations per level, but we could use Levene's test for testing equality of variances, which was illustrated in Section 6.1.2 and was used in the example in Section 12.2.1.2.

In the Morrow et al. (1998) case study mentioned earlier in this chapter, the authors found that observations were not independent within a day. This can create a problem in the following way. Assume that our experiment consists of using two levels of a single factor, so that we could perform the analysis either as a t-test or by using ANOVA. The easiest way to see how lack of independence within a factor level can cause a problem is to consider the t-test. Recall from Section 6.1.2 that the denominator of the t-statistic estimates the standard deviation of $\bar{x}_1 - \bar{x}_2$. Under the assumption of independence *between* the two levels, the variance of the difference of the two means is equal to the sum of the variances. This assumption will generally be met but if the data are collected over time, the assumption of independence *within* a level may not be met.

It follows from Section 4.2.2 that the variance of the ith sample mean is equal to σ_i^2/n_i *only* if the observations that comprise the mean are independent. If only consecutive observations are correlated (i.e., "autocorrelation" exists) and the correlation is ρ, the standard deviation in the denominator of the t-statistic (i.e., the estimated standard deviation of the difference of the two means) will be the square root of $(2\sigma^2/n)\,[1 + (n - 1)\rho]$, rather than the assumed value of the square root of $2\sigma^2/n$. Thus, when ρ is positive, the usual case, the value of the t-statistic will be inflated, with a false signal of significance more likely to occur than under independence. In the less likely case of ρ being negative, the t-statistic would be deflated, with significant mean differences possibly being camouflaged.

12.2.3 ANOVA and ANOM

One thing to keep in mind about an ANOVA table is that the units represented by the table are in the square of the original unit of measurement since we are working with sums of squares.

Generally, the square of the original unit doesn't have any physical meaning (e.g., what does "temperature squared" mean). Because of this shortcoming, Ellis Ott invented a method of analysis that he could use in his consulting work with engineers. It is called *Analysis of Means* (ANOM) and has some clear advantages over ANOVA, although it is not as well known as ANOVA, so the latter is used much more frequently. The two techniques test slightly different hypotheses when $k > 2$, however, as with ANOM one tests whether each population mean is equal to the average of the k population means. (When $k = 2$, ANOM, ANOVA, and the independent sample t-test are all equivalent.)

Of course, if each of $k > 2$ means were equal to the average of the means, then the means would have to all be equal, so the set of hypotheses that are tested are equivalent to the single hypothesis in ANOVA. Unlike ANOVA, however, with ANOM the individual tests are performed, which necessitates the use of different theory and methodology. Accordingly, it is possible for the results to differ, although there will usually be agreement.

If H_0 is rejected using ANOVA, it is necessary to perform further testing to determine why it was rejected, whereas with ANOM a sample mean that stands apart from the overall average is identified as part of the (single-stage) ANOM procedure.

The MINITAB ANOM display that corresponds to the output in Section 12.2.1.3 is given in Figure 12.1.

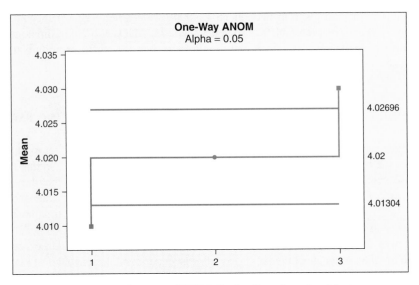

Figure 12.1 One-way ANOM display for a three-level factor.

The α *decision lines*, as they are called, are computed as

$$\overline{y}_{\text{all}} \pm h_{\alpha,k,v}\, s \sqrt{(k-1)/kn}$$

with s, k, and n as previously defined and $h_{\alpha,k,v}$ is a tabular value, given in Table G. (Note that $h_{\alpha,k,v}$ is replaced by $t_{\alpha/2,v}$ when two means are compared.)

In every ANOM display the means are displayed by vertical lines extending from the overall mean to the values of the individual means. From the ANOVA output it was apparent that the first and third means differed. (The other mean might appear to be missing, but that is because it is equal to the midline.) We receive essentially the same message from ANOM as these two means are judged to differ from the overall mean, and since they are on opposite sides of the overall mean, there is strong evidence that they differ from each other.

12.2.3.1 ANOM with Unequal Variances

There is also an Analysis of Means procedure, due to Nelson and Dudewicz (2002), that can be used when there is evidence of unequal variances. This might be viewed as being analogous to the t-test given in Section 6.1.2 that can be used when the pooled t-test cannot be used because of evidence of (highly) unequal variances.

Nelson and Dudewicz (2002) dubbed their procedure HANOM (heteroscedastic analysis of means). Before this procedure would be used, the variances should be tested, and in the one-factor case it is quite possible that there will be enough observations per factor level for such a test to have reasonable power for detecting unequal variances.

One possible test would of course be Levene's test that was discussed earlier in the chapter, or if one wanted to stick strictly with analysis of means procedures, the Analysis of Means for Variances (ANOMV), due to Wludyka and Nelson (1997), might be used.

If the results from one of these methods provided evidence that the variances are more than slightly unequal, HANOM could be used. It should be noted that another option would be to use heteroscedastic ANOVA (e.g., see Bishop and Dudewicz, 1981), that was mentioned pheviously. A second option would be to use Kruskal–Wallis ANOVA, a nonparametric method given in Section 16.4.1 that assumes neither normality nor equal variances. It would be preferable to use heteroscedastic ANOVA if approximate normality seemed to exist, however, since that would be a more powerful procedure than the Kruskal–Wallis procedure under (approximate) normality.

12.3   ONE TREATMENT FACTOR AND AT LEAST ONE BLOCKING FACTOR

Often there will be only one factor of interest, but there will be other factors that may influence the response values, and thus should be adjusted for, if possible, so that they do not cause erroneous conclusions to be drawn. By "adjusted for" we mean that the residual (error) sum of squares should not be inflated by real effects. Unduly and unnecessarily inflating the error sum of squares can lead to failure to conclude that a treatment effect exists when in fact it does exist.

Design of experiments had its beginnings in agriculture, and one of the best examples of the need for at least one blocking factor would be an experiment designed to test three crop fertilizers, with an eye toward selecting the best fertilizer. This is a classical experimental design problem, the type of problem that was considered by early contributors to the field, dating from the 1920s.

It would be reasonable to suspect possible differences in soil fertility in either or both directions, especially if the plot of land used for the experiment were reasonably large. We will first illustrate "either" and will then illustrate "both."

12.3.1   One Blocking Factor: Randomized Block Design

We will first assume that the variation in soil fertility is horizontal, so we want to adjust for this. We could then use a *randomized block design* with vertical strips of soil serving as the blocks. The levels of the factor, such as different fertilizers, would be randomly assigned to the plots within each vertical strip, and this would be done for each vertical strip (block). If all of the fertilizers appeared in each block, the design would be a *randomized complete block design* (RCBD); if not, it would be an incomplete block design, and such designs should be at least partially balanced.

■ EXAMPLE 12.1

Problem Statement and Data

Assume that we have three fertilizers to be tested, the decision is made to use four blocks, and the data below contain a measure of crop growth for each combination of block and fertilizer when the experiment is conducted.

Fertilizers	Blocks			
	1	2	3	4
A	5.6	6.4	6.6	5.8
B	5.1	6.2	6.4	5.7
C	5.0	6.1	5.8	5.5

Assumption

Since this is a two-way arrangement of the data, it would be analyzed using two-way analysis of variance. There is, however, the restriction that treatments and blocks are independent, so there will be no interaction term. (Interaction is illustrated in Section 12.5.1.)

In the context of this problem, interaction would occur if the different fertilizers didn't maintain approximately their relative positions (in terms of the magnitude of the response values) across the blocks. We can see just by eyeballing the data that the assumption of no interaction appears to be met, but of course we would prefer to take a more formal approach. As with all assumptions, this assumption should be checked using graphical and/or statistical methods and it can be checked by plotting the data, as the reader is asked to do in Exercise 12.24. In order to do a parametric analysis with hypothesis tests and other inferences, we also must assume that the errors within each cell have a normal distribution with a common variance, but obviously we can't test this with only one observation per cell.

We could analyze these data using hand computation, but doing so is time consuming and unnecessary, so the formulas needed for hand computation will not be given. They can be found in any book on experimental design.

Blocking

If the blocking has been both necessary and effective, the block totals will differ considerably, with "considerably" being a relative term. Here the block totals are 15.7, 18.7, 18.8, and 17.0, respectively, and that is not small variation since the observations are all in the relatively narrow range from 5.0 to 6.6. Therefore, we would expect the blocking to have a material effect on the analysis, and we will see that by analyzing the data with and without the blocking.

If we ignore the blocking, we simply have a one-way analysis to perform, and the output is as follows:

```
One-way ANOVA: response versus fertilizers

Analysis of Variance for response
Source        DF        SS        MS        F        P
Fertilizer     2     0.500     0.250     0.96     0.420
Error          9     2.350     0.261
Total         11     2.850

                        Individual 95% CIs For Mean
                        Based on Pooled StDev
Level    N     Mean     StDev      ---------+---------+---------+-
1        4   6.1000    0.4761          (-----------*-----------)
2        4   5.8500    0.5802      (-----------*-----------)
3        4   5.6000    0.4690      (-----------*-----------)
                                   --------+---------+---------+-----
Pooled StDev = 0.5110               5.50      6.00      6.50
```

We see that although the means differ noticeably, they do not differ sufficiently relative to the variability in the data, so the *p*-value far exceeds .05.

The analysis using the blocking factor is as follows:

```
Two-way ANOVA: response versus blocks, fertilizers

Analysis of Variance for response
Source          DF          SS          MS          F           P
Blocks           3      2.2033      0.7344       30.05       0.001
Fertilizer       2      0.5000      0.2500       10.23       0.012
Error            6      0.1467      0.0244
Total           11      2.8500
```

We observe that the blocking was effective, as evidenced by the small p-value for blocks. Notice also that the sum of squares for blocks is 77% of the total sum of squares, so isolating the block component has a strong effect on the p-value for fertilizer, and that value is now small.

Conclusion

We thus conclude that there is a difference between fertilizers, with blocking being necessary to see the difference. The question then arises as to which fertilizers are different. Rather than select one multiple comparison procedure from the many that have been proposed, a simple, approximate method would be to compare a difference in means against $t_{\alpha/2,v}\widehat{\sigma}\sqrt{2/n}$, as this is the expression that is used in constructing a confidence interval for a difference in population means. If the absolute value of a difference in treatment means exceeds this value, then the confidence interval would not include zero. For this example we obtain 0.27 as the value of the expression if we use $\alpha = .05$. The difference between the first and third means exceeds this number and the other differences are 0.25. Thus, the difference between the first two fertilizers might possibly be due to chance, but it would not be unreasonable to select the first one.   ∎

Another approach to determining which fertilizers differ is to use ANOM, although there is apparently no software that will provide an ANOM display for a randomized block design without a macro being written. Consequently, we will not give the ANOM display for this dataset.

Example 12.1 illustrated a typical scenario when a randomized block design should be used. A single blocking factor can be used indirectly when there is one hard-to-change factor. If a factor has two levels and all of the experimental runs (assume there are three other factors) with the factor at one level are run first followed by all of the runs with the factor at the other level, the design has been blocked, perhaps inadvertently.

■ EXAMPLE 12.2

Experiment

As another example, but this time without numbers, consider the following civil engineering problem. An analyst wishes to assess the effect of three different maintenance procedures, A, B, and C, on bridge deck life. The analyst, with cooperation from the local authorities, has 90 bridges available in which to assess the three different maintenance procedures.

Would it be appropriate for the procedures to be randomly assigned to the bridges? What would happen if the bridges differed considerably in terms of traffic volume? Obviously, a maintenance procedure that was quite good could erroneously be declared as inferior to the other two procedures if it was assigned to bridges that as a group had considerably more traffic volume than the bridges to which the other two procedures were assigned.

Blocking

The solution would be to classify the bridges by traffic volume categories (e.g., light, medium, and heavy) and use traffic as the blocking factor. If, for example, there were 30 bridges in each category, the random assignment of maintenance procedures to categories would result in each maintenance procedure assigned to 10 bridges within each block. The statistical analysis would parallel the analysis in Example 12.1 except that in this case there are multiple occurrences of a factor level (maintenance procedure type) within each block. ∎

12.3.2 Two Blocking Factors: Latin Square Design

In Example 12.1, differences in soil fertility were assumed to be only in one direction. What if the differences existed for both directions? Then two blocking factors would have to be used. This leads to consideration of a *Latin square design*. As the name implies, this is a square configuration, meaning that the number of levels of the factor and of each of the two blocking factors must be the same. The size of a Latin square is called the order of the design, and a Latin square of order 4 is given below, with the letters denoting the levels of the factor.

A	B	C	D
B	A	D	C
C	D	A	B
D	C	B	A

If this particular Latin square were used and an experiment conducted, the values of the response variable could be inserted in the appropriate cells and the data could then be analyzed. This would be a 3-way analysis: treatments (e.g., fertilizers), rows, and columns. The analysis can be performed with virtually all statistical software; it is not a matter of having a special Latin square analysis capability. An ANOM display could also be constructed, analogous to the way that an ANOM display would be constructed for a randomized block design.

It is also possible to construct Latin square-type designs (Graeco-Latin, etc.), for which more than two blocking factors can be used.

12.4 MORE THAN ONE FACTOR

Typically, there is more than one factor that relates to the response variable, just as regression models generally contain multiple predictors. As we extend the discussion of designed experiments and consider more than one factor, it seems logical to pose the following question: Why not study each of the factors separately rather than simultaneously?

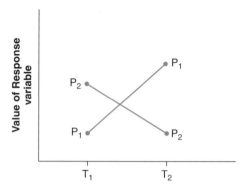

Figure 12.2   Interaction profile. P_1, low pressure; P_2, high pressure; T_1, low temperature; T_2, high temperature.

Figure 12.2 provides an answer to that question. Assume that a company's engineering department is asked to investigate how to maximize process yield, with the engineers generally agreeing that temperature and pressure have a profound effect on yield. Three of the engineers are given this assignment, and these will be represented by Bob, Joe, and Sarah, respectively. Each engineer conducts his or her own experiment. Assume that Bob and Joe each investigate only one factor at a time, whereas Sarah decides to look at both factors simultaneously. Assume further that Figure 12.2 depicts what can be expected to result when both factors are studied together.

If Bob had used the low temperature (T_1) and varied the pressure from low to high, he would conclude that the best way to increase the yield is to increase the pressure, whereas he would have reached the opposite conclusion if he had used the high temperature. Similarly, if Joe had set the pressure at the high level (P_2), he would have concluded that the best way to increase yield would be to reduce the temperature, whereas he would have reached the opposite conclusion if he had used the low pressure level.

Sarah, on the other hand, would be in the proper position to conclude that interpreting the effects of the two factors would be somewhat difficult because of the *interaction effect* of the two factors. Interaction effects are depicted graphically by the lack of parallelism of lines, as in Figure 12.2.

This type of feedback is not available when factors are studied separately rather than together. These "one-at-a-time" plans have, unfortunately, been used extensively in industry. They *do*, however, have value for certain objectives, as discussed in Section 12.15, so a blanket indictment of these designs is unwise. For example, when screening designs are used there is generally interest in identifying factors that seem important, so at that stage interactions are not important. In fact, as is discussed in Section 12.15, one-at-a-time designs do have value as screening designs.

12.5   FACTORIAL DESIGNS

The examples in Section 12.2.1 each had three levels. Designs with three levels can obviously require a large number of points when there are more than a few factors, however, so designs with two levels are used much more extensively.

12.5.1 Two Levels

The simplest design with two factors is one that has two levels. We represent factorial designs as s^k, with s denoting the number of levels and k the number of factors. Therefore, for a two-factor design with two levels, that is, a 2^2, there are four design points. For three factors the design would be written as a 2^3, for four factors it would be a 2^4, and so on.

The designation of the two levels of each factor is mathematically arbitrary, but frequently "-1" is used for the "low" level of a quantitative factor, such as the lower of two temperatures, and "1" is used for the "high" level, such as the higher of two temperatures. Thus, the actual temperatures are not used; rather, the temperature values are "coded." For example, if the two temperatures were 380 °C and 340 °C, the 1 and -1 would be obtained as $(380 - 360)/20$ and $(340 - 360)/20$, respectively. That is, the denominator is half the difference between the two values and the number that is subtracted is the average of the two values. Coding is discussed in more detail in Section 12.12.

The four combinations for a 2^2 design, often referred to as *treatment combinations*, are thus $(-1, -1)$, $(1, -1)$, $(-1, 1)$ and $(1, 1)$, which form a square with a center of zero. If the numbers within each set of parentheses are multiplied together and the products added, the sum is zero. Similarly, if this set of four products is in turn multiplied by the corresponding first number in each set of parentheses and those products are added, the sum is also zero, and similarly if the second number within each set of parentheses is used instead of the first number.

This indicates that the design is *orthogonal*, which means that the effects of each factor and the interaction of the two factors can be estimated independently of each other. The concept of interaction is illustrated and explained later in this section.

Although we shall continue to deemphasize hand computation and emphasize computer-aided analyses, it is helpful to go through the calculations for a simple 2^2 design, which although not of practical value, does have illustrative value. Therefore, consider Figure 12.3,

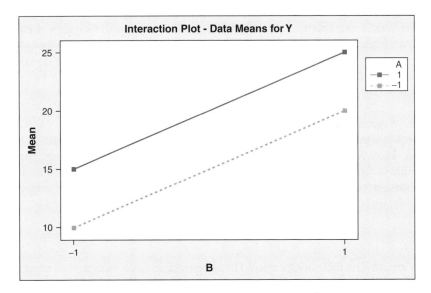

Figure 12.3 Interation plot: data display for a 2^2 design.

which can be produced in MINITAB by using either the `FFINT` command or the `INTERACT` command.

In this case we can see the data values just as easily as if they were presented in tabular form. Although the plot is labeled an interaction plot, representing the (statistical) interaction (or not) between the two factors, we can also use it to obtain the A effect and B effect estimates, in addition to the AB (interaction) effect estimate. (The latter is analogous to a cross-product term in regression.)

From the plot we can see that changing A from the low level to the high level does have an effect on the response, and the effect is the same (5 units) at both levels of B. Therefore, in this instance we would estimate the effect of factor A, termed a *main effect* in factorial designs, as 5. If the differences were not the same, then we would average the two differences. Another correct and very intuitive way to estimate the A effect is to subtract the average response at the low level from the average response at the high level. Thus, the effect would be estimated by $(15 + 25)/2 - (10 + 20)/2 = 20 - 15 = 5$.

We have to look at Figure 12.3 a bit differently to see the B effect estimate in the same two ways. The two lines being parallel means that the change in the response as B is increased from the low level to the high level is the same at each level of A. Thus, the B effect is 10. Viewing the B effect estimate by using average responses, analogous to what was done for factor A, we thus obtain the B effect estimate as $(20 + 25)/2 - (10 + 15)/2 = 22.5 - 12.5 = 10$.

The term "interaction" does not mean, for example, anything like a chemical interaction. Instead, for a two-factor interaction it simply means that the effect of a factor differs over the two levels of the other factor, which, as indicated in Section 12.4, means that the lines in an interaction plot will not be parallel. In Figure 12.3 this is not the case, however, as we have seen. Thus, the estimate of the interaction effect, written as AB, is zero. In general, the estimate of the AB interaction effect will be $\frac{1}{2}[A_2B_2 - A_2B_1 - (A_1B_2 - A_1B_1)]$, with the subscript "2" denoting the high level and subscript "1" denoting the low level.

When there are more than two factors, we may still compute a main effect estimate as the average response at the high level minus the average response at the low level for each factor; interaction effect estimates cannot be explained as easily.

In this simple example the interaction effect estimate being zero makes it easier to determine the main effect estimates unambiguously, but an analysis with hypothesis tests as was performed in Section 12.2.1 is not possible because there aren't enough observations to provide an estimate of σ^2 when there is only one observation for each factor level combination. The analysis for such designs is discussed in Section 12.5.3.

12.5.1.1 Regression Model Interpretation

The effect estimates are used to develop the regression equation for the data in the experiment. The coefficients in the regression model are one-half the effect estimates; this relationship is derived in Appendix B at the end of this chapter. For example, for an unreplicated 2^2 design with effect estimates for A, B, and AB being 5, -3.8, and -1.9, respectively, and the average of the four data points being 24.6, the regression equation is

$$\widehat{Y} = 24.6 + 2.5A - 1.9B - 0.95AB$$

How do we interpret these coefficients? The interpretation is the same as the interpretation of the coefficients in any regression model developed from orthogonal data. That is, if A is increased by one coded unit (staying within the range of values for that factor used in the

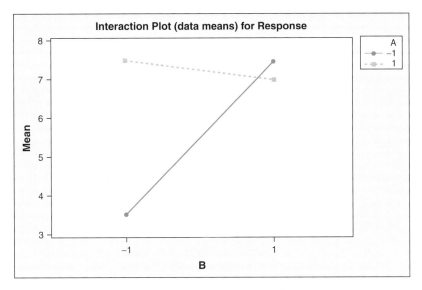

Figure 12.4 Interaction plot for a 2^2 design.

experiment), we would predict that the response would increase by 2.5 units. Similarly, if B is increased by one coded unit, we would predict that the response would decrease by 1.9 units. The interaction coefficient is trickier to interpret because the product AB could be either positive or negative. Let the current coded level for both A and B be -0.5. If A and B are each increased by one coded unit, $\widehat{Y}$ increases by 0.6, with the change having no effect on the contribution of the interaction term. Thus, we cannot interpret interaction coefficients the same way that we interpret coefficients for main effects.

12.5.1.2 Large Interactions
The analysis can be complicated by sizable interactions, as will now be illustrated. Consider Figure 12.4.

At the high level of B, the A effect is small and negative, whereas at the low level of B the A effect is much larger and positive. Note that if we average these two effect estimates, we will have an effect estimate that does not well-represent what happened at either of the two levels of factor B. As a more extreme example, if we average 100 and 300 to obtain 200, the latter is not representative of either 100 or 300. The magnitude of the interaction is indicated by the extent to which the lines deviate from being parallel, and when the lines cross the interaction is extreme.

In such cases it is not possible to estimate main effects in the usual way. Instead, it is necessary to look at what happens at levels of factors involved in large interactions. These effects have been called *simple effects*, although it would be better to call them *conditional effects* because they are computed by conditioning on one level of another factor (i.e., they are computed using half of the data when factors have two levels). The average of the conditional estimates is the main effect estimate. For an unreplicated 2^2 design, the conditional effects of factor B are the effects of that factor at each of the two levels of factor A. That is, they are the average at the high level of factor B minus the average at the low level of B for each of the two levels of A, with the conditional effects of A defined analogously.

Large interaction effects can cause the conditional effects to differ greatly. Unfortunately, there is no well-accepted rule stating when conditional effect estimates should be computed instead of main effect estimates, although Daniel (1976) stated that either or both of the main effect estimates should not be reported when a two-factor interaction effect estimate is more than one-third of either or both of the two main effects whose factors comprise the interaction. For example, if the AC interaction effect estimate is 6.32 and the A and C main effect estimates are 11.34 and 31.76, respectively, the main effect estimate for A should not be reported. Although Daniel (1976) did not state what should be done instead, the implication was that conditional effects should be reported.

12.5.2 Interaction Problems: 2^3 Examples

In this section we look at three experiments in which a 2^3 design was used. This is a design with 3 factors each at 2 levels, so there are $2^3 = 8$ design points, with the configuration of points forming a cube:

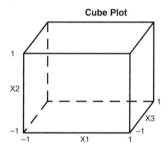

The discussion at the end of the preceding section showed how a large interaction can cause problems in assessing main effects in a two-factor design. In this section we look at an actual experiment that has a large 3-factor interaction, and then consider a 2^3 experiment given by Daniel (1976) that has a large two-factor interaction.

■ **EXAMPLE 12.3**

Experiment and Data

Although 3-factor interactions are rare, they do occur so it is important to understand their deleterious effects. In their article "Sequence-Leveled Experimental Designs at Work" (*Quality Engineering*, **14**(2), 223–230, 2001–2002), C. W. Carter and T. B. Tripp described an experiment involving capacitors with the response variable being the direct-current leakage of capacitor-grade tantalum powder with the anodized pellets sintered at 1600 °C. The experimental variables are formation temperature, formation voltage, and phosphorus concentration, all at two levels, with each factor-level combination replicated nine times so that the design is a 2^3 with nine replications. The data, in the usual coded form for the design, are in 3WAYINT.MTW.

The initial analysis of the data is as follows:

```
Factorial Fit: leakage versus phosphorus,voltage,formation

Estimated Effects and Coefficients for leakage (coded units)

Term                     Effect    Coef    SECoef      T        P
Constant                           1506    168.6     8.93    0.000
phosphorus                2653      1326    168.6     7.86    0.000
voltage                   -789      -394    168.6    -2.34    0.023
formation                -2517     -1258    168.6    -7.46    0.000
phosphorus*voltage        -936      -468    168.6    -2.78    0.007
phosphorus*formation     -2615     -1308    168.6    -7.75    0.000
voltage*formation         1088       544    168.6     3.23    0.002
phos.*volt.*formation      980       490    168.6     2.90    0.005

Unusual Observations for leakage

Obs      leakage       Fit     SE Fit     Residual      St Resid
11       16500.0    7294.4      477.0       9205.6         6.82R
16        2970.0    7294.4      477.0      -4324.4        -3.21R

R denotes an observation with a large standardized residual
```

Analysis

We immediately observe that every effect is significant, and this should raise a red flag because this will rarely occur. This could be caused by a bad data point, and indeed there is an observation that looks very suspicious: the 16,500 value.

There is another problem, however. This analysis assumes that the variances for each of the eight treatment combinations are the same. That assumption will clearly be violated when the response values differ by an order of magnitude across the treatment combinations, and that is the case here. One solution to the problem is to try to transform the dependent variable to correct the problem. Carter and Tripp took a different approach, however. Specifically, they used the expression for the contrast of the three-factor interaction and constructed a confidence interval. Since the interval did not include zero, they concluded that there is a 3-factor interaction. They then examined the formation temperature × formation voltage interaction in light of the significant 3-factor interaction, by looking at the two-factor interaction over the different levels of phosphorus. We might call these *conditional interaction effects*. Similarly, one would want to look at the conditional main effects. ∎

∎ EXAMPLE 12.4

Experiment

As a second example illustrating interaction problems, Daniel (1976, p. 54) described an experiment in which the objective was to investigate the effect of stirring (A), temperature (B), and pressure (C) on the thickening time of cement. This was an unreplicated 2^3 design. The data were also analyzed in Ryan (2000), and the analysis was complicated by the fact that there was a large BC interaction. Because of this, the conditional main effects of B and C differ considerably, and it was observed that the conditional effects for B were -180 and -80, whereas the conditional effects for C were -121 and -26. These numbers differ so much that reporting the average for each pair would be misleading. ■

■ EXAMPLE 12.5

Experimental Design

As a final example of a 2^3 design, we consider the experiment, with a 2^3 design with three centerpoints, reported by E. Palmqvist, H. Grage, N. Q. Meinander, and B. Hahn-Hägerdal in their paper "Main and Interaction Effects of Acetic Acid, Furfural, and p-Hydroxybenzoic Acid on Growth and Ethanol Productivity of Yeasts" (*Biotechnology and Bioengineering*, **63**(1), 46–55, 1999).

Objective

Although there were three response variables used in the experiment, interest centered on one of these: Baker's yeast. The objective was to examine the effect of the three factors in the article title, acetic acid (HAc), furfural (F), and p-hydroxybenzoic acid (POH), on the ethanol yield of Baker's yeast. The data from the experiment were as follows:

HAc	F	POH	Baker's yeast yield
0	0	0	0.38
10	0	0	0.39
0	2	0	0.31
10	2	0	0.35
0	0	2	0.37
10	0	2	0.38
0	2	2	0.33
10	2	2	0.40
5	1	1	0.38
5	1	1	0.41
5	1	1	0.41

Of concern is the fact that the runs were apparently not randomized and, in particular, it appears as though the three centerpoints were run consecutively. If so, this invalidates any analysis using ANOVA because those centerpoint response values will not measure experimental error at that point, let alone be representative of the experimental error at the other design points.

Conclusion

The authors reported that there were no significant effects and indeed this is what happens when ANOVA is used. Underestimating the experimental error would exaggerate rather than diminish the importance of the various effects, so that can't be the reason. There are, however, large interaction effects involving HAc and F that are approximately two-thirds the size of the main effects. Consequently, a conditional effects analysis would have to be performed, without clear evidence of a good estimate of the experimental error. ▪

12.5.3 Analysis of Unreplicated Factorial Experiments

An unreplicated factorial, such as the 2^2 used for illustration in Section 12.5.1, cannot be analyzed in the same way that the data from the one-factor design with three levels were analyzed in Section 12.2.1. The problem is that an unreplicated factorial will have $n - 1 = 2^k - 1$ degrees of freedom for estimating effects, but the number of effects to be estimated is also $2^k - 1$. Consequently, there will not be any degrees of freedom left for estimating σ^2. Thus, F-statistics and/or t-statistics cannot be computed.

Therefore, a different method of analysis must be used. Recall a normal probability plot, as illustrated in Section 6.1.2. When we test for the significance of effects, the null hypothesis is that the mean of the effect being tested is zero. Since normality and a common variance are assumed, if those assumptions are valid and no effects are real, the effect estimates would all be distributed as $N(0, \sigma^2)$. Therefore, a normal probability plot of the effect estimates would be a reasonable approach, with effects judged to be real when the effect estimates plot well off the line, as such points would likely represent effects whose mean is not zero.

Figure 12.5 is an example of such a plot. Many years ago, judgment would be used to determine whether or not a plotted point was far enough off the line drawn through the main cluster of points to warrant calling the effect that the point represented a significant effect. Such methods are now used very infrequently as they have been superseded by better methods.

Figure 12.5 is a normal probability plot of estimates of the seven effects that are estimable when a 2^3 design is used: the three main effects, three two-factor interactions, and the one three-factor interaction. Points that are labeled on the plot, which was produced by

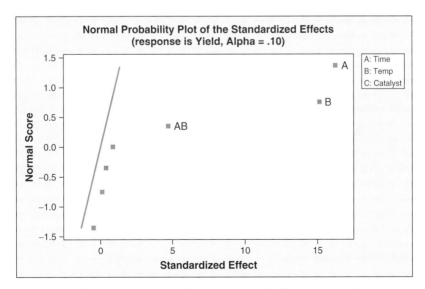

Figure 12.5 Normal probability plot of standardized effect estimates.

MINITAB, are those that are judged to be significant. There are various methods for determining significant effects, all of which use a pseudoerror term, obtained by treating small effects as error. The method incorporated by MINITAB is due to Lenth (1989).

The plot is produced by using the `GEFFECTS` subcommand of the `FFACTORIAL` command, with the latter used to specify the model. Notice that the line is constructed through the point $(0, 0)$, *not* through the center of the four points that almost form a straight line. Since the AB interaction effect is declared significant, great care would have to be exercised in trying to assess the magnitude of the A and B effects.

Although a normal probability plot of effect estimates is very useful, many practitioners have pointed out that it is not useful for trying to explain results to management when managers have not had any training in design of experiments. Clearly, there is indeed a problem of the same type that motivated the development of Analysis of Means. Consequently, various users favor a Pareto chart (see Section 1.7) of effect estimates, which might be used to complement the normal probability plot. Thus, the latter might be used by the analyst to determine effects that seem to be statistically significant, with the analyst's conclusion conveyed to management in the form of a Pareto chart.

12.5.4 Mixed Factorials

Although two-level designs predominate, there are many experiments conducted for which some of the factors have more than two levels. This is undoubtedly due to the influence of Genichi Taguchi (see Section 12.17) and later researchers who have worked with such designs. These designs are called *mixed factorials*, the simplest of which is a 3×2 design—a design for which the first factor has three levels and the second factor has two levels. If a factor has three levels, both the linear and quadratic effects of the factor can be obtained. This means that the interaction between a 3-level factor and a 2-level factor can be represented as two interaction plots: linear $\times$ linear and quadratic $\times$ linear. The latter can be somewhat difficult to interpret, however. Furthermore, the analysis by software will not be forthcoming unless special steps are taken to produce it. Specifically, the quadratic component of a 3-level factor has level designations of $(1\ -2\ 1)$, corresponding to the linear component designation of levels as $(-1\ 0\ 1)$. A quadratic $\times$ linear interaction in MINITAB, for example, cannot be obtained unless it is produced manually.

Other aspects of the analysis are straightforward, however, and in MINITAB the analysis can be performed using either the `ANOVA` command or `GLM` command.

12.5.5 Blocking Factorial Designs

There will often be constraints that prohibit all of the runs of a factorial design from being made under the same set of conditions (e.g., it may not be possible to make all of the runs in one day). When this is the case, it is necessary to make the experimental runs in blocks, with the case study that follows being an example of this.

■ EXAMPLE 12.6 (CASE STUDY)

Problem Statement

Young (1994) described an experiment in which blocking was used. The problem addressed was the unstable effort required to close the hatch of the hatchback version of a new design of a car. There was a desire to investigate the influence on process stability of a component of the hatch gasket called the stuffer. It was believed that variation relative to the nominal values for length (L), firmness (F), and position (P) might cause problems. A 2^3 design was used, with the low level for each of the three factors being the low end of normal variation and the high level being the high end of the normal range.

Blocking

Car-to-car variation might be expected, so it would be desirable to block on cars. There is the obvious question of how many cars to use. It was decided that two cars would not be representative of production, so four cars were used, with four of the eight runs being made on each of two cars. Thus, the design was both blocked and replicated, with the blocking performed on the three-factor interaction. The response variable was the closing effort (CE) required and the data were as follows, with the presence of a letter in the treatment combination indicating that the factor is at the high level and absence of a letter indicating the low level. The treatment combination (1) represents all factors being at the low level.

Treatment Combination	CE	Block
(1)	7.0	1
lp	7.0	1
lf	6.5	1
pf	7.0	1
(1)	7.0	3
lp	7.5	3
lf	6.5	3
pf	7.0	3
l	5.0	2
p	5.0	2
f	3.5	2
lpf	4.5	2
l	14.0	4
p	11.0	4
f	13.0	4
lpf	13.5	4

The first thing we notice is that there is a considerable difference in the measurements between the fourth block and the other three blocks, so that blocking was indeed necessary. The analysis performed by Young (1994) showed no effect to be significant other than the blocking factor, which the reader is asked to verify in Exercise 12.85. Thus, it appears as though the fears regarding excess variation were unfounded. ∎

12.6 CROSSED AND NESTED DESIGNS

It is important to recognize that factors are *crossed* in a typical factorial design. That is, each level of each factor occurs in combination with all possible combinations of levels of the other factors. When factors are not crossed, they are *nested*; that is, all possible combinations are not used. For example, assume that there are six similar but not identical machines, each of which has a head that could not be used on any of the other five machines. If we wanted to investigate the effect of head and machines, we would have to recognize that we have a *nested design*. That is, the head effect is nested under the machine effect. Nested designs are not covered in this text but are covered in any book on design of experiments. It is important for engineers and other users of experimental designs to recognize when factors are nested rather than crossed so that the proper analysis can be performed.

12.7   FIXED AND RANDOM FACTORS

When there are at least two factors, it is necessary to determine whether the factors are *fixed* or *random*, as the analysis will depend on this classification. A factor is fixed if the levels of a factor are predetermined and the experimenter is interested only in those particular levels (e.g., temperature at 250 °C, 300 °C, and 350 °C). A factor is classified as random if the levels are selected at random from a population of levels (e.g., 254 °C, 287 °C, 326 °C), and the inference is to apply to this population of levels rather than to the particular levels used in the experiment, as is the case with fixed factors.

For example, consider the case of a single factor A, with the model written

$$Y_{ij} = \mu + A_j + \varepsilon_{ij} \tag{12.3}$$

with j denoting the jth level of the single factor, and i denoting the ith observation. Thus, Y_{ij} represents the ith observation for the jth level. Furthermore, μ is a constant that is the value that we would expect for each Y_{ij} if there was no factor effect, A_j represents the effect of the jth level (as a departure from μ), and ε_{ij} represents the error term (with the errors assumed to be random), analogous to the error term in a regression model. If there was no factor effect, there would be no need for an A_j term in the model, so the model could then be written

$$Y_{ij} = \mu + \varepsilon_{ij} \tag{12.4}$$

The F-test given in Eq. (12.2) is then a test of H_0: all $A_j = 0$ against H_a: not all $A_j = 0$, with the outcome determining whether the appropriate model is given by Eq. (12.3) or Eq. (12.4).

If the factor is random, we would test $H_0: \sigma_A^2 = 0$ against $H_a: \sigma_A^2 \neq 0$. If the factor were temperature and we rejected H_0, we would be concluding that there is a temperature effect within the range of temperatures from which the ones used in the experiment were randomly selected, whereas if we fixed specific temperatures and rejected H_0, we would conclude that at least one of the temperatures used has an effect.

Regarding the numerical analysis of the data, with two factors, A and B, there are obviously four possibilities: (1) both could be fixed, (2) both could be random, (3) A could be fixed and B random, or (4) A could be random and B fixed.

For the first case when both factors are fixed, the main effects and the interaction (if separable from the residual) are tested against the residual. When both factors are random, the main effects are tested against the interaction effect, which in turn is tested against the residual. When one factor is fixed and the other one random, the fixed factor is tested against the interaction, the random factor is tested against the residual, and the interaction is tested against the residual. (By "tested against" we mean that the mean square for what follows these words is used in producing the F-statistic.)

These testing procedures result from what the expected mean square (EMS) for each effect is under each of the four scenarios. For example, when both factors are random and have two levels and there is one observation for each AB_{ij} combination (i.e., an unreplicated 2^2 design), the expected mean square for effect A [i.e., EMS(A)] is $\sigma_\epsilon^2 + \sigma_{AB}^2 + 2\sigma_A^2$. To test $H_0: \sigma_A^2 = 0$, we need to construct an F-test such that the denominator of the F-statistic has the first two variance components in common with EMS(A), so that the EMS expressions differ only in terms of the variance component that is being tested.

This is the guiding principle behind the construction of the tests. Textbooks on design of experiments have often given rules for determining the EMS expressions, but very simple rules can be given, as discussed by Lenth (2001) and Dean and Voss (1999, p. 616).

Although it is desirable to know how the tests are constructed, engineers and others need not be concerned with remembering such rules, as it is simply a matter of specifying the fixed and random factors with whatever software is used. For example, the default option for the ANOVA command in MINITAB is that the factors are fixed. If one or more factors are random, they would have to be specified as such using the RANDOM subcommand.

The failure to designate B as a random factor could make a considerable difference on whether A was judged significant or not, especially when the interaction effect is large. This is because A is tested against the interaction effect when B is declared a random factor, but is tested against the error (residual) when B is a fixed effect. The proper designation of factor B could, somewhat paradoxically, "penalize" A twice because the A effect can be small simply because the interaction is large, and then having to test A against the interaction further reduces the chance that A could be significant. Nevertheless, factors should be properly designated.

12.8 ANOM FOR FACTORIAL DESIGNS

One of the primary advantages of ANOM is that it allows the user to easily see the magnitude of interaction effects relative to the magnitude of main effects. This is because for any two-level factorial design there will be a "high" and a "low" for each interaction effect. That is, two averages are computed, each computed from half of the data, and the interaction effect estimate is the difference of these two averages, just as is true for main effect estimates. For example, for a 2^2 design the observations that would be used in computing one of the averages would be those for which the factors are both at their low levels or both at their high levels, with the other average computed from the remaining observations. This same principle would apply to two-factor interactions when there are more than two factors, and the two averages for a three-factor interaction would be computed using the average at $+1$ and the average at -1, these being the products across the three factors. (Of course, this $+1$ and -1 approach also applies to two-factor interactions.)

An ANOM display with main effects and interactions displayed in this manner would allow easy comparison of the magnitude of interaction effects and the corresponding main effects. As stated previously, Daniel (1976) recommended that a main effect not be reported if an interaction effect estimate was more than one-third of the main effect estimates of the factors that comprise the interaction. Unfortunately, widely available software with ANOM capability doesn't permit this comparison, except for the macro that can be downloaded at the Minitab, Inc. website (see http://www.minitab.com/

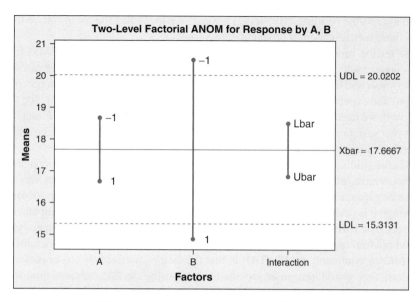

Figure 12.6   ANOM display for a 2^2 design.

support/answers/answer.aspx?ID=1755), which has this capability for a 2^2
design. Either the design must be replicated or an estimate of the standard deviation must
be entered along with the degrees of freedom. Of course, an arbitrarily large value for
the latter could be entered if the estimate is not obtained directly from data. An example of the display is given in Figure 12.6, which shows the B effect being significant at
$\alpha = .05$.

Anyone wanting to use MINITAB with interaction effects shown on the same display
with the main effect estimates for other designs would have to write a macro to accomplish
that. For at most a moderate number of factors, all of the main effects and interaction
effects could be shown on one graph. Then the user could apply rules of thumb such as
the one given by Daniel (1976) to determine whether it is practical to report main effect
estimates.

12.8.1   HANOM for Factorial Designs

As indicated in Section 12.2.3.1, Nelson and Dudewicz (2002) presented heteroscedastic
analysis of means (HANOM) for the case of unequal variances, this being the counterpart
to heteroscedastic ANOVA.

The first step with their approach is to test for equality of variances, and in order to do this
there must be more than a few observations per treatment combination. Otherwise, a test
for equality of variances, such as Analysis of Means for Variances (ANOMV) by Wludyka
and Nelson (1997), will not have sufficient power for detecting unequal variances. Only in
applications of experimental design for which observations are relatively inexpensive will
there be more than a few observations per treatment combination, and the main thrust of
the coverage of multifactor designs in this chapter is for unreplicated designs. Therefore,
HANOM will not be illustrated and the reader is referred to Nelson and Dudewicz (2002)
for details of their approach.

12.9 FRACTIONAL FACTORIALS

A 2^k design can be impractical and expensive to run when k is not small. For example, an experimenter may be able to afford 32 points, but 64 might be impractical, and 128 points would be impractical in almost all types of experiments. Not only would the latter be impractical, it would also be highly wasteful. The total number of effects that *could* be estimated is $2^k - 1$, as stated previously. This follows from the fact that there are two possibilities for each factor—either being in an effect or not—and if they were all absent, there would be no effect. For each additional data point we are able to estimate another effect, but we would hardly ever have a need to estimate 127 effects.

Therefore, when there are more than a few factors to be investigated, we will almost always want to use a fraction of the 2^k possible runs. We shall focus attention on two-level fractional factorial designs as this type of fractional factorial design has been used extensively in practice. Fractional factorials were first presented by Finney (1945) and popularized in the landmark papers of Box and Hunter (1961a,b), and we shall adopt the notation of the latter throughout this section.

When we use a fractional factorial design, we generally want to use a design that has the maximum *resolution* for a given number of design points. Designs of resolution III, IV, and V, written with Roman numerals, are the ones that are most frequently used. The first of these allows main effects (only) to be estimated, in the absence of interactions. With a resolution IV design, only main effects are estimable, but main effects can be estimated even if two-factor interactions are significant. With a resolution V design, both main effects and two-factor interactions are estimable, in the absence of higher-order interactions.

In order to distinguish between designs that have the same resolution, other criteria can be employed.

12.9.1 2^{k-1} Designs

A two-level fractional factorial can be written in the general form 2^{k-p}, with, as before, k denoting the number of factors, and the fraction of the full 2^k factorial that is to be run is $1/2^p$. Thus, a 2^{3-1} design would be a 1/2 fraction of a 2^3. Sixteen-point designs (so that $k - p = 4$) are the ones that have been used the most often in industry.

For simplicity, however, we shall first illustrate a 2^{3-1} design, which, although of limited usefulness, does have value for illustrative purposes. We should first recognize that with only four design points (here we are assuming that the design is not replicated), we will have only 3 degrees of freedom so we can estimate only 3 effects. Which three effects do we choose to estimate? Although in rare instances a two-factor interaction might be of more interest to an experimenter than a main effect, we would generally choose to estimate main effects over interactions, if we had to select one over the other. Thus, the logical choice would be to estimate the three main effects: A, B, and C. Before we can undertake that task, however, we must determine what four design points to use. We cannot randomly select four treatment combinations from the eight that are available. For example, we obviously could not estimate the main effect of A if we happened to select four treatment combinations in which A was at the high level and none in which A was at the low level; similarly for B and C. Thus, we would clearly want to have two treatment combinations in which A is at the high level and two in which A is at the low level, and the same for B and C.

With a little trial and error, we could obtain four treatment combinations that satisfy this property without too much difficulty, but it would obviously be preferable to use some systematic approach. Whenever a 1/2 fraction is used, we have to select one effect to "confound" with the difference between the two fractions. This means that the chosen effect would be estimated by the difference of the averages of the treatment combinations in each fraction (which of course is the way that we would logically estimate the difference between the two fractions). If we have to "give up" the estimate of one effect in this way (which is obviously what we are doing since we will run only one of the two fractions), it would be logical to select the highest-order interaction to relinquish, as this is the effect that is least likely to be significant. For a 2^3 design, that is the ABC interaction.

It is important for students and practitioners to know how fractional factorials are constructed and also to know exactly what risks are taken, but we want to use software to construct the designs, and the output from MINITAB is given below. Note that the design with the factor levels is not part of the output; this has to be stored by the user and can then be displayed.

```
Factorial Design

Fractional Factorial Design

Factors:     3        Base Design:      3, 4      Resolution: III
Runs:        4        Replicates:       1         Fraction: 1/2
Blocks:      none     Center pts (total):              0

*** NOTE *** Some main effects are confounded
             with two-way interactions

Design Generators: C = AB

Alias Structure

I + ABC

A + BC
B + AC
C + AB
```

The reason we see $C = AB$ in the output is that a common way of constructing a 2^{k-p} design is to first construct the full factorial for $k - p$ factors and then obtain the columns for the p factors that have not been accounted for yet by letting them represent products of the columns in the full factorial that has been constructed. Here $p = 1$, so we just need one additional column, and $C = AB$ means that we are going to confound C with AB. Thus, the (main) effect of C will be estimated in the same way that the AB interaction would be estimated (i.e., the same linear combination of response values), but of course we are not going to try to estimate this interaction. The term *alias* is also frequently used to signify effects that are confounded.

We can see this relationship by looking at the treatment combinations.

```
Data Display
```

Treatment Combination	A	B	C
c	−1	−1	1
a	1	−1	−1
b	−1	1	−1
abc	1	1	1

Notice that if we multiply corresponding numbers in the A and B columns together, we obtain the C column. Thus, the C effect is estimated in the same way that the AB interaction would be estimated (if it were estimated). Notice also that if we multiply all three columns together, which would give us the ABC interaction, we obtain a column in which all of the numbers are one. This is what is called the *principal fraction* and is the default option in MINITAB. The ABC interaction in this case is the *defining contrast*. This is often written $I = ABC$. This states that the ABC interaction is *confounded (aliased)* with the mean estimate as the latter would also be obtained by adding all of the observations and dividing by the total number of them. If the other fraction had been used, the defining contrast would have been $I = -ABC$, which states that the estimate of the mean would be equal to minus the estimate of the ABC interaction.

We have two fractions to choose from when we use a 1/2 fraction, and the obvious question is: Does it make any difference which fraction we use? It *could* make a difference because, for example, it might be impractical or even impossible to use the high level of each factor in combination with the low levels of the other factors, as the principal fraction requires. Furthermore, having, for example, the three factors at all high or all low levels might have an extreme and unexpected effect on the response that would not be detected if the fraction that would produce such a response value was not used. This is where engineering knowledge would come into play in suspecting something of this sort.

The analysis of data from a fractional factorial design proceeds along the same lines as the analysis of an unreplicated factorial. We might hope that the analysis of a half-fraction would be straightforward—and often it is—but it will often be necessary to additionally run the other half-fraction in order to dealias interactions. A more economical approach, however, would be to add only the additional treatment combinations to dealias certain significant effects.

■ **EXAMPLE 12.7**

Problem Statement

In their paper "Optimization of Tool Life on the Shop Floor Using Design of Experiments" (*Quality Engineering*, **6**(4), 609–620, 1994), D. Watkins, A. Bergman, and R. Horton gave a case study that detailed the use of design of experiment techniques to increase tool life in a manufacturing facility. Scrap rates were high and this was attributable to tool failures. Therefore, due to high production rates, even a small increase in tool life could yield significant operating cost savings.

Experimental Design

Five factors were investigated in the experiment using a 2^{5-1} (resolution V) fractional factorial with three replicates at each setting. The response of interest was tool life as measured in the number of quality ends cut before tool failure. The experiment was performed on both spindles (right and left) of a single threading machine. Approximately half the runs were performed on each end. The spindle blocking effect was not expected to be important. This turned out to be an incorrect assumption and was considered to be an important experimental discovery.

Analysis

Data analysis was then performed, ignoring third- and fourth-order interactions, using both Pareto and F-statistic analyses of the effects. An empirically determined corrective factor was applied to accommodate the obvious spindle effect. The data were analyzed both with and without this corrective factor. Two main effects, chaser type and coolant, were identified as well as a significant interaction. Using these results, factor settings were determined that should improve tool life.

Prior to experimentation, approximately 60 parts were produced before tool failure. After this preliminary experimentation, considered incomplete in terms of process optimization by the experimenters, the number of parts produced prior to tool failure was approximately 170 (about a 200% increase).   ■

In general, a 2^{k-1} design should almost always be constructed by letting I equal the highest-order interaction, as indicated previously. For example, if we had four factors, we would choose $I = ABCD$ and this would produce a resolution IV design as the main effects would not be confounded with two-factor interactions, but certain two-factor interactions would be confounded with other two-factor interactions.

As stated, we generally want to let software construct designs for us. It is useful, however, to know that all of the treatment combinations for one fraction have an odd number of letters in common with the defining contrast, with the treatment combinations in the other fraction having an even number of letters in common with the defining contrast. (Zero is treated as even.) For example, with a 2^{5-1} design and $I = ABCDE$, treatment combinations ab and abc would be in different fractions. This is a useful spot check before an experiment is run to see if all of the treatment combinations have been written properly.

12.9.2   Highly Fractionated Designs

A 2^{k-1} design will often require more experimental runs than can be afforded when k isn't small. For example, 128 runs would be necessary when $k = 8$. The 127 effects that could be estimated would include many high-order interactions that almost certainly would not be significant, so most of those experimental runs would, in essence, be wasted. Therefore, designs of the form 2^{k-p} should be considered, with p selected to be as large as necessary so that a practical number of design points can be used.

More highly fractionated designs than a 2^{k-1} do come with a price, however, as the aliasing becomes more complex. For example, each effect is confounded with three other effects when a 2^{k-2} design is used, as compared with one effect when a 2^{k-1} design is employed. Consequently, it is important to understand the alias structure for these designs and to make sure that effects that might be significant are not confounded with each other.

Of course, we want to use software for design construction, as indicated previously, and the alias structure is available, and with MINITAB it is printed out, as in Example 12.7 in Section 12.9.1. Following is the output when a 2^{6-2} design is constructed using

MINITAB:

```
Factorial Design

Fractional Factorial Design

Factors:     6         Base Design:       6,16      Resolution: IV
Runs:        16        Replicates:          1       Fraction: 1/4
Blocks:      none      Center pts (total): 0

Design Generators: E = ABC F = BCD

Alias Structure
    I + ABCE + ADEF + BCDF

    A + BCE + DEF + ABCDF
    B + ACE + CDF + ABDEF
    C + ABE + BDF + ACDEF
    D + AEF + BCF + ABCDE
    E + ABC + ADF + BCDEF
    F + ADE + BCD + ABCEF
    AB + CE + ACDF + BDEF
    AC + BE + ABDF + CDEF
    AD + EF + ABCF + BCDE
    AE + BC + DF + ABCDEF
    AF + DE + ABCD + BCEF
    BD + CF + ABEF + ACDE
    BF + CD + ABDE + ACEF
    ABD + ACF + BEF + CDE
    ABF + ACD + BDE + CEF
```

The display shows that each effect is confounded with three other effects. The design is resolution IV because the shortest "word," excluding I, in "$I + ABCE + ADEF + BCDF$," is four.

The most highly fractionated design is one for which the number of experimental runs is equal to one more than the number of factors. Such designs are used for screening important factors and are called *orthogonal main effect plans*, as only main effects are estimable with these designs. These are simply 2-level fractional factorials when the number of runs is a power of 2, with Plackett–Burman designs available when the number of runs is not a power of 2, but is a multiple of 4 (e.g., 20).

Certain orthogonal main effect plans have come to be known as "Taguchi designs" but can also be constructed as fractional factorials. This is discussed further in Section 12.17.

12.10 SPLIT-PLOT DESIGNS

Many industrial experiments are split-plot experiments, although they may not be recognized as such and consequently are often analyzed improperly. Since design of experiment applications were originally in agriculture, we can use an agricultural example to explain the design. Assume that we have a few acres of land that we want to use to test several fertilizers to try to determine which one seems to be the best. For the sake of illustration, we will assume that differences in soil fertility are suspected for the east–west direction, but not for the north–south direction (analogous to the randomized block design illustration in Section 12.3.1).

If we have four fertilizers, we might use four blocks, which would be rectangular arrays of land, and define plots within each block. The fertilizers would then be assigned randomly to the plots within each block. (Note that if north–south variability were also suspected, the treatments would be assigned to the blocks in a Latin square configuration, not assigned randomly.) If we were interested simply in comparing the fertilizers, we would have a randomized block design. But assume further that there are two types of seeds that we also wish to test. It would be logical to split each plot into two sections and then randomly assign the two types of seeds to the two subplots within each of the whole plots.

This would be a *split-plot design*. How does this differ from a 2×4 factorial for the two factors, run with four replicates? With the latter, the eight treatment combinations would be randomly assigned to the experimental units. Then this process would be repeated three times. That is clearly quite different from what happens with the split-plot layout.

An important point is that there is a whole-plot error term and a subplot error term. If this isn't recognized and the data are analyzed as if there were only one error term, the results could be misleading. The importance of using the proper error term is emphasized in various articles, including Box and Jones (2000–2001).

Split-plot and related designs, such as split-split plot designs and split-lot designs, are covered in books on design of experiments. For example see Ryan (2007) for further reading on split-plot and split-split plot designs. (*Note*: There is a movement to replace the term "split-plot" with the more general term "split-unit" since these designs are now used in many applications for which there is no plot involved, such as in engineering applications.)

12.11   RESPONSE SURFACE DESIGNS

The emphasis in this chapter has been on designs for factors with two levels because those are the designs used most frequently in practice. It is necessary to use more than two levels to detect curvature, however, and, in general, to determine the shape of the "response surface." That is, how does the response vary over different combinations of values of the factors? In what region(s), if any, is the change approximately linear? Are there humps, and valleys, and saddle points, and, if so, where do they occur? These are the types of questions that response surface methodology (oftentimes abbreviated as RSM) attempts to answer.

The most frequently used response surface designs are *central composite designs* (*CCDs*). These designs permit the estimation of nonlinear effects and are constructed by starting with a two-level full factorial and then adding center points (i.e., at the center of the full factorial) and axial (star) points that lie outside the square formed from connecting the factorial points. The design for two factors is shown in Figure 12.7.

The value of α would be selected by the experimenter. Desirable properties of the design include orthogonality and rotatability. A design is rotatable if $Var(\widehat{Y})$ is the same for all points equidistant from the center of the design. The term "orthogonal" was, of course, used in Section 12.5.1. In the case of two factors, rotatability is achieved when $\alpha = 1.414$. If there are not tight constraints on the number of center points that can be used, both orthogonality and rotatability can be achieved by selecting the number of center points to achieve orthogonality since the number of center points does not affect rotatability.

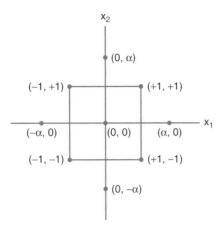

Figure 12.7   Central composite design for two factors.

12.12   RAW FORM ANALYSIS VERSUS CODED FORM ANALYSIS

It was stated in Section 8.5 that regression methods are used in analyzing data from designed experiments and this is illustrated in various sources, including Ryan (2000, Section 13.6). In regression the data usually are not coded but rather are used in raw form. Therefore, that raises the question as to whether coding is essential for the analysis of data from designed experiments or is simply a convenience.

As long as we have a single factor, the analysis, in terms of the conclusions that are reached, is invariant to whatever coding, if any, is used. This is also true when there is more than one factor, provided that the model does not contain any interaction terms. To see this, assume that we have two factors, X_1 and X_2, and an unreplicated 2^2 design. In coded form, X_1, X_2, and X_1X_2 are of course pairwise orthogonal. That is, the dot product of the following columns is zero:

A	B	AB
−1	−1	1
1	−1	−1
−1	1	−1
1	1	1

Assume that the two levels of factor A are 10 and 20 and the two levels of factor B are 20 and 30. The raw-form representation of the two factors and their interaction is then

A	B	AB
10	20	200
20	20	400
10	30	300
20	30	600

Notice that the last column of numbers cannot be transformed to the corresponding set of coded-form numbers. Furthermore, if the analysis were performed in raw form, there would be a major multicollinearity problem as the correlation between A and AB is .845 and the correlation between B and AB is .507. If we let the four response values be 6.0, 5.8, 9.7, and 6.3, respectively, it can be shown that the least squares coefficient of A is positive when the raw form is used and negative when the coded form is used.

Following the example at the beginning of Section 12.2.1, we can see that the A effect is negative because $6.0 + 9.7$ exceeds $5.8 + 6.3$. Specifically, the effect estimate is $(6.0 + 9.7)/2 - (5.8 + 6.3)/2 = 1.8$. Therefore, since the effect estimate is positive, the regression coefficient should also be positive. Thus, we have a wrong sign problem, which unlike the "wrong sign" problem discussed in Section 9.2, really *is* a wrong sign problem!

12.13 SUPERSATURATED DESIGNS

A *supersaturated design* is a design for which there are fewer degrees of freedom for estimating effects than there are factors. Thus, not even all of the main effects are estimable. These designs are useful for examining main effects when faced with a very large number of factors. Supersaturated designs are not orthogonal designs, so considerable care must be exercised when using such designs. Since there are more effects to be estimated than can be estimated, a variable selection technique is generally used to determine which effects to estimate. Variable selection methods can perform poorly when variables are highly correlated, however, so it is important that the correlations between the columns in the design not be very large. Lin (1995) considered the maximum number of factors that can be used for a specified number of runs and when the degree of nonorthogonality is also specified.

Despite the necessary caveats, supersaturated designs do warrant considerable attention. When classical designs are used, the number of significant effects is generally less than half the number of effects that can be estimated with the design that is employed. This is somewhat wasteful of design points unless the "extra" design points are used to estimate σ^2, which in essence is what is done when normal probability plot methods are employed. Those methods for identifying the real effects work well when there is not a large fraction of effects, and such methods are needed when a design is not replicated. We thus might view a normal probability plot approach as a type of variable selection method.

Therefore, a variable selection technique used with supersaturated designs and a normal probability plot when used with classical designs have essentially the same objective; the difference is simply in the number of design points that are used.

The price we pay with supersaturated designs is the loss of orthogonality, which could cause a failure to identify the real effects. When a design is selected with only a small loss of orthogonality, however, the cost savings in terms of fewer design points can easily offset the risk. See Lin (1995) for guidance on use of these designs.

12.14 HARD-TO-CHANGE FACTORS

One thing to keep in mind about experimental designs is that it may not be possible to freely change and reset the levels of certain factors, and certain factor-level combinations may not even be possible. For example, assume that there are three factors to be investigated, each at two levels, and factor C is a hard-to-change factor. When the design combinations are listed in standard order, the first four factor-level combinations have factor C at its low level and the next four have that factor at its high level. If we run the design with the treatment combinations in this order, however, we lose potential benefits that result from randomization, and test statistics (such as F-tests) will be biased. See Ganju and Lucas (1997) for a discussion of the latter. See also Joseph (2000), who considers hard-to-change factors in the context of a particular example. In particular, the author pointed out that changing the copper level in a solder bath took about 3 hours and was extremely difficult to do.

Undoubtedly, there are countless examples of hard-to-change factors, but until the past several years the subject has received very little attention in the literature. Ju and Lucas (2002) considered the case of one factor not being reset in L^k designs. They make the subtle distinction between a randomized run order and a completely randomized design. With the latter, each factor would be reset before each experimental run, even if the factor level will be the same for the next experimental run. For example, if a furnace temperature is to be 370 °F on the fourth and fifth experimental runs, after the fourth run it would, under complete randomization, be cooled down and reheated to 370 °F, just as it would have to be cooled down and then reheated to 330 °F if this were to be the setting for the fifth experimental run.

There is evidence to suggest that failure to reset factors probably occurs in most experiments, and undoubtedly even occurs very often when factors are easy to change.

It might seem that simply using a randomized run order (such as 8, 5, 1, 6, 3, 4, 2, 7 for a 2^3 design) would be sufficient, but failure to reset creates problems for inferences, such as F-tests. Problems also result when the level of a hard-to-change factor is changed only once in an experiment (e.g., half of the runs at the high level followed by half of the runs at the low level). Ju and Lucas (2002) showed that it is preferable to use a split-plot design when there is one hard-to-change factor, but this is not necessarily the case when there are at least two factors.

Of course, one of the objectives of a study will often be to determine the best levels of each factor to use, so of course the frequent changing of factor levels would occur only in the experiment, not in practice. Therefore, it should be stressed to industrial personnel and others who will be involved in conducting experiments that the changing of levels is just for the purpose of experimentation that will provide important feedback. Because bias will be created if randomization is restricted, this should generally be avoided by making necessary changes of factor levels in experimentation, if even doing so is troublesome, but not impossible.

Whereas hard-to-change factors can create problems due to the way that the experiment must be run, combinations of factor levels that are not possible, or at least practical, create a different type of problem. If a full factorial design were to be used, and one treatment combination was not possible, then not using that combination would convert the design from an orthogonal design to a near-orthogonal design. If a fractional factorial design could be used and only one combination was impossible, then a 1/2 fraction might be used with the fraction utilized that did not contain the impossible combination. If more than one combination was impossible, then a 1/4 fraction might be considered, and so on.

12.15   ONE-FACTOR-AT-A-TIME DESIGNS

As illustrated in Section 12.4, varying one factor at a time can produce misleading results when interactions exist. One-factor-at-a-time (OFAT) designs do have value when used as screening designs, however, as screening designs generally do not provide for the estimation of interactions. Furthermore, it is possible to construct OFAT designs that are resolution V and thus permit the estimation of interactions. One particular advantage of these designs is that they have much fewer factor-level changes than do fractional factorial designs, and thus can be good designs when there are hard-to-change factors.

The simplest type of OFAT design is the following, called a standard OFAT design. Assume that there are k factors. The first run has all of the factors at the low level and the last run has all of the factors at the high level (or the reverse). The second through $(k + 1)$ runs have $(k - 1)$ factors at the low (high) level and the other factor at the high (low) level, with the next k runs being the complete reversal of these k runs. The last $(k - 1)(k - 2)/2$ runs are obtained by changing the factors in pairs, starting from the point of all of the factors being at the low level and using all pairs (i, j) for which $1 \leq i < j \leq k - 1$.

Although the individual two-factor interactions are not estimable with this design, sums of interaction effects are estimable, in addition to the main effects. Thus, if a significant interaction exists, the group that it is in may stand out.

Strict OFAT designs have the minimum number of factor-level changes. See Qu and Wu (2005) for details.

Although these designs are appealing in the presence of hard-to-change factors, the designs are not orthogonal—a trait they share with supersaturated designs, discussed in Section 12.13. This doesn't automatically render the designs of secondary importance, however, because what is lost in orthogonality is offset, at least partially, by having a smaller number of runs. [Just having a small number of runs doesn't automatically make the design worthy of consideration, however. Designs developed by Dorian Shainin have a very small number of runs relative to the number of factors, but Ledolter and Swersey (1997) showed that these designs are not worthy of consideration.]

The extent of nonorthogonality in OFAT designs is usually small, however, so these designs should be given consideration.

12.16 MULTIPLE RESPONSES

Most practical problems have more than one response variable that is of interest. Despite this, there is essentially no mention of designs for multiple responses in engineering statistics books, and very little mention in books on design of experiments.

Analyzing data with multiple responses is more complicated than the analysis of data with a single response variable due to the fact that it is more difficult to analyze multivariate data than univariate data. A frequent objective when multiple responses are used is to try to determine settings for the factors that will produce values for the response variables that are optimal in some sense. Unfortunately, this cannot be done, in general, so multiple response optimization really involves compromises rather than optimization. That is, values for the response variables that are considered to be optimal generally cannot all be attained, due to correlations between the response variables. Accordingly, a "desirability function" approach (Derringer and Suich, 1980) is often used.

Unfortunately, the complexity of the multiresponse optimization problem has resulted in some rather unrealistic assumptions being made in the literature. One such assumption is that all response variables are functions of the same independent variables, and additionally that the model for each response variable has the same general form. It is very unlikely that either of these assumptions will be met.

Appropriate software must be used for multiple response optimization. The MROPTI-MIZER in Minitab can be used for optimization, based on the desirability function, for one or more response variables, but it is preferable to use this capability in menu mode (starting from "DOE") rather than in command mode. JMP can also be used for multiple response optimization.

Certain aspects of the multiresponse optimization problem remain unsolved. The reader is referred to Section 6.6 of Myers and Montgomery (1995) or a later edition of that book for additional information on the subject.

12.17 TAGUCHI METHODS OF DESIGN

In the 1980s and 1990s, engineers and other industrial personnel in the United States and United Kingdom, in particular, in their quest for quality improvement moved toward what is generally referred to as "Taguchi methods of design." The book by Taguchi (1987) contains the essentials of the Taguchi approach. Since Taguchi and his colleagues are entrepreneurs, they made a major effort to market this design approach, with engineers being part of their target market. The result was that many engineers and other industrial personnel soon began to equate experimental design with Taguchi, and some writers even gave Taguchi credit for inventing design of experiments. This occurred despite the fact that only a few of the designs attributed to Taguchi were actually invented by Taguchi. In particular, some of the orthogonal arrays presented by Taguchi are actually Plackett-Burman designs, which as stated previously are two-level designs with the runs not being a power of 2 (e.g., 12, 20, and 24). Certain other orthogonal arrays presented by Taguchi were equivalent to suboptimal fractional factorial designs (i.e., the design resolution was less than what it could have been with a fractional factorial design with the same number of design points). This is illustrated with an example from the engineering literature in Exercise 12.35. Other Taguchi designs are equivalent to fractional factorial designs constructed in the usual (optimal) manner, as the reader is asked to show in Exercise 12.75.

In general, there was much false information promulgated. It is important to note that the "Taguchi movement" was not all bad, however, as there were some positive ramifications. Most notable among them was the realization of the need to identify factors that contribute to variability. Heretofore, (classical) designs were for determining the magnitude of main effects and interaction effects on the mean of the response variable, not on the variance of the response variable.

This resulted in the consideration of designs that incorporated "noise factors," which could be controlled during an experiment, such as a laboratory experiment, but could not otherwise be controlled. As has been discussed in the literature, many of the successful applications of the Taguchi design approach have been in industries characterized by high-volume, low-cost manufacturing. In such industries, designs with a larger number of points may not be a problem. Therefore, in those cases, designs that would otherwise be declared inefficient, such as Taguchi's product arrays, might not entail extra cost that would be prohibitive. In other applications, however, it is important that efficient designs be used, as well as designs that are easy to understand and to use. Engineers who have been trained in the use of Taguchi design methods have been trained to follow certain prescribed steps. The construction of experimental designs and analysis of the resulting data cannot be regimented, however, and as with other areas of statistics, the use of experimental designs and analysis of data obtained from designed experiments is very much an art.

12.18   MULTI-VARI CHART

A multi-vari chart is a graphical tool that is in keeping with Taguchi's idea of looking at factors that affect variation, as the chart shows variation over one factor or multiple factors—hence the name of the chart. It can be used in conjunction with design of experiments or with passive observation of data.

Unfortunately, there is much misinformation in the literature and elsewhere about multi-vari charts. For example, MINITAB calls them "Shainin charts," which is understandable since Dorian Shainin is most frequently associated with the chart, but he did not develop the chart. MINITAB plots means in its multi-vari chart routine, and although this does show the variability of the means, it is not the way the chart is generally viewed. The chart can be correctly constructed using JMP, as will be illustrated shortly.

Such confusion is understandable since there are not very many papers on multi-vari charts in the literature, and the general belief is that the tool has been underutilized. Indeed, the title of the Zaciewski and Nemeth (1995) paper seems to say it all: "The Multi-Vari Chart: An Underutilized Quality Tool."

Assume that we have a 2^3 design with four replications. One simple way to display the variability for each treatment combination would be to plot the largest observation and the smallest observation for each treatment combination and then connect these points, thus creating for each treatment combination what Juran (1974) termed a "basic vertical line." Graphical displays of stock market prices include connecting the high and low points of a stock price on each day and Juran may have derived the method from this application. In any event, the use of the general idea of a multi-vari chart seems to predate Seder (1950), who is generally credited with its invention, and since the idea of connecting extreme points with a vertical line is a very simple method, its use may go back much earlier than is known.

There is no one way that a multi-vari chart must be constructed, so there is room for creativity. If the chart were used as stated for the replicated 2^3 design, it would provide a visual check on whether the constant variance assumption appeared to be at least approximately met. If the engineering tolerances were displayed on such a chart, certain treatment combinations that could pose problems relative to the tolerances might then be avoided.

A multi-vari chart is available in JMP and Figure 12.8 is the chart for the replicated 2^3 experiment mentioned in Section 12.5.2. These are leakage data and the magnitude of the variability for two of the treatment combinations dwarfs the variability for the other six treatment combinations. Of course, ANOVA could not be applied to these data because of the differences in variability, and consequently, the authors were forced to use a different approach.

Such a chart might seem to be similar to an ANOM display since vertical lines are used, but with an ANOM display means are plotted and there is no intent to display variability.

For additional reading on multi-vari charts, see de Mast, Roes, and Does (2001).

12.19   DESIGN OF EXPERIMENTS FOR BINARY DATA

Binary data, such as go/no go data, occur frequently in industry. Since "no go" is undesirable (such as "no, the product cannot be shipped yet"), it is certainly desirable to ascertain the factors that affect the outcome. Thus, there is a need for designed experiments with a binary response variable.

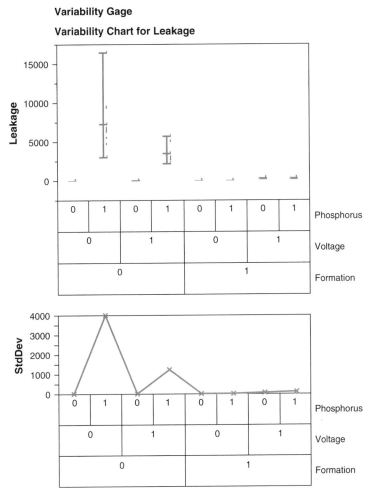

Figure 12.8 Multi-vari chart.

12.20 EVOLUTIONARY OPERATION (EVOP)

Despite the words of Box (1966) that were repeated in the first paragraph of this chapter, management is often reluctant to perturb processes that are performing reasonably well. It isn't necessary to make major changes in the settings of process variables in order to glean important information about processes, however, as data resulting from small changes in the variables can also be quite helpful.

Such small changes can be implemented in the context of design of experiments by using Evolutionary Operation (EVOP). The latter can be performed using either factorial designs or simplex EVOP, in which there is quickly a move away from an unfavorable condition, although a move into an unfavorable region can also occur since movements are determined by a single observation, unlike the use of factorial designs in EVOP, which has been termed Box-EVOP.

Although intended by Box (1957) to be a pencil-and-paper tool that can be used by plant personnel, software for EVOP is available, as developed by some small software companies. See Box and Draper (1969) and Ryan (2000, Chapter 15) for additional details on EVOP.

12.21 MEASUREMENT ERROR

The validity of the analyses given in this chapter hinge on good measurements of the response variable, in particular, being taken, as measurement problems can increase the variability of the response and this can drown out what would otherwise be significant effects. In fact, measurement error is often a significant problem, as indicated by the quote from Bisgaard (1999) at the beginning of Chapter 13. Extensive study of this issue by statisticians has occurred only in the past ten to fifteen years, however.

Because this subject is so important, a separate chapter is devoted to it; specifically, Chapter 13.

12.22 ANALYSIS OF COVARIANCE

No chapter on experimental design and design of experiments would be complete without at least brief mention of analysis of covariance. Consider the following experiment: a one-factor ANOVA is to be performed to determine if there is any difference in three types of feeds on the weight gain of pigs during a specified period of time. Thirty pigs are to be used in the experiment but unfortunately the pigs differ noticeably in weight. Therefore, any difference in the feeds could be confounded with different amounts of weight gain that is strongly related to different initial weights of the pigs. So the different initial weights of the pigs (with weight of course a quantitative variable) should be adjusted for in trying to determine if there are differences among the feeds. This adjustment can be handled with analysis of covariance, with initial weight in this example being the single covariate. Analysis of covariance can be performed in MINITAB with the ANCOVA command. See Snedecor and Cochran (1980, Chapter 18) for a discussion of how the adjustment is performed. The paper by Silknitter, Wiskowski, and Montgomery (1999) is a useful tutorial on analysis of covariance.

12.23 SUMMARY OF MINITAB AND DESIGN-EXPERT® CAPABILITIES FOR DESIGN OF EXPERIMENTS

Although it is important to see how effect estimates are obtained, just as in multiple regression one should use software for computations. All major statistical software packages have extensive capability for constructing experimental designs and analyzing the resultant data, and there are many specialized software products that are exclusively for design of experiments, such as Design-Expert®.

MINITAB can also be used to construct experimental designs for a single factor or for many factors and to analyze data from experiments. Of course, design construction is more involved when there are multiple factors, especially when fractional factorial designs are constructed. Most of the designs discussed in this chapter can be constructed with MINITAB, although designs with blocking and split-plot designs must be analyzed somewhat indirectly. Information on the use of MINITAB for experimental designs is given at the textbook website, in addition to information on downloadable macros.

Design-Expert$^{\circledR}$ and the introductory version of it (Design-Ease$^{\circledR}$) are software devoted exclusively to design of experiments. The former provides capability for almost all of the design methods presented in this chapter, as well as multiple response optimization, with one notable exception being that ANOM is not included.

12.23.1 Other Software for Design of Experiments

In addition to the software listed and illustrated in this chapter, various other software is available, including JMP from SAS Institute, Inc.

12.24 TRAINING FOR EXPERIMENTAL DESIGN USE

The statistical design of experiments is very important in engineering, but unfortunately it has not been used to anywhere near the fullest extent possible. As the late statistician Bill Hunter often said, one should "do statistics," and this certainly applies to engineering students. An excellent way to learn experimental design, especially some of the nuances that don't come through strongly in textbooks, is to conduct actual, simple experiments that do not require much in the way of materials or time. The primary motivation for students conducting such simple experiments was the paper by Hunter (1977), and there have been various other articles on the subject written since then, especially articles appearing in the *Journal of Statistics Education*, an online publication (see `http://www.amstat.org/publications/jse`).

One example of a hands-on exercise that I have used in training industrial personnel in experimental design is commonly referred to as the catapult experiment, with a small catapult used to catapult a ping-pong ball, which is measured for distance traveled. Factors that can be varied include the rubber band type, position of the arm, and type of ball used. Data are generated from the experiment, which would then be analyzed using the methods of this chapter. Of course, an important part of the exercise is trying to figure out why certain effects (possibly including interactions) are significant. A good description of an actual experiment of this type is given in the paper "Training for Design of Experiments Using a Catapult" by J. Anthony (*Quality and Reliability Engineering International*, **18**, 29–35, 2002). Anthony states the following: "The engineers in the company felt that the experiments were useful in terms of formulating the problem, identifying the key control factors, determining the ranges of factor settings, selecting the experimental layout, assigning the control factors to the design matrix, conducting the experiment, and analyzing and interpreting the results of the experiment."

12.25 SUMMARY

The statistical design of experiments is very useful in engineering. The basic concepts and designs were presented in this chapter; readers are referred to books on experimental design, of which there are dozens, for additional information.

Designed experiments will not automatically be beneficial as appropriate designs must be thoughtfully selected and the appropriate experiments carefully conducted. Care must be exercised and even well-planned experiments can be undermined by out-of-control processes and measurement error. Consequently, process control methods, such as those given in Chapter 11, should always be used in conjunction with designed experiments, and measurement systems should also be checked. Methods for doing the latter are given in Chapter 13.

Well-planned experiments can also be undermined by missing data, and a decision must be made as to whether or not the missing data should be imputed, for which various methods have been proposed. This becomes a particularly vexing problem when the missing data are on treatment combinations that would seem to be possibly the best in terms of maximizing or minimizing the response variable (whatever the objective is). See Cooley, Franklin, and Elrod (1999–2000) for a discussion of this for an actual experiment.

Some of the material presented in this chapter (e.g., Section 12.14) has not previously been presented in engineering statistics textbooks, nor has it been presented in books on design of experiments. Unfortunately, some very important aspects of experimental design have for the most part escaped attention, but they must be considered when experiments are conducted. Engineering students, practicing engineers, and others should be aware of these nuances.

APPENDIX A. COMPUTING FORMULAS

Although the emphasis in this chapter and in the book is on using software for computation, some readers may wish also to perform some computations so as to add to their insight and obtain a feel for how the numbers are obtained. Hand computation for more than two factors in ANOVA would be cumbersome, so the computing formulas will be given for certain sections in the chapter up through two factors.

For Section 12.2.1.2

The computer output was shown for the first of the two examples in Section 12.2.1. It would not be difficult to perform that computation by hand (calculator) and computing formulas are given herein for accomplishing that.

In order to reproduce the first part of the computer output, we need to compute the sum of squares for between (the columns), within (the columns), and the total sum of squares. The latter is always computed the same way, regardless of the experimental design or the number of factors. In words, first all of the observations are squared and summed. From this sum is subtracted the square of the sum of all of the observations divided by the number of observations.

In terms of a computing formula, let Y_{ij} denote the jth observation in the ith column. Then the total sum of squares, denoted by SS_{total}, is given by

$$SS_{\text{total}} = \sum_{i=1}^{3} \sum_{j=1}^{7} Y_{ij}^2 - \left(\sum_{i=1}^{3} \sum_{j=1}^{7} Y_{ij} \right)^2 \bigg/ 21$$

$$= 339.3712 - (84.42)^2/21$$
$$= 339.3712 - 339.3684$$
$$= 0.0028$$

for this example, with of course 21 being the total number of observations, 3 the number of columns, and 7 the number of observations within each column.

The approach to obtaining the between (columns) sum of squares can be computed as

$$
SS_{\text{between}} = \sum_{i=1}^{3} \frac{T_i^2}{7} - \frac{\left(\sum_{i=1}^{3} \sum_{j=1}^{7} Y_{ij} \right)^2}{21} \tag{A12.1}
$$

$$
= \frac{787.925 + 791.860 + 795.804}{7} - \frac{(84.42)^2}{21}
$$

$$
= 339.3698 - 339.3684
$$

$$
= 0.0014
$$

where T_i represents the total of the observations for the ith level of temperature, and, as indicated, 7 represents the number of observations for the ith level. For ANOVA formulas in general, whenever a number is squared, it is always divided by the number of observations that were summed to produce that number.

The within (columns) sum of squares, also called "error," is generally obtained by subtraction. That is, $SS_{\text{total}} - SS_{\text{between}} = SS_{\text{within}} = 0.0028 - 0.0014 = 0.0014$ for this example. Thus, we have obtained the values for the sums of squares that were given in the computer output in Section 12.2.1.2.

For Section 12.3.1

The analysis of data from a randomized block design is similar to the analysis of data from any design with two factors. The value for $SS_{\text{treatments}}$ is obtained in the same manner as the value for SS_{between} with the T_i in Eq. (A12.1) representing, as in that equation, the treatment totals. The value of SS_{blocks} is also obtained in essentially the same way if we now let T_i in that equation represent the block totals. SS_{total} is obtained in the same way as for any design and SS_{residual} is obtained by subtraction. The degrees of freedom for Treatments and Blocks are the number of treatments minus one and the number of blocks minus one, respectively. The degrees of freedom for Total is always the number of observations minus one, and the degrees of freedom for Error is of course obtained by subtraction.

For Other Sections

It is not particularly instructive or enlightening to perform hand computations when there are more than two factors in a designed experiment. In the case of a Latin square design, the computations are not difficult, however, and are just an extension of the computations for a randomized block design as there is a second blocking factor and the sum of squares for that blocking factor is computed the same way as the computations for the first blocking factor.

APPENDIX B. RELATIONSHIP BETWEEN EFFECT ESTIMATES AND REGRESSION COEFFICIENTS

It was stated in Section 12.5.1.1 that the regression coefficients are half the effect estimates (for two-level designs). This can be explained as follows. Define a matrix $\mathbf{X}$ such that the first column is a column of ones (which provides for estimation of b_0) and the other columns contain the coded values (i.e., $+1$ and -1) of the effects to be estimated. Since all but the first column sum to zero and pairwise dot products of all columns are zero, $\mathbf{X'X}$ is a diagonal matrix with n as each element on the main diagonal. Thus, $(\mathbf{X'X})^{-1}$ is a diagonal matrix with each diagonal element $1/n$. The regression coefficients are obtained from $(\mathbf{X'X})^{-1}\mathbf{X'Y}$ with $\mathbf{Y}$ a vector that contains the response values. The vector obtained from $\mathbf{X'Y}$ then has the sum of the response values as the first element and every other element of the form $\sum_{j=1}^{n}(X_{ij}Y_j)/n$ for each effect X_i that is being estimated. Since each X_{ij} is either $+1$ or -1, each sum is thus of the form $\left(\sum y_{(+)} - \sum y_{(-)}\right)/n$, with $y_{(+)}$ denoting a value of Y for which X_{ij} is positive, and similarly $y_{(-)}$ denoting a value of Y for which X_{ij} is negative. Note that the divisor is n instead of $n/2$, with the latter being the number of terms in each of these last two sums. Thus, this simplifies to $\frac{1}{2}(\overline{y}_{(+)} - \overline{y}_{(-)}) = \frac{1}{2}$ (effect estimate).

REFERENCES

Bisgaard, S. (1999). " Quality Quandries". *Quality Engineering*, **11**(4), 645–649.

Bishop, T. A. and E. J. Dudewicz (1981). Heteroscedastic ANOVA. *Sankhya B*, **43**, 40–57.

Box, G. E. P. (1957). Evolutionary operation: A method for increasing industrial productivity. *Applied Statistics*, **6**(2), 81–101.

Box, G. E. P. (1966). Use and abuse of regression. *Technometrics*, **8**, 625–629.

Box, G. E. P. (1990). George's Column. *Quality Engineering*, **2**(3), 365–369.

Box, G. E. P. (1992). George's Column. *Quality Engineering*, **5**(2), 321–330.

Box G. E. P. and J. S. Hunter (1961a). The 2^{k-p} fractional factorial designs, Part I. *Technometrics*, **3**, 311–351.

Box G. E. P. and J. S. Hunter (1961b). The 2^{k-p} fractional factorial designs, Part II. *Technometrics*, **3**, 449–458.

Box, G. E. P. and S. Jones (2000–2001). Split plots for robust product and process experimentation. *Quality Engineering*, **13**(1), 127–134.

Box, G. E. P. and N. R. Draper (1969). *Evolutionary Operation*. New York: Wiley.

Box, G. E. P., S. Bisgaard, and C. Fung (1990). *Designing Industrial Experiments*. Madison, WI: BBBF Books.

Brown, M. B. and A. B. Forsythe (1974). Robust tests for the equality of variances. *Journal of the American Statistical Association*, **69**, 364–367.

Clark, J. B. (1999). Response surface modeling for resistance welding. *Annual Quality Congress Proceedings*, American Society for Quality, Milwaukee, WI.

Cooley, B. J., L. A. Franklin, and G. Elrod (1999–2000). A messy, but instructive, case study in design of experiments. *Quality Engineering*, **12**(2), 211–216.

Daniel, C. (1976). *Applications of Statistics to Industrial Experimentation*. New York: Wiley.

de Mast, J. , K. C. B. Roes, and R. J. M. M. Does (2001). The multi-vari chart: a systematic approach. *Quality Engineering*, **13**(3), 437–447.

Dean, A. and D. Voss (1999). *Design and Analysis of Experiments*. New York: Springer.

Derringer, G. and R. Suich (1980). Simultaneous optimization of several response variables. *Journal of Quality Technology*, **25**, 199–204.

Finney D. J. (1945). Fractiorial replication of factorial arrangements. *Annals of Eugenics*, **12**, 291–301.

Ganju, J. and J. M. Lucas (1997). Bias in test statistics when restrictions in randomization are caused by factors. *Communications in Statistics—Theory and Methods*, **26**(1), 47–63.

Hunter, W. G. (1977). Some ideas about teaching design of experiments with 2^5 examples of experiments conducted by students. *The American Statistician*, **31**(1), 12–17.

Joseph, V. R. (2000). Experimental sequence: A decision strategy. *Quality Engineering*, **13**(3), 387–393.

Ju, H. L. and J. M. Lucas (2002). L^k factorial experiments with hard-to-change and easy-to-change factors. *Journal of Quality Technology*, **34**(4), 411–421.

Juran, J. M. (1974). Quality improvement. Chapter 16 in J. M. Juran, F. M. Gryna, Jr., and R. S. Bingham, Jr., eds. *Quality Control Handbook*. New York: McGraw-Hill.

Ledolter, J. and A. Swersey (1997). Dorian Shainin's variable search procedure: A critical assessment. *Journal of Quality Technology*, **29**, 237–247.

Lenth, R. V. (1989). Quick and easy analysis of unreplicated factorials. *Technometrics*, **31**, 469–473.

Lenth, R. V. (2001). Review of *A First Course in the Design of Experiments: A Linear Models Approach* by Weber and Skillings. *The American Statistician*, **55**, 370.

Levene, H. (1960). Robust tests for the equality of variances. In *Contributions to Probability and Statistics: Essays in Honor of Harold Hotelling* (I. Olkin et al., eds.) Stanford, CA: Stanford University Press, pp. 278–292.

Lin, D. K. J. (1995). Generating systematic supersaturated designs. *Technometrics*, **37**(2), 213–225.

Morrow, M. C., T. Kuczek, and M. L. Abate (1998). Designing an experiment to obtain a target value in the chemical process industry. In *Statistical Case Studies: A Collaboration Between Academe and Industry* (R. Peck, L. D. Haugh, and A. Goodman, eds.). Philadelphia: Society for Industrial and Applied Mathematics and Alexandria, VA: American Statistical Association.

Myers, R. H. and D. C. Montgomery (1995). *Response Surface Methodology: Process and Product Optimization Using Designed Experiments*. New York: Wiley.

Nelson, P. R. and E. J. Dudewicz (2002). Exact analysis of means with unequal variances. *Technometrics*, **44**(2), 152–160.

Qu, X. and C. F. J. Wu (2005). One-factor-at-a-time designs of resolution V. *Journal of Statistical Planning and Inference*, **131**, 407–416.

Ryan, T. P. (2000). *Statistical Methods for Quality Improvement*, 2nd ed. New York: Wiley.

Ryan, T. P. (2007). *Modern Experimental Design*. Hoboken, NJ: Wiley.

Seder, L. A. (1950). Diagnosis with diagrams—Part I. *Industrial Quality Control*, **6**(4), 11–19.

Silknitter, K. O., J. W. Wisnowski, and D. C. Montgomery (1999). The analysis of covariance: A useful technique for analysing quality improvement experiments. *Quality and Reliability Engineering International*, **15**, 303–316.

Snedecor, G. W. and W. G. Cochran (1980). *Statistical Methods*, 7th ed. Ames, IA: Iowa State University Press.

Taguchi, G. (1987). *System of Experimental Design*, Volumes 1 and 2. White Plains, NY: Unipub/Kraus International Publications.

Wludyka, P. S. and P. R. Nelson (1997). An analysis of means type test for variances from normal populations. *Technometrics*, **39**, 274–285.

Young, J. C. (1994). Why the results of designed experiments are often disappointing: A case study illustrating the importance of blocking, replication & randomization. IIQP Research Report RR-94-10, University of Waterloo.

Zaciewski, R. D. and L. Nemeth (1995). The multi-vari chart: An underutilized quality tool. *Quality Progress*, **28**(10), 81–83.

EXERCISES

12.1. Use the following applet, due to Christine Anderson-Cook, to do a one-factor two-level test (left hand versus right hand): `http://www.amstat.org/publi cations/jse/java/v9n1/anderson-cook/BadExpDesignApplet. html`.

12.2. Assume that a one-factor design is to be used and there are two levels of the factor.
   (a) What is the smallest possible value of the F-statistic and when will that occur?
   (b) Construct an example with 10 observations at each level that will produce this minimum value.

12.3. Assume that data from a 2^3 design have been analyzed, and one or more of the interactions are significant. What action should be taken in investigating main effects for factors that comprise those interactions?

12.4. Six similar (but not identical) machines are used in a production process. Each machine has a head that is not interchangeable between machines, although there are three different types of heads that can be used on a given machine. A study is to be performed to analyze process variability by studying the six machines and three heads. Could the data be analyzed as a 3×6 factorial design? Why or why not?

12.5. Explain why residual plots should be used with data from designed experiments.

12.6. How many degrees of freedom will be available for estimating σ^2 if a 2^3 design is run with four replicates?

12.7. Oftentimes the design of an experiment is lost and all that is available are the treatment combinations and the values of the response variable. Assume that four factors, each at two levels, are studied with the design points given by the following treatment combinations: (1), ab, bc, abd, acd, bcd, d, and ac.
   (a) What is the defining contrast?
   (b) Could the design be improved using the same number of design points?
   (c) In particular, which three main effects are confounded with two-factor interactions?

12.8. Assume that a 2^{3-1} design has been run using the treatment combinations (1), ab, ac, and bc. What was the defining contrast? What would you recommend if the treatment combination bc is an impossible combination of factor levels?

12.9. Assume that ANOVA is applied to a single factor with four levels and the F-test is significant. What graphical aid could be used to determine which groups differ in terms of their means?

12.10. The following data are from an experiment described by B. N. Gawande and A. Y. Patkar in their paper "Application of Factorial Designs for Optimization of Cyclodextrin Glycosyltransferase Production from *Klebsiella pneumoniae pneumoniae* AS-22" [*Biotechnology and Bioengineering*, **64**(2), 168–173, 1999]. A 2^{5-1} design was used to investigate the effect of dextrin, peptone, yeast extract, ammonium dihydrogen orthophosphate, and magnesium sulfate on enzyme production. These five factors are denoted by A, B, C, D, and E, respectively. Analyze the data.

A	B	C	D	E	Response
−	−	−	−	−	3.97
−	−	−	+	+	5.99
−	−	+	−	+	4.13
−	−	+	+	−	5.59
−	+	−	−	+	5.18
−	+	−	+	−	6.47
−	+	+	−	−	5.12
−	+	+	+	+	6.53
+	−	−	−	−	5.39
+	−	−	+	+	5.25
+	−	+	−	+	5.39
+	−	+	+	−	6.06
+	+	−	−	+	4.98
+	+	−	+	−	6.74
+	+	+	−	−	5.66
+	+	+	+	+	8.42

12.11. An experimenter decides to use a 2^{6-2} design and elects to confound ABD and CEF in constructing the fraction.

(a) Determine what effect(s) would be aliased with the two-factor interaction AB.
(b) Is there a better choice for the two defining contrasts?

12.12. Analysis of Means is applied to data from a 2^3 design with three replicates. Assume that $SS_{residual} = 22$ (from ANOVA) and that the B main effect was found to be significant using $\alpha = .05$. The length of the line segment corresponding to B in the ANOM display must have exceeded _____.

12.13. Assume that a 2^4 design is to be run. Explain to the person who will conduct the experiment what the treatment combination ab means.

12.14. Given the following results when a 2^2 design is used, what is the estimate of the A effect?

	B_{low}	B_{high}
A_{low}	25	30
A_{high}	40	20

12.15. How many effects can be estimated when an unreplicated 2^3 design is used, and what are those effects? (List them.)

12.16. Fill in the blanks in the following output for data from a completely randomized, one-factor design with three levels.

One-Way Analysis of Variance

Analysis of Variance for Y

Source	DF	SS	MS	F	P
A	—	—	192.3	—	0.005
Error	12	—	22.9		
Total	—	658.9			

12.17. Assume a one-factor experiment with the factor being random. Explain, *in words*, the null hypothesis that is tested.

12.18. Consider a 2^6 design and a 2^{6-2} design.
  (a) Explain what is gained by using the latter relative to the former. Is anything lost? In particular, what would you say if an experimenter believes that some two-factor interactions may be significant but doesn't have any a priori information on which of these interactions might be significant? (Assume that $I = ABCD = CDEF$.)
  (b) Would this be a good design if the experimenter felt that the only two-factor interactions that could be significant were AB and EF? Explain.

12.19. When Analysis of Means is applied to certain full or fractional factorial two-level designs, the UDL will be above the midline on the chart by an amount equal to $t_{\alpha/2, v} \, s/4$. Name one of the designs for which this is true.

12.20. Analysis of Means is to be applied to data from a 2^3 design with three replicates. The estimate of σ using s thus has 16 degrees of freedom. If any main effect or interaction effect is to be significant using $\alpha = .01$, the *difference* between the two means for any of the effects must exceed _____ when $s = 4$.

12.21. Assume that Analysis of Means is being applied to data from a one-way classification (i.e., one factor) with five levels. State the general form of one of the five hypothesis tests that would be performed.

12.22. Explain why the decision lines used in Analysis of Means could not be applied to data obtained in real time (i.e., what is called Phase II for control charts).

12.23. The following statement is paraphrased from an Internet message board: "I have two factors, each at two levels. When I multiply the factor levels together to create the levels for the interaction, the interaction levels are $+1$ and -1, but when I look at the combinations of the factors in the original units I see four combinations. How can $+1$ and -1 each represent two different combinations of factor levels?" Explain the faulty reasoning in these statements.

12.24. Construct an interaction plot of the data given in Example 12.1 to show that there is no evidence of an interaction between blocks and fertilizers.

12.25. The data given in ENGSTAT1224.MTW are from a statement by Texaco, Inc. to the Air and Pollution Subcommittee of the Senate Public Works Committee on June 26, 1973. It was stated that an automobile filter developed by Associated Octel Company was effective in reducing pollution, but questions were raised in regard to various matters, including noise. The president of Texaco, John McKinley, asserted that the Octel filter was at least equal to standard silencers in terms of silencing properties. The data in the file are in the form of a 3×2^2 design with SIZE of the vehicle being 1 = small, 2 = medium, and 3 = large; TYPE being 1= standard silencer and 2 = Octel filter, and SIDE being 1= right side and 2 = left side of the car. The response variable is NOISE.

(a) Are these factors fixed or random, or can that be determined? (Remember that the classification of a factor as fixed or random affects the way the analysis is performed when there is more than one factor.)

(b) Since a 3×2^2 design (a mixed factorial) was not covered explicitly in the chapter, initially ignore the SIZE factor and analyze the data as if a replicated 2^2 design had been used, bearing in mind that the objective of the experiment is to see if the Octel filter is better than the standard filter.

(c) Now analyze the data using all three factors, using appropriate software such as the general linear model (GLM) capability in MINITAB, and compare the results. In particular, why is one particular effect significant in this analysis that was not significant when the SIZE factor was ignored? (*Hint*: Look at the R^2 values ignoring and not ignoring SIZE.)

(d) Is there an interaction that complicates the analysis? If so, what approach would you take to explain the comparison between the standard filter and the Octel filter?

12.26. Critique the following statement: "I'm not worried about the effect of large interactions because if I have such an interaction, I will simply include in the model the factors that comprise large interactions."

12.27. If you are proficient in writing MINITAB macros, write a macro that could be used for a 2^3 design such that the main effects and interaction effects are all shown on the same display, with the line that connects the points for each interaction computed as discussed in Section 12.7. Then use your macro to construct an ANOM display for the data in Section 9.3, using the center-point runs to estimate σ_ϵ.

12.28. Conduct an actual experiment using a 2^3 design [perhaps doing one of the experiments described in Hunter (1977)] and do a thorough analysis of the data. If data points are both inexpensive and easy to obtain, conduct a replicated experiment.

12.29. When should a randomized block design be used? If an experiment were run and the block totals were very close, what would that suggest?

12.30. (Harder problem) Construct a standard OFAT design for four factors and compute the correlations between the columns of the design. How many effects can be estimated with this design? Compare this design to a 2^4 design. What is lost and what is gained with the OFAT relative to the 2^4? Would you recommend that the OFAT design be used? Why or why not?

12.31. Construct a standard OFAT design for five factors and compute the correlations between the columns of the design. Do you consider the latter to be unacceptably large? Why or why not?

12.32. Explain how a large three-factor interaction can interfere with the interpretation of main effects in a 2^3 design.

12.33. Usher and Srinivasan [*Quality Engineering*, **13**(2), 161–168, 2000–2001] described a 2^4 experiment for which there were six observations per treatment combination, but the data were not analyzed as having come from a replicated experiment. Specifically, the authors stated "Note that these six observations do not represent true replicates in the usual sense, because they were not manufactured on unique runs of the process." Consequently, the average of the six values at each treatment combination was used as the response value, and the data were thus analyzed as if they had come from an unreplicated experiment. Explain how analyzing the data as if the data had come from a replicated experiment could produce erroneous results.

12.34. Explain a statement such as "the best time to design an experiment is after it has just been run."

12.35. A. Gupta [*Quality Engineering*, **10**(2), 347–350, 1997–98] presented a case study involving antibiotic suspension products, with "separated clear volume" (the smaller the better) being the response variable. The objective was to determine the best level of each of five two-level factors so as to minimize the response variable, with the selection of a particular level of a factor being crucial only if the factor is significant. An L_{16} orthogonal array [see part (b)] was used, and that design and the accompanying response values are as follows:

Response	A	B	C	D	E
47	8	50	0.2	0.4	Usual
42	8	50	0.4	0.4	Modified
50	8	60	0.4	0.4	Usual
51	8	60	0.2	0.4	Modified
40	16	50	0.4	0.4	Usual
44	16	50	0.2	0.4	Modified
46	16	60	0.2	0.4	Usual
40	16	60	0.4	0.4	Modified
28	8	50	0.2	0.6	Usual
23	8	50	0.4	0.6	Modified
30	8	60	0.4	0.6	Usual
38	8	60	0.2	0.6	Modified
26	16	50	0.4	0.6	Usual
31	16	50	0.2	0.6	Modified
32	16	60	0.2	0.6	Usual
33	16	60	0.4	0.6	Modified

(a) Should the data be analyzed in the original units of the factors (with of course some numerical designation for the two levels of factor E)? Why or why not?

(b) Show that this L_{16} array is equivalent to a suboptimal 2^{5-1} design. [*Hint:* The interaction that is confounded with the mean can be determined using MINITAB, for example, by trying to estimate all the effects. That is, by using a command such as FFACT C1= (C2-C6) 5, with the response values being in C1 and the factors being in C2-C6.]

(c) Explain the consequences of using this L_{16} array rather than the 2^{5-1} design with maximal resolution in terms of the estimation of the two-factor interactions.

(d) Estimate as many effects as you can with this design and determine the significant effects. With the second level (60) of the second factor considered to be preferable for nonstatistical reasons, what combination of factor levels appears to be best? Is it necessary to qualify your answer in any way since the L_{16} design is a resolution IV design whereas a 16-point resolution V design could have been constructed?

12.36. (a) Analyze the following data for two groups using both an exact t-test and an F-test. Show that $F = t^2$.

1	2
1.43	1.48
1.62	1.66
1.35	1.36
1.46	1.49
1.49	1.53
1.54	1.58
1.60	1.64
1.57	1.62

(b) Do the samples seem to be independent? Comment.

(c) Construct a confidence interval for the difference between the two means (see Section 6.1.2). Will the confidence interval result (regarding significance or not) have to agree with the F-test? Explain.

12.37. The following interaction profile shows the results of an unreplicated 2^2 design:

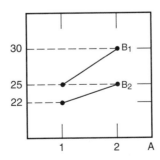

Estimate the A effect.

12.38. Given below is part of a dataset (all of the predictor variables but only one of the response variables) originally given by Czitrom, Sniegowski, and Haugh (in *Statistical Case Studies: A Collaboration Between Academe and Industry*, 1998).

Selectivity	Bulk Gas Flow (sccm)	CF_4 Flow (sccm)	Power (Watts)
10.93	60	5	550
19.61	180	5	550
7.17	60	15	550
12.46	180	15	550
10.19	60	5	700
17.50	180	5	700
6.94	60	15	700
11.77	180	15	700
11.61	120	10	625
11.17	120	10	625

The objective of the experiment was to maximize selectivity by using appropriate settings of the three factors that are given above.

(a) How would you describe the design that was used?

(b) Would you expect the experiment to be run using the sequence of treatment combinations as given in the table? Why or why not?

(c) Determine which effects, if any, are significant.

(d) Can we determine the treatment combinations to use to maximize selectivity just from inspection of the table?

12.39. L. B. Hare (*Journal of Quality Technology*, Vol. 20, pp. 36–42, 1988) reported the use of a 2^{5-1} design in an experiment for which the primary goal was to determine the effect of five factors on fill weight variation for a dry mix soup filling process. The largest effect estimate was the main effect estimate of factor E (3.76), and the largest two-factor interaction involving this factor was the BE interaction, for which the effect estimate was 3.24. This was the second largest effect.

(a) How would you recommend that the influence of factor E be reported?

(b) What effect is the E effect confounded with, assuming that the design was constructed in an optimal manner? Should this confounding be of any concern?

(c) Similarly, what effect is the BE interaction confounded with and should that confounding be of any concern?

12.40. Explain why it is important to properly classify each factor as fixed or random when at least two factors are involved in an experiment. The default for statistical software is for fixed factors. Explain the "double whammy" effect of improperly analyzing a two-factor experiment with both factors assumed random when in fact both factors are fixed in the presence of a large AB interaction.

12.41. Assume that Analysis of Variance calculations have been performed for a problem with two levels of a single factor. This would produce results equivalent to an independent sample t-test provided that the alternative hypothesis for the t-test is (choose one): (a) greater than, (b) not equal to, or (c) less than.

12.42. The following data are from an actual experiment and are analyzed in Ryan (2000, Section 13.8). The objective was to investigate the effect of stirring (A), temperature (B), and pressure (C) on the thickening time of a certain type of cement. Analyze the data.

Treatment Combination	Response
(1)	297
a	300
b	106
ab	131
c	177
ac	178
bc	76
abc	109

12.43. Analyze the following data from a 2^2 design with three replicates:

	A Low	A High
B Low	10 12 16	8 10 13
B High	14 12 15	12 15 16

12.44. A 2^{4-1} design is run with $ABCD$ confounded with the difference between the two fractions. Give five of the treatment combinations for one of the two fractions (your choice).

12.45. A 2^2 design with two replicates is run and the following results are obtained. The response totals are: 20 when both factors are at the low level, 30 when both factors are at the high level, 15 when factor A is at the high level and factor B is at the low level, and 18 when factor A is at the low level and factor B is at the high level. Determine the estimate of the A effect.

12.46. Critique the following statement: "The coefficients obtained from use of a 2^2 design may be unreliable because of multicollinearity."

12.47. Consider a 2^2 design. Construct an interaction plot for which the estimate of the interaction effect is zero, being careful to do the appropriate labeling.

12.48. Close-to-orthogonal designs, such as supersaturated designs, are sometimes used. When this is done, it is necessary to consider the correlations between the factors. Explain why these are not true correlations.

12.49. An unreplicated 2^{5-1} design is used and a normal probability plot is subsequently used. If all effects that can be estimated are used for the plot:
   (a) How many points will the plot contain?
   (b) What would you suspect if the plot showed the vast majority of the effects to be significant?

12.50. A 2^3 design is run and the estimate of the *ABC* interaction effect is 4.82. If the response values for four of the treatment combinations are: *b* (2.94), *c* (6.86), *a* (9.12), and *abc* (4.14), what must have been the sum of the other four response values?

12.51. The issue of analyzing data from a designed experiment in coded form versus raw form was covered and illustrated numerically in Section 12.12. Consider a model with two main effects and an interaction term.

    (a) Show algebraically why the main effect model coefficients will be different for the two forms, and thus the *p*-values will differ.

    (b) Use 12, 10, 21, and 28, respectively, and 23, 26, 18, and 14, respectively, as the two sets of response values for the treatment combinations in the example in Section 12.12 with the two sets of values being replicates of the treatment combinations used and show numerically that the *p*-values for the two main effects differ between the raw-form data and the coded-form data.

12.52. Explain how you would design an experiment if you had four factors to examine in the experiment and one of the four was a hard-to-change factor.

12.53. Explain, in words, what is being tested when an experiment has two fixed factors and their possible significance is to be tested with hypothesis tests.

12.54. Would you ever use stepwise regression in trying to determine the effects that are significant in analyzing data from a designed experiment in which the columns of factor settings were orthogonal? If you did so, would the results you obtained likely differ from the results obtained if you looked at the *p*-values for the *F*-statistics as in Section 12.3, for example? If so, which results would you go by? Would your answers be different if a supersaturated design were used (i.e., nonorthogonal columns)? Explain.

12.55. Construct an example for a single (fixed) factor with three levels and five observations per level for which the overall *F*-test shows a significant result but the averages for the three levels are 18.1, 18.2, and 18.3, respectively, and all 15 numbers are different.

12.56. Assume that you have encountered some computer output for a 2^2 design with both factors being random. The sums of squares for each of the two factors are small but the interaction is large. Since both main effects are tested against the interaction effect because the factors are random, what would be the first step you would take as a data analyst before you passed these results on to your boss?

12.57. Sopadang, Cho, and Leonard [*Quality Engineering*, **14**(2), 317–327, 2001] described an experiment in which 15 factors were examined in 16 experimental runs. What was the resolution of the design and for what purpose would you suggest that the design be used?

12.58. Explain why a 2^{5-1} design should not be run with *ACDE* confounded with the difference between the two fractions.

12.59. There are sixteen treatment combinations used when a 2^{5-1} design is run with $I =$ *ABCDE*. Which one of the following treatment combinations is used in error with the other twelve (correct) treatment combinations: a, bcd, ac, bce, ace?

12.60. Assume that ANOM is used for data from a 2^3 design. How would you suggest that the effects be arranged (ordered) in a single ANOM display so as to permit an easy visual comparison of interactions relative to main effects?

12.61. Explain to a person unfamiliar with experimental design what (1) denotes when used to represent a treatment combination and explain why it is used.

12.62. What does it mean to "dealias" effects?

12.63. Explain why a person who runs an experiment using a 2^3 design does not want to see a significant *ABC* interaction effect.

12.64. Assume that a 2^4 design is run but the observation from one of the 16 treatment combinations is lost. What action would you suggest be taken regarding the remaining 15 observations?

12.65. Construct an example of data from a 2^2 design for which the estimate of the *A* effect is 0 and the estimate of the *B* effect is 5.

12.66. Is there a finite upper bound on the amount by which the conditional effects of *A* could differ for data from a 2^2 design? Explain.

12.67. What course of action would you recommend if two interaction effects for a 2^4 design exceeded two of the main effects for factors that comprise the interaction effects?

12.68. Which fractional factorial design will have the highest resolution, a 2^{8-3} design or a 2^{6-1} design, each of which has 32 design points? Explain.

12.69. Explain the condition(s) under which the selection of a 1/4 fraction of a full factorial design should not be chosen randomly.

12.70. Consider a 2^2 design with four replications.
   (a) How many degrees of freedom will the error term have if all effects that can be estimated are in fact estimated?
   (b) Will it be necessary to use a normal probability plot to determine the effects that are significant? Why or why not?
   (c) If the estimate of the conditional effect of factor *A* at the low level of factor *B* is 15 and the estimate of the interaction effect is 30, what is the estimate of the *A* effect?

12.71. To see the importance of maintaining processes in a state of statistical control when experiments are performed, consider the following. An industrial experiment is conducted with temperature set at two levels, 350 °F and 450 °F. Assume that no attempt at process control is made during the experiment and consequently there is a 3-sigma increase in the average conductivity from the original value of 13.2, due to some cause other than the change in temperature, and the increase occurs right when the temperature is changed. Assume that σ is known to be 4.5 and that 20 observations were made at each of the two temperature levels.

(a) What is the expected value of the appropriate test statistic for this scenario and the corresponding expected p-value?

(b) Answer these same two questions if the process went out of control, by the same amount, when the first 10 observations had been made at the second temperature. What argument do these two sets of numbers support?

12.72. After reading a few articles on experimental design, but not really being knowledgeable in the subject, a scientist proclaims that he is going to use a mixed factorial: a 1×2 design. Explain why this is really not a mixed factorial. What type of design is it? Would it be of any practical value if the design were unreplicated? Explain. Would your answer be the same if the design were replicated? Explain.

12.73. An experiment is run using a 2^3 design. The eight response values are as follows:

| | C_{low} | | C_{high} | |
	B_{low}	B_{high}	B_{low}	B_{high}
A_{low}	16	18	22	24
A_{high}	19	14	21	27

(a) Construct the BC interaction plot.

(b) Based solely on this plot, would you recommend that the B and C main effects be reported? Why or why not?

12.74. A scientist wishes to investigate the heat loss for three types of thermal panes but, for reasons of convenience, makes only three observations on each type.

(a) Would you recommend that an ANOVA go forward, or would you suggest that more observations be obtained?

(b) If the analysis does proceed with the present number of observations, what is the smallest value of the F-statistic that would result in rejection of the null hypothesis if the experimenter decides that the p-value must be less than .01?

12.75. If you have access to MINITAB, show that a Taguchi orthogonal array design constructed for seven factors, each at two levels, with eight experimental runs (i.e., the L_8 in Taguchi's notation), is the same design as the 2^{7-4} design produced by MINITAB. Comment.

12.76. L. Huang and J. C. Chen [*Journal of Industrial Technology*, **17**(2), 1–8, 2001; available at http://nait.org/jit/Articles/huang020801.pdf] stated that they used an experimental design for the purpose of developing a regression model "that was capable of predicting the in-process surface roughness of a machined work piece using a turning operation." Read and critique their article, paying particular attention to their experimental design and the multiple regression equations that they obtain [especially Eqs. (4) and (5)]. What would you have done differently, if anything, in terms of the way the experiment was designed and carried out, the results that the authors obtained, and the presentation of those results?

12.77. Assume that there is one quantitative variable that is believed to relate to the response variable and four qualitative variables that are arranged in the form of an experimental design. How would you suggest that the data be analyzed?

12.78. Consider a 2^2 design with three observations per treatment combination. If the AB interaction is very close to zero, what will be the relationship between the conditional effects of factor A?

12.79. Consider the MINITAB output at the beginning of Section 12.9.2. How would you explain to a manager who has never studied experimental design what $B + ACE + CDF + ABDEF$ means?

12.80. If a randomized block design would be useful in your field of engineering or science, give an example. If such a design would not be useful, explain why.

12.81. Consider the article referenced in Exercise 12.76. Can the main effect of the first factor, S, be interpreted unambiguously for that dataset? Explain.

12.82. An experimenter runs a resolution V design, constructs various types of graphs of the data, and notices that two main effects appear to be strongly significant. What must be assumed (about interactions) before any action should be taken regarding the two factors?

12.83. A design with five factors, each at two levels, was run and the alias structure involving factor E was $E = ABCD$. What is the resolution of the design?

12.84. If possible, read the following paper: "The Stability of the Adsorption of Some Anions on Chemical Manganese Dioxide" by D. F. Aloko and K. R. Onifade [*Chemical Engineering and Technology*, **26**(12), 1281–1283, 2003]. A 2^2 design was used in an experiment that the authors described, with the two main effects and interaction effect used as the terms in a regression model for each of three response variables. The value of R^2 was given for each model and the values for one of the three response variables were 12.80, 7.86, 10.24, and 4.97, with the treatment combinations for the design in the order given in the third paragraph of Section 12.5.1. The value of R^2, which was stated as being the "regression coefficient," was .84. Do you disagree with any of these results? If so, how would you explain this to the authors?

12.85. Given the data in Example 12.6, verify the statement that there are no significant effects other than the block effect.

12.86. In the article "Characterization and Optimization of Acoustic Filter Performance by Experimental Design Methodology" by R. Gorenflo et al. [*Biotechnology and Bioengineering*, **90**(6), 746–753, 2005], the authors stated the following relative to the designed experiment that they described: "Interactions higher than two-factor interactions were assumed to be negligible." Do you agree with their decision? Would it be necessary to read the article to comment on their decision? Explain.

CHAPTER 13

Measurement System Appraisal

Ideally, we want measurement error, if it exists, to be small relative to the natural variation in a process. Unfortunately, in engineering and in other areas of science, measurement error can often be so great as to wreak havoc on designed experiments and other statistical analyses. Indeed, in referring to industrial experiments, Bisgaard (1999) stated: "If I were to provide a list of the most frequently encountered reasons why experiments fail, trouble with the measurements would come in as the clear 'winner'." Undoubtedly, other experts would make similar statements, so measurement system appraisal is of paramount importance.

There are various factors that can cause whatever is being measured to not be measured accurately. For example, a measurement instrument might provide relatively poor results. If the instrument has a constant bias, this can be estimated and accounted for, whereas poor precision of the measurement instrument without a constant bias is more of a problem. Measurement error can also be caused by operators and even by what is being measured, as well as interactions between these factors. External factors can also cause problems. For example, Fluharty, Wang, and Lynch (1998) discussed the impact of environmental interference on measurement systems and provided a simulation exercise that can be used for illustration.

As an extreme but simple example, consider the task of measuring the width of a desk in a classroom using a ruler. Assume that the width is known and also assume that the ruler is shorter than the desk. How much variability in measurements would you expect among a class of 20 students? How much variability would you expect if the ruler (or some other instrument) was longer than the width of the desk and the students had gained some experience in making measurements?

The point is that the measuring instrument, the item being measured, and the person making the measurement can all have a considerable effect on the recorded measurement. Consequently, since measurement error can have such a profound effect on the results of designed experiments, as Bisgaard (1999) stated, it is frequently almost imperative to perform a study to identify the magnitude of the components of the error. Such a study is discussed in the next section.

13.1 TERMINOLOGY

Whenever a designed experiment is used, there is an underlying model with σ_ϵ^2 as one of the terms in the model. Ideally, this term represents only variability due to replication. Even if this were the case (i.e., the model is correct, which it won't be), the replication error could be sizable. Therefore, it is important to assess the replication variability in a study.

The term *reproducibility* refers to the variability (or relative lack thereof) that results from making measurements under different conditions of normal use, such as the variability of averages of measurements by different operators on the same parts. *Repeatability* refers to, as the term implies, the variation (or lack thereof) of repeated measurements under identical, repeatable measurement conditions, such as measurements made on the same part by the same operator.

A study to assess these components of measurement variability is, in industry, termed a *gage reproducibility and repeatability* (gage R&R) study. In the context of experimental design, it is desirable to assess the relative magnitudes of these variance components through the use of designed experiments, with the data analyzed using Analysis of Variance (ANOVA) techniques.

Thus, the design of the experiment is quite important. The standard approach is to use a two-factor design, with the factors being "parts" and "operators." The former is a random factor, with the parts that are used being selected from a collection of parts, but operators may be either a random factor or a fixed factor. If the operators that will be used in the experiment are the same ones that are used to monitor the process, then operators is a fixed factor. (Recall the discussion in Section 12.7 regarding fixed and random factors.)

In addition to the proper classification of the operator factor as fixed or random, there are other important issues to be addressed such as determining the number of parts to use and the number of measurements to be made on each part. Burdick, Borror, and Montgomery (2003) in their review paper recommended using many parts with few measurements per part, rather than the reverse.

Thus, experiments to assess measurement system capability must be designed with the same care as experiments in general, and the experiment described next is an example of one for which the design could have been better.

■ **EXAMPLE 13.1**

Study Objective

Buckner, Chin, and Henri (1997) described a study that related to the objective of improvement of wafer-to-wafer and within-wafer sheet resistance uniformity for the Novellus Concept One-W tungsten deposition tool, as part of the SEMATECH equipment improvement project.

Study Design

It was necessary to determine whether the total measurement precision of the Prometrix RS35e four-point prober was adequate to monitor progress toward the objective. Five factors were examined, four at two levels and one at three levels, with "operator" being the three-level factor. The design had 16 points, so it was an irregular fraction of a mixed factorial. Both sheet resistance and uniformity were used as response variables, with a separate analysis conducted for each.

Conclusion

No factor was found to have a significant effect in either analysis, although the analysis for uniformity appears to be incorrect. It was also concluded that the prober was adequate for the stated objective, but it is interesting that no main effects were significant. Whenever this occurs, the design of the experiment should be rethought to try to determine if the design was flawed. After all, an experiment won't be run unless the experimenter believes that at least some of the factors significantly influence the response variable.

Indeed, it was admitted that the design could have been improved as the authors stated that the decision to run the experiment over five days could cause variation in workday cycles, which was not an isolated factor and could thus inflate the error term and consequently cause the factors to not be significant.

That may indeed have happened. Furthermore, large interactions can cause main effects to not be significant, and there were some sizable interactions in the data.

Although it is desirable that there not be any significant effects so that no corrective action is needed, in this instance it appears as though more work (and perhaps another experiment) is needed before such a conclusion can be drawn. ■

In general, experiments to assess measurement system capability must be as carefully designed as other types of experiments.

13.2 COMPONENTS OF MEASUREMENT VARIABILITY

If we let σ_y^2 denote the total measurement variability, we might think of modeling this variance in the same general way that we would model a response variable Y, or model the variance when a weighted least squares method is used in regression analysis. That is, we would try to think of the variance components that would add, apart from a random error, to σ_y^2. One logical way to proceed would be to use ANOVA with all factors treated as random, provided that the factors really are random.

Montgomery (1997, p. 473) described a study for which the objective was to assess the variability due to operators and the variability due to the gage. Three randomly selected operators measured each of 20 selected parts twice with a gage. The objective was to assess variability due to operators, parts, and gage. Letting σ_α^2 and σ_β^2 denote the variability due to operators and parts, respectively, and $\sigma_{\alpha\beta}^2$ represent the variability due to the interaction between operators and parts, it follows that

$$\sigma_y^2 = \sigma_\alpha^2 + \sigma_\beta^2 + \sigma_{\alpha\beta}^2 + \sigma^2 \tag{13.1}$$

with σ^2 denoting the variance that results when a part is repeatedly measured by the same operator using the same gage (and, in general, under the same conditions). Thus, σ^2 measures the gage repeatability, with $\sigma_\alpha^2 + \sigma_{\alpha\beta}^2$ measuring the reproducibility. Therefore, we could also represent Eq. (13.1) as $\sigma_y^2 = \sigma_{\text{measurement}}^2 = \sigma_{\text{reproducibility}}^2 + \sigma_{\text{repeatability}}^2 + \sigma_{\text{parts}}^2$. In terms of ANOVA jargon, repeatability is a measure of the "error" and reproducibility corresponds to the operators. Of course, if we are to measure reproducibility, we must identify all of the factors and extraneous factors must not have an influence. Similarly, processes must be in statistical control, as when any designed experiment is used.

We can view the breakdown of σ_y^2 using an equation such as Eq. (13.1), but we don't need to literally model σ_y^2, although this could be done. Rather, we can use standard experimental designs for location (as in Ryan, 2007) and simply use expected mean square expressions to allow us to estimate variance components in the presence of at least one random factor. (If there were no random factors—see Section 12.7 for a discussion of fixed and random factors—then the only variance component would be σ^2.)

To illustrate, assume that only one factor is used in a particular experiment and that factor is random (say, operators, with a group selected at random). It can be shown that the expected mean square for operators is $\sigma^2 + n\sigma_{\text{operators}}^2$, with n denoting, in this example, the number of measurements made by each operator. Since the expected mean square error is σ^2 (under the assumption that the model is correct), it follows that we would estimate $\sigma_{\text{operators}}^2$ as $\widehat{\sigma}_{\text{operators}}^2 = (MS_{\text{operators}} - MS_{\text{error}})/n$, with the components of the numerator obtained in the usual way (see Chapter 12). The numerator could easily be negative if there were no operator effect. More specifically, if the F-statistic for testing the operator effect were less than 1, the numerator, and hence the fraction, would be negative since the F-statistic is the ratio of the two terms in the numerator. We might proceed by interpreting a negative estimate as signifying that there is no operator effect and not being concerned with estimating the variance component, or we could use a more sophisticated approach to obtain an estimate of the variance component, one that forces the estimate to be positive.

In general, the negative estimate problem will occur for the model in Eq. (13.1) when one of the F-statistics is less than 1. This is not uncommon; see Table 4 of Inman, Ledolter, Lenth, and Niemi (1992) for an example of a negative variance component estimate in the measurement system case study they presented, although they indicated that the wrong model was fit. (Technically, however, we cannot expect to have the right model.)

If the objective is primarily to identify significant factors, estimating variance components may be of little interest. In other applications, both point estimates and confidence intervals may be desired. Burdick and Larsen (1997) reviewed methods for obtaining confidence intervals for variance components and Burdick et al. (2003) stressed the importance of computing confidence intervals and gave additional references. They also emphasized the importance of using more than the usual three operators in a study because this causes confidence intervals to be rather wide.

We would be happy if a starting model of Eq. (13.1) could be reduced to simply $\sigma_y^2 = \sigma^2$, as this would mean that no corrective action (such as retraining certain operators) would apparently be needed.

Khattree (2003) gave nine possible models for a gage R&R study, discussed each, and provided estimating equations for the variance components under various methods of estimation.

13.2.1 Tolerance Analysis for Repeatability and Reproducibility

Just as tolerances (i.e., engineering specifications) are used in assessing process capability (see Section 11.1), tolerances are also used relative to repeatability and reproducibility. In particular, with σ, as in Eq. (13.1), estimated from either a designed experiment or from a study leading to a control chart analysis, the "equipment variation" (EV) might be defined for a normal distribution as $EV = 2(2.575)\widehat{\sigma} = 5.15\widehat{\sigma}$, with 5.15 representing the limiting value of the number of standard deviations of the difference between tolerance limits that cover 99% of a normally distributed population (for any degree of confidence since this is a limiting value). The equipment variation as a percentage of the tolerance $(USL - LSL)$ would then be given by $100[(EV)/(USL - LSL)]$. Obviously, a small value for the latter is desired, with what would be an acceptable value depending on the circumstances.

The same type of computations could be performed for operators, that is, using σ_α, the square root of the variance component for operators in Eq. (13.1). Letting $OV = 5.15\sqrt{\widehat{\sigma}_\alpha^2 + \widehat{\sigma}_{\alpha\beta}^2}$ denote "reproducibility variation," the reproducibility variation as a percentage of the tolerance is given by $100[(OV)/(USL - LSL)]$.

Combining repeatability and reproducibility into a single measure produces the gage R&R tolerance percentage. In terms of the notation in Eq. (13.1), this would be $100[\sqrt{(OV)^2 + (EV^2)}/(USL - LSL)]$. This is also called the *precision over tolerance* (P/T) *ratio* and could equivalently be written $5.15\widehat{\sigma}_{\text{gageR\&R}}/(USL - LSL)$ and/or multiplied by 100 so as to express the number as a percentage. Alternatively, 6 could be used in place of 5.15, motivated by Six Sigma considerations. Another, slightly different name for this ratio is the "precision-to-tolerance ratio" (PTR), as used by Burdick et al. (2003).

The acceptable percentage for each of these measures can vary, but under 10% is certainly acceptable for the gage R&R tolerance percentage, as indicated by, for example, Montgomery and Runger (1993) and the Automotive Industry Action Group's (1995, p. 60) *Measurement Systems Analysis* manual, as well as in the 2002 edition. It has also been claimed that under 30% is acceptable (Czitrom, 1997), while others say that 10–30% is in the gray area. Over 30% is considered to be unacceptable (Automotive Industry Action Group, 2002). Clearly, we want the percentage to be as small as possible and the largest percentage that can be tolerated will depend on various factors, including whether or not designed experiments are being used without a measurement system analysis.

13.2.2 Confidence Intervals

The statistical measures given in Section 13.2.1 are all point estimates of the corresponding population quantities, which are functions of the unknown variance components. Accordingly, confidence intervals would be highly desirable, just as confidence intervals always provide more information than point estimates. The construction of a confidence interval for σ_y^2 is complex and beyond the level of this text, however, and depends on the model that is assumed, such as Eq. (13.1), for example. The interested reader is referred to Burdick et al. (2003) for details.

13.2.3 Examples

■ **EXAMPLE 13.2**

Consider a Gage R&R study that involves 3 parts and 2 operators, with repeat measurements made on each part–operator combination and the measurement being the diameter of a manufactured product. The data are given below and also at the textbook website. Part and Operator should be considered random factors in this study.

Part	Operator	Diameter
12	3	34
12	4	36
12	3	34
12	4	36
11	3	30
11	4	32
11	3	30
11	4	31
12	3	36
12	4	35
12	4	34
12	3	36
11	3	31
11	4	32
11	3	31
11	4	32
10	3	24
10	4	25
10	3	24
10	3	25
10	4	25
10	3	24
10	4	25
10	4	26

Performing a Gage R&R study (with a specified tolerance $= USL - LSL = 60$) using the ANOVA method and using MINITAB produces the following numerical output. [*Note*: MINITAB assumes that both Part and Operator are random and there is no option to specify otherwise. If, for example, Part is random but Operator is fixed, then the user would have to manually compute the F-statistic and p-value, following the rules given in Section 12.7. It should be noted, however, that the model given in Eq. (13.1) would not apply if both factors are not random since there is no variance component for a fixed factor. In general, a Gage R&R study without both factors being random would be highly atypical, although possible.]

Gage R&R Study - ANOVA Method

Two-Way ANOVA Table With Interaction

Source	DF	SS	MS	F	P
Part	2	438.083	219.042	404.385	0.002
Operator	1	4.167	4.167	7.692	0.109
Part * Operator	2	1.083	0.542	0.975	0.396
Repeatability	18	10.000	0.556		
Total	23	453.333			

Two-Way ANOVA Table Without Interaction

```
Source          DF     SS         MS        F        P
Part             2   438.083   219.042   395.263   0.000
Operator         1     4.167     4.167     7.519   0.013
Repeatability   20    11.083     0.554
Total           23   453.333
```

Gage R&R

```
                               %Contribution
Source             VarComp     (of VarComp)
Total Gage R&R      0.8552          3.04
  Repeatability     0.5542          1.97
  Reproducibility   0.3010          1.07
    Operator        0.3010          1.07
Part-To-Part       27.3109         96.96
Total Variation    28.1661        100.00
```

```
                             Study Var   %Study Var   %Tolerance
Source           StdDev (SD)  (6 * SD)     (%SV)      (SV/Toler)
Total Gage R&R    0.92477     5.5486       17.42         9.25
  Repeatability   0.74442     4.4665       14.03         7.44
  Reproducibility 0.54867     3.2920       10.34         5.49
    Operator      0.54867     3.2920       10.34         5.49
Part-To-Part      5.22599    31.3559       98.47        52.26
Total Variation   5.30718    31.8431      100.00        53.07
```

Number of Distinct Categories = 7

Gage R&R for Diameter

The output shows, in particular, that the Total Gage R&R is under 10% of the tolerance, so that is acceptable. There should be concern, however, about the part-to-part variability, which is large and accounts for the vast majority of the total variation.

The Number of Distinct Categories in the output is computed in a nonintuitive manner. Specifically, it is defined as the part standard deviation divided by the Total Gage R&R standard deviation, with that quotient multiplied by 1.41 and the decimal fraction dropped. So for this example, the computation is $(5.22599/0.92477) \times 1.41 = 7.97 = 7$. [The multiplier 1.41 is used by the Automotive Industry Action Group; the number is actually $\sqrt{2}$. The latter results from dropping a "1" while simplifying the Discrimination Ratio, which in terms of the notation used here is defined as $\sqrt{(\sigma_y^2 + \sigma_\beta^2)/(\sigma_y^2 - \sigma_\beta^2)}$.]

The number of distinct categories has nothing to do with the number of categories in the data (and the computed value, before truncation, will generally not be an integer), but it is an important, if somewhat crude, measure because obviously repeatability and reproducibility cannot be assessed unless there is more than slight variability in the data attributable to something else. The Automotive Interest Action Group has stated that the number of distinct categories be at least 5 for the measurement system to be declared adequate. Here that requirement is met.

Graphical output is also automatically produced in MINITAB, and the graphical output for this example is shown in Figure 13.1.

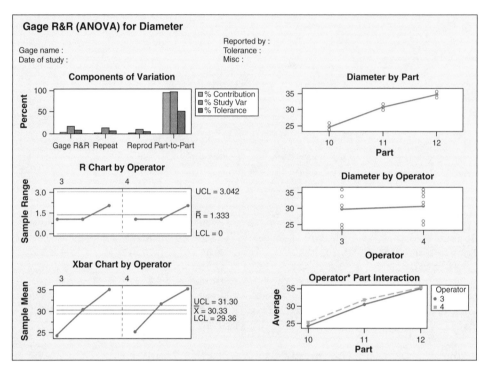

Figure 13.1 Graphical Gage R&R output.

Of course, the graph shows that the part-to-part variation dwarfs the other components of variation. Although $\overline{X}$ and R control chart limits are given, these charts should not be given any attention unless the limits are computed from a large amount of data (at least 100 observations), whereas in this example there are only 24 observations. Furthermore, in this example the $\overline{X}$-chart for operator is potentially quite misleading since it is the part variability that has caused two of the three points for each operator to be outside the control limits.   ■

A Gage R&R study can of course also be performed with other popular statistical software. The output from JMP is somewhat different from the MINITAB output as the user must select either the "QC" pull-down menu or the "Graph" pull-down menu and then select "Variability/Gage chart." The initial output is in the form of a multi-vari chart (see Section 1.6), but numerical output, including ANOVA tables and variance component percent breakdowns, can also be produced.

13.3 GRAPHICAL METHODS

Although a general model such as that given in Eq. (13.1) might be expected to hold over time in many applications, the sizes of the variance components might be expected to change over time. Therefore, some type of monitoring device should be used, and a control chart is ideal for this purpose.

For example, repeatability could be monitored with a control chart for standard deviations, and there are various ways in which this could be done. If a traditional control chart approach is taken, σ would be estimated from historical data and, preferably, probability limits constructed. Individual standard deviations, s, would then be plotted in real time. Alternatively, the *NIST/SEMATECH e-Handbook of Statistical Methods* (2002) computes an upper limit only in Section 2.2.3.1 and computes that limit in an unusual manner, by using $UCL = s_1\sqrt{F_{\alpha,\,j-1,k(j-1)}}$, with s_1 denoting the pooled standard deviation, j denotes the number of current measurements, and k denotes the number of points in time when the data were collected. The second degrees of freedom for F is the degrees of freedom associated with the historical data, and the first degrees of freedom corresponds to the current measurements.

The degrees of freedom for F are of the same general form as obtained when ANOVA is used, but in that case the data are obtained at the same point in time.

Thus, this is neither a standard control chart approach, for which the Phase I estimation of the historical data leads to the assumption of known σ for Phase II, nor a traditional ANOVA approach.

13.4 BIAS AND CALIBRATION

If σ^2 is large, that poses a problem since this could cause σ_y^2 to be unacceptably large. Serious measurement system problems could still exist even if σ^2 were small, however, as the measuring instrument could be biased. Constant bias would not present a problem, however, as an adjustment could be made for this. As discussed in Section 8.4.1, inverse regression or the classical method of calibration could be used to adjust for the bias. This could be done if an accurate measuring instrument were available, or if standards were available, for example, from the National Institute of Standards and Technology.

However the adjustment is made, it is important that the measurement system remain in a state of statistical control during the time that the data are collected. Accordingly, control chart procedures should be used, and the importance of using such procedures for this purpose is stressed in the literature. For example, Pankratz (1997) stated:

> The calibration procedure is based on the assumption that data being collected are representative of measurements that will be made on a daily basis. Therefore, SPC is crucial during the data collection step. No other tool provides the insight into the stability of the apparatus while the calibration data are being generated.

The recommendation to use SPC (statistical process control) procedures actually applies whenever *any* type of experimentation is performed, so the same advice in the context of designed experiments was given and emphasized in Section 12.1.

The monitoring of simple linear regression functions that are used in inverse regression and in the classical approach to calibration is an emerging field. The quality of the fit as measured by R^2, the magnitude of which shows how constant the bias is if in fact there is bias, and the slope and intercept should be monitored with process control/control chart techniques, and the issue should be addressed for both the analysis of historical data and real-time monitoring.

Kim, Mahmoud, and Woodall (2003) addressed the real-time process monitoring issue and proposed using three exponentially weighted moving average charts—one for monitoring the slope, one for monitoring the intercept, and one for monitoring the stability of the variation about the regression line by using residuals. Other methods have also been proposed.

The analysis of historical data might proceed by using a bivariate chart to simultaneously monitor the slope and intercept and a Shewhart chart to monitor the stability of the variation about the regression line.

When regression is used, the parameters are typically estimated from a single sample, but in the approach of Kim et al. (2003) the use of multiple samples is assumed in the analysis of historical data, and this was also assumed in other papers. One advantage of using multiple samples is that stability can be checked over time, which of course would not be possible if only one large sample were used. For real-time monitoring, samples of size 4 were used by Kim et al. (2003), following the work of Kang and Albin (2000), but this issue should be examined. We generally prefer to have more observations than this to fit a simple linear regression line. The sample size issue has been under investigation.

Other graphical methods might also be employed, and Buckner, Chin, Green, and Henri (1997) used boxplots to detect the existence of a warm-up effect in their measurement system study, the data for which are given in Exercise 13.8.

13.4.1 Gage Linearity and Bias Study

Similar to the discussion in the preceding section, one way to assess bias and determine whether or not the bias is a function of the magnitude of the measurement (using reference values) is to perform a gage linearity and bias study. The following example is from the *Measurement Systems Analysis Reference Manual* (Chrysler, Ford, General Motors Supplier Quality Requirements Task Force) and is also included as the sample datafile GAGELIN.MTW with MINITAB.

The setting is as follows. A plant foreperson selected five parts that represented the expected range of the measurements. Each part was measured by layout inspection to determine its reference (master) value, and one operator then randomly measured each part 12 times. The following data are given and are also at the textbook website:

Part	Master	Response	Difference (Bias)
1	2	2.7	0.7
1	2	2.5	0.5
1	2	2.4	0.4
1	2	2.5	0.5
1	2	2.7	0.7
1	2	2.3	0.3
1	2	2.5	0.5
1	2	2.5	0.5
1	2	2.4	0.4

1	2	2.4	0.4
1	2	2.6	0.6
1	2	2.4	0.4
2	4	5.1	1.1
2	4	3.9	−0.1
2	4	4.2	0.2
2	4	5.0	1.0
2	4	3.8	−0.2
2	4	3.9	−0.1
2	4	3.9	−0.1
2	4	3.9	−0.1
2	4	3.9	−0.1
2	4	4.0	0.0
2	4	4.1	0.1
2	4	3.8	−0.2
3	6	5.8	−0.2
3	6	5.7	−0.3
3	6	5.9	−0.1
3	6	5.9	−0.1
3	6	6.0	0.0
3	6	6.1	0.1
3	6	6.0	0.0
3	6	6.1	0.1
3	6	6.4	0.4
3	6	6.3	0.3
3	6	6.0	0.0
3	6	6.1	0.1
4	8	7.6	−0.4
4	8	7.7	−0.3
4	8	7.8	−0.2
4	8	7.7	−0.3
4	8	7.8	−0.2
4	8	7.8	−0.2
4	8	7.8	−0.2
4	8	7.7	−0.3
4	8	7.8	−0.2
4	8	7.5	−0.5
4	8	7.6	−0.4
4	8	7.7	−0.3
5	10	9.1	−0.9
5	10	9.3	−0.7
5	10	9.5	−0.5
5	10	9.3	−0.7
5	10	9.4	−0.6
5	10	9.5	−0.5
5	10	9.5	−0.5
5	10	9.5	−0.5
5	10	9.6	−0.4
5	10	9.2	−0.8
5	10	9.3	−0.7
5	10	9.4	−0.6

The analysis of these data in MINITAB produces the output shown in Figure 13.2.

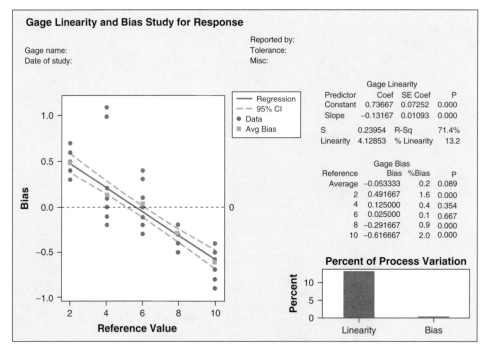

Figure 13.2 Gage linearity and bias results.

The Gage Linearity part of the output shows the simple linear regression equation that results from regressing the bias (difference) against the master (reference) value. The graph shows a clear relationship between the bias and the master value, as does the numerical output. Such a result should of course not be unexpected. The Gage Bias portion of the output shows the average bias for each reference value and the p-value for each t-test for the hypothesis that the mean bias is equal to the stated reference value (or, equivalently, a test that the bias is equal to zero for each reference value). We can see that the hypothesis is rejected for three of the five reference values. The displayed p-values for each of the five reference values are those that result from using the measurements that correspond to each of the indicated reference values and performing a one-sample t-test, with each reference value being the hypothesized value. The first p-value is obtained by performing a pooled-t test (see Section 6.1.2), with the variance pooling performed by pooling the bias variances over the reference values (call it s_p^2), and computing $t = \overline{\text{BIAS}}(s_p/\sqrt{n})$. The (two-sided) p-value is then twice the tail area for this t-value.

13.4.2 Attribute Gage Study

An attribute gage study calculates the amount of bias and repeatability of a measurement system when the data are in attribute form (i.e., a part being either accepted or not accepted). Given below are data from the MINITAB sample file AUTOGAGE.MTW. (Although not stated, these values are apparently departures from a nominal value since the numbers are negative.)

```
Partnumber  Reference  Acceptances

     1        -0.050         0
     2        -0.045         1
     3        -0.040         2
     4        -0.035         5
     5        -0.030         8
     6        -0.025        12
     7        -0.020        17
     8        -0.015        20
     9        -0.010        20
    10        -0.005        20
```

The lower tolerance limit is −0.02 and the upper tolerance limit is 0.02, with the Acceptances showing the number of parts out of 20 that were declared acceptable (i.e., within tolerance) for each reference value. The output for these data, using MINITAB, is shown in Figure 13.3.

Ideally, all parts with a reference value below −0.02 should of course be rejected. The fact that this did not occur shows some bias. The question is then whether or not the amount of bias is of any concern and whether the bias is statistically significant. The hypothesis test result shows a p-value that is quite small, so the bias is significant. In particular, we can see from the small graph that there is about a 50–50 chance of accepting a measurement of −0.0275.

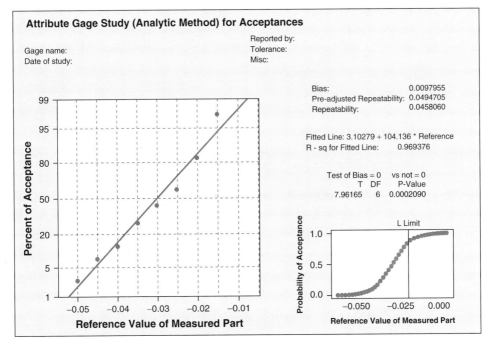

Figure 13.3 Attribute gage study (analytic method) results.

13.4.3 Designs for Calibration

Somewhat analogous to the experimental designs that were presented in Chapter 12, there are many calibration designs that have been developed for various applications, such as for mass weights, roundness standards, and ankle blocks. These designs are not well known but are catalogued and described in detail in Section 2.3.3 of the *NIST/SEMATECH e-Handbook of Statistical Methods* (http://www.itl.nist.gov/div898/handbook/mpc/section3/mpc33.htm).

These designs do not require the use of regression methods, unlike inverse regression and the classical method of calibration. As a simple example, single-point calibration would proceed as follows. A test item and a reference standard would be used, with the true measurement of the latter being known. Let Y denote the measurement of the test item with a particular instrument. This measurement is comprised of the true (unknown) value, bias (if it exists), and random error. That is, $Y = Bias_1 + Y^* + error_1$, with Y^* denoting the true value. A reference standard, with a known value, is measured with the same instrument. Letting R denote the measured value for the reference standard and R^* the known value, we then have $R = Bias_2 + R^* + error_2$. Of course, we don't know $Bias_2$ unless we know $error_2$, and we are not going to know the latter. If we make the plausible assumption that the bias is the same on the test item and the standard item, we would estimate Y^* as $Y^* = Y - R + R^*$. That is, the value that we are trying to estimate is estimated as the test item measurement minus the reference standard measurement plus the known value of the reference standard.

Of course, in statistics we don't make decisions based on samples of size one (i.e., $Y-R$ is a single difference), so this example has more illustrative value than practical value. It would obviously be preferable to use multiple differences obtained from multiple measurements, and this is where calibration designs come into play. These designs, which are not well known outside NIST, specify how the differencing and other computations are to be performed when a test item and a reference standard are used. See Section 2.3.4 of the *NIST/SEMATECH e-Handbook of Statistical Methods* for a catalog of calibration designs.

13.5 PROPAGATION OF ERROR

To this point in the chapter, measurement error and measurement variability have been considered for a single response variable, Y. Now assume that there are two response variables, X and Y, and a third response variable is defined as $Z = X + Y$. The question is then how measurement error in X and Y "propagates" to measurement error in Z. We can't say anything about Z and the magnitude of the measurement error in Z without knowing the magnitude and direction of the average measurement error in X and Y, and knowing whether or not X and Y are independent. Thus, if we wanted to obtain some idea of the measurement error of the perimeter of a constructed object that is difficult to measure, and the perimeter is computed from measurements made on the length, we would have to make certain assumptions.

The same is true for other functional forms, such as estimating the area of an object from measurements made on the length and width. Of course, here the functional form is $Z = XY$, with Z denoting the area and X and Y representing the length and width, respectively.

A propagation of error analysis can be applied to more than just measurement error. For example, Tiffany, Thayer, Coehlo, and Manning (1976) did such an analysis involving systematic (measurement) error and random error.

In general, some propagation of error formulas are based on assumptions that may be unreasonable. For example, when the propagation of error approach is used to obtain an expression for the standard deviation of area as a function of length and width, the approximate formula

$$s_{\text{area}} = \sqrt{\text{width}^2 \times s_{\text{length}}^2 + \text{length}^2 \times s_{\text{width}}^2}$$

is based on the assumption that length and width are independent, an assumption that is unlikely to be satisfied in very many applications. The exact formula is much more complex and is the variance of a product (e.g., length times width), as originally given by Goodman (1960).

Error propagation and measurement theory in general are discussed, with engineering applications, in Gertsbakh (2003), and there is also considerable discussion of error propagation in the *NIST/SEMATECH e-Handbook of Statistical Methods* (`http://www. itl.nist.gov/div898/handbook`), especially Section 2.55 and the case study in Section 2.64. See also Section 2.3.6.7.3.

13.6 SOFTWARE

Software for the methods discussed in this chapter is readily available because the methods employed in a gage R&R study, for example, are commonly used general purpose methods such as Analysis of Variance (ANOVA), as well as control charts—methods that have been discussed in detail in previous chapters.

13.6.1 MINITAB

A Gage R&R study can be performed in MINITAB for both crossed and nested factors, and this was illustrated for crossed factors. The hypothesis tests in the ANOVA table that is produced are constructed under the assumption that the factors are random and there is appropriately no option for specifying anything else. When the response variable is binary (such as "go/no go" data), an Attribute Gage study can be performed, and this was also illustrated. For the latter, two methods are available for determining whether or not the bias is zero: (1) the method advocated by the Automotive Interest Action Group (AIAG) and (2) linear regression. A Gage Linearity and Gage Bias study (for measurement data) can also be performed, with the former assessing the performance (precision) of the instrument over the range of measurements and the latter determining the degree of bias, if any. Finally, a gage run chart is essentially a control chart that shows the extent of departure from the overall mean of measurements made by the different operators on different parts, with part numbers and operator displayed for each plotted point.

13.6.2 JMP

JMP provides certain options that MINITAB does not provide. In particular, JMP allows up to three factors in estimating variance components, whereas MINITAB does not permit more than two factors. The Gage R&R analysis is available only for the two-factor crossed model, however.

13.7 SUMMARY

The intent of this chapter was to provide a brief but important overview of measurement issues. Statistical methods and designed experiments can be undermined by substantial measurement error, as indicated in the quote from Bisgaard (1999). Consequently, measurement error must be carefully considered and evaluated, if necessary, before relying on statistical results.

Experimental designs can be used for the purpose of estimating variance components, which can pinpoint the source(s) of measurement error problems. Once sources have been identified, they might be monitored individually, as discussed by Yashchin (1994) and Woodall and Thomas (1995). Methods for performing measurement system capability analyses were reviewed by Burdick et al. (2003) and calibration and measurement issues and methods were reviewed by Iyer (2003). Other useful sources include Automotive Industry Action Group (2002) and Horrell (1991), and especially Taylor (1997). A very short (48-page), simplified introduction to Gage R&R studies was given by Barrentine (1991). See also Section 2.4, Gage R&R Studies, and Section 2.5, Uncertainty Analysis, of the *NIST/SEMATECH e-Handbook of Statistical Methods* (Croarkin and Tobias, 2002).

REFERENCES

Automotive Industry Action Group (1995). *Measurement Systems Analysis*, 2nd ed. Detroit, MI: Chrysler, Ford, General Motors Supplier Quality Requirements Task Force.

Automotive Industry Action Group (2002). *Measurement Systems Analysis*, 3rd ed. Detroit, MI: Chrysler, Ford, General Motors Supplier Quality Requirements Task Force.

Barrentine, L. B. (1991). *Concepts for R&R Studies*. Milwaukee, WI: ASQ Quality Press.

Bisgaard, S. (1999). "Quality Quandries." *Quality Engineering*, **11**(4), 645–649.

Buckner, J., B. L. Chin, and J. Henri (1997). Prometrix RS35e gauge study in five two-level factors and one three-level factor. In *Statistical Case Studies for Industrial Process Improvement* (V. Czitrom and P. D. Spagon, eds.). Philadelphia: Society for Industrial and Applied Mathematics, and Alexandria, VA: American Statistical Association.

Buckner, J., B. L. Chin, T. P. Green, and J. Henri (1997). Revelation of a microbalance warm-up effect. In *Statistical Case Studies for Industrial Process Improvement* (V. Czitrom and P. D. Spagon, eds.). Philadelphia: Society for Industrial and Applied Mathematics, and Alexandria, VA: American Statistical Association.

Burdick, R. K. and G. A. Larsen (1997). Confidence intervals on measures of variability in R&R studies. *Journal of Quality Technology*, **29**(3), 261–273.

Burdick, R. K., C. M. Borror, and D. C. Montgomery (2003). A review of methods for measurement systems capability analysis. *Journal of Quality Technology*, **35**(4), 342–354.

Croarkin, C. and P. Tobias, technical editors. (2002). *NIST/SEMATECH e-Handbook for Statistical Method.* (http://www.itl.nist.gov/div898/handbook/mpc/section3/mpc33.htm.)

Czitrom, V. (1997). Introduction to gauge studies. Chapter 1 in *Statistical Case Studies for Industrial Process Improvement* (V. Czitrom and P. D. Spagon, eds.). Philadelphia: Society for Industrial and Applied Mathematics, and Alexandria, VA: American Statistical Association.

Fluharty, D. A., Y. Wang, and J. D. Lynch (1998). A simplified simulation of the impact of environmental interference on measurement systems in an electrical components testing laboratory. Chapter 15 in *Statistical Case Studies: A Collaboration Between Academe and Industry* (R. Peck, L. D. Haugh, and A. Goodman, eds.). Philadelphia: Society for Industrial and Applied Mathematics, and Alexandria, VA: American Statistical Association.

Gertsbakh, I. (2003). *Measurement Theory for Engineers*. New York: Springer.

Goodman, L. (1960). On the exact variance of products. *Journal of the American Statistical Association*, **55**, 708–713.

Horrell, K. (1991). *Introduction to Measurement Capability Analysis*. SEMATECH report 91090709A-ENG.

Inman, J., J. Ledolter, R. V. Lenth, and L. Niemi (1992). Two case studies involving an optical emission spectrometer. *Journal of Quality Technology*, **24**(1), 27–36.

Iyer, H. (2003). Statistical calibration and measurements. Chapter 20 in *Handbook of Statistics 22* (R. Khattree and C. R. Rao, eds.). Amsterdam: Elsevier Science B.V.

Kang, L. and S. L. Albin (2000). On-line monitoring when the process yields a linear profile. *Journal of Quality Technology*, **32**, 418–426.

Khattree, R. (2003). Repeatability, reproducibility, and interlaboratory studies. Chapter 22 in *Handbook of Statistics 22* (R. Khattree and C. R. Rao, eds.). Amsterdam: Elsevier Science B.V.

Kim, K., M. A. Mahmoud, and W. H. Woodall (2003). On the monitoring of linear profiles. *Journal of Quality Technology*, **35**, 317–328.

Montgomery, D. C. (1997). *Design and Analysis of Experiments*, 4th ed. New York: Wiley.

Montgomery, D. C. and G. C. Runger (1993). Gauge capability and designed experiments. Part I: Basic methods. *Quality Engineering*, **6**, 115–135.

Pankratz, P. C. (1997). Calibration of an FTIR spectrometer for measuring carbon. Chapter 3 in *Statistical Case Studies for Industrial Process Improvement* (V. Czitrom and P. D. Spagon, eds.). Philadelphia: Society for Industrial and Applied Mathematics, and Alexandria, VA: American Statistical Association.

Ryan, T. P. (2007). *Modern Experimental Design*. Hoboken, NJ: Wiley.

Taylor, J. R. (1997). *An Introduction to Error Analysis: A Study of Uncertainties in Physical Measurements*, 2nd ed. Sausalito, CA: University Science Books.

Tiffany, T. O., P. C. Thayer, C. M. Coehlo, and G. B. Manning (1976). A propagation of error analysis of enzyme activity. A model for determining the total system random error of a kinetic enzyme analyzer. *Clinical Chemistry*, **22**(9), 1438–1450.

Van Den Heuvel, E. R. (2000). Gage R&R studies for nonstandard situations. *Annual Quality Congress Proceedings*, Milwaukee, WI: American Society for Quality.

Woodall, W. H. and E. H. Thomas (1995). Statistical process control with several components of common cause variability. *IIE Transactions*, **27**, 757–764.

Yashchin, E. (1994). Monitoring variance components. *Technometrics*, **36**, 379–393.

EXERCISES

13.1. If a negative estimate of a variance component means that the corresponding factor or interaction has no effect, does a positive estimate of a variance component likely mean that the corresponding effect is significant? Explain.

13.2. Assume that a study is to be performed with the intention of determining what factors affect a particular response variable (as in Chapter 12).

(a) Would a nearly constant bias in the response variable values create problems relative to trying to determine what effects are significant?

and/or

(b) Would there be a problem only in terms of estimating the magnitude of the significant effects? Explain.

13.3. A company performs a study to see if there is a machine effect that is contributing to its problems with excessive variability in a particular product line. Three machines are picked at random for the study and this is the only factor that is examined in the study. If $SS_{machines} = 72.4$, $MS_{error} = 12.4$, and there are 10 measurements made for each machine, what is the numerical value of $\hat{\sigma}^2_{machines}$?

13.4. Inman et al. (1992) (see References) described a measurement system study in which the data were recorded to two decimal places, which the authors suspected was probably more precise than the capability of the measuring instrument. Explain why this is a problem.

13.5. A useful class experiment would be the one mentioned at the beginning of the chapter. The students in a class (of at least moderate size) could each measure the width of the desk used by the instructor, using both a ruler or other measuring instrument that is at least as wide as the desk and one that is not as wide as the desk. Of course, repeatability could be assessed by having each student make two or three measurements using the same measuring instrument.

13.6. E. R. van den Heuvel (2000) (see References) gave the following data, which were from training material for Six Sigma as an aid in explaining a Gage R&R study.

| Product | Operator 1 Measurements | | | Operator 2 Measurements | | | Operator 3 Measurements | | |
	1	2	3	1	2	3	1	2	3
1	7.1	7.3	7.2	7.6	7.5	7.5	7.0	7.1	7.1
2	13.2	13.3	13.4	13.4	13.7	13.6	13.0	13.3	13.3
3	2.1	2.1	2.1	2.6	2.5	2.6	2.2	2.0	2.0
4	27.2	27.5	27.5	28.2	28.1	28.2	27.6	27.8	27.4
5	18.9	18.8	18.5	19.1	19.4	19.1	19.0	18.8	18.8
6	3.1	3.1	3.3	3.6	3.6	3.7	3.2	3.2	3.3
7	21.3	21.5	21.6	22.2	22.0	22.0	21.7	21.7	21.4
8	6.4	6.4	6.6	7.2	6.9	6.8	6.5	6.7	6.7
9	38.2	38.5	38.6	38.6	38.8	39.1	38.7	38.6	38.5
10	21.8	21.8	21.6	21.6	21.9	22.0	21.6	21.9	21.8

Perform a Gage R&R analysis, using software and assuming that product and operator are both random factors. Draw appropriate conclusions.

13.7. Assume, relative to Exercise 13.6, that a foreperson states that the three operators involved in the study are the only ones who will be doing this work. Does the model given in Eq. (13.1) still apply? Explain.

13.8. Buckner et al. (1997) (see References) described a Gage R&R study that was performed to assess instrument repeatability and measurement system reproducibility. Three operators were supposed to weigh ten sample wafers, weighing each one twice. The data are as follows:

Wafer	Operator 1 First	Operator 1 Second	Operator 2 First	Operator 2 Second	Operator 3 First	Operator 3 Second
1	54.0083	54.0084	54.0098	54.0041	54.0143	54.0096
2	53.7703	53.7693	53.7708	53.7601	53.7670	53.7685
3	54.2083	54.2084	54.2094	54.1975	54.2079	54.2073
4	54.1110	54.1099	54.1111	54.1007	54.1110	54.1083
5	54.0347	54.0341	54.0350	54.0240	54.0360	54.0333
6	54.1152	54.1130	54.1132	54.1017	54.1141	54.1117
7	54.2192	54.2194	54.2195	54.2081	54.2176	54.2189
8	54.1555	54.1561	54.1561	54.1462	54.1516	54.1550
9	54.4405	54.4399	54.4405	54.4293	54.4381	54.4391
10	54.2291	54.2284	54.2290	54.2179	54.2252	54.2278

(a) Perform a Gage R&R analysis, using only the information that is given here.

(b) The authors stated that the first and second measurements taken by each operator on each wafer were at least half a day apart. Does this necessitate a change in the method of analysis? Indeed, can a Gage R&R analysis still be performed? Explain. If not, can an alternative analysis be performed? If so, perform the analysis.

(c) If possible, read the article and see if your approach to part (b) agrees with what the authors did.

CHAPTER 14

Reliability Analysis and Life Testing

The terms "survival data," "reliability data," and "lifetime data" are all quite similar. The first term is generally applied to people, such as survival data in years after treatment for cancer patients who follow a specified treatment. The second and third terms refer to electrical or mechanical components, products, or systems but are used in somewhat different ways. For example, we might refer to the reliability of an electrical system, whereas we would be most concerned with the lifetime, not the reliability, of a light bulb since light bulbs function normally until they burn out and are not repaired.

As consumers, we are all interested in product reliability. Similarly, manufacturers are interested in product reliability, which directly relates to warranty costs. Various definitions of reliability have been given, but the one stated below should suffice.

DEFINITION

Reliability is the probability of a product fulfilling its intended function during its life cycle.

A more elaborate definition, which says essentially the same thing, was given by Barlow (1998): "*Reliability* is the *probability* of a device performing its purpose adequately for the period of time intended under operating conditions encountered."

As contrasted with quality control and quality improvement, we might say that a product "has quality" if it is still functioning as intended at a given point in time and the extent of its useful life is what the public considers acceptable. Not everyone has come to view quality as extending beyond the point of manufacture, however, as there are still some outdated views of quality. For example, Rausand and Høyland (2004, p. 6) stated, "Reliability is therefore an extension of quality into the time domain."

Reliability theory and methodology developed as a tool to help nineteenth century maritime and life insurance companies compute profitable rates to charge their customers (*NIST/SEMATECH e-Handbook of Statistical Methods*, Section 8.1.2). Today we would

---

refer to this as survival analysis since people are involved, but the general principles of reliability analysis and survival analysis are certainly related.

To the layperson, product reliability probably means "when can I expect product X to give out completely, at what earlier point in time can I expect to have to start making repairs, and once the repairs begin, how slowly or how rapidly can I expect the product to deteriorate?" This, in essence, is what we mean by reliability. In other words, reliability essentially refers to a continuum of time, whereas there is no mention of "rate of degradation," or similar terminology, in any common definition of quality. The possible use of degradation data in designed experiments is discussed in Section 14.8 and using degradation data in conjunction with acceleration models is discussed in Section 14.3.1.

14.1 BASIC RELIABILITY CONCEPTS

Reliability data are measurements of the time to failure, with possibly other measurements made on sample items. One goal in reliability work is to estimate the percentage of units of a product that will still be functioning after a particular period of time. This could then be translated into a probability statement, for example, "the probability is .24 that a newly designed light bulb will last longer than 25,000 hours." In order to be able to make such a statement, it is necessary to select a probability distribution that will facilitate the construction of reasonably precise probability statements of the type that one wishes to make.

Probability distributions that have frequently been used in reliability work include the Weibull distribution, which was covered in Section 3.4.6. Early work in reliability utilized the exponential distribution to a considerable extent, but it was eventually realized that the Weibull distribution generally provides a much better model. The exponential distribution does have some important applications, however, as was discussed in Section 3.4.5.2. (See the discussion of the Weibull distribution vis-à-vis the exponential distribution in Section 14.5.3.) A good discussion of the exponential distribution and its limited usefulness in reliability work—though it is still widely used—can be found at http://www.reliasoft.com/newsletter/4q2001/exponential.htm. Limitations of the exponential distribution in reliability applications are discussed in Section 14.5.2.

One restriction generally imposed on the selection of a distribution is that the distribution must be defined only for positive values, since the random variable is time, which of course cannot be negative. This doesn't mean that distributions for which the random variable can be negative cannot be used in reliability applications, or other applications, however.

In particular, this alone would not rule out the normal distribution as a possible model because there are many random variables (e.g., height, weight, aptitude test scores) that cannot be negative, yet a normal distribution (which of course has a range of $-\infty$ to $+\infty$) is a good model. As long as the mean is well above zero, this may not be a problem. A more important issue is whether the random variable can be expected to be skewed, and we certainly expect failure times to be skewed.

Since reliability data are, at least for the most part, *failure data*, the definition of a "failure" in a given application must be carefully considered. In the example cited previously, is it reasonable to say that a light bulb has failed if it hasn't lasted for 25,000 hours? Often the threshold value for determining failure is chosen subjectively. In many applications, however, an item does not need to have ceased functioning for the item to be declared a "fail." Specifically, failure might be declared if an item can no longer deliver a specified proportion of its functionality. For example, if a TV picture tube is becoming weak, there may still be a picture, but the picture quality is probably unacceptable. Similarly, a cable TV connection may be poor at the company's end, but the picture may be viewable.

In both cases we would have what is termed a *repairable system*. An item that is repairable, such as a car, may eventually last a long time. Items that are not repairable would include a light bulb that has burned out and an extremely old car that is "beyond repair." These would obviously be termed *nonrepairable systems* and there is a different set of statistical methods applicable to such systems. Most reliability data are collected on nonrepairable populations, rather than on repairable populations, for which units that fail can be repaired and returned to the population.

The term *failure mode* is also used in reliability. A failure mode is simply a cause of failure for a component or unit. There can be different causes of failure and each failure mode has a probability distribution. A *failure mode and effects analysis* (FMEA) is often conducted for the purpose of outlining the expected effects of each potential failure mode. This consists of examining as many components, assemblies, and subsystems as necessary to identify failure modes (i.e., how faults are observed) and the causes and effects of the failures, noting that each failure mode can be caused by several different failure causes.

An example of an FMEA is given in Section 14.7. A related concept is a *failure mode, effects, and criticality analysis* (FMECA), which occurs if priorities (criticalities) are assigned to the failure mode effects.

Under the *competing risk model*, the component or unit fails when one of the possible causes of failure (i.e., failure mode) occurs. Each failure mode thus has its own reliability, $R(t)$, defined in the next section, with the reliability for the component or system under the competing risk model being the product of the reliabilities for the failure modes, and the component or system failure rate being the sum of the individual failure mode failure rates. The latter holds for models in general as long as independence holds, and as long as component failure results when the first mechanism failure occurs.

These results also hold for a *series system*, for which the system fails if any one of its components fails. In contrast, a *parallel system* is one for which the system continues to operate until *all* of the components fail. In between these two extremes is an "*r* out of *n*" system, which will continue to function as long as any *r* out of *n* components are functioning. See Section 8.1.8.4 of the *NIST/SEMATECH e-Handbook of Statistical Methods* for additional information about this type of system.

Which of these models is applicable of course depends on how the system was constructed.

In many applications it isn't practical to wait until failures occur. For example, a piece of machinery may have some critical components whose reliability needs to be quickly and precisely assessed, as the reliability of the machinery strongly depends on the reliability of these components. Such a situation would almost dictate the use of *accelerated testing*, which would be performed in a laboratory. In particular, it isn't possible to wait ten years or so until an automobile gives out to determine whether or not it functioned as designed during its useful life. As discussed by Fluharty, Wang, and Lynch (1998), automotive engineers are designing products with anticipated lives that exceed ten years and 150,000 miles, so accelerated testing is used to simulate high mileage and aging. The failure rate in an accelerated test would then have to be converted to a failure rate under normal use conditions, and this would entail the use of models. Clearly, the modeling aspect is important because the results of accelerated tests are of little value if the results cannot be converted to normal use failures.

14.2 NONREPAIRABLE AND REPAIRABLE POPULATIONS

By a "nonrepairable population" we mean one for which units that fail are not repaired. Models that are frequently used for nonrepairable populations include the exponential, Weibull, gamma, lognormal, and extreme value models given in Chapter 3, in addition to a few other models (see Croarkin and Tobias, 2002, Section 8.1.6). Models that are used for repairable systems include the homogeneous Poisson process, with the cumulative number of failures to time T given by the Poisson distribution with parameter λT. (The Poisson distribution was covered in Section 3.3.4.) This model is widely used in industry but the most commonly used model is the Duane model. Duane (1964) presented failure data of different systems during their development programs and showed that the cumulative *mean time between failures* (*MTBF*) plotted against cumulative operating time was close to a straight line when plotted on log–log paper. This model is also often called the *power law model* and has also been referred to as the Army Materials System Analysis Activity model.

The *pdf* of the model can be written

$$P(N(T) = k) = \frac{a^k T^{bk} e^{-\lambda t}}{k!}$$

with $N(T)$ denoting the cumulative number of failures from time 0 to time T, a and b are constants, and λ is the parameter to be estimated. [See Croarkin and Tobias (2002, Section 8.1.7.2) for further details.]

Most books on reliability are books for nonrepairable systems, which is understandable because most reliability data that are collected are on nonrepairable systems. A notable exception is Rigdon and Basu (2000), which is devoted to repairable systems. See also Section 8.1.7 of Croarkin and Tobias (2002), which discusses some of the models that are used for repairable systems.

14.3 ACCELERATED TESTING

An accelerated failure model relates the time-to-failure distribution to the stress level. The general idea is that the level of stress is only compressing or expanding time and not changing the shape of the time-to-failure distribution. With this assumption, projecting back to stress under normal use is not as difficult as it might seem as one has to simply handle the change of scale in regard to the time variable. That is, changing stress is equivalent to transforming the time scale used to record the time at which failures occur, so it is a matter of "decelerating" back to normal use by using an appropriate acceleration factor. This is accomplished by using *acceleration models*. These models are mechanistic models that are based on the physics or chemistry that underlies a particular failure mechanism. Acceleration models of course use time as a variable but generally do not contain interaction terms. Models without such terms seem to work well.

The general idea is that for two stress levels, $S_2 > S_1$, the time to failure at stress level S_1 can be represented as

$$T(S_1) = a(S_2, S_1)T(S_2) \tag{14.1}$$

with $T(S_i)$ denoting the time to failure at stress S_i, $i = 1, 2$. In order to be able to project back to obtain $T(S_1)$, the acceleration rate $a(S_2, S_1)$ must obviously be known. In order to easily solve for $a(S_2, S_1)$ for a specified model, such as the Arrhenius equation presented in Section 14.3.1, it is necessary to use the equivalent form

$$\log[T(S_1)] = \log[a(S_2, S_1)] + \log[T(S_2)] \tag{14.2}$$

If we let $\log[T(S_1)]$ correspond to the probability of failure by time t with parameters $\mu(S_1)$ and $\sigma(S_1)$, it follows that

$$\mu(S_1) = \log[a(S_2, S_1)] + \mu(S_2) \tag{14.3}$$

and $\sigma(S_1) = \sigma(S_2) = \sigma$. In words, Eq. (14.3) states that the mean time to failure at stress S_1 is equal to the log of the acceleration rate plus the mean time to failure at stress S_2.

In order for accelerated testing to be successful, an acceleration factor must be identified, if one exists. Unfortunately, acceleration factors don't always exist. For example, Wu and Hamada (2000, p. 548) pointed out that a factor for accelerating the failure of hybrid electronic components has not been found. There are some acceleration factors that do work well, however, including temperature and voltage.

There are also some major obstacles to the successful use of accelerated testing. The first is that the relationship between time to failure and stress must be modeled, and it may be difficult to check the adequacy of the model. In essence, what is needed is a mechanistic model (Chapter 10), and as Hooper and Amster (2000) pointed out, one must understand the physics of failure.

An obvious question (or two) to ask is: Why is a model necessary? Can't we just obtain all of the data that we need from accelerated testing? The answer is "no" because most of the data will come from high-stress testing, so it is necessary to extrapolate the failure results at high-stress levels to lower levels, including "no stress."

Since accelerated testing is performed because of relative scarcity of failure data under normal stress conditions, we would like to have some assurance that accelerated testing will be successful. A simulation procedure for arriving at a test design that will increase the chances of success is given in Chapter 8 of the *NIST/SEMATECH e-Handbook of Statistical Methods* (Croarkin and Tobias, 2002), and specifically at `http://www.itl.nist.gov/div898/handbook/apr/section3/apr314.htm`.

Even though selecting an appropriate accelerated test model can be difficult and model validation even more so, there is some "history" that can be relied on in zeroing in on an appropriate model. For example, Spence (2000) points out that the Arrhenius model has been used to calculate acceleration factors for semiconductor life testing for many years. This model is considered in the next section.

14.3.1 Arrhenius Equation

We would expect that as temperature is increased, a reaction that is affected by temperature, such as a chemical reaction, will occur more rapidly. The Arrhenius equation embodies this idea and is named after S. A. Arrhenius (1859–1927). The equation when applied to accelerated testing is given by

$$\mu(S) = \log(R_0) + \frac{E_a}{k(S + 273)} \qquad (14.4)$$

with R_0 denoting the rate constant, E_a is the activation energy, k is Boltzmann's constant, S is the stress (temperature in degrees Celsius in the Arrhenius equation), and $\mu(S)$ is a function of the time to failure at stress S. Equation (14.4) does not by itself give the acceleration rate, which is obtained by using two temperatures, S_1 and S_2. That rate must be solved for by using Eqs. (14.3) and (14.4), and doing so produces

$$a(S_2, S_1) = \exp\left[\left(\frac{E_a}{k}\right)\left(\frac{1}{(S_1 + 273)} - \frac{1}{(S_2 + 273)}\right)\right] \qquad (14.5)$$

as the reader is asked to verify in Exercise 14.3. The value obtained from this equation after specifying S_1 and S_2 would then be used in Eq. (14.1) to solve for (estimate) $T(S_1)$.

Regarding computation, Section 8.4.2.2 of the *NIST/SEMATECH e-Handbook of Statistical Methods* shows how to use JMP for Arrhenius parameter estimation.

14.3.2 Inverse Power Function

The Arrhenius equation is for temperature as the accelerating variable. When voltage is the acceleration variable, the inverse power function is frequently used, and the use of the function leads to

$$\mu(S) = \beta_0 + \beta_1 \log(S)$$

and an acceleration factor of

$$a(S_2, S_1) = \left(\frac{S_1}{S_2}\right)^{\beta_1}$$

Since $S_2 > S_1$, β_1 must obviously be negative since the acceleration rate must be greater than one. Of course, β_0 and β_1 would have to be estimated, and the method of maximum likelihood is a common choice for this purpose.

As in regression analysis, once the parameters are estimated, residual analysis would be used to check the fit of the model. Of course, in regression analysis normality is assumed, but normality is generally not part of reliability work. Therefore, a probability plot of the residuals would be constructed using the distribution that is assumed under the model that is fit, with the objective to see if the points plot on a straight line. If they do so, this would validate the implied distributional assumption, which in turn would essentially validate the selected stress function.

14.3.3 Degradation Data and Acceleration Models

It is possible to predict when failures will occur if failure can be related directly over time to changes in the values of a measurable product random variable. Then degradation data can be used without having to wait for failures, with failure predicted by extrapolating from the degradation data. In order for this to work, a variable must be identified that changes slowly over time and eventually reaches a value that causes the unit to fail. Unfortunately, it is often difficult or impossible to find such a variable. Consequently, although the use of degradation data is an attractive alternative to having to wait for failures, it could be difficult to implement the idea.

If such a variable is found and the approach is used, it is highly desirable to obtain an actual failure or two to compare the actual failure time with what is predicted using the acceleration model, and thus obtain a "residual" or two that can be used for model criticism.

For more information on the use of degradation data, see Section 8.4.2.3 of Croarkin and Tobias (2002). For additional information on accelerated testing the reader is referred to Nelson (1990).

14.4   TYPES OF RELIABILITY DATA

Answers to questions about the reliability of a particular product are complicated by the fact that we generally cannot take, say, 100 units of product X and see how long it takes for each unit to become inoperable because of practical considerations. For example, let's say we start with 100 light bulbs and after a certain length of time, 98 of these have burned out. We wait for the other two bulbs to burn out ... and wait ... and wait, but the bulbs defy us and continue functioning. So we grow weary of this waiting game and terminate the test. What have we done? For one thing, we have created *censored data*.

Now what do we do since we don't have data on all 100 light bulbs? We could compute the average lifespan for the 98 bulbs, but we would be underestimating what the average would have been if we had the time and patience to wait for the other two light bulbs to burn out.

This scenario is typical of what occurs with life testing in that the data are censored. That is, we do not have measurements on every item in the sample. We thus have to estimate "what would have been" so we need statistical techniques that are different (and more involved) than what has been presented in previous chapters.

Not surprisingly, software must be used to handle this situation. We recognize the added complication of censoring at this point in the chapter, but we will wait until near the end of the chapter to return to it after first considering more simplified scenarios.

Much has changed during the last forty years regarding views toward product reliability and desirable product lifespans. It was argued decades ago that it would be bad for the U.S. economy for products to last a long time as then consumers wouldn't need to purchase replacements. So it was viewed as desirable to have "built-in obsolescence" and to have products with not-too-long lifespans.

That view changed when it became apparent that Japanese cars, in particular, had a long lifespan, so consumers began to demand this. Now, in the early years of the twenty-first century, consumers have a much different view of product reliability and the desired lifespan of products. In particular, wouldn't it be nice if light bulbs lasted "almost forever" so that those of us who live in houses with high ceilings and light fixtures near the ceiling wouldn't have to risk life and limb to replace them? Perhaps with such considerations in mind, companies like General Electric, Cree Research, and Emcore have worked on building a better light bulb—one whose lifespan is a dramatic improvement over the light bulbs that we have used for decades. Subsequent research accomplishments will present new challenges for modeling product lifespan.

Unfortunately, product reliability generally commands the most attention when there is loss of life, as happened when the combination of Firestone tires on Ford Explorer vehicles gained national attention during 2001 because of the deaths that were apparently caused by either the vehicle or the tires (or both), and which led to the dissolution of the long-standing business relationship between Firestone and Ford.

In this chapter we survey reliability and life testing, focusing more on concepts and applications than on theory. We also want to recognize that reliability is inextricably linked to nonconformities, as units with nonconformities at the beginning of use will likely have a shorter lifespan than those units without any nonconformities.

14.4.1 Types of Censoring

Since censored data predominate in reliability studies, it is important to understand the different types of censored data. There are three basic types of censoring: right censored, left censored, and interval censored, in addition to the exact failure time.

With *exact failure* data, the time that an item fails is recorded, and the testing does not terminate until the item has failed. Thus, there is "zero censoring" and obviously exact failure data is the easiest data with which to work. Unfortunately, we will not often encounter such data (exclusively) as this would suggest that the quality of what is being examined is probably poor.

With *right censoring*, an item is removed from test when it is still functioning, so the time that it would have failed if testing on the item had continued is unknown. When this type of censoring is used, one must be able to assume that items that are removed from test at a given time are representative of all of the items that are on test at that time. This type of censoring is also called *Type I censoring*.

With *left censoring*, items are removed from the test at a certain time if they are not functioning. The time of failure is of course unknown; all that is known is that it failed sometime between the start of the test and the time at which it was removed. This type of censoring is also called *Type II censoring*. This type of censoring is rarely used, however.

With *interval censoring*, an item that is put on test is found to be functioning at a certain time thereafter (say, t_l), and then found to be not functioning when checked at a later time (say, t_u). All that is known is that the item failed sometime between t_l and t_u. This type of data is sometimes referred to as "readout data."

14.5 STATISTICAL TERMS AND RELIABILITY MODELS

We will let $f(t)$ denote an arbitrary failure distribution, with t denoting time (of failure). Since time is continuous, we will consider only continuous distributions. Therefore, we will not be concerned with $f(t)$ for a given value of t, but rather will be concerned with $F(t)$, the probability of failure by time t, and simple functions of it. Although the data that are collected are failure data, we will naturally focus attention on the positive side of the coin: nonfailure. Accordingly, we define $R(t) = 1 - F(t)$ as the *reliability function*, which is the probability that a unit is still functioning at time t. (In survival analysis this is referred to as the *survival function*.) The *failure rate* is defined as $h(t) = f(t)/R(t)$, which is sometimes referred to as the instantaneous failure rate.

Depending on the situation, the mean time to failure ($MTTF$) can also be of interest. It is defined as $MTTF = \int_0^\infty t f(t)\,dt$. Since $f(t)$ in a particular application may be highly skewed, the median life, defined as t_m such that $R(t_m) = .50$, may sometimes be preferable. If $MTTF$ is to be estimated, which of course would be necessary if $f(t)$ were not known, care should be exercised in using sample data for the estimation as early failures could greatly overestimate *MTTF* if the time interval for the sample is not reasonably long. Conversely, the *MTTF* could be underestimated if early failures are rare but failures frequently occur just beyond the sampling interval.

The mean time between failures (*MTBF*), mentioned in Section 14.2, is also a useful measure for repairable systems, although its usefulness is limited by the fact that most systems are nonrepairable, as stated previously. It is estimated by the total time that a system has been in operation divided by the total number of failures. The general form of a confidence interval for the *MTBF* is given in many books on reliability [e.g., see, Von Alven (1964) or Dhillon (2005)].

The mean residual life—*MRL(t)*—is also sometimes used. It is defined as

$$MRL(t) = \int_{o}^{\infty} R(x|t) \, dx = \frac{1}{R(t)} \int_{t}^{\infty} R(x) \, dx$$

with $R(x|t)$ denoting the *conditional survivor function* at time t. The mean residual life can be written in terms of the failure rate, as has been shown by Gupta and Bradley (2003).

These are all important functions in reliability work.

DEFINITION

The *reliability function* is given by $R(t) = 1 - F(t)$ and is the probability that a unit is still functioning at time t. The failure rate, also called the *hazard function*, is given by $h(t) = f(t)/R(t)$ and measures susceptibility to failure as a function of the age of the unit.

In words, $h(t)$ is the probability of failure at time t divided by the probability that the unit is still functioning at time t. It has been stated that the failure rate is the rate at which the population survivors at any given instant are "falling over the cliff." As mentioned previously, however, we technically cannot speak of probability of failure at a particular point in time because we use only continuous time-to-failure distributions. Therefore, borrowing a calculus argument, we should view $h(t)dt$ as approximately equal to the probability of a failure occurring in the interval $(t, t + dt)$, given that the unit has survived to time t, with $f(t) = (d/dt)F(t)$.

This is somewhat analogous to saying that, at a particular point in time, a car is traveling 50 miles per hour. If the driver doesn't drive for an hour, she will not travel 50 miles. But she *would* have traveled 50 miles if she had maintained that constant speed for an hour, so her interval of driving time would be $(t, t + 1$ hour).

We illustrate these concepts in the next few sections for some commonly used failure distributions.

14.5.1 Reliability Functions for Series Systems and Parallel Systems

As stated in Section 14.1, with a series system, failure occurs when one of the components of the system fails. If independence of the components is assumed, which may or may not be a plausible assumption, depending on the application, the system reliability is the product of the component reliabilities: that is, $R_s(t) = \prod_i R_i(t)$, with $R_s(t)$ denoting the system reliability at time t and $R_i(t)$ denotes the reliability for the ith series component at time t.

Some series systems consist of identical parts, such as cells of a battery. For k identical, independent components, $R_s(t) = [R_i(t)]^k$.

Series systems do not always have independent components, however, and determining the system reliability for dependent components is generally complicated and will not be pursued here. The interested reader is referred to Nelson (1982, p. 172). One approach that will sometimes work for a complex multicomponent system is to try to reduce it to a single-component system with a known reliability function, or at least a known cumulative distribution function.

The system reliability for independent components can be illustrated as follows.

■ **EXAMPLE 14.1**

Light-emitting diodes (LEDs) are part of the move toward brighter and more efficient lights. Assume that a system consists of two different types of LEDs, one whose lifetime is believed to be about 12,000 hours with a standard deviation of 900 hours and to be (approximately) normally distributed, and for the other LED a normal distribution with a mean of 11,000 hours and a standard deviation of 800 hours is believed to provide a good fit. (The question of whether or not a normal distribution would seem to be a viable model for this application is addressed in Exercise 14.29.)

The reliability for the two-component system is given by $R_s(t) = [1 - F_1(t)][1 - F_2(t)]$. For example, the system reliability at $t = 13{,}000$ is given by $R_s(13{,}000) = [1 - F_1(13{,}000)][1 - F_2(13{,}000)] = (1 - .8665)(1 - .9938) = .0008$. Thus, it is highly unlikely that the system will be functioning after 13,000 hours. ■

For parallel systems, the system fails when the last component fails, so the probability that the system is still functioning at time t is equal to one minus the probability that all components have failed by time t: that is, $R_s(t) = 1 - [F_1(t)][F_2(t)] \cdots [F_k(t)]$ for a parallel system of k independent components, and thus $R_s(t) = 1 - [F_i(t)]^k$ when the distributions are identical.

14.5.2 Exponential Distribution

Although the exponential distribution is no longer as heavily used in reliability applications as was once the case, it is a base distribution that has illustrative value. From Section 3.4.5.2, if we let $\lambda = 1/\beta$ in Eq. (3.14), we obtain $f(t) = \lambda \exp(-\lambda t)$ for the exponential distribution. It follows that $R(t) = \exp(-\lambda t)$, so that $h(t) = \lambda$. Thus, the hazard rate is constant for the exponential distribution and does not depend on t. (The exponential distribution is the only distribution that has a constant failure rate.)

Thus, an unused item would have the same reliability as a used item. Does this sound reasonable? It obviously doesn't apply to cars and people, for example. If it did, an infant would be just as likely to die while an infant as a 95-year-old would be to die at that age. Similarly, a new car isn't going to fail at the same rate as a 15-year-old car. And so on.

The "bathtub" curve is generally considered to be a reasonable representation of the failure rate for many, if not most, products, and is shown in Figure 14.1. That is, if a

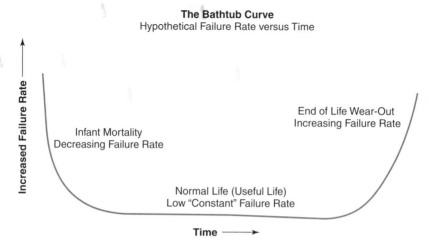

Figure 14.1 "Bathtub" curve of product life: hypothetical failure rate versus time.

very large sample of failure times was available for a particular product, a graph of the frequencies of the different failure times would at least roughly resemble the form of a bathtub. In words, the failure rate would be high when the product is first used (since the product could be defective); then the rate would level off and be flat for a long period of time, and finally the rate would increase as the product neared the end of its life.

Since this obviously sounds reasonable, we might restrict our attention to probability models whose failure rate would graphically correspond roughly to a bathtub. Such models are discussed in the following sections. (The bathtub model applies to both nonrepairable and repairable systems.)

14.5.3 Weibull Distribution

The two-parameter Weibull distribution, given in Section 3.4.6 (there is also a three-parameter Weibull distribution), is generally a better model for reliability applications than the exponential distribution and indeed is one of the most frequently used distributions in reliability work. This is partly due to the flexibility in fitting data that is afforded by a distribution that has two or three parameters. The distribution is named after Walodie Weibull, a Swedish engineer who proposed the distribution in 1939 as an appropriate distribution for the breaking strength of materials, although the distribution is used just about as often in analyzing lifetime data.

Extreme value theory can be used to justify the use of the Weibull distribution to model failure when there are many similar "competing" failure processes and the first failure causes the mechanism to fail. With $f(t) = (\beta/\alpha)(t/\alpha)^{\beta-1}\exp[-(t/\alpha)^\beta]$, it can easily be shown, as the reader is asked to do in Exercise 14.1, that $R(t) = \exp[-(t/\alpha)^\beta]$. It follows that $h(t) = (\beta/\alpha)(t/\alpha)^{\beta-1}$. Notice that $h(t)$ depends on t unless $\beta = 1$. (Recall from Section 3.4.6 that the Weibull distribution reduces to the exponential distribution when $\beta = 1$.) The shape of the curve when $h(t)$ is graphed against t will depend on the combination of values for α and β.

Therefore, for simplicity assume that $\alpha = 1$. We see that $h(t)$ is a constant (namely, $1/\alpha$) when $\beta = 1$. When $\beta > 1$, $h(t)$ is an increasing function of t and is a decreasing function of t when $\beta < 1$. Since β, if not equal to one, must obviously be either greater than one or less than one, we cannot capture the entire "bathtub." This is clearly true for any value of α. Therefore, what is frequently done is to use $\beta < 1$ and concentrate on modeling the portion of the bathtub that corresponds to the major portion of a product's life. As such, the Weibull distribution has been successfully used in modeling failure rates for material strengths, capacitors, and ball bearings, just to name a few.

The natural logarithm (base e) of a Weibull random variable has an extreme value distribution (Section 14.5.5), and the log of data well-fit by a Weibull is often used as that simplifies the analysis somewhat.

14.5.4 Lognormal Distribution

Another model worthy of consideration in reliability applications is the lognormal distribution. Recall from Section 3.4.8 that this distribution arises if the natural logarithm of a random variable has a normal distribution. Unlike the exponential and Weibull distributions, the lognormal distribution does not have a failure rate that can be written in a simple form. The lognormal has been used successfully in many applications to model the failure rates of semiconductor failure rate mechanisms such as corrosion, diffusion, metal migration, electromigration, and crack growth. It is also commonly used as a distribution for repair time, as well as being commonly used in the analysis of fatigue failures caused by fatigue cracks.

14.5.5 Extreme Value Distribution

Assume that a system consists of n identical components that are aligned in a series, and the system fails when the first of these components fails. System failure times then are the minimum of n random component failure times. Since the n components are identical, they would have identical distributions.

The *pdf* of the smallest extreme (i.e., the minimum) is given by

$$f(x) = \frac{1}{b} \exp\left(\frac{x-a}{b}\right) \exp\left[-\exp\left(\frac{x-a}{b}\right)\right] \qquad -\infty < x < \infty$$

with a the location parameter and b the scale parameter.

Since the extreme value distribution is the limiting distribution of the minimum of a large number of unbounded identically distributed random variables, this distribution would be a logical choice for modeling system failures. If n is large, the Weibull distribution might also be considered since the Weibull is the limiting distribution of the extreme value distribution. (An extreme value variant of the Weibull distribution is referred to as a Type III extreme value distribution, with the extreme value distribution itself labeled Type I.)

14.5.6 Other Reliability Models

The *gamma distribution* (Section 3.4.5) is infrequently used as a reliability model, although it is appropriate when standby backup units are in place that have exponential lifetimes. Although the proportional hazards model of Cox (1972) is frequently used in survival analysis, it has very limited applicability in engineering and thus won't be covered in detail here.

The Birnbaum–Saunders distribution (not covered in Chapter 3) results from a physical fatigue process for which crack growth (e.g., a crack in cement) causes failure. For more information on the model, see Birnbaum and Saunders (1969) and Section 8.1.6.6 of Croarkin and Tobias (2002).

14.5.7   Selecting a Reliability Model

As with modeling in general, there are various approaches that can be used to select a reliability model. As implied previously, there may be physical aspects that could suggest the choice of a particular model. If not, probability plots (illustrated in Section 6.1.2) and other plots can be performed when the data are failure data from nonrepairable populations, although probability plots for distributions other than the normal distribution are not available in some statistical software, and this is especially true for distributions such as the extreme value distribution. These plots are available in MINITAB, however, and Figure 14.2 is a panel of probability plots for some of the distributions that are available under MINITAB's reliability/survival menu.

Figure 14.2 shows that a Weibull distribution gives by far the best fit of the four fitted distributions, and in fact the data were simulated from a two-parameter Weibull distribution. (For the other distributions for which MINITAB automatically produces a probability plot, a three-parameter Weibull distribution gave a slightly better fit, as did a three-parameter lognormal distribution. This is not surprising as a model other than the model used in generating the data will often give a slightly better fit. Here, however, model parsimony

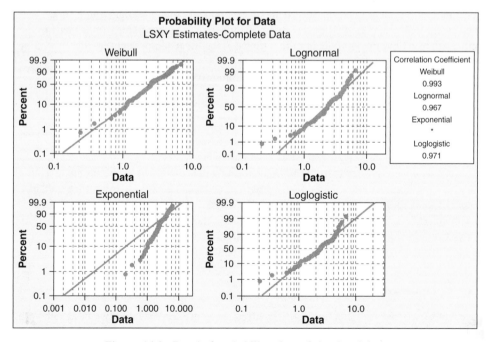

Figure 14.2   Panel of probability plots of simulated data.

would mandate that the simpler two-parameter model be fit since there is hardly any difference in the fit of the three models.)

Various statistical tests have also been given for determining whether or not a particular distribution provides a good fit to reliability data, and in particular, certain tests have been developed for the exponential distribution because of the central role that it has played in reliability work.

For repairable populations, trend plots and Duane plots can be used (see Croarkin and Tobias, 2002, Section 8.2.2.3), provided that appropriate software or computer paper is available, although use of the latter is an antiquated approach.

Selection of a life distribution model can be made using theoretical considerations or for other reasons such as a model simply providing a good fit. In general, the more parameters a distribution has, the greater modeling flexibility the distribution provides. Accordingly, we would expect the two-parameter Weibull distribution to be particularly useful, and the three-parameter Weibull distribution to provide even greater flexibility.

Furthermore, use of the Weibull distribution can be supported by extreme value theory, as was stated previously. That is, if multiple (competing) failure mechanisms exist and failure of the unit occurs at the occurrence of the first failure mechanism, the `Weibull distribution should be a prime candidate for the life distribution model.

The lognormal distribution will, in general, be a good candidate when failures result from chemical reactions. As stated previously, a gamma distribution will rarely give a good fit to failure time data as its failure rate generally does not represent the failure rates of products.

14.6 RELIABILITY ENGINEERING

Reliability engineering is an activity rather than a branch of engineering like mechanical engineering or civil engineering. As such, reliability engineering can be performed by design engineers, quality engineers, or systems engineers, in addition to reliability engineers. The objective is to determine the expected or actual reliability of a product, process, or service and of course to try to improve reliability. The American Society for Quality (ASQ) offers certification in reliability engineering in the form of its certified reliability engineer (CRE) designation. Various other certification programs are offered, including the one offered by the University of Tennessee. Reliability engineering is viewed as consisting of reliability, maintainability, and availability. (Maintainability refers to the field/activity of estimating the distribution of time to repair, and availability is a commonly used measure of performance of repairable systems.)

14.6.1 Reliability Prediction

Reliability prediction is part of reliability engineering, and by this we mean predicting future failures. Kvam and Miller (2002) stated "Engineering practitioners have been apprehensive about applying predictive inference to problems that involve the estimation of future failures." Reasons for this could certainly be debated, but it is inarguably true that prediction intervals outside the area of regression analysis have been almost completely ignored in courses and texts on engineering statistics. One impediment to the widespread use of prediction and prediction intervals in reliability engineering is that such intervals (e.g., for multiple failure models) lack the simplicity of the prediction interval form given in Eq. (7.1) and can be computer-intensive.

14.7 EXAMPLE

Here is a typical failure mode effects analysis (FMEA).

```
         Failure Mode of Circuit Breakers, Percentage of Total
                     Failures in Each Failure Mode

Failure Characteristics  Percentage of Total Failures, All Voltages

Back-up protective equipment required,                              9
  failed while opening

Other circuit breaker failures:

    Damaged while successfully opening                              7
    Failed while in service                                       32
    Failed to close when it should                                 5
    Damaged while closing                                          2
    Opened when it shouldn't                                      42
    Failed during testing or maintenance                           1
    Damage discovered during testing or maintenance                1
    Other                                                          1

        Total Percentage                                         100
```

[*Source*: J. C. Das, "Industrial Power Systems," in *Wiley Encyclopedia of Electrical and Electronics Engineering Online* (J. Webster, ed.). Used with permission of John Wiley & Sons, Inc.]

14.8 IMPROVING RELIABILITY WITH DESIGNED EXPERIMENTS

Just as the goal should be to improve quality, as emphasized throughout Chapter 11, improving reliability should also be a goal. Measuring reliability and predicting reliability are obviously important, but that isn't enough if product reliability is judged unacceptable. Wu and Hamada (2000, Chapter 12) discuss the use of experimental design for improving reliability. This does not require the use of esoteric experimental designs, as designs presented in Chapter 12 of this book can be used, but the manner in which the values of the dependent variable are determined does dictate how the analysis is to be performed.

Specifically, if the data are not censored, standard analysis methods can be used, unless one is concerned over the fact that the error terms in the linear model with failure time as the dependent variable cannot have a normal distribution since the failure times cannot be normally distributed because they are nonnegative. To adjust for this, Wu and Hamada (2000) advocated using a transformation, such as a log transformation, so that the transformed times can be negative. Although theoretically proper, regression models are usually applied without transforming the dependent variable when the latter can only assume positive values. Since normality doesn't exist in practice anyway, one might simply check to see if the residuals appear to be approximately normally distributed. The residuals will usually have closer to a normal distribution than will the errors when the errors are not normally distributed, but approximate normality for the residuals may be good enough. Otherwise, a log transformation would be advisable.

■ **EXAMPLE 14.2 (CASE STUDY)**

Problem Statement

Textbook problems are often much simpler than problems encountered in real life, with the consequence that students may fail to understand that the use of statistics to solve problems will not necessarily be easy. A good example of this is a problem that was faced by the Controls Division of the Eaton Corporation in Athens, Alabama. During 1988, the failure rate of an industrial thermostat assembly inexplicably rose to as high as 7% in certain batches. The employees decided to conduct an experiment, and subsequently conducted a second experiment that was guided by what they learned from the first experiment. Bullington, Lovin, Miller, and Woodall (1993) described only the results of the second experiment.

Experimental Design

Eleven factors were chosen for study and the experimenters decided to use the Taguchi L_{12} array, which is better known as a 12-run Plackett–Burman design (see Section 12.9.2). There were 10 observations for each of the 12 treatment combinations, and the 120 thermostats were cycled until failure, with testing stopped after 7,342,000 cycles. Thus, the data were right censored. The data are in CH14ENGEATON.MTW at the text website.

It should be noted that this is *not* a Plackett–Burman design with 10 replications, as each set of 10 thermostats manufactured at each treatment combination was made in the same batch of production.

This is an important point, as we would expect the experimental error to be less with this manufacturing arrangement than if, say, 120 different batches were used. The standard analysis of the data would be based on the assumption that the experiment was replicated, and of course the estimate of the error variance helps determine what effects are significant when a replicated design is used. So an underestimate of the error variance could cause certain effects to be falsely declared significant. (This relates to the discussion in Chapter 12 of multiple readings versus replication and the importance of determining the correct error term when trying to ascertain the significance of effects.)

Data Analysis

Indeed, the analysis given by Bullington et al. (1993) showed all 11 of the main effects (the only effects that are estimable) to be significant, with the largest p-value being .0025. Even though the selection of these 11 factors was based on the outcome of a previous experiment, it is unlikely that all of the effects are really significant. In an attempt to circumvent the analysis problem caused by each set of 10 observations being from the same batch, Bullington et al. (1993) used the average of the 10 numbers as a single response value. The data must then be analyzed as having come from an unreplicated design.

Analysis Problems

The Eaton experimenters decided to use only the two shortest failure times for each treatment combination in their analysis. Doing so discards 80% of the data and biases the results, however. In particular, for several treatment combinations the two shortest failure times differed considerably from some of the other eight failure times. Thus, this analysis is clearly objectionable.

The fact that all 11 main effects are significant when the 120 observations are used suggests that the error variance has probably been underestimated, but we can't draw that conclusion without first considering the (complex) alias structure of a Plackett–Burman design. Each main effect is partially aliased with every two-factor interaction that does not involve the main effect letter. For example, the main effect of A is partially aliased with every two-factor interaction that does not contain the letter A. That's $\binom{11}{2} = 55$ interactions! (The term "partially aliased" means that the coefficient of the effect in the alias structure is less than one.)

The main effect estimates are given in Table 14.1.

TABLE 14.1   Main Effect Estimates for Thermostat Experiment Data

Effect	Estimate
A	−0.62
B	0.44
C	−0.64
D	0.57
E	−2.05
F	0.46
G	−0.78
H	−1.11
I	−0.66
J	−0.55
K	−0.71

Clearly, E and H stand out, with all of the other effect estimates being below 1.00 in absolute value. [Factor E was the grain size of the beryllium that was used in the diaphragm and factor H was the length of time of the heat treatment. The various factors are described in detail in Bullington et al. (1993).] Wu and Hamada (2000) analyzed these data and noted that the coefficients (i.e., half the effect estimates) for the other nine effects are in a narrow range (.22 to .39 in absolute value). This suggested to them that these nine effects being significant could be due to their partial aliasing with the EH interaction. To check this, they removed factor B from the analysis (one factor had to be removed since the model was saturated). Their subsequent analysis showed, interestingly, that E, H, and EH were the only significant effects. Thus, their suspicion does seem to have been correct, and a meaningful analysis resulted despite the shaky nature of the multiple values that were not true replicates.

In MINITAB, reliability data are analyzed using the LREGRESSION command, which stands for "life regression." The sequence of MINITAB commands that produced Table 14.1 and other statistics are given at the textbook website. A distribution would have to be assumed for the error distribution, and of course this assumption would have to be checked.

One way to do so would be to construct a probability plot of the standardized residuals. We are immediately faced with the question of what should be the distribution of the standardized residuals. In this case study, the failure times were believed to follow a lognormal distribution. A distribution must be specified because the parameters are estimated using maximum likelihood. The standardized residuals could not have a lognormal distribution, even approximately, however, as this would mean that the failure times were never underestimated since the values of a lognormal random variable can never be negative. Clearly, this would be nonsensical.

We can avoid this dilemma by taking the logarithm of the failure times and using those transformed values in the analysis. Then the error distribution is normal, under the initial assumption of a lognormal distribution, and a normal probability plot of the standardized residuals could be constructed. The reader is asked to do this in Exercise 14.2 and interpret the plot.

Whether we construct a normal probability plot or simply look at the list of the standardized residuals, there are certain observations that stand out. In particular, the first observation is much smaller than the other observations for that treatment combination, and the same can be said for observation #51. Similarly, observations #101and #102 stand out, as does #120. Since there is considerable variability among the failure times for a given treatment combination, these may be valid numbers, but they should definitely be checked out.

14.8.1 Designed Experiments with Degradation Data

With improvement in quality, as has been the objective particularly during the past fifteen years or so, failure data is harder to come by because there may be very few failures during normal testing intervals. This can even be a problem with accelerated testing since a good accelerating factor may be hard to find.

Consequently, it may be necessary to use product degradation data in designed experiments rather than failure data. Wu and Hamada (2000, Section 12.8) gave a procedure that might be used with a designed experiment. As stated in Section 14.3.3 of this book, a detailed discussion of the use of degradation data, including its advantages and disadvantages, can be found in Section 8.4.2.3 of Croarkin and Tobias (2002). See also Nelson (1990, pp. 521–544) and Tobias and Trindade (1995, pp. 197–203).

14.9 CONFIDENCE INTERVALS

Various confidence intervals can be constructed in a reliability analysis, including confidence intervals on parameter values, but we are generally most interested in confidence intervals on percentiles. In confidence intervals such as those given in Chapter 5, the width of the confidence interval varies inversely with $\sqrt{n}$. This does not hold true for confidence intervals constructed in reliability analysis, however. Rather, the width of the interval varies inversely with the square root of the number of *failures*. Thus, having a large sample size isn't necessarily helpful, as we also need at least a moderate number of failures.

For the Bullington et al. (1993) data analyzed in Example 14.2, we would certainly be interested in constructing confidence intervals on selected percentiles. For example, we might be interested in the interval for the 50th percentile of the number of cycles before failure, and also, say, the 80th percentile, as well as the 95th and 99th percentiles, for a given combination of (coded) values of factors E and H, with the EH interaction also in the model. The results when factor E is at the high level (i.e., $+1$) and factor H is at the low level (i.e., -1) are as follows.

Table of Percentiles

Percent	Predictor Row Number	Percentile	Standard Error	95.0% Normal CI Lower	Upper
50	1	145.6535	20.6414	110.3296	192.2870
80	1	279.9207	41.9363	208.6951	375.4549
90	1	393.8559	62.9653	287.9100	538.7881
95	1	522.1653	89.0778	373.7653	729.4862

From casual inspection of the data we can see that the best settings would be both factors at the low level, so when those settings are used we obtain the following results:

Table of Percentiles

Percent	Predictor Row Number	Percentile	Standard Error	95.0% Normal CI Lower	Upper
50	1	6860.876	1166.750	4916.156	9574.885
80	1	13185.41	2490.279	9106.057	19092.24
90	1	18552.22	3760.132	12470.27	27600.44
95	1	24596.12	5304.854	16116.87	37536.39

Both the percentile estimates and the standard errors are very large numbers, with the latter resulting partly from the fact that most of the data for this treatment combination were censored, so this creates considerable uncertainty. Since the censoring value was 7342 and the estimate of the 90th percentile value is more than twice this value, there is obviously considerable extrapolation beyond the censoring value.

It is best that tables of this type be produced in menu mode in MINITAB because of the number of subcommands that would have to be entered if command code were used.

14.10 SAMPLE SIZE DETERMINATION

A simple formula for determining sample size, based on minimal assumptions, was given by Meeker, Hahn, and Doganaksoy (2004). Analogous to the Section 14.5 notation, let $R(t) = R$ denote the desired reliability for a specified value of t, and let α have the usual designation, as in a $100(1 - \alpha)\%$ confidence interval. Then

$$n = \frac{\log(\alpha)}{\log(R)} \tag{14.6}$$

■ **EXAMPLE 14.3**

To illustrate the application of the formula, assume that the reliability of a certain type of light-emitting diode (LED) is to be .99 for a specified number of hours of operation. If the degree of confidence is 95%, $n = \log(\alpha)/\log(R) = \log(.05)/\log(.99) = 298$, so this number of units would need to be used in the test.

If a Weibull distribution is assumed, Meeker and Escobar (1998) gave the sample size expression as

$$n = \frac{1}{k^\beta} \left(\frac{\log(\alpha)}{\log(R)} \right) \tag{14.7}$$

with β the Weibull shape parameter and k a constant to be selected. It is worth noting that Eq. (14.7) reduces to Eq. (14.6) if $\beta = 1$ with k chosen to be 1. ∎

It is also possible to use software such as MINITAB for such purposes as determining the sample size for a confidence interval on the reliability after a certain length of time and a specified value for the distance between the estimate of the reliability and a lower or upper confidence bound. Below we use the same time and parameters as in the preceding example and specify a distance of 0.05. This results in a sample size of $n = 44$, as can be seen.

```
Estimation Test Plans

Uncensored data

Estimated parameter: Reliability at time = 48
Calculated planning estimate = 0.965571
Target Confidence Level = 95%

Planning distribution: Weibull
Scale = 448.3, Shape = 1.5
```

	Sample	Actual Confidence
Precision	Size	Level
0.05	44	95.1233

14.11 RELIABILITY GROWTH AND DEMONSTRATION TESTING

Reliability growth testing is performed for the purposes of (1) assessing current reliability, (2) identifying and eliminating faults, and (3) forecasting future reliability. This is somewhat similar to the stages in the use of control charts, as discussed in Chapter 11. Reliability demonstration testing is used to see if a desired reliability level has been achieved after faults have been identified and removed.

For example, assume that a Weibull distribution with a scale parameter of 448.3 and a shape parameter of 1.5 (α and β, respectively, in the *pdf* for the Weibull distribution in Section 14.5.3 and also in Section 3.4.6) is used as the reliability model and no failures will be allowed. How many items must be tested for a period of 48 hours to demonstrate a reliability of 90% with a confidence level of 95%? This can be determined using appropriate software and only using software because the hand computations would be quite complicated. The MINITAB output for this is given below and we see that $N = 86$ items must be tested.

```
Demonstration Test Plans

Substantiation Test Plan
Distribution: Weibull, Shape = 1.5
Scale Goal = 448.3, Target Confidence Level = 95%


                            Actual
Failure  Testing  Sample  Confidence
   Test     Time    Size      Level
      0       48      86    95.0859
```

Reliability demonstration testing is related to demonstrating a mean time to failure (*MTTF*).

14.12 EARLY DETERMINATION OF PRODUCT RELIABILITY

As discussed by Mann, Schafer, and Singpurwalla (1974) and Meeker and Escobar (1998), traditional reliability methods fail when data are sparse. When test data are sparse, PREDICT (Performance and Reliability Evaluation with Diverse Information Combination and Tracking) can be used to estimate reliability. This is a formal, multidisciplinary process that combines all available sources of information. It is described in detail by Booker, Bement, Meyer, and Kerscher (2003).

14.13 SOFTWARE

14.13.1 MINITAB

MINITAB has excellent reliability capabilities, some of which were illustrated in the chapter. Capabilities not illustrated include accelerated life testing, with the Arrhenius equation being one of four equations from which the user can select. There are many other methods available, including parametric and nonparametric distribution analyses and growth curve analyses.

14.13.2 JMP

JMP has very limited capability for reliability analyses, and what is there is really survival analysis methods playing a dual role. Although it is true that the survival of a patient is roughly tantamount to a product still functioning after an extended period of time, obviously patients aren't "accelerated" so there are certain tools in reliability analysis that are not used in survival analysis.

14.13.3 Other Software

Other general purpose statistical software also have reliability analysis capability, in addition to software that is specifically for reliability analysis. One list of such software is given at `http://www.asq.org/reliability/resources`.

14.14 SUMMARY

Reliability engineering and reliability analysis are important parts of engineering. Reliability is distinct from quality (improvement) in that with reliability we are essentially trying to model and depict time to failure, whereas with quality improvement the objective is to prevent bad product from occurring very often and to reduce the variability of the values of process characteristics over the units of production. There is no attempt to try to determine how long an item will last.

Although the definition of quality implies quality over time, quality control/improvement methods do not require monitoring when the product is in customers' hands. Such monitoring would be of no value anyway in terms of quality improvement since the quality is set once it leaves the production process.

In order to model failures, there of course must generally be failures. If product reliability is so good that reliability testing doesn't produce practically any failures, we then have a problem similar to the problem caused by an extremely small percentage of nonconforming units when a p-chart (Section 11.9.1) is used.

The usual approach when regular testing is insufficient is to use accelerated testing. The success of accelerated testing depends very strongly on whether or not accelerating factors can be found that lead to failure. Such factors don't always exist.

A model that ideally fits the failure times well is necessary for both normal stress testing and accelerated testing. The Weibull distribution and lognormal distribution often work well as a model for normal testing, with the model for accelerated testing determined by the choice of the acceleration factor(s).

A good introductory-level presentation of the analysis of reliability data is Nelson (1983). Another useful reference not previously mentioned is Meeker and Escobar (1998), and Blischke and Murthy (2003) is a book that consists entirely of case studies, 26 of them. A good, short (199-page) introduction to engineering reliability, although at a higher level than this chapter, is Barlow (1998).

REFERENCES

Barlow, R. E. (1998). *Engineering Reliability*. Philadelphia: Society for Industrial and Applied Mathematics and Alexandria, VA: American Statistical Association.

Birnbaum, Z. W. and S. C. Saunders (1969). A new family of life distributions. *Journal of Applied Probability*, **6**, 319–327.

Blischke, W. R. and D. N. Prabhakar Murthy (2003). *Case Studies in Reliability and Maintenance*. Hoboken, NJ: Wiley.

Booker, J. M., T. R. Bement, M. A. Meyer, and W. J. Kerscher (2003). PREDICT: A new approach to product development and lifetime assessment using information integration technology. Chapter 11 in *Handbook of Statistics 22: Statistics in Industry* (R. Khattree and C. R. Rao, eds.). Amsterdam: Elsevier Science B.V.

Bullington, R. G., S. Lovin, D. M. Miller, and W. H. Woodall (1993). Improvement of an industrial thermostat using designed experiments. *Journal of Quality Technology*, **25**(4), 262–270.

Cox, D. R. (1972). Regression models and life tables. *Journal of the Royal Statistical Society, Series B*, **34**, 187–220.

Croarkin, C. and P. Tobias, technical editors (2002). *NIST/SEMATECH e-Handbook of Statistical Methods* (http://www.itl.nist.gov/div898/handbook/mpc/section3/mpc33.htm).

Dhillon, B. S. (2005). *Reliability, Quality and Safety for Engineers*. Boca Raton, FL: CRC Press.

Duane, J. T. (1964). Learning curve approach to reliability monitoring. *IEEE Transactions on Aerospace*, **2**, 563–566.

Fluharty, D. A., Y. Wang, and J. D. Lynch (1998). A simplified simulation of the impact of environmental interference on measurement systems in an electrical components testing laboratory. Chapter 15 in *Statistical Case Studies: A Collaboration Between Academe and Industry* (R. Peck, L. D. Haugh, and A. Goodman, eds.). Philadelphia: Society for Industrial and Applied Mathematics and Alexandria, VA: American Statistical Association.

Gupta, R. C. and D. M. Bradley (2003). Representing the mean residual life in terms of the failure rate. *Mathematical and Computer Modeling*, **37**, 1271–1280.

Hooper, J. H. and S. J. Amster (2000). Analysis and presentation of reliability data. In *Handbook of Statistical Methods for Engineers and Scientists*, 2nd ed. (H. M. Wadsworth, ed.). New York: McGraw-Hill.

Kvam, P. H. and J. G. Miller (2002). Discrete predictive analysis in probabilistic safety assessment. *Technometrics*, **34**(1), 106–117.

Mann, N. R., R. E. Schafer, and N. D. Singpurwalla (1974). *Methods for Statistical Analysis of Reliability and Life Data*. New York: Wiley.

Meeker, W. Q. and L. A. Escobar (1998). *Statistical Methods for Reliability Data*. New York: Wiley.

Meeker, W. Q., G. J. Hahn, and N. Doganaksoy (2004). Planning life tests for reliability demonstration. *Quality Progress*, **37**(8), 80–82.

Nelson, W. (1982). *Applied Life Data Analysis*. New York: Wiley.

Nelson, W. (1983). *How to Analyze Reliability Data*, ASQC Basic References in Quality Control: Statistical Techniques, Vol. **6**. Milwaukee, WI: American Society for Quality Control.

Nelson, W. (1990). *Accelerated Testing: Statistical Models, Test Plans, and Data Analyses*. New York: Wiley.

Rausand, M. and A. Høyland (2004). *System Reliability Theory: Models, Statistical Methods, and Applications*. Hoboken, NJ: Wiley.

Rigdon, S. E. and A. P. Basu (2000). *Statistical Methods for the Reliability of Repairable Systems*. New York: Wiley.

Spence, H. (2000). Application of the Arrhenius equation to semiconductor reliability predictions. Manuscript.

Tobias, P. A. and D. C. Trindade (1995). *Applied Reliability*, 2nd ed. New York: Chapman and Hall.

Von Alven, W. H. ed. (1964). *Reliability Engineering*. Englewood Cliffs, NJ: Prentice Hall.

Wu, C. F. J. and M. Hamada (2000). *Experiments: Planning, Analysis, and Parameter Design Optimization*. New York: Wiley.

EXERCISES

14.1. Derive the reliability function for the Weibull distribution that was given in Section 14.5.3.

14.2. Referring to Example 14.2, construct a normal probability plot of the standard-ized residuals, after first taking the (base e) log of the failure values. (If you use MINITAB, you can use the SPplot subcommand with the LREG main command in MINITAB to produce the plot.) Interpret the plot. In particular, are there any unusual values? If the standardized residuals appear to be ap-proximately normally distributed, then the untransformed failure times must be approximately lognormally distributed. Does the latter appear to be the case? Explain.

14.3. Verify Eq. (14.5).

14.4. A complex electronic system is built with a certain number of backup components in its subsystems. One subsystem has three identical components, each of which has a probability of .25 of failing within 1100 hours. The system will operate as long as at least one of the components is operating (i.e., a parallel system). What is the probability that the subsystem is operational after 1100 hours?

14.5. Suppose that the time to failure, in minutes, of a particular electronic component subjected to continuous vibrations can be approximated by a Weibull distribution with $\alpha = 50$ and $\beta = 0.40$. What is the probability that such a component will fail in less than 5 hours?

14.6. A freight train requires two locomotives for a one-day run. It is important that the train not arrive late, so the reliability of the locomotives should be determined and will hopefully be adequate. Past experience suggests that times to failure for such locomotives can be approximated by an exponential distribution with a mean of 42.4 days. Assume that the locomotives fail independently and determine the reliability, $R(t)$, for the train. Should the company be concerned about the reliability number? Explain.

14.7. It is known that the life length of a certain electronic device can be modeled by an exponential distribution with $\lambda = .00083$. Using this distribution, the reliability of the device for a 100-hour period of operation is .92. How many hours of operation would correspond to a reliability of .95?

14.8. Although a normal distribution is not one of the most frequently used distributions for modeling failure, it has been used to model certain types of failure behavior. Assume that the time to failure of an item is approximately normally distributed with a mean of 90 hours and a standard deviation of 10 hours.

(a) What is the numerical value of $R(t)$ for $t = 105$?
(b) What value of t corresponds to a reliability of .95?

14.9. Assume that an electronic assembly consists of three identical tubes in a series system and the tubes are believed to function (and fail) independently. It is believed that the function $f(t) = 50t \exp(-25t^2)$ adequately represents the time to failure for each tube. Determine $R_s(t)$. Looking at your answer, does $f(t)$ appear to be a realistic function? Explain.

14.10. The design of an electrical system is being contemplated, with an eye toward using enough components in a parallel system to provide acceptable reliability. Assume that the time to failure for each component is approximately normally distributed with a mean of 4500 hours and a standard deviation of 400 hours, and that the components are independent. If the probability of failure within the first 5,000 hours is to be less than .005, how many components should be used?

14.11. What is the failure rate for the time to failure distribution given in Exercise 14.10, if it can be obtained? If it cannot be obtained, explain why that is the case.

14.12. Determine the failure rate when the time to failure for a particular type of component is believed to be reasonably well-represented by a Weibull distribution with $\alpha = 40$ and $\beta = 0.35$. Graph the failure rate. Does this resemble a bathtub curve, as in Figure 14.1? Does the curve look reasonable relative to what we might expect for the hazard rate for a company whose products have a good reputation for good reliability? Explain.

14.13. Obtain the expression for the failure rate for a normal distribution if it is possible to do so. If it is not possible, explain why not.

14.14. Obtain the general expression for $R(t)$ when time to failure is assumed to have the extreme value distribution given in Section 14.5.5.

14.15. Show that the failure rate for the extreme value distribution given in Section 14.5.5 is not a constant.

14.16. A parallel system consists of three nonidentical components with failure probabilities during the first 100 hours of .052, .034, and .022, respectively. What is the probability that the system is still functioning at 100 hours?

14.17. Consider the following statement: "For a series system comprised of independent components, the reliability of the system is less than or equal to the lowest component reliability." Does that seem counterintuitive? Show why the statement is true.

14.18. Show that the reliability for a parallel system is greater than or equal to the highest component reliability.

14.19. If possible, give three examples of engineering applications for which a constant failure rate seems appropriate, and thus the exponential model would seem appropriate as a model for time to failure. If you have difficulty doing so, state how the failure rates would deviate from being constant in engineering applications with which you are familiar.

14.20. Critique the following statement: "Making normal stress estimates using accelerated testing is as risky as extrapolation in regression since in both cases the user is extrapolating from the range of the data."

14.21. A system need not be strictly a parallel or a series system, as many other combinations are possible and are in use. Assume that the Weibull model is being used and the system is such that three of the components are arranged as in a parallel system but the fourth component stands alone and must function for the system to function. Give the general expression for $R_s(t)$, assuming independence for the components.

14.22. The time to failure of a choke valve can be modeled by a Weibull distribution with $\beta = 2.25$ and $\alpha = 0.87 \times 10^4$ hours.
  (a) Determine $R(4380)$, the reliability at 4380 hours.
  (b) Determine the *MTTF*.

14.23. Determine the *MTTF* for the lognormal distribution in terms of the parameters of the distribution.

14.24. A machine with constant failure rate λ has a probability of .50 of surviving for 125 hours without failure. Determine the failure rate λ if $R(t) = \exp(-\lambda t)$.

14.25. Consider an exponential distribution with $\lambda = 2.4 \times 10^{-5}$ (hours) and determine the *MTTF*.

14.26. Consider $R(t) = 1 - (x/\beta)^\alpha$, $0 \leq x \leq \beta$, α, $\beta > 0$. Determine $f(t)$ and the failure rate.

14.27. Determine the failure rate for a continuous uniform distribution defined on the interval (a, b). Does this seem to be a reasonable failure rate? Are there any restrictions on t, relative to a and b? Explain.

14.28. Using the results of Exercise 14.26, determine the *MTTF*.

14.29. Consider Example 14.1 on LEDs. Do you think a normal distribution would generally be a viable model for such applications? Why or why not?

14.30. A computer has two independent and identical central processing units (CPUs) operating simultaneously. At least one CPU must be operating properly for the computer to function successfully. If each CPU reliability is .94 at a certain point in time, what is the computer reliability at that point in time? Does this reliability seem acceptable? Explain.

14.31. A safety alarm system is to be built of four identical, independent components, with the reliability of each component being p. Two systems are under consideration: (1) a 3-out-of-4 system, meaning that at least 3 of the components must be functioning properly, and (2) a 2-out-of-4 system. Determine the system reliability for each as a function of p. If the cost of the second system is considerably less than the cost of the first system, would you recommend adoption of the second system if $p \geq .9$? Explain.

14.32. Assume that a system has four active, independent, and nearly identical units that form a series system. If the reliability of two of the units after 1,000 hours is .94 and .95 for the other two units, what is the system reliability at 1,000 hours?

14.33. The first ten failure times for a repairable system are (in hours), 12,150, 13,660, 14,100, 11,500, 11,750, 14,300, 13,425, 13,875, 12,950, and 13,300. What is the estimate of the *MTBF* for that system?

14.34. Could a *MTBF* be computed for a nonrepairable system? Why or why not?

CHAPTER 15

Analysis of Categorical Data

Not all engineering data are in the form of continuous measurements, as count data are frequently obtained, with counts recorded of the observations in a sample that fall in each of two or more categories. In this chapter we consider the analysis of such data, using the approximate and/or exact methods of analysis that are available.

15.1 CONTINGENCY TABLES

In this section we consider an engineering-related example to illustrate a contingency table analysis.

■ **EXAMPLE 15.1**

Experiment Objective

Assume that operators in a particular plant are rotated among shifts and there is interest in seeing whether or not there is any relationship between operators and shifts in terms of performance. That is, do one or more operators perform better on certain shifts? Categorical data are often in the form of count data, and the company in question might decide to conduct an experiment involving three operators as they are rotated among the three shifts. The number of nonconforming parts that the operators are directly responsible for is recorded for each worker over a period of three months.

Data

The results are given in the table below, which is called a *contingency table*.

		Operator		
		1	2	3
Shift	1	40	35	28
	2	26	40	22
	3	52	46	49

Modern Engineering Statistics   By Thomas P. Ryan
Copyright © 2007 John Wiley & Sons, Inc.

Null Hypothesis

The null hypothesis for this problem is that the two classification variables, shift and operator, are independent. If that were true, the operators' performance *relative to each other* would be constant over the machines. That is, if the first operator has 30% more nonconformities on the graveyard shift (shift #3) than on the first shift, as we observe, then the other operators would perform (approximately) the same under the assumption of independence. And similarly for comparisons regarding the second shift.

We observe that these relationships do not hold in this example, so the question is whether there is enough of a departure from the numbers that would indicate independence to lead to the rejection of the hypothesis of independence.

In order to make that assessment, we need to compare the observed frequencies with the expected frequencies under the assumption of independence. As should be intuitively apparent, the expected frequencies are those that give the proper proportionate relationships for the columns and rows. For example, consider the following table:

1	2	5
2	4	10
3	6	15

The numbers in the second row are twice the numbers in the first row, and the numbers in the third row are three times the numbers in the first row and 1.5 times the numbers in the second row. This simultaneously causes the columns to also be proportional.

Viewed in another way, if the classification variables were independent, the probability of an observation falling in a particular cell would be equal to the probability of the observation being in the column times the probability of the observation being in the row. For example, let A_i denote the ith row and let B_j denote the jth column. Then $P(A_i \cap B_j) = P(A_i)P(B_j)$, under the assumption of independence. Let N denote the total count for all observations. The expected count for cell (i, j) is then $N \times P(A_i)P(B_j)$. We can simplify this by recognizing that $P(A_i) = n(A_i)/N$ and $P(B_j) = n(B_j)/N$, with $n(\cdot)$ denoting the number of observations in the indicated row or column. Therefore, letting E_{ij} denote the expected frequency for the (i, j)th cell, we have

$$E_{ij} = N \left(\frac{n(A_i)}{N} \right) \left(\frac{n(B_j)}{N} \right)$$

$$= \frac{n(A_i)n(B_j)}{N}$$

(15.1)

Preliminary Analysis

In words, the expected count in cell (i, j) under the assumption of independence is equal to the count for the ith row times the count for the jth column divided by the total number of observations. Applying Eq. (15.1) to the operator–shift example, we obtain the following results, with the expected frequencies in parentheses:

Operator

		1	2	3
	1	40 (35.96)	35 (36.87)	28 (30.17)
Shift	2	26 (30.72)	40 (31.50)	22 (25.78)
	3	52 (51.32)	46 (52.62)	49 (43.06)

The reader who wishes to do so can confirm that the ratios between the expected row numbers are constant across the columns, and thus the ratios between the expected column numbers are constant across the rows.

Note that these two classification variables are on a nominal scale rather than, say, an ordinal scale. That is, shift #3 is not higher than shift #1. When ordinal variables are involved, a contingency table approach will not be the best method of analysis, as a method should be used that incorporates the ordinal nature of the variables. See books on categorical data analysis such as Agresti (2002) or, at a lower mathematical level, Agresti (1996).

Assumption

Note also that an assumption of independence is required for the analysis. In general, this means that there should be random sampling, and for this example we may state further that there should be no dependencies in the data that comprise the counts. For example, whether a unit of production that is handled at a particular point in time by operator #2 on shift #1 is nonconforming or not should not depend on whether the previous unit handled by that operator on that shift was nonconforming or not.

Test Statistic

The next step is to develop a statistic that utilizes the differences between the observed and the expected values. One such statistic is

$$ \chi^2 = \sum_{i=1}^{m} \frac{(O_i - E_i)^2}{E_i} \tag{15.2} $$

with the summation extending over the cells in the table. For a large sample size, this statistic has approximately a χ^2 distribution with $(r - 1)(c - 1)$ degrees of freedom, with r denoting the number of rows and c denoting the number of columns. (The symbol χ^2 is used to denote the test statistic because that is what most writers use. This is analogous to stating that a t-statistic has a t-distribution.) Here $(r - 1)(c - 1)$ literally represents the number of entries that are free to vary. Specifically, with the column and row totals both fixed by design, as will be the case in some applications, if we know $(r - 1)(c - 1)$ of the entries in $(r - 1)$ rows and $(c - 1)$ columns, then the other entries can be determined by subtraction. Thus, $(r - 1)(c - 1)$ of the numbers are "free to vary."

This argument does not apply when the row and column totals are not both fixed, however. For a 2 × 2 table, the four | observed − expected | values are the same except for the sign (as the reader is asked to explain in Exercise 15.15), so they are related through the absolute value function. Thus, it can be claimed, as R. A. Fisher did, that there is only one independent value.

This argument does not apply to larger tables, however, as the deviations will generally all be different. The deviations must sum to zero, however, since the row totals of expected values must equal the row totals for the observations, and similarly for the columns. Therefore, the number of deviations that are free to vary in any column is one less than the number of entries in the column (i.e., the number of rows), and similarly for the rows. Thus, the degrees of freedom when the row and column totals are not fixed is still $(r-1)(c-1)$.

DEFINITION

An $R \times C$ contingency table has $(R-1)(C-1)$ degrees of freedom, with the χ^2 statistic in Eq. (15.2) having this number of degrees of freedom when the analysis is performed using Eq. (15.2).

Computer Output

Given below is (edited) output for this example, which was produced using the CHIS command in MINITAB.

```
                    Chi-Square Test

      Expected counts are printed below observed counts

                          Operator

                1           2           3         Total
Shift    1      40          35          28          103
                35.96       36.87       30.17

         2      26          40          22          88
                30.72       31.50       25.78

         3      52          46          49          147
                51.32       52.62       43.06

         Total  118         121         99          338

         Chi-Sq = 0.454 + 0.095 + 0.156 +
                  0.726 + 2.292 + 0.553 +
                  0.009 + 0.834 + 0.821 = 5.939

              DF = 4,   P-Value = 0.204
```

Conclusion

Since the p-value, which is $P(\chi_4^2 > 5.939 \mid \text{independence})$, is well above .05, we would logically not reject the null hypothesis of independence and thus conclude that there is not sufficient evidence to enable us to reject the null hypothesis. The output shows that there is a mixture of observations that seem to more or less conform to the independence hypothesis and those that don't conform, with the greatest difference between the observed and expected values occurring in the middle cell.

Note that a difference between the observed and expected values need not be great in order for a cell to make a substantial contribution to χ^2 if the denominator is small. For example, compare the contribution to χ^2 of the middle cell versus the contribution directly beneath it (i.e., in the third row, second column), relative to the respective differences between the observed and expected values. The latter are 8.50 and 6.62 for the middle cell and the other cell, respectively, but the contribution to χ^2 of the middle cell is more than 2.5 times the contribution of the other cell. ∎

Much more extreme comparisons can result when one cell has an expected frequency that is small (5 or less), but the other expected frequency in the comparison is not small. Accordingly, there are various rules-of-thumb that have been proposed regarding when this type of test can be used. One commonly applied rule is that the expected frequencies should all be at least 5. Other rules permit one expected frequency under 5, but none smaller than 3.

This raises the question of what should be done when a selected rule is violated. One possibility would be to collapse the table so as to combine cells, with this performed in such a way so as to remove the problem. Of course, this would not be possible if a table had only two rows and two columns.

Fortunately, there are exact tests that can be used, under certain conditions, when appropriate software and/or tables are available and the use of the χ^2 test would be questionable. These are discussed in the next section.

15.1.1 2 × 2 Tables

In this section we consider the smallest contingency table possible—a 2 × 2 table. For tables of this size it is easy to perform an exact analysis, rather than having to rely on the chi-square approximation. We will look at examples using each method of analysis and compare the results.

■ **EXAMPLE 15.2**

Study Objective and Data

Assume that someone searches through some clinical records of 50 smokers and 100 nonsmokers and wishes to test whether there is any relationship between smoking and lung cancer, this being a classical example. The data that are obtained might appear as follows:

	S	NS	
LC	40	20	60
NCL	10	80	90
	50	100	150

Here *S*, *NS*, *LC*, and *NCL* denote smoke, not smoke, lung cancer, and not lung cancer, respectively.

Without performing any statistical analysis, we can easily see from the data that there does appear to be a relationship between the two classification variables, and we will see that we will easily reject the hypothesis of independent classification variables.

One possible approach would be to again use the test statistic given by Eq. (15.2), as we want to again compare observed frequencies with expected frequencies. Since 50 smokers and 100 nonsmokers were sampled, we would expect the number of nonsmokers with lung cancer to be about twice as many as the number of smokers with lung cancer, if the variables were independent. Since 60 were observed with lung cancer, this means that 40 of the nonsmokers would be expected to have lung cancer, and 20 to not have it—just the reverse of what was observed. Once we have these expected frequencies, the other two are determined since, in particular, the column totals are fixed.

Analysis and Conclusion

Thus, we easily obtain the expected frequencies, and those frequencies and the observed frequencies are as follows, with the expected frequencies shown in parentheses:

	S	NS	
LC	40 (20)	20 (40)	60
NCL	10 (30)	80 (60)	90
	50	100	150

In a 2×2 table, all of the $(O_i - E_i)^2$ values will be the same. This follows from the fact that if one of the $O_i - E_i$ values in a given column is a, then the other one must be $-a$. Since the observed and expected frequencies in a given column must add to the same number—namely, what was observed—the differences must add to zero, and similarly for the rows.

Thus, in this example,

$$\chi^2 = \frac{400}{20} + \frac{400}{40} + \frac{400}{30} + \frac{400}{60} = 50$$

The degrees of freedom for a 2×2 table is 1 because once we have computed one expected frequency, the other three can be obtained by subtraction, as indicated previously. That is, only one number is free to vary. So, as in Example 15.1, it is easy to "see" what the term "degrees of freedom" means in this context. The computed value of χ^2 is quite large, so we know what the outcome of the hypothesis test will be. If we go through the formality of comparing the calculated value with the tabular value, we would be comparing the 50 against, say, $\chi^2_{.05,1} = 3.84$. Obviously, the observed value far exceeds the calculated value, so we reject the hypothesis and conclude that the two classification variables are dependent.   ■

15.1.2 Contributions to the Chi-Square Statistic

It is generally helpful to know which cell or cells are making the greatest contribution(s) to the value of the chi-square statistic. This was illustrated in Example 15.1 and will also be illustrated in this section for Example 15.2. The fact that the four components of the numerator of Eq. (15.2) are all the same (which as stated always occurs with a

2 × 2 contingency table) makes discussion of the contributions to the chi-square statistic easier than when the numerators differ.

MINITAB (edited) output for that example is as follows:

```
          Chi-Square Test

Expected counts are printed below observed counts

            C1           C2          Total
    1       40           20            60
          20.00        40.00

    2       10           80            90
          30.00        60.00

  Total     50          100           150

  Chi-Sq = 20.000 + 10.000 +
           13.333 + 6.667 = 50.000

  DF = 1, P-Value = 0.000
```

The output shows how highly dependent the value of the chi-square statistic is on the expected frequencies, as in this case all of the observed frequencies differ from their corresponding frequencies by the same amount. It should be apparent that a contribution could be large even for a relatively small difference between an observed and expected frequency when the expected frequency is small. This is why the test should certainly not be used when an expected frequency only slightly exceeds 1, with a commonly used rule of thumb, as stated previously, that no expected frequency be less than 5.

So what does a practitioner do when there are small expected frequencies? This issue is addressed in the next section.

15.1.3   Exact Analysis of Contingency Tables

The analysis given in the preceding section is approximate since the test statistic given by Eq. (15.1) does not have a chi-square distribution. Indeed, it should be apparent that if we take any function of count data, the result will not be continuous data, so the statistic given by Eq. (15.1) could not have a continuous distribution, and thus could not have a chi-square distribution.

Whenever any approximation is used, the expected adequacy of the approximation must be addressed, but this of course isn't known for all possible configurations of counts in a contingency table. Rather than worry about this problem, we can use an exact approach *if* the row and column totals are both fixed by design. Generally, this will not be the case, however, as either the column totals or the row totals (or both) won't be fixed. For example, if we decide to sample 50 smokers and 100 nonsmokers, as was done in Example 15.2, the number of people who have lung cancer and the number who do not have lung cancer will be random variables. Conversely, if we fix the number who have lung cancer and the number who do not, the number of smokers and the number of nonsmokers will then be random variables.

In the case of a 2×2 contingency table with fixed column and row totals, *Fisher's exact test* can be used. Fisher (1934) actually recommended that the test be used whether the marginal totals are fixed or not, but that view is not presently held. The test is performed by using the hypergeometric distribution (Section 3.3.3) to compute the probability of the observed configuration of counts, in addition to the probability of all possible configurations that are more extreme than what was observed. [With Tocher's (1950) modification, one does not include the observed configuration.] For example, if the first row of the table in Example 15.2 had been 41 and 19 and the second row had been 9 and 81, that would have been a (slightly) more extreme configuration, relative to the null hypothesis of independence, than what was observed.

It would be highly impractical to try to perform these computations by hand. Fortunately, software such as *StatXact* is readily available. Tables are also available when the number of counts in the different cells of the table are small.

An exact test can also be performed for larger tables, but again the marginal totals should be fixed. Of course, the use of exact tests is not as important for large cell counts, as then the chi-square approximation should be satisfactory.

■ **EXAMPLE 15.3**

Consider the following 4×3 contingency table, which is *StatXact* Example 9 at the Cytel website (http://www.cytel.com/Products/StatXact/example_09.asp). This is (likely hypothetical) minority discrimination data of firefighter exam scores with the column headings denoting "Pass," "No Show," and "Fail," respectively. The row labels are, in order, White, African-American, Asian-American, and Hispanic.

P	NS	F
5	0	0
2	1	2
2	0	3
0	1	4

There is no way that a chi-square computation could be performed since the observed frequencies are so small. The row totals are all 5 so they were apparently fixed. The column totals are clearly random, however, since they depend on the exam scores. This makes an exact analysis a bit risky, but that is about all that can be done, as clearly there is no way to collapse rows and columns to create a smaller table and use a chi-square analysis as there would still be expected frequencies less than 5, and the table would have also lost its original identity.

So risky though an exact analysis may be, it is about all that can be done with this example if a larger sample size cannot be obtained. The *StatXact* output for this example given at the Cytel website showed a *p*-value of .0727 for the chi-square analysis and a *p*-value of .0398 for the exact analysis. Thus, if .05 is used as the threshold value, the two approaches lead to different conclusions.

Since exact analysis of contingency tables larger than a 2×2 is generally impractical without the appropriate software, we will defer analysis to the next example. ■

■ **EXAMPLE 15.4**

Consider the following 2×2 table, with arbitrary classification variables.

A

		1	2
B	1	28	9
	2	4	11

In view of the extreme reversal for the elements of the two columns (28 is much larger than 9 but 4 is much less than 11), we would expect to reject the hypothesis of independence, and in fact the p-value is .002, as determined using statistical software. For example, the appropriate part of the output from *StatXact* is

```
Inference:

Hypergeometric Prob. of the table:                    0.001348

                              P-Value
    Type       DF    Tail    1-Sided      2-Sided      Point Prob.

Asymptotic    1     .GE.    0.0006119    0.001224
Exact               .GE.                 0.001622      0.001348
```

Since the one count that is below 5 is only slightly below 5 and the other counts are well above 5, we would expect the signal from the chi-square test to also strongly signal a lack of independence, but to have at least a slightly different p-value. In addition to the *StatXact* output given above, MINITAB gives the p-value as .001 when using the chi-square approximation, so there is only a slight difference. MINITAB does not use the continuity correction, however, and when that is used the p-value is.003. In general, continuity corrections should be used because there is a need to compensate for the "mismatch" that exists when a continuous distribution is applied in determining probabilities and p-values for a discrete random variable. ■

In this example there was no harm in using the (approximate) chi-squared test since it leads to the same conclusion as results from application of the exact test. Even though one of the observed frequencies is less than 4, the expected frequency for that cell is 9.23, and as stated previously, it is the expected frequencies that determine whether or not a chi-square approach should be used.

We can see the deleterious effect of small expected frequencies with the following example, which has three small expected frequencies (shown in parentheses along with the other expected frequency):

	A_1	A_2	Totals
B_1	13 (10.91)	2 (4.09)	15
B_2	3 (5.09)	4 (1.91)	7
Totals	16	6	22

With such small expected frequencies, we would expect the chi-square approximation to be poor. That approximation produces a p-value of .032, whereas the use of Fisher's exact test produces a p-value of .054. Thus, if .05 is used as a threshold value, the decisions reached with each approach would differ, as was the case with Example 15.3.

Of course, Fisher's exact test would be rather difficult to justify if the column and row totals were not fixed by design. Indeed, Fisher's test has been controversial over the years, and the analysis of contingency tables in general has sparked much debate.

One important point should be made regarding the exact test versus the chi-square test. Assume a True–False test with 100 questions, with 50 being true and the other 50 false. A student supposedly guesses on each question and obtains 64 correct answers, with "true" being correctly guessed 32 times and "false" correctly guessed the other 32 times. How likely is this to happen. If we run this problem through MINITAB or other statistical software, we obtain a p-value of .005. We would also, however, obtain the same p-value if only 18 questions of each type were answered correctly. If we hypothesized "guess," we would reject it in both cases, but we wouldn't expect a person to do worse than 50% accuracy by guessing, so we would view the second result much differently than the first result. The bottom line is that the test of interest here is one-sided but we can't select the side when the chi-square test is used and obtain the correct p-value in computer output relative to the alternative hypothesis. Although the latter is generally "not independent," we are usually interested in one side or the other, such as good performance or bad performance.

This is unlike the exact test of a contingency table, for which only the appropriate direction would be considered. There are many applications (such as in Exercise 15.26) for which the appropriate direction is clear. For example, assume that a basketball referee trainee is shown forty 10-second sequences of a game in which a foul has been committed or not committed, but the call is not an easy one, and the trainee is expected to make the right call on each one. If there are 20 sequences in which a foul was committed and 20 in which a foul was not committed, it would be ideal for the trainee to obtain the correct decision for all 40 sequences. Let's assume that 18 of the fouls were correctly identified and 16 of the non-fouls were properly identified. Since this is a good score, we would expect the null hypothesis of independence to be rejected. We would obtain the same p-value for the chi-square test, however, if we have the reverse: 2 of the fouls properly identified and 4 of the non-fouls correctly identified. Since we want to declare "good performance," we want the p-value to be small when this occurs. For each of these two sets of numbers the p-value is 0.000 to three decimal places, whereas for the second set the p-value should be 1.000.

With the exact test, the p-value is computed as the probability associated with tables at least as extreme, relative to the alternative hypothesis, as the observed configuration (ignoring Tocher's modification). Only with extremely small cell counts would it be practical to do this by hand, using the hypergeometric distribution (Section 3.3.3) to compute the probability for each set of counts.

Tables are available for small values of N and software packages for exact tests are available in SAS Software, *StatXact*, and Fisher's exact test in MINITAB. There are also Java applets for exact tests that are available on the Internet. For example, as of this writing one such applet can be found at `http://www.matforsk.no/ola/fisher.htm` and the use of it for the example in this section gives a p-value of .0016, in agreement with the .002 value that was given earlier.

15.1.4   Contingency Tables with More than Two Factors

It should be noted that there will often be a need to consider more than two classification factors, especially when one of the factors can be viewed as a dependent variable, with the other factors being independent variables. One might design an experiment with a categorical random variable as the dependent variable (as discussed in the next section), with all of the other factors also being categorical variables. In that case, a multiway contingency table would result, with the data perhaps analyzed in that manner, or with a log-linear model, which would be a more flexible approach. Log-linear models are beyond the scope of this text; they are covered extensively in Christensen (1997) and also in other books, including Lindsey (1995).

15.2   DESIGN OF EXPERIMENTS: CATEGORICAL RESPONSE VARIABLE

Example 15.1 at the beginning of this chapter could be viewed as a designed experiment of sorts in that the objective was to see whether operator and shift were independent in terms of nonconforming units. This can be couched in experimental design terms by stating that we are testing for interaction. Indeed, if we construct a graph of the points in the table that is in the form of an interaction profile (see Section 12.5.1), we see that the lines cross, thus suggesting that there is more than moderate interaction.

This says nothing about the existence of main effects, however, although if we followed the general advice of putting in our model the terms that comprise each interaction, we would include the two main effects in this case. The shift effect, as evidenced by the disparate row totals, may be nonremovable; the column totals suggest that it might be desirable to monitor the operators, perhaps by the use of control charts (see Chapter 11).

There may be other factors that affect the rate of nonconforming units, however, so what is needed is a designed experiment. Bisgaard and Gertsbakh (2000) pointed out that often the only reliable response data in the production of many items such as cathode-ray tubes, metal castings, and plastic moldings is the classification of a production unit as either conforming or nonconforming. If the factors that cause a unit to be conforming or nonconforming can be identified, then those factors might be monitored with control charts with the goal of keeping the factors at their respective levels that seem best for maximizing the proportion of conforming items. Indeed, Bisgaard and Gertsbakh (2000) stated: "Consequently it ought to be standard practice to experiment with processes to screen out the most important factors influencing the rate of defectives, and when those have been identified, engage in optimization experiments . . . .".

Management is often reluctant to conduct such experiments, however, fearing that substantially more nonconforming units may be produced during the experimentation. If the percentage of nonconforming units is to be reduced, though, the factors that contribute to the rate of nonconforming units must be identified. One solution to this problem is to make very small changes to the process conditions, in the spirit of EVOP (see Section 12.20). One benefit of the variable sampling scheme of Bisgaard and Gertsbakh (2000) is that there is a movement away from process conditions that are unfavorable in terms of increasing the likelihood of producing nonconforming units.

All things considered, it is difficult to rationalize the avoidance of designed experiments that will lead to a reduction in the percentage of nonconforming units. Nevertheless, if it is deemed impractical to use designed experiments for this purpose, but there is a goal of identifying factors that affect the categorical response variable, *logistic regression* can be used for this purpose, with the conditions that produce the values of the response variable noted (e.g., conforming or nonconforming), with the conditions representing values of the independent variables. Logistic regression was covered briefly in Section 9.10.

15.3   GOODNESS-OF-FIT TESTS

Categorical data also result when distributions are fit to data and a comparison of fitted values with observed values is performed, using a chi-square test or other test. More specifically, for a discrete distribution there are natural categories, namely, the possible values of the random variable if the latter is discrete and the number of possible values is finite. For a continuous random variable, intervals are created, similar to what is done when a histogram is constructed, and a goodness-of-fit test is performed that uses the differences between the observed and fitted values, just as is done in the discrete case.

■   **EXAMPLE 15.5**

Study Objective

To take a simple example, let's assume that you find a single die (as in "a pair of dice") in an old trunk, with the die appearing to be affected somewhat by age. You would like to know if the die could still be used in games of chance. Since the die appears to be somewhat scuffed up, you decide to test the hypothesis that the die is still usable (i.e., balanced). Translated into statistical terms, this means that you are testing the hypothesis that the number of spots showing on the die when it is rolled has a discrete uniform distribution for the six possible outcomes. That is, the distribution is $f(x) = 1/6$, $x = 1, 2, 3, 4, 5, 6$.

Experiment

So we conduct an experiment and roll the die 300 times, record the outcomes, and try to determine if the extent of the departure of the data from the theoretical values under the discrete uniform distribution is sufficient to lead to rejection of that distribution: that is, to reject the hypothesis that the die is balanced. Assume that the results are 44, 55, 51, 49, 53, and 48, for 1–6, respectively. Of course, each theoretical value is 50, so it is a matter of determining whether or not the extent of the observed departures from the theoretical values could have occurred by chance.

Exact Analysis

The exact analysis would entail determining the probability of obtaining a configuration that is at least as extreme as the observed configuration when the null hypothesis is true. That would be computationally complex but a chi-square goodness-of-fit test should be a good substitute when none of the expected frequencies are small (i.e., less than 5), as in this example.

Chi-Square Test

The chi-square statistic given in Eq. (15.2) is the appropriate statistic and the computations yield $\chi^2 = 1.52$, with each E_i in the formula equal to 50 for this example. (Since each $E_i = 50$, this number can be seen as the sum of the squared deviations from 50 divided by 50. That produces $76/50 = 1.52$.)

Conclusion

The degrees of freedom = number of categories $-1 = 5$, and the p-value can then be shown to be .0892. Since the p-value is not small, there is not sufficient evidence to reject the hypothesis that the distribution is the one that was fit. (Note that in this case there were no parameters in the fitted distribution to estimate. In general, the degrees of freedom for the chi-square statistic is the number of categories minus one minus the number of estimated parameters.) ■

In this example the categories were naturally defined, whereas this generally won't be the case for a continuous distribution. The comment made in Section 1.5.1 about needing to know the shape of the unknown distribution in order to construct the classes for a histogram in such a way that the histogram will be likely to reflect the distribution shape of course also applies here. Therefore, a hypothesized distribution could provide a less-than-excellent fit simply because of a bad choice for the number of classes coupled with inherent random variation in the data.

As an extreme example, assume that a normal distribution is hypothesized and only three class intervals are used. Could a histogram constructed from these three intervals have tails? That is, could the histogram "tail off" as one moves away from the mean as a normal distribution does? Obviously, the answer is "no," so this would be a very poor choice for the number of class intervals to test the assumption of normality. From a practical standpoint, one could use a number of classes that would seem to be a good choice for the distribution that is being tested (see Scott, 1979) and, especially if the hypothesis is rejected, one could vary the number of classes slightly to see if this affects the hypothesis test conclusion.

■ **EXAMPLE 15.6**

Assume that we have a sample of 98 observations and we hypothesize that the data have come from a population with a $N(100, 25)$ distribution. We will apply the power-of-2 rule given in Section 1.5.1.

With 98 observations, the power-of-2 rule specifies seven intervals. A decision must be made regarding how the intervals should be constructed: Should they be constructed from the data or from the fitted distribution? The customary procedure is to construct the intervals based on the fitted distribution. Remember that when a chi-square statistic is computed, we don't want to have any small expected frequencies; determining the class intervals from the fitted distribution (and perhaps having equal probability intervals) will prevent that from happening.

Therefore, assume that we want each interval to be a 1/7 probability interval. It can be shown that the intervals are then $(-\infty, 94.662]$, $(94.662, 97.170]$, $(97.170, 99.100]$, $(99.100, 100.900]$, $(100.900, 102.830]$, $(102.830, 105.338]$, and $(105.338, \infty)$, with "]" indicating that the end value is included in the interval and "(" or ")" indicating that the value is not included in the interval. The 98 simulated observations lead to interval counts of 18, 11, 16, 12, 7, 13, and 21, respectively, for the seven classes. This results in a value of 9.2429 for the chi-square statistic and an associated p-value of .1604. Therefore, despite a major difference from the expected frequencies in two of the classes, there is not sufficient evidence to reject the hypothesis of a $N(100, 25)$ distribution. Thus, the appropriate decision is reached. (Chi-square goodness-of-fit tests can be performed in MINITAB, but the user has to perform them manually by doing the appropriate computations.) ■

There are other goodness-of-fit tests, including the Kolmogorov–Smirnov test. Whereas the chi-square goodness-of-fit test is essentially for discrete distributions but is also used for continuous distributions, the Kolmogorov–Smirnov test can be applied only to continuous distributions. The test uses the cumulative distribution function (*cdf*) (see Section 3.3), and the test statistic is the largest absolute value of the difference between the empirical *cdf* (i.e., the *cdf* for the data) and the theoretical *cdf*.

For testing the fit of a normal distribution, the Shapiro–Wilk test is widely favored because it has good properties in testing against a variety of alternatives.

Of course, we should bear in mind that fitting a normal distribution and using a chi-square test is an alternative to simply constructing a normal probability plot—which is much easier when the software that is being used has that capability. Furthermore, in MINITAB the result of the Anderson–Darling goodness-of-fit test is given, and the results of two other goodness-of-fit tests, the Ryan–Joiner and Kolmogorov–Smirnov tests, can also be displayed, if desired.

Some statistical software has the capability for probability plots for many distributions, and MINITAB has this capability for 14 distributions. This reduces somewhat the need for goodness-of-fit testing.

15.4 SUMMARY

Categorical data frequently occur in engineering, especially binary data that is in the form of go/no-go data. The analysis of such data can be performed in various ways, including using contingency tables. Only two-way tables, corresponding to two classification variables, were illustrated in this chapter, but three-way tables and higher order tables can also be used.

It should be kept in mind that a contingency table analysis will not necessarily be the appropriate method of analysis whenever there are two classification factors (variables). Oftentimes, a measure of association between the two variables is needed, not a test to determine whether or not the variables are independent.

One important but relatively untapped area is in trying to determine the cause of no-go data by modeling the data.

REFERENCES

Agresti, A. (2002). *Categorical Data Analysis*, 2nd ed. Hoboken, NJ: Wiley.

Agresti, A. (1996). *An Introduction to Categorical Data Analysis*. New York: Wiley.

Bisgaard, S. and I. Gertsbakh (2000). 2^{k-p} experiments with binary responses: inverse binomial sampling. *Journal of Quality Technology*, **32**(2), 148–156.

Christensen, R. (1997). *Log-Linear Models and Logistic Regression*, 2nd ed. New York: Springer-Verlag.

Fisher, R. A. (1934). *Statistical Methods for Research Workers*, 5th ed. Edinburgh, UK: Oliver & Boyd.

Fisher, R. A. (1935). *The Design of Experiments*. Edinburgh, UK: Oliver & Boyd.

Lawrence, W. F., W. Liang, J. S. Mandleblatt, K. F. Gold, M. Freedman, S. M. Ascher, B. J. Trock, and P. Chang (1998). Serendipity in diagnostic imaging: Magnetic resonance imaging of the breast. *Journal of the National Cancer Institute*, **90**(23), 1792–1800.

Lindsey, J. K. (1995). *Modeling Frequency and Count Data*. Oxford: Clarendon Press.

Scott, D. W. (1979). On optimal and data-based histograms. *Biometrika*, **66**, 605–610.

Tocher, K. D. (1950). Extension of the Neyman–Pearson theory of tests to discontinuous variates. *Biometrika*, **37**, 130–144.

EXERCISES

15.1. The following survey data have been listed by Western Mine Engineering, Inc. at their website:

	Number of Mines (Union)	Number of Mines (Non-Union)
Wages increased in past 12 months	58	70
Wages decreased in past 12 months	0	3
No change in wages in the past 12 months	18	44

Can the hypothesis that wage changes are independent of whether a mine is a union or a non-union mine be tested with a χ^2 test? If not, can the table be modified in a logical way so that the test can be performed? If so, perform the modification and carry out the result. What do you conclude if you performed the test?

15.2. Given the following 2×2 contingency table,

$$A$$

B	25	35
	20	30

assume that the marginal totals are random and determine the expected frequency that corresponds to the observed frequency of 30 in testing the hypothesis that A and B are independent.

15.3. On February 12, 1999 the United States Senate voted on whether to remove President Clinton from office. The results on the charge of perjury are given in the following table:

	Democrat	Republican	All
Not guilty	45	10	55
Guilty	0	45	45
All	45	55	100

(a) Assume that someone decided (naively) to do a chi-square test on these data to test whether party affiliation and decision are independent. From a purely statistical standpoint, would it make any sense to perform such a test since the U.S. Senate was not sampled, and indeed this is the vote tabulation for all senators? Critique the following statement: "There is no point in performing a chi-square test for these data since we would be testing the hypothesis that Democrats and Republicans vote the same and I can see from inspection of the table that this is not true."

(b) Assume that the following counter argument is made. "No, this *is* a sample from a population since this is not the first time that the Senate has voted on whether or not to remove a president, and it probably won't be the last time. Therefore, we can indeed view these data as having come from a population—for the past, present, and future." Do you agree with this position?

15.4. Consider a 4×5 contingency table. How many degrees of freedom are associated with the table and what does "degrees of freedom" mean for a contingency table?

15.5. Give an example of a 3×3 contingency table for which the value of χ^2 is zero.

15.6. Compute the value of χ^2 for the following 2×2 table and test whether classification variables A and B are independent:

$$A$$

B	25 36
	20 54

15.7. A company believes it has corrected a problem with one of its manufacturing processes and a team that worked to correct the problem decides to collect some data to test the belief. Specifically, the team decides to compare the results for the week before the correction with data they will collect during the week after the correction. The team collects that data and the results are summarized in the following table:

	Conforming	Nonconforming
Before	6217	300
After	6714	12

(a) Could a contingency table approach be used to analyze the data? Why or why not?

(b) If so, will the results agree with a test of the hypothesis of two equal proportions, using methodology given in Chapter 6? Explain.

(c) Perform a contingency table analysis, if feasible.

(d) Would it be possible to use Fisher's exact test? Why or why not?

15.8. The following table was constructed using data from the Spring 2002 Fraternity and Sorority Report of Purdue University:

	Men	Women
Freshman	4,160	2,836
Sophomore	4,204	3,047
Junior	4,010	2,919
Senior	4,240	3,056

(a) Assume that we are interested in determining if the ratio of men to women is independent of class standing. If a contingency table approach were used, what would be the population to which the inference were applied?

(b) The table given here is a 4×2 table. Because of the hypothesis to be tested, must the data be arranged in a 2×4 table instead of a 4×2 table before the analysis can be performed? Explain.

(c) State the null hypothesis for the contingency table analysis, perform the appropriate analysis, and draw a conclusion.

15.9. Assume a 2×2 contingency table and state the null hypothesis in terms of proportions.

15.10. Generate 98 observations using a $N(50, 25)$ distribution, use the number of classes specified by the power-of-2 rule given in Section 1.5.1, and use equal probability classes as in Example 15.6. Test the hypothesis that the data came from the $N(50, 25)$ distribution. What do you conclude? Repeat this exercise using one more class than you just used, and then one fewer class. Is your conclusion sensitive to the number of classes used? Explain.

15.11. Generate 98 observations from the exponential distribution (see Section 3.4.5.2) with $\beta = 10$. Test the hypothesis that the data are from a population with a $N(10, 2)$ distribution. Use the power-of-2 rule to determine the number of equiprobability classes. What do you conclude? Would you anticipate that the result would be sensitive to the number of classes used? Explain.

15.12. Give a 2×2 contingency table for which the numerical value of χ^2 is zero.

15.13. Fill in the missing value in the following 2×3 contingency table such that the value of χ^2 is zero.

20	35	18
24	42	

15.14. Assume that you want to test if attendance at a particular high school multiclass reunion differs across the classes, but all you have are the attendance figures. What else would you need, if anything, in order to be able to conduct the test? If you had that information and were able to perform the test, what would be the hypothesis that you were testing, and for that matter would it make any sense to perform the test?

15.15. Consider a 2×2 contingency table and explain why | observed − expected | is the same for each cell.

15.16. Consider a 3×3 contingency table. What is the value of χ^2 if every number in the second row of the table is 1.7 times every number in the first row and every number in the third row is 2.4 times every number in the second row? If the third number in the third row is increased by 10%, can the value of χ^2 be determined? Why or why not?

15.17. The faculty of a small midwestern university is surveyed to obtain their opinion of a major university restructuring proposal. The results are summarized in the following table:

	Support	Opposed	Undecided
Men	116	62	25
Women	84	78	37

(a) Assume that every faculty member was surveyed. Would you perform a contingency table analysis of these data? Why or why not? If the analysis would be practical, perform the analysis and draw a conclusion.

(b) Now assume that a random sample of the faculty was used and the results in the table are for that sample. Now would you perform a contingency table analysis? Why or why not? If the analysis would be practical, perform the analysis and draw a conclusion.

15.18. Explain in detail why a 3×4 contingency table has 6 degrees of freedom, with "degrees of freedom" in this context literally meaning how many observations are free to vary, as was explained in Section 15.1.

15.19. Consider a 2×2 contingency table and the difference in each cell between the observed and the expected value. What is the sum of those differences over the four cells? What is the sum of the absolute values of those differences if one of the differences is 8?

15.20. Generate 200 observations from an exponential distribution with a mean of 5, which can be done in MINITAB in command mode with the RANDOM command and the appropriate subcommand, or by using menu mode. (The exponential distribution was covered in Section 3.4.5.2.) Read the paper by Scott (1979) (see References) and construct a histogram in accordance with the number of intervals suggested by the paper, then test for goodness of fit when an exponential distribution with a mean of 5 is fit to the data. Comment.

15.21. Agresti (2002) (see References) gave a 2 × 2 table in which the entries are proportions that add to 1.0 across the rows. Is this a contingency table? Why or why not?

15.22. Consider a 2 × 4 contingency table with every value in the second row equal to 1,000 times the corresponding value in the first row (admittedly a hypothetical situation). What will be the numerical value of the χ^2 statistic?

15.23. Assume that a 2 × 2 table is included in a report with the cell proportions given. The report sustains water damage, however, and only the row totals and one of the cell values is discernible. Assume that the proportions are proportions relative to fixed column totals of 1.0, with the row totals not summing to 1.0. Can the numerical value of the χ^2 statistic be computed? Why or why not?

15.24. Agresti (2002) (see References) gave the following 2 × 2 table, with the data discerned from Lawrence et al. (1998) (see References).

Diagnosis of Test

Breast Cancer	Positive	Negative	Total
Yes	0.82	0.18	1.0
No	0.01	0.99	1.0

(a) Could Fisher's exact test be applied to these data? Why or why not?
(b) Is there any other test that could be applied, using only the table? If so, apply the test and draw a conclusion. If not, explain what additional information would be needed.
(c) If any additional information is needed, obtain it from the journal article, if possible, and perform an appropriate test.

15.25. Consider the following 2 × 3 contingency table for arbitrary factors A and B:

A

		1	2	3
B	1	30	58	67
	2	24	50	55

Without doing any calculations, explain why the use of Eq. (15.2) will not result in the rejection of the null hypothesis that the factors are independent.

15.26. Although there will not be many applications in which Fisher's exact test will be strictly applicable since it is unusual for both the column and row totals of a 2 × 2 contingency table to be fixed, the test was motivated by an experiment, the famous tea-tasting experiment, for which the test does strictly apply. The experiment, described in Fisher (1935) (see References) involves one British woman who was a colleague of Fisher. When drinking tea, she claimed to be able to determine whether milk or tea was added to the cup first. The experiment involved 8 cups of tea, with the tea being added first in four of the cups and milk added first in the other four cups. The results of the experiment are given below.

| | Added First | | |
Guess	Milk	Tea	Total
Milk	3	1	4
Tea	1	3	4
Total	4	4	8

(a) To this point, not enough information has been given to allow Fisher's exact test to be strictly applied. What other condition must be met, and in fact was met?

(b) Perform the test, using either appropriate software, one of the many Java applets that can be found on the Internet, or hand calculation since the latter is practical for this dataset. Do the data support the woman's claim? Explain.

15.27. The following 2×2 contingency table was given by Agresti (2002) (see References), with the source of the data being the Florida Department of Highway Safety and Motor Vehicles and the data from accident records in 1988.

| | Injury | |
Safety Equipment in Use	Fatal	Nonfatal
None	1,601	162,527
Seat belt	510	412,368

(a) What is the hypothesis that would be tested?

(b) Would Fisher's exact test be applicable here? Why or why not?

(c) Perform the appropriate test and draw a conclusion.

CHAPTER 16

Distribution-Free Procedures

Since "all models are wrong but some are useful," as G. E. P. Box was quoted in Chapter 10, the question naturally arises as to how good our results are when we use methods based on certain distributional assumptions that we know are not valid. We often invoke the Central Limit Theorem and argue that this theorem ensures that we will have approximate normality; but how good the approximation is depends on the application, as has been observed in previous chapters, and it also depends on the shape of the distribution for the population from which the sample was obtained.

For example, using a sample of size 25 to construct a confidence interval for the mean of a population that we assume to have a normal distribution won't cause a major problem if there is not more than a slight-to-moderate deviation from normality. On the other hand, constructing an $\overline{X}$ control chart with a subgroup size of 5, a common choice as indicated in Chapter 11, can create problems if there is a moderate departure from normality.

16.1 INTRODUCTION

When we use a *distribution-free procedure* (also called a nonparametric procedure), we do not make any distributional assumptions. It follows that these procedures do not utilize test statistics of the form that have been presented in the preceding chapters, as those have known distributions. In general, when a nonparametric procedure is used, the data are not used in the same way that they would be used with a parametric approach. Specifically, the actual observed values are not used, and in many procedures the data are replaced by ranks, with the data being ranked from smallest to largest. In certain other nonparametric procedures, continuous data are converted to binary data.

What happens when we thus "throw away data"? The answer is that we do have some loss of information. For example, if we eschew both types of two-sample t-tests that were described in Section 6.1.2 because we are not sure that the distributions are even unimodal, the question arises as to what is lost, if anything, if the two populations actually had (approximately) normal distributions. More specifically, if the two means differ by a certain amount, how much is lost with a nonparametric procedure when the corresponding parametric procedure is appropriate? The answer is that the distribution-free procedure is, for very large sample sizes, roughly 95% as efficient as the corresponding parametric procedure. This means, for example, that if 95 observations are required under normality to properly reject the null hypothesis under a specified condition (i.e., the probability of properly concluding that the means are different when they are in fact different by a specified amount), then 100 observations will be required under the nonparametric procedure to have the same probability of properly rejecting the hypothesis that the means are equal when in fact they differ by a specified amount.

These numbers are termed *asymptotic relative efficiency* (A.R.E.) values and they essentially tell us how much is lost when the corresponding parametric procedure is appropriate. Of course, such numbers must be interpreted in light of the fact that we never have normality in practice (and similarly for other distributions), so that should be kept in mind. Nevertheless, A.R.E. values are useful, and we might be concerned by values that are much under 100%, especially if we are planning to use a small sample size, as often happens with nonparametric procedures.

We should also think about what is gained when a nonparametric procedure is used, as such gains are the motivation for using nonparametric methods. Our other option would be to use methods specifically for certain distributions, such as testing that the means of two Poisson populations are the same. Such methods are not easily accessible by practitioners, however, so it would be much simpler to use a nonparametric procedure when there is evidence that the parametric method assumptions are seriously violated.

In subsequent sections we start with basic nonparametric methods and progress through various methods that correspond to parametric methods presented in previous chapters. We will carefully point out these correspondences, when they exist, and we will also point out the A.R.E. values for methods for which the values are known.

16.2 ONE-SAMPLE PROCEDURES

One of the basic assumptions of one-sample procedures, as given for example in Chapter 5, is that the sample is a random sample. This is not always stated, but the consequences of a strong violation of the assumption can be pronounced. This can be illustrated as follows. Assume that we wish to test, by way of a confidence interval or hypothesis test, that a population mean is equal to 50. When we use $\sigma_{\bar{x}} = \sigma_x / \sqrt{n}$, we are tacitly assuming that we have a random sample because this relationship between $\sigma_{\bar{x}}$ and σ_x does not otherwise hold. If consecutive observations are correlated rather than independent, we might substantially underestimate $\sigma_{\bar{x}}$. [The explanation of this is beyond the intended scope of this book; the interested reader is referred to Cryer (1986).] The result would be that the confidence interval is too narrow and/or the hypothesis test may give a significant result only because $\sigma_{\bar{x}}$ was underestimated.

Not all samples contain independent data, and data collected over time will generally not be independent. Such data are said to be *autocorrelated*.

DEFINITION

Since "auto-" means "self," *autocorrelated data* are data that are "self-correlated." That is, there is a relationship between consecutive observations and, in general, observations that are k units apart. This is referred to as *autocorrelation*.

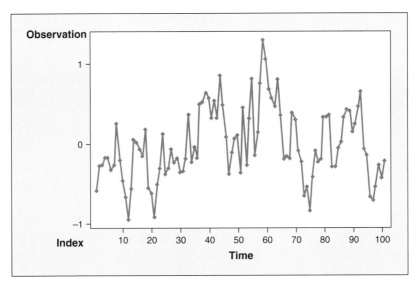

Figure 16.1   Time sequence plot of autocorrelated data.

So how does one test data that may be correlated over time? Several methods are given in the next section.

16.2.1   Methods of Detecting Nonrandom Data

The simplest way of detecting nonrandom data is to just plot the data. If the data have been collected at different points in time, then the data should be plotted against time. If, for example, there is a high correlation between consecutive observations, this should be apparent simply from the graph. For example, consider Figure 16.1. This is a graph of 100 observations with the correlation between consecutive values equal to .67.

The correlation between observations is apparent, especially for the last 30 numbers or so, but it is desirable to use a quantitative test to confirm the impression one receives from the visual test. Figure 16.2 contains a graph of the autocorrelations of observations k units apart for $k = 1, 2, \ldots, 25$. The dotted lines are essentially two standard deviation boundaries, so an autocorrelation that lies outside those boundaries is declared statistically significant. Here we see that the first two autocorrelations lie outside the boundaries.

In addition to a time sequence plot and an autocorrelation test, there is a simple non-parametric test, the *runs test*, that is frequently used to assess possible nonrandomness.

16.2.1.1   Runs Test

In Section 1.3 the concept of a random variable was motivated by a sequence of heads and tails when a coin was tossed 16 times. If the coin is a fair coin, there should not be long runs of heads or tails, nor should heads and tails have a tendency to alternate. We could test for nonrandomness by computing the number of runs above and below the average value. A large number of runs would occur if the data were negatively autocorrelated. That is, a graph of the observations against time would have something resembling a sawtooth pattern. A small number of runs would suggest that the data are positively autocorrelated. For the data represented by Figure 16.1, the number of runs is 27, which is far below

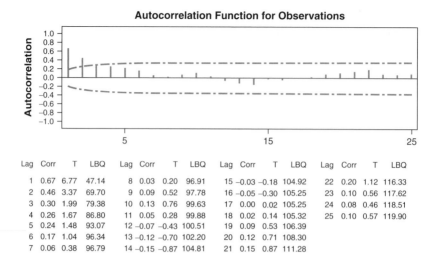

Figure 16.2   Autocorrelation function with 95% confidence limits.

the expected number of runs, 50.3861, under the assumption of a random sample. [The expected number of runs is a function of the total number of observations and the number of observations above (n_1) and below (n_2) the mean. A large sample approximation to the expected number is $2n_1n_2/(n_1 + n_2) + 1$.]

The probability of obtaining 27 or fewer runs when the expected number is 50.3861 is virtually zero, so the conclusion from the runs test is that the data are not random. The small number of runs strongly suggests that the data are positively autocorrelated, and it was stated at the start of Section 16.2.1 that the autocorrelation is .67.

There is no corresponding parametric test for detecting the randomness of a sequence of numbers.

16.2.2   Sign Test

It was stated at the beginning of this chapter that when nonparametric procedures are used, the data are not used in the manner in which they are used with a parametric procedure. Assume that we want to test the hypothesis that a population mean is at most 20, this being our null hypothesis. Nothing is known, or guessed, about the distribution of the population, so with a sample of 25 observations we could not proceed on the assumption that the sample mean would have approximately a normal distribution, as 25 will not be enough observations to result in an approximate normal distribution for the sample mean if the population were highly skewed.

Therefore, we must proceed with a nonparametric approach. Assume that 18 of the sample values are larger than 20. Does this suggest that the mean is probably greater than 20, so the hypothesis of at most 20 is false? If the sample is random and the observations are independent (this could be checked by graphing the data and computing the autocorrelation function, as illustrated in Section 16.2.1), the appropriate distribution would be the binomial distribution (see Section 3.3.2). If the mean were at most 20, we would not expect the vast majority of the observations to be larger than 20. If the mean were exactly 20, our best guess would be that half of the observations would be greater than 20, provided that we additionally knew that the distribution was symmetric. Accordingly, viewing the problem as one having two possible outcomes for each observation, at most 20 and greater than 20, we would compute the probability of having at least 18 binomial "successes" in 25 trials when the probability of a success is .5. That probability is .0216. Since this is a small probability, we might be tempted to conclude that the mean must be greater than 20.

However, what must we assume, which has not been stated previously, in order to use the sign test in this way and reach a conclusion regarding the mean? (The reader is asked to address this question in Exercise 16.1.) What if instead of focusing on the mean, we wanted to test the assumption that the median of the population was 20 against the alternative hypothesis that it is greater than 20? (Of course, the alternative hypothesis would be selected before the sample was obtained, as discussed in Section 5.3.1.) The binomial probability would be the same as given previously and we would conclude that the median was not 20.

Note that it is better to simply give the p-value when using a discrete distribution such as the binomial because with a discrete random variable we cannot have an exact, prespecified significance level, and if n is small we might not even be able to come close to a desired significance level.

Another computational problem arises when we have a sample size that is at least moderate. This is because software or special tables are needed and such tables do not go beyond small-to-moderate values of n. (Of course, we think of nonparametric procedures as being small sample procedures for the most part, as with large samples we would be inclined to try to invoke the Central Limit Theorem and use a parametric procedure.)

For example, if $n = 52$, we would be hard pressed to find a table that would allow us to use the sign test. We know, however, that the normal approximation to the binomial distribution (mentioned in Section 5.4) works quite well when $p = .5$, so a normal approximation could be used. We should also keep in mind that good statistical software is plentiful and it is much easier to use software than to use tables or hand computations.

It should be clearly understood that when we use tests such as the sign test and the tests that are presented in the following sections, we still must assume the absence of autocorrelation. We are simply not assuming normality. Indeed, as Box (1999) stated, "For example, so-called "distribution-free" tests recommended as substitutes for the t-test are just as disastrously affected by failure of the *distributional assumption* of zero serial correlation as is the t-test itself."

16.2.3 Wilcoxon One-Sample Test

When the sign test is used, data are discarded (as is true for *all* nonparametric procedures) because the distance that each point is from the median is not used. A step toward recovering some of the lost data would be to rank the distances that the data points are from the median. That is, compute $| x_i - \text{median} |$ for $i = 1, 2, \ldots , n$, and rank the absolute values from smallest to largest, while keeping track of which values of $(x_i - \text{median})$ are positive and which are negative. Let T_+ denote the total of the ranks for the positive deviations and let T_- denote the total of the ranks for the negative deviations. Typically, the smaller of the two totals is used as the test statistic, but that doesn't really matter.

If the sample data support the hypothesized median, the two totals should be about the same. If, however, the data do not support the hypothesized median, we would expect the totals to differ considerably. Specifically, if the hypothesized median appears to be too small, we would expect T_+ to be much larger than T_-, and conversely when the median seems too large. For the latter scenario we would expect most of the sample data to be less than the median. What if half of the data are below the median but T_- is much smaller than T_+? The reader is asked to explain what this suggests in Exercise 16.2.

■   **EXAMPLE 16.1**

Given below is MINITAB output for the Wilcoxon one-sample test applied to 100 random numbers generated from the standard normal distribution. As the output shows, a test of the median = 0 against median ≠ 0 is performed, and of course the hypothesis should not be rejected since the median *is* 0, and of course it isn't rejected since the *p*-value is large.

```
Wilcoxon Signed Rank Test: C1

Test of median = 0.000000 versus median not = 0.000000

                N for      Wilcoxon                    Estimated
          N     Test      Statistic        P             Median
C1      100      100        2841.0    0.278            0.09436          ■
```

The Wilcoxon test can be viewed as a median test and also as an alternative to a paired *t*-test (Section 6.2) since the latter reduces to a single sample because differences are obtained. When used as an alternative to the paired *t*-test, the Wilcoxon test has an A.R.E. value of .955 when the differences have a normal distribution. Thus, not much is lost when the differences have a normal distribution, but much is gained when the differences are considerably nonnormal.

16.3   TWO-SAMPLE PROCEDURES

There are also two-sample nonparametric procedures that are alternatives to two-sample parametric procedures. For example, assume that there are two independent samples but each of the two populations is believed to be highly skewed. Neither of the two *t*-tests given in Section 6.1.2 could be used. One nonparametric test that could be used instead is the Wilcoxon two-sample test, also referred to as the *Mann–Whitney test* and probably better known by that name. Therefore, we will use this name in subsequent sections.

16.3.1   Mann–Whitney Two-Sample Test

As in the one-sample Wilcoxon test, the data are ranked. With two samples, the samples are combined for the ranking, with the ranking being from the smallest to the largest. If the numbers differ considerably between the samples, then the ranks will differ considerably and the sum of the ranks for the two samples will be much different. The null hypothesis is that the two populations have identical distributions with the same median; the alternative hypothesis is that the populations differ only in terms of their medians.

■ **EXAMPLE 16.2**

Objective

As has been pointed out, for example, by James (1989), many quality characteristics such as diameter, roundness, mold dimensions, and customer waiting time do not have normal distributions. If the distributions were known, then appropriate distribution theory could be applied in drawing inferences regarding such characteristics, but the correct distribution is always unknown.

Study Objective and Data

Assume that a company has two plants that manufacture a particular product, and of course it is desired that the diameters of the product coming from the two plants be the same. A sample of 12 items is obtained from each plant and the data are as follows:

1: 21.82 21.73 21.85 21.79 21.83 21.80 21.84 21.86 21.77 21.79 21.78 21.69
2: 21.86 21.84 21.89 21.88 21.87 21.79 21.88 21.80 21.79 21.83 21.86 21.75

Statistical Test

Since the samples are obviously independent, the Mann–Whitney test is an appropriate test. The output is

Mann-Whitney Test and CI: 1,2

```
1          N = 12      Median =      21.795
2          N = 12      Median =      21.850
Point estimate for ETA1-ETA2 is -0.040
95.4 Percent CI for ETA1-ETA2 is (-0.080,0.000)
W = 114.5
Test of ETA1 = ETA2 vs ETA1 not = ETA2 is significant at 0.0433
The test is significant at 0.0425 (adjusted for ties)
```

Conclusion

The output shows that the two plants may differ in terms of their medians as the p-value is less than .05. The reader is asked to consider this example further in Exercise 16.19.   ■

The parametric counterpart to the Mann–Whitney test is the independent sample t-test and the A.R.E. is .955 when the two populations have a normal distribution.

16.3.2 Spearman Rank Correlation Coefficient

There are nonparametric counterparts to the (Pearson) correlation coefficient in Section 8.3. There will often not be a linear relationship between two random variables, yet we may want some measure of the degree of association between them. Furthermore, the data may come in the form of ranks, which would also preclude the use of the Pearson correlation coefficient. Some years ago I was asked by a national organization to pass judgment on the work of a consultant in a particular study regarding worker safety. One set of data was in the form of ratings (1–5) by two raters, and the objective was to see whether there was agreement between the two people.

This was clearly an instance in which the Pearson correlation coefficient was inappropriate since, for one thing, the data are discrete rather than continuous and the range for each variable is limited. The consultant had used Spearman's rank correlation coefficient and found a high degree of correlation, which caused the consultant to conclude that the raters were in general agreement. The very small sample size made this claim questionable, however.

Spearman's correlation coefficient is computed by first ranking the data on each variable from smallest to largest, and then applying the formula for Pearson's correlation coefficient to the ranks. Thus, if the ranks are the same, the correlation is 1; if the ranks are just the opposite, the correlation is -1. Such values could easily occur with a very small sample, however, and this could simply signify a (possible) monotonic relationship. For example, the (y, x) values of (4, 13), (16, 14), (18, 26), and (19, 27) would graph similar to a graph of the log function (i.e., highly nonlinear), yet the Spearman correlation is 1.

There are many other measures of association in use, although some are most useful in applications in the social sciences and for that reason they will not be given here.

16.4   NONPARAMETRIC ANALYSIS OF VARIANCE

There are also nonparametric counterparts to analysis of variance. As with other nonparametric procedures, the data are converted to ranks if the data are not already in the form of ranks. In the following sections we consider one-way and two-way nonparametric analysis of variance.

16.4.1   Kruskal–Wallis Test for One Factor

Just as with the Mann–Whitney test, when we have more than two levels of a factor the data are merged, then ranked and the ranked data returned to their respective groups. Recall that in one-way analysis of variance the general idea is to see if the means from the groups differ significantly from the overall mean, and Eq. (12.2) is used to measure this difference. It stands to reason that we should do essentially the same thing with ranks; that is, see the extent to which the average ranks for the groups differ from the overall average rank. The latter will always be $(N + 1)/2$, which follows from the fact that the sum of the ranks, which is the same as the sum of the first N integers, is $N(N + 1)/2$.

Thus, the obvious choice for a statistic is one that measures the sum of the squared deviations of the average ranks for the groups from $(N + 1)/2$, and the Kruskal–Wallis test statistic does that. Analogous to the notation used in Section 12.2.1.3, we will let $\overline{R}_i$ denote the average rank for group i and n_i denote the number of observations in group i, and of course $\overline{R}_{\text{all}} = (N + 1)/2$ is the average of all the ranks. The expression that corresponds to the sum of squares between groups in analysis of variance is then $\sum_{i=1}^{k} n_i (\overline{R}_i - (N + 1)/2)^2$.

In one-way analysis of variance, the model can be written as $Y_{ij} = \mu_i + \epsilon_{ij}$ and one proceeds to compute the mean square between groups and divide that number by the mean square error. In nonparametric one-way analysis of variance, we can think of a representation of the observed ranks as $R_{ij} = \overline{R}_{\text{all}} + e_{ij}$. Then, analogous to analysis of variance, we could compute the variance of the ranks, which would be

$$\sum_{ij} e_{ij}^2 / (N-1) = \sum_{ij} \left(R_{ij} - \frac{N+1}{2} \right)^2 \Big/ (N-1) \qquad (16.1)$$

It can be shown, as the reader is asked to do in Exercise 16.3, that this variance, which we will denote by S^2, is $N(N+1)/12$.

The Kruskal–Wallis test statistic is then the sum of squares between groups divided by S^2. Specifically, the statistic is given by

$$H = \frac{12}{N(N+1)} \sum_{i=1}^{k} n_i \left(\bar{R}_i - \frac{N+1}{2} \right)^2 \qquad (16.2)$$

Note that this is very similar to what occurs in analysis of variance, the difference being that with the latter it is the mean square between groups that is divided by S^2.

As in analysis of variance, there is a more efficient way to compute the value of H. Specifically, it can be shown, as the reader is asked to do in Exercise 16.4, that the test statistic can be written

$$H = \frac{12}{N(N+1)} \sum_{i=1}^{k} \frac{R_i^2}{n_i} - 3(N+1) \qquad (16.3)$$

Of course, we should prefer not to do this computation by hand. Sample MINITAB output is

Kruskal-Wallis Test: C1 versus C2

Kruskal-Wallis Test on C1

C2	N	Median	Ave Rank	Z
1	10	22.50	10.6	−2.16
2	10	18.50	12.7	−1.23
3	10	38.50	23.2	3.39
Overall	30		15.5	

H = 11.76	DF = 2	P = 0.003	
H = 11.76	DF = 2	P = 0.003	(adjusted for ties)

In this particular example there were no ties. If there had been any repeat observations, there would have been repeated ranks and it would have been necessary to divide the value of H by $1 - \sum_{m=1}^{a} (T_i^3 - T_i)/(N^3 - N)$, with T_i denoting the number of tied observations in the ith group of ties from the group of a ties. Since there were no ties with these data, there was no adjustment, so the two values of H in the computer output are the same.

The Z-value for any particular group is computed by subtracting the average rank, $(N+1)/2$, from the average rank for each group and then dividing by $s_{pr}/\sqrt{n}$, with s_{pr} denoting the square root of the pooled standard deviation of the ranks, and n is the number of observations in a group (which in this example is constant over the groups).

The null hypothesis that the populations are the same (i.e., equal means) is easily rejected since the *p*-value is very small. As in analysis of variance, what would logically follow would be a determination of the group(s) that produces the large *H* value, and the *Z*-values can be helpful in that regard. Here it is apparent that group 3 differs considerably from the other two groups.

The parametric counterpart is one-factor analysis of variance, and the A.R.E. value relative to the *F*-test is .955 when the populations are normal.

It is worth noting that this test becomes the equivalent of the Mann–Whitney test when there are only two groups, just as Analysis of Variance becomes the equivalent of the independent sample *t*-test when there are two groups in a parametric analysis.

16.4.2 Friedman Test for Two Factors

The distribution-free counterpart to two-way analysis of variance with one factor being a blocking factor is due to Friedman (1937). As with the Kruskal–Wallis test, ranks are used, and the data within each block are ranked from smallest to largest. For simplicity, we will use *F* to denote Friedman's statistic, although this is not what he used. With this notation, the test statistic is

$$F = \frac{12}{wk(k+1)} \sum_{i=1}^{k} (R_i)^2 - 3w(k+1)$$

with k denoting the number of levels of the factor of interest, w denotes the number of blocks, and R_i is the total of the ranks, summed over the blocks, for the ith level of the factor of interest. Provided that neither the number of levels of the factor nor the number of blocks is small, F is approximately χ^2_{k-1}. Of course, we generally do not want to use hand computation.

In two-way analysis of variance there is interest in looking at the effects of each factor as well as their interaction effect. Friedman's test is not designed to be the nonparametric counterpart to this parametric scenario, however. Indeed, with a randomized block design (see Section 12.3.1), one assumes that there is no interaction between the treatment factor and the blocking factor. There is, however, interest in determining if the blocking has been successful, whereas this is not done with Friedman's test. Thus, there is not a direct correspondence.

There is, however, an analogy between the Kruskal–Wallis test vis-à-vis Friedman's test relative to blocking or no blocking in analysis of variance with a single factor. It was shown in Section 12.3.1 how important blocking can be in analysis of variance with a single factor. Similarly, blocking via Friedman's test can also be important when a nonparametric approach is used, as is illustrated in the following example.

■ **EXAMPLE 16.3**

Problem Setting

Because of difficult economic times, a company is considering phasing out one or more of its product lines. It has plants in the United States, Germany, and the United Kingdom, and it has four lines of a particular type of product. Its marketing department makes customer satisfaction surveys and the data in Table 16.1 below are composite satisfaction scores for the four product lines (designated as A, B, C, and D), and the three countries.

TABLE 16.1   **Satisfaction Scores**

Country	A	B	C	D
United States	6.4	6.7	7.3	5.9
Germany	6.1	6.8	7.1	6.0
United Kingdom	7.0	7.4	7.5	6.9

Study Objective

The company would like to determine if the satisfaction scores differ significantly across the product lines, with a high satisfaction score of course being best. Note that even though we have a two-way table, we can't analyze this as if it were a contingency table problem because the entries in the table are not counts. Since this is not a designed experiment with randomization, we cannot analyze this as a randomized block design, and similarly randomization would be needed to analyze the data as a general two-way analysis of variance problem, in addition to the assumptions of constant variance and normality within each cell. Accordingly, a nonparametric analysis is appropriate.

Computer Output

The MINITAB output obtained by first using the Kruskal–Wallis approach and then the Friedman approach is as follows:

Kruskal-Wallis Test: C1 versus C2

Kruskal-Wallis Test on C1

C2	N	Median	Ave Rank	Z
1	3	6.400	5.0	-0.83
2	3	6.800	7.3	0.46
3	3	7.300	10.3	2.13
4	3	6.000	3.3	-1.76
Overall	12		6.5	

H = 6.38   DF = 3   P = 0.094

* NOTE * One or more small samples

Friedman Test: C1 versus C2, C3

Friedman test for C1 by C2 blocked by C3

S = 9.00   DF = 3   P = 0.029

```
                            Est         Sum of
C2          N             Median         Ranks
1           3             6.2875          6.0
2           3             6.8125          9.0
3           3             7.1625         12.0
4           3             6.0875          3.0

Grand median = 6.5875
```

Conclusions

We can see that the results differ with the two approaches, as the Kruskal–Wallis p-value is .094 whereas the Friedman p-value is .029. Thus, the conclusions would differ if we used .05 as the threshold value.

We can easily see why that is the case. Notice that the UK scores are greater than the corresponding scores for Germany and the United States. This would contribute to a block effect that would be ignored if a Kruskal–Wallis approach were used. This becomes apparent when we view the ranks with and without blocking.

When the Friedman test is performed, the ranks are assigned within each block and the ranks are then added over the blocks for each level of the factor. For this example, the ranks in each row will be 2, 3, 4, 1 in this order for the four columns. Thus, the rank sums for the four product lines will be three times this (i.e., three blocks): namely, 6, 9, 12, 3. The Kruskal–Wallis ranks are given in Table 16.2.

TABLE 16.2   Ranks Ignoring Blocks

Country	A	B	C	D
United States	4	5	10	1
Germany	3	6	9	2
United Kingdom	8	11	12	7

The extent to which the sums of the ranks for the columns differ from 15, 24, 33, and 6, respectively (the sums of 4, 5, and 6; 7, 8, and 9; 10, 11, and 12; and 1, 2, and 3), is a measure of the extent to which the two analyses will differ. Here, the primary difference is in the D column, and of course this causes differences elsewhere (here in the second and third columns) since the sum of the ranks must be the same in both cases.

Thus, we can easily see why different conclusions are drawn, and in this instance we see why blocking was needed. We conclude that satisfaction differs and so the company should consider eliminating D.

We should note before leaving this example that an exact Friedman test can be performed and is available in the *StatXact* software, but the controversies regarding exact tests should of course be kept in mind.   ■

The parametric counterpart to the Friedman test is of course two-way analysis of variance. Unlike some other nonparametric tests, the A.R.E. is not a single number but rather depends on the number of columns (also called blocks in the literature) in the two-way layout.

16.5   EXACT VERSUS APPROXIMATE TESTS

Software for the nonparametric methods given in the preceding sections is readily available and can be applied to obtain exact p-values, so there is no need to resort to approximations, such as the chi-square analysis of a contingency table given in Section 15.1. The computer output to this point in the chapter has mostly been MINITAB output, but exact tests are also available in *StatXact*, in particular, and in other software.

16.6   NONPARAMETRIC REGRESSION

Within the past fifteen to twenty years, there has been considerable interest in using alternatives to traditional regression methods. Regression models were discussed in Chapters 8 and 9, with the models intended to cover a full range of data, specifically the range of data in the sample that was used in constructing the model. Often, however, different models will be needed for different areas of the data space. Constructing such models would be difficult with a small amount of data, however, and there would be other complications.

Consequently, a better approach would be to "let the data model themselves," and in this section we will consider only a single predictor. The most popular method for doing so is *loess*, which is also called *lowess*. This is an acronym that represents "local regression"; it was introduced by Cleveland (1979) and was called *locally weighted regression*. The "local" refers to the fact that a separate regression function, usually a straight line, is fit for each point, x_0, in the sample, using a specified number of points on each side of x_0. The word "weighted" results from the fact that a weight function is used, with the weights reflecting the distance that each x_i is from x_0. That is, the points that are the farthest from x_0 receive the least weight. Therefore, the fit in each "neighborhood," as it is called, is actually a weighted least squares fit.

The objective of local regression is to identify the model (possibly with polynomial terms) that is appropriate for each data segment.

Although *loess* is in some ways an attempt to let the data model itself, the user still must make certain decisions, such as determining the neighborhood size (i.e., the fraction of observations used for each model fit). Ideally, the size should be small when there is very little vertical scatter at and near x_0, and large when there is considerable vertical scatter. That is, if an appropriate function seems to be easily determinable at x_0, there would be no need in using observations that are very far from x_0, and in particular in areas where there may be considerable vertical scatter.

The determination of the neighborhood size roughly corresponds to determining the number of regressors in linear regression, and Cleveland and Devlin (1988) apply the concept of Mallows C_p statistic (see Section 9.6.4.1) to local regression for the purpose of determining the neighborhood size.

The mechanics of local regression are somewhat beyond the intended level of this book and therefore will not be given. The interested reader is referred to Loader (1999).

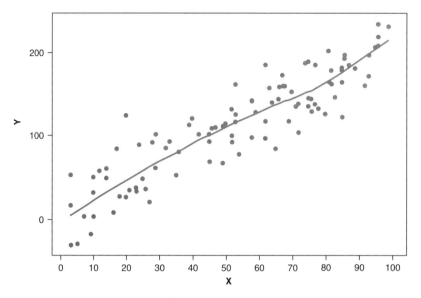

Figure 16.3   Example of LOWESS smooth.

Various statistical software have *loess* capability, including MINITAB, which refers to it as "LOWESS". Figure 16.3 was produced using MINITAB, with the commands that produced it given at the textbook website.

Note the bend of the line in the upper right portion of the graph, as the line attempts to accommodate the curvature of the points in that area. Of course, we don't want the line to accommodate outliers, so it is possible to use a robust version that is insensitive to outliers. Cleveland (1985) indicated that two robust steps should adequately smooth outlier effects for most data sets.

Although locally weighted regression is considered to be a nonparametric regression approach, with *loess* the errors can be specified either to be normally distributed or to have a symmetric distribution.

As with any fitting procedure, it is important to recognize that data may be too sparse in certain areas of the data space to support any type of fit in those areas. In regard to *loess*, it would be better either to not construct the smooth in such areas or at least to indicate (using colors, as has been suggested) areas in which a smooth is not well-supported.

■ **EXAMPLE 16.4**

In their article "Regression Model for Daily Maximum Stream Temperature" in the *Journal of Environmental Engineering* (July 2003; available online at `http://cadswes.colorado.edu/PDF/RiverWare/NeumannASCE_JEE2003.pdf`), D. W. Neumann, B. Rajagopalan, and E. A. Zanoga showed scatter plots of maximum daily stream temperature at Reno, Nevada against various possible predictors, including maximum daily air temperature at Reno, and each scatter plot showed a nonlinear relationship and was subsequently fit with a *loess* smooth. What is interesting about this application is that the authors used a stepwise *linear* regression procedure to select the important predictors, which resulted in the selection of two predictors. The experimenters fit a multiple linear regression model in these two predictors and also fit a "local nonlinear regression model" (i.e., a *loess* model) and decided to use the former because it was simpler.

16.7 NONPARAMETRIC PREDICTION INTERVALS AND TOLERANCE INTERVALS

Nonparametric tolerance intervals were covered in Section 7.3 and nonparametric prediction intervals were covered in Section 7.4.7; the reader is referred to that material for details. Those methods are generally used when a normal probability plot or other test indicates that the distribution of the population is not well-approximated by a normal distribution, although it is also possible to construct tolerance intervals for other specified distributions, as was discussed in Section 7.5.

16.8 SUMMARY

Introductory statistics texts and courses emphasize parametric procedures. The late Gott-fried Noether, author of *Introduction to Statistics the Nonparametric Way*, believed that nonparametric procedures should be emphasized. Since we almost always do not know the distribution of population values from which our sample was obtained, one can indeed make a strong argument for using, even emphasizing, nonparametric procedures. A useful approach is to apply both a parametric procedure and the corresponding nonparametric procedure, if one exists, and compare the results. If the results are conflicting, then the results of the nonparametric procedure should be followed.

Exact methods of analyzing categorical data, such as the exact analysis of contingency tables, have become popular during the past ten years. These methods will often be prefer-able to approximate methods, such as the analysis of contingency tables using chi-square. Care must be exercised, however, as exact methods do tend to be conservative. See Agresti (2001) for a summary and critique of exact methods.

REFERENCES

Agresti, A. (2001). Exact inference for categorical data: recent advances and continuing controversies. *Statistics in Medicine*, **20**(17–18), 2709–2722.

Box, G. E. P. (1999). Statistics as a catalyst to learning by scientific method. Part II. A discussion. *Journal of Quality Technology*, **31**(1), 16–29.

Cleveland, W. S. (1979). Robust locally weighted regression and smoothing scatterplots. *Journal of the American Statistical Association*, **74**, 829–836.

Cleveland, W. S. (1985). *The Elements of Graphing Data*. Pacific Grove, CA: Wadsworth.

Cleveland, W. S. and S. J. Devlin (1988). Locally weighted regression: an approach to regression analysis by local fitting. *Journal of the American Statistical Association*, **83**, 596–610.

Cryer, J. (1986). *Time Series Analysis*. Boston: Duxbury Press.

Friedman, M. (1937). The use of ranks to avoid the assumption of normality implicit in the analysis of variance. *Journal of the American Statistical Association*, **32**, 675–701.

James, P. C. (1989). C_{pk} equivalences. *Quality*, **28**(9), 75.

Loader, C. (1999). *Local Regression and Likelihood*. New York: Springer-Verlag.

EXERCISES

16.1. Consider the sign test in Section 16.2.2 and the example that was given. Explain why the test could not be used to test the mean of a population if nothing were known about the distribution for the population.

16.2. Consider the Wilcoxon one-sample test given in Section 16.2.3 and explain what a small value of T_- would suggest when half of the sample data values are less than the hypothesized median.

16.3. Show that the reciprocal of the variance of the ranks is the constant in Eq. (16.2).

16.4. Show that Eq. (16.2) can be simplified to Eq. (16.3).

16.5. Describe, in general, the condition(s) under which the Friedman test should be used instead of the Kruskal–Wallis test.

16.6. Consider two ways of classifying students at a university: (1) male or female and (2) engineering major or nonengineering major. Assume that you have the grade point averages for a given semester and you want to test the independence of those two classification variables for that semester.
  (a) Would it be possible to test this hypothesis using a contingency table approach? Why or why not?
  (b) If it were possible, what would be the population to which the inference would apply (remember that you have the "population" for a given semester)?

16.7. The data in CLINTON.DAT contain a conservative score/rating that had been assigned to each U.S. Senator at the time of the Senate votes on perjury and obstruction of justice alleged against then President Clinton. Assume that you want to test the hypothesis that the scores for the Democrats and the Republicans have the same distribution.
  (a) Test this hypothesis using the appropriate test (as if we didn't know what the conclusion must be!).
  (b) Having done this, you have concluded, subject to a certain probability of having reached the wrong conclusion, that the distributions either are or are not different. But don't you know whether the distributions are the same or not in terms of location just by inspecting the data? Can the data for each party be viewed as sample data from some population? Why or why not?

16.8. Assume a 2×2 contingency table.
  (a) If the difference between the observed and expected value (i.e., $O - E$) is -5 for one of the cells, what must be the sum of these differences over the other three cells?
  (b) Construct an example for which the absolute value of all four differences is 5.

16.9. Assume that a sample of 60 nonsmokers and 40 smokers is obtained and whether or not each person has any type of lung disorder is recorded. Would you apply Fisher's exact test in analyzing the data? Why or why not? If you would not do so, what would you do instead?

16.10. Assume a single sample from which an inference is to be made about the median of the distribution from which the data were obtained. Of the two distribution-free methods that might be used, a sign test and the one-sample Wilcoxon test, which test is more sensitive to outliers? Explain.

16.11. Consider the data in `3WAYINT.TXT` that was used in Example 12.3 in Section 12.5.2. Could that data be analyzed using one of the methods presented in this chapter? Explain.

16.12. You are given the data in `SCATTER.MTW`.
  (a) Construct a scatter plot.
  (b) Would you suggest fitting a simple linear regression model to the data, or do you believe that a *loess* fit would be more appropriate? Explain.
  (c) Construct a *loess* smooth and a simple linear regression line and compare them, overlaying the lines if possible. What does this comparison suggest as being the best fit to the data?

16.13. Consider the data in `TABLE102.MTW`. Use MINITAB or other software to construct a *loess* smooth. Do the data support a smooth for the entire range of X?

16.14. A process engineer collects 100 data points from past records and plots the historical data on a chart. You are given the chart and you notice that 70 of the points plot above the midline. Use the sign test to determine if this seems plausible. What did you assume when you made the test? Can such assumptions be defended in the absence of additional information?

16.15. You are given the following bivariate data:

X	23.5	23.8	24.2	25.1	26.2	27.8	26.1	27.2	24.2	25.1	24.2	23.3	24.6	27.0	25.2
Y	17.9	18.3	16.9	17.1	19.9	20.1	18.4	18.2	17.5	17.3	17.0	18.9	18.4	19.2	19.1

Compute the value of the Spearman rank correlation coefficient.

16.16. Construct an example with 15 bivariate observations for which the value of the Spearman rank correlation coefficient is -1, with the first and last Y-values being 16.8 and 19.9, respectively, and the first and last X-values being 22.1 and 17.2, respectively.

16.17. Consider a sequence of 14 data points with 10 points above the assumed mean for the population from which the sample was drawn. Does this suggest that the mean may not be the value that is assumed? Perform the appropriate test and draw a conclusion.

16.18. Consider Example 16.2 in Section 16.3.1. If you were the quality control manager for the two plants, would you consider additional tests beyond the one in that section? Explain.

16.19. Like the independent sample t-test, the Mann–Whitney test can be applied when there are unequal sample sizes. One such example is given in *StatXact User Manual Volume 1*: a blood pressure study with 4 subjects in a treatment group and 11 subjects in a control group. The data are as follows:

Treatment	94	108	110	90							
Control	80	94	85	90	90	90	108	94	78	105	88

With only this information, it would not be practical to use the independent sample t-test as there is no way to test for normality when a sample contains only four observations, and similarly a test of equality of variances would be impractical. Therefore, apply the Mann–Whitney test and reach a conclusion.

16.20. Read the article referred to in Example 16.4 and especially look at the smooths of the scatter plots for the two selected predictors. Comment on the decision that was reached.

CHAPTER 17

Tying It All Together

Students all too often complete a statistics course without having any real idea of how to apply what they've learned. They remember a few concepts and methods but have very little understanding of when and how to apply important statistical methods that have broad applications. This can be traced to the way that statistics courses are structured, as there are only a few methods that one must choose from in working chapter exercises because the exercises are usually at the end of each section and require the use of techniques presented in those sections.

Consequently, the student is hardly ever forced to think about all that has been presented in the course and draw from that material in working problems and/or other class assignments. Furthermore, it may not be apparent that the analysis of data requires a combination of techniques, not just one or two.

There are exercises at the end of this chapter for which it may not be obvious how to proceed, but this parallels the real world as problems encountered in the latter do not come with suggestions as to how they should be solved!

17.1 REVIEW OF BOOK

Data can be collected either as observational data or from a designed experiment. The graphical methods given in Chapter 1 can be used in both cases: the examples were from observational studies and certainly the vast majority of graphs that most of us encounter are graphs of data from observational studies (or simply from just plain "observation").

Statistical distributions (also known as probability distributions), serve as models of reality. The major distributions were covered in Chapter 3 and a few other distributions were mentioned in Chapter 14. There are hundreds of other distributions. The general idea is to select a distribution that will be an adequate substitute for the unknown distribution of a random variable that is of interest. General knowledge of the distributions of certain types of random variables (height, weight, part dimensions, etc.) can be helpful in expediting the search for a suitable distribution, which of course should be computer-aided.

After a distribution has been selected as a model, the parameters of the distribution must be estimated, with interest generally centering on estimating the mean, standard deviation, and/or variance. For a normal distribution, the mean and variance are parameters in the probability density function (*pdf*), whereas at the other extreme, neither the mean nor the variance of the binomial distribution is part of the *pmf*. The Poisson distribution has the same mean and variance, and these are equal to the only parameter in the *pmf*, λ. There are various methods used for estimating parameters and one of the most frequently used methods is the method of maximum likelihood.

Oftentimes, sample statistics such as the mean and standard deviation are calculated without any attempt at estimating specific parameters, as was presented in Chapter 2. There is certainly nothing wrong with stopping at that point since description is one of the uses of statistics.

Although description is useful, the field of statistics is really built on inference. Much space in Chapters 5 and 6 was devoted to confidence intervals for various parameters, including the relationship between confidence intervals and the corresponding hypothesis tests, as it is important to recognize that these are essentially two sides of a coin. The reader should bear in mind that confidence intervals are inherently more informative than hypothesis tests.

Prediction intervals and tolerance intervals have many engineering applications and there are many different types of such intervals, but very little time is usually devoted to these intervals in statistics courses, a custom that should be changed (as has been stated by others in the statistics profession), and Chapter 7 was intended to be a step in that direction. An upsurge in interest in these intervals would motivate the major statistical software companies to include them in their software, which in turn would make the intervals easier to use and increase their popularity.

Regression analysis, in its varied forms, and analysis of variance (ANOVA) models associated with designed experiments are the two most frequently used statistical techniques. They are intertwined in that analysis of variance tables are part of standard regression output and regression methods are very useful in supplementing the information contained in analysis of variance tables when designed experiments are performed.

Subject-matter knowledge can often be used to suggest a particular regression model, or to suggest a mechanistic model (Chapter 10), that might later be modified using regression tools so as to produce an empirical–mechanistic model. This is a relatively untapped area in which statisticians can work with engineers and others in the development of improved models.

Analysis of Means (ANOM) has been underutilized as a complement to ANOVA, or as a possible substitute for it. Developed by Ellis Ott as a communicative aid for his consulting work with engineers, it is an intuitive and powerful tool that, unlike ANOVA, does not stray from the original unit of measurement. Therefore, ANOM was given a moderate amount of space in Chapter 12.

Statistically designed experiments can provide important information, provided that good designs are used and measurement error is not a problem. Unfortunately, measurement error is often a problem (e.g., see the quote from Bisgaard in the first paragraph of Chapter 14). Therefore, it is important to conduct studies, such as gage reproducibility and repeatability (R&R) studies with designed experiments, so as to estimate variance components and identify any measurement problems.

Quality and reliability would seem to be related semantically, but quality improvement methods, which include control charts, process capability indices, and designed experiments, are used for assessing and improving quality, whereas in reliability work the emphasis is on assessing reliability, although reliability improvement is also of interest. The choice of appropriate statistical distributions is an important part of reliability analysis, whereas distributions are often not specified in quality improvement work. Instead, control chart users have been a bit lax in using standard methods and not attempting to gain insight into the nature of the distributions of the random variables that they are charting. The state-of-the-art is constantly evolving and today's process monitoring needs and application opportunities certainly could not have been envisioned by Walter Shewhart, who passed away four decades ago. One of the hindrances to progress in control chart usage in engineering is excessive reliance on the basic charts that Shewhart outlined over 80 years ago.

Measurement data are used extensively in both reliability and quality improvement, with the latter also concerned with control charting attributes data in the form of nonconformities and nonconforming data. Methods for analyzing categorical data arranged in contingency tables can also be employed, and binary engineering data (such as go/no-go data) can be modeled with an eye toward determining what factors are related to the binary variable. The latter is a relatively untapped area.

Regardless of the methods that are used, it is imperative that assumptions be checked, and the user should understand which assumption–method combinations are critical and which are of less importance. This was stressed in Chapters 5 and 6, in particular, and also discussed in subsequent chapters.

Since "all models are wrong; some are useful," one can argue for the use of nonparametric methods, which have only general distribution assumptions. A good strategy is to use a parametric approach, if desired, and then apply the corresponding nonparametric approach to the same data. If the results agree, the assumptions may not have been seriously violated. If the results greatly differ, however, the parametric assumptions have probably been violated, so either a nonparametric approach should be used or the data should be transformed in an effort to meet the assumptions of the parametric method.

Nonparametric methods are undoubtedly underutilized, as are robust methods. In listing areas of progress in the field of statistics during 1950–1980, Efron (2001) listed nonparametric and robust methods first. (The latter are methods that are relatively insensitive to extreme observations, such as the trimmed mean given in Section 2.1, and/or insensitive to violations of assumptions.)

Some of the 18 exercises given at the end of the chapter contain information that suggests a particular method should be used; the wording of other exercises is not as helpful. If you are able to work all of the exercises successfully, or at least the vast majority of them, you have undoubtedly *learned* a good measure of statistical methodology, rather than merely being *trained* in it.

17.2 THE FUTURE

Students of statistics as well as infrequent users of statistical methodology should bear in mind that the field is ever-changing, as are most fields. Many promising new methods of today will be replaced by superior, or at least alternative, methods of tomorrow, some of which will be developed in response to new data-analytic needs. Therefore, in addition to reading textbooks, it is also important to read articles such as Efron (2001) that summarize the developments of the past and look toward the future.

17.3   ENGINEERING APPLICATIONS OF STATISTICAL METHODS

There are, of course, a huge number of successful engineering applications of statistical methods that are described in the literature. One excellent source is the Institute for Scientific Information's Web of Science (http://scientific.thomson.com/products/wos). Of course, there are many other sources, one of which is NCHRP 20–45 "Scientific Approaches to Transportation Research," an online publication available at http://onlinepubs.trb.org/onlinepubs/nchrp/cd-22/start.htm. The references list application papers that use analysis of variance, linear regression, contingency table analysis, goodness of fit, logistic regression, before and after studies, plus a few other statistical methods not covered in this text. These are organized within a statistical method by application areas including pavements, environment, bridges, traffic, safety, and planning, and most of the papers are available as *pdf* files.

REFERENCE

Efron, B. (2001). Statistics in the 20th century, and the 21st. Invited lecture at the Statistiche Woche 2001 in Vienna, Austria, Oct. 16–19, 2001. Published in *Fetschrift 50 Jahre Österreichische Statistische Gesellschaft*, Seiten 7-19. This is available online at http://www.statistik.tuwien.ac.at/oezstat/festschr02/papers/efron.pdf.

EXERCISES

17.1. An experimenter decides to test the equality of two population means by taking two small samples when the population standard deviations are unknown. He performs the *t*-test but later remembers that he forgot to check the distributional assumption. When he does so he sees that the two normal probability plots suggest considerable nonnormality. What would you recommend be done at that point?

17.2. A construction company institutes an accident prevention program and a year later would like to determine if the program has been successful. It has ten crews that each work on a very large construction project that so far has taken two years for each project. The company has recorded the number of accidents for the year before the program was started and the year after the program was started and the results are as follows.

Before	4	8	3	7	2	1	6	7	4	3
After	2	5	3	6	3	0	5	5	2	2

Use an appropriate analysis and determine whether or not the program had been successful.

17.3. Given the following graph, would you fit a simple linear regression model to the data represented by these points, or would you use some other approach? Explain.

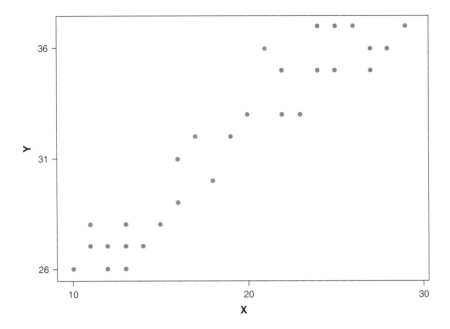

17.4. Respond to the following statements: "I am going to run an analysis of variance with three levels of a single factor. I know that I should check for normality of each population, but how can I check for normality when I have only five observations per factor level? For that matter, how can I check the equal variances assumption with each variance computed from only five observations?"

17.5. What is the purpose in studying Bayes' rule in Chapter 3 and other probabilistic ideas and methods when these methods are not used directly in subsequent chapters?

17.6. A control chart user constructs a Phase I control chart for individual observations. Explain why the probability of a randomly selected point falling outside the 3-sigma limits will (almost certainly) not be .0027 even if the population from which the data came had been a normal distribution.

17.7. A small company suspects that one of its two plants is considerably more efficient than the other plant, especially in regard to nonconforming units. There is one particular product that has been especially troublesome for the first plant to produce, with the percentage of units that are nonconforming running at over 2% for that plant. The number of nonconforming units of the type in question is tabulated for a period of two weeks, with the following results:

Plant	Number Inspected	Number of Nonconforming Units
1	6420	130
2	7024	102

(a) A company employee will perform a test of the equality of the percentage of nonconforming units, which of course is the null hypothesis. What can be said about the inherent truth or falsity of the hypothesis for this scenario?

(b) There are three possible ways of performing the analysis. Name them. Which method would you suggest be employed?

(c) Use the method that you selected in part (b) and perform the analysis.

17.8. A college basketball team makes 7 of 16 3-point field goal attempts in a particular game and makes 28 field goals overall, for an overall field goal percentage of 40%. What was the team's percentage of successful 2-point field goal attempts? Can the standard deviation of the latter be computed? If so, what is it? If not, explain why it can't be computed.

17.9. A process engineer plots points on a control chart and observes that six consecutive points plot above the chart midline, which is the target value for the process characteristic that is being plotted. She immediately concludes that the process must be out of control since the probability of six points plotting above the midline when the process is in control is $(1/2)^6 = 1/64 = .016$. What is the fallacy in that argument?

17.10. A histogram was presented in Chapter 1. It is worth noting that a histogram has variability just as does any random variable. Which type of histogram will have greater variability—one constructed using the $\sqrt{n}$ rule or one constructed using the power-of-2 rule? Generate 10 sets of 100 observations from a $N(50, 10)$ distribution and then construct a histogram of each type for each of the 10 data sets. This can be done in MINITAB using the RAND command. Specifically,

```
MTB>LET K1 = SQRT(100)
MTB>RAND 100 C1;
SUBC>NORM 50 K1.
```

will generate one set of 100 random observations from the $N(50, 10)$ distribution. Compare the variability of the histograms of each type and comment.

17.11. You are presented with data from a 2^3 experiment that was conducted to determine the factors that seemed to be significant and to estimate the effect of those factors. You notice that the AC and BC interactions are large, relatively speaking, in that each exceeds the main effect estimate for one of the factors that comprise the interaction. Explain how you will analyze the data.

17.12. P.-T. Huang, J. C. Chen, and C.-Y. Chou [*Journal of Industrial Technology*, **15**(3), 1–7, 1999 (available at http://www.nait.org/jit/Articles/huan0699.pdf)] used an experimental design for the purpose of developing a multiple linear regression model for detecting tool breakage, as a tool can continue functioning even when it is broken. Read and write a critique of their paper, paying attention to the presentation of their experimental design and the analysis of the resultant data.

17.13. A practitioner has data on six variables that are believed to be related, with one clearly being "dependent," in a regression context, on the other five. All of the pairwise correlations for the six variables are between .50 and .92, and the dependent variable is binary. The practitioner develops a multiple linear regression model using all five of the available predictor variables, then constructs a confidence interval for the parameter associated with the third variable and makes a decision based on the endpoints of that interval. Two mistakes were made: what were they?

17.14. A scientist returns to his company after attending a three-day short course on design of experiments. In the course, the importance of having the levels of quantitative variables equispaced was stressed. He wishes to run an experiment with a single factor but is reluctant to do so because the levels of a quantitative factor that he wants to use are not equispaced. Should he be concerned about this? Explain.

17.15. An experimenter favors nonparametric methods, claiming that they are much easier to use than parametric methods because she doesn't have to worry about any assumptions, and thus doesn't have to spend time checking assumptions. Do you agree that assumptions can be dispensed with when nonparametric methods are used?

17.16. A student computes the average field goal percentage for all basketball teams in a particular conference by computing the average of the field goal percentages of the various teams.

(a) Explain why this will not give the correct answer.
(b) Under what conditions will the computed number be close to the actual average field goal percentage?

17.17. The Dean of Students at a small university is interested in determining if fraternity or sorority membership is having an adverse effect on grades. He is interested in looking at grade point averages of at least 3.0 and below 3.0, with the latter in his mind being the value that separates good from not-so-good performance. The university's data recording system and computer systems made it easy to compile the following numbers, which are for the entire university, for one semester.

Status	≥ 3.0	< 3.0
Member	564	838
Nonmember	1239	1473

(a) Is this a sample from some population? If so, what is the population?
(b) *If* this is a sample from some population, perform the appropriate analysis and draw a conclusion, justifying the analysis that you performed.
(c) What analysis would you have recommended if the grade point averages had not been dichotomized?

17.18. Safety engineers at a particular company are having difficulty obtaining tread wear information on certain brands of tires that they are considering for use on their company vehicles. Consequently, they decide to conduct an experiment to compare tread wear for three brands of tires. For some unknown reason, the variability in the tread wear differed considerably over the three brands. They didn't know any reason why this should have occurred.

(a) Would you recommend that they repeat the experiment or simply analyze the data that were obtained?

(b) Assume that they used two blocking variables in the experiment (e.g., as in Section 12.3.2). Can the data be analyzed using any methodology that was presented in the text? Explain.

Answers to Selected Exercises

CHAPTER 1

1.1. Below is the time sequence plot that is actually in the form of a scatter plot. There is considerable nonrandomness in this plot, especially the strong upward trend since about 1970 as well as the monotonicity during certain intervals of years (e.g., strictly decreasing from 1938 to 1943).

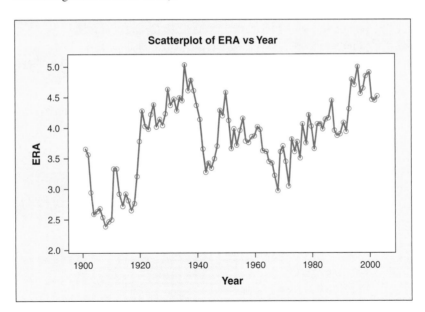

Scatter plot of ERA versus year.

1.2. The plots will look the same. A linear transformation is being made and software will similarly "transform" the vertical axis scale. MINITAB, for example, can be used to show that the time sequence plots are identical.

Modern Engineering Statistics   By Thomas P. Ryan
Copyright © 2007 John Wiley & Sons, Inc.

1.5. (a) The percentages are not additive since neither is based on 100% of the population of men and women combined.

(b) The risk will always be 100%, regardless of the amount of vitamin C that is ingested.

1.6. (a) A bar chart, ordered by salary, with the job titles on the horizontal axis.

(b) The estimate would be $8(51,211) = \$409,688$ for the company with 8 civil engineers. If the number of civil engineers were not known, it would be best not to provide an estimate because the variability in the number of civil engineers per company is not given.

1.7. (a) No, the slopes (of a line fit through the points, for example) will not be the same because the axes are reversed.

(b) No, there are no points on the graph that stand out as being unusual.

(c) We would guess that the point with the higher acceptance rate would be the public university (which it is—University of California–Berkeley).

1.9. (a) Dotplot. **(b)** Dotplot or histogram. **(c)** Histogram.

1.14. (a) The population would be his or her career batting average.

(b) This would logically be defined as the population.

(c) The population would be the total votes cast for the country.

(d) The population would be the set of test scores for the class.

(e) The population is the collection of all transistors of that type produced on a given day.

1.15. There is no time order involved in a dotplot, histogram, stem-and-leaf display, scatter plot, or boxplot. A time sequence plot would be another possibility.

1.20.

	Number of Classes	
n	Power-of-2 Rule	Square Root Rule
50	6	7
100	7	10
200	8	14
300	9	18
400	9	20
500	9	23

1.21. There variables were displayed in Figure 1.5. A fourth variable could be displayed by having a separate graph for each value of that variable. And so on.

1.23. 23.1.

1.25. No, a histogram would not be particuarly useful. Since the data are obtained over time, a display that incorporates time should be used.

1.27. A Pareto chart could be a good choice, depending on how many countries are represented. A bar chart would be another possibility, and for political reasons might be the preferred alternative.

1.33. **(a)** The histograms are similar in that they both exhibit extreme right skewness. Beyond that, a finer comparison would not be practical since each histogram is constructed for only 12 observations, so very large histogram variation would be observed in repeated sampling.

(b) Although both histograms exhibit right skewness, the two large values for TR90 coupled with the fact that there are only 12 observations make it impossible to construct a meaningful histogram.

1.39. 52.

1.42. **(a)** For negative correlation, a line fit to the points will likely have a negative slope, depending on the degree of the negative correlation.

(b) The plot should simply be a random scattering of points.

1.47. It is not possible for any quantity to drop by more than 100%.

1.48. It is improper to use the expression "four times less," as "times" should be used only to signify something greater than the original quantity.

1.51. Since they add to 100%, if we know one percentage then we also know the other one.

1.55. The plots could be compared since they were constructed the same way, but either plot alone could be misleading because of the nonconstant time increment.

1.63. **(a)** Far more than five classes should be used. A reasonable number would be 11, which is the number suggested by the power-of-2 rule.

(b) Left-skewed data are rare, so that should have been a tipoff that the histogram with five classes was misleading. We would logically expect the data to be right skewed, and the information given here, albeit limited, also suggests that.

1.71. Pareto chart.

1.75. A pie chart with 50 slices would be virtually impossible to read. A Pareto chart should be constructed instead.

CHAPTER 2

2.1. **(a)** The average is 97.040 and the standard deviation is 0.078.

(b) The average could easily be computed without a calculator since the numbers all deviate only slightly from 97. Treat the decimal fraction part as integer devistions from 97 (−80, 118, 34, etc.). Then add the deviations sequentially in your head, divide by 6, then divide by 1000, and add the result to 97. This produces 97 + (241/6)/1000 = 97.0402.

(c) Using a similar approach to compute the variance would not be practical for this example, due in part to the fact that the average is not an integer.

2.4. 2.

2.7. (a) The average is 100.85.

(b) There is no parameter that is being estimated with this average.

(c) The standard deviation is 7.69. The unit of measurement for the standard deviation is the original unit of measurement: cost index.

2.10. No, the average salaries for starting pitchers and relief pitchers differ by such a large amount that it would be inappropriate to compare standard deviations, as we would expect the salaries of starting pitchers to have the larger standard deviation simply because their average salary is so much higher. A comparison using the coefficient of variation would be a logical alternative.

2.13. (a) The standard deviation for the new numbers is 100 times the standard deviation of the original numbers.

(b) The coefficient of variation is unaffected because the mean of the new numbers is also 100 times the mean of the original numbers.

2.15. A typical salary is best represented as the median salary. The need to use the median instead of the average shouldn't depend greatly on company size since the larger the company, the more personnel with high salaries.

2.18. 2.

2.19. (b).

2.23. (a) $12,245.

(b) Cost per share is $12,245/450 = $27.21. The cost would remain unchanged if an additional 200 shares were purchased at this price since each of the two components in the weighted average would be $27.21.

2.27. A very large sample should be used unless a large estimation error can be tolerated.

2.30. 4.12.

2.33. 6.32428.

2.37. 14,240.

2.40. The variance of the new numbers is the multiplier squared times the variance of the old numbers. Since $(-1)^2 = 1$, the variance is unchanged by multiplying by −1.

2.45. The transformation must have been of the form $Y = 0.1X + c$, with c an arbitrary constant. Thus, there is essentially an infinite number of possible transformations as c could be any real number.

2.47. 3.1 should be -3.1.

2.49. 84.

2.51. 623.

2.53. No, not enough information is given.

2.56. 0.42.

2.60. **(a)** The average cannot be determined from the information that is given. All that can be determined are the five deviations from the average.
(b) 17.5.

2.65. The variance of all of the numbers is 691.6; the variance after 10% trimming is 288.1. The numbers differ greatly because of the extreme observations.

2.67. 104.

2.69. **(a)** Yes, November.
(b) Actually, this cannot be determined because the numbers are obviously not correct within rounding.

2.71. This can occur in only two ways: in the trivial case where both the numbers are the same so that the standard deviation and variance are both zero, and in any of an infinite number of examples for which the variance is one. One such example is $3 + \sqrt{0.5}, 3 - \sqrt{0.5}$.

2.75. Eighty percent of the employees are men and 20% are women.

CHAPTER 3

3.1. $P(B \mid A) = 0$ because B cannot occur if A has occurred since the events are stated as being mutually exclusive.

3.3. **(a)** .9.
(b) Yes, .9.

3.5. 2/3.

3.9. .42.

3.11. **(a)** .0918.

(b) The answer in part (a) is just an approximation since normality was not assumed and the Central Limit Theorem must thus be invoked.

3.12. 33.4.

3.13. 63/64.

3.17. The possible values of a binomial random variable are 0 through n. The possible values for each Bernoulli trial are 0 and 1.

3.18. .15.

3.21. $E(a + bX) = E(a) + E(bX) = a + bE(X)$.
$Var(a + bX) = Var(bX) = b^2 Var(X)$ since an additive constant does not affect the variance but a multiplicative constant does.

3.23. .326.

3.25. 1.25.

3.28. 2.

3.31. A distinction must be made between a population distribution and an empirical distribution. Although the range of *possible* values has decreased, the variance can still be essentially unchanged if the largest values are well within the upper boundary.

3.33. 800.

3.34. 1/64.

3.35. **(a)** .1056. **(b)** Zero. **(c)** $n > 36$.

3.39. The relationship will not be of any value unless the distribution of CMMH is known.

3.40. .35.

3.41. The first moment.

3.43. $P(A \mid B) = P(A)$ results from the formula for conditional probability only if $P(A \cap B) = P(A) P(B)$. It then follows from this that $P(B \mid A)$ must simplify to $P(B)$ since $P(B \cap A) = P(A \cap B)$.

3.47. $E(X) = 4.58$ and $Var(X) = 0.28$.

3.49. The probabilities given could not be correct because they add to more than one.

3.51. .281.

3.52. $npa^{(n-1)/n}$.

3.53. 1/128. There are many possible solutions since n and p are both unknown.

3.55. 1/6.

3.57. **(a)** The mean is 32; the variance is 12.
    (b) Normality or nonnormality has no effect on the computations for the mean and variance in part (a).

3.59. **(a)** $a = 1$. **(b)** The distribution is not symmetric.

3.61. .9406. This is an approximation because there is a finite population, but n/N is small enough to permit the correction factor to be ignored.

3.65. It is not a *pdf* because it integrates to 2/3 instead of 1.

3.67. .254.

3.69. (a) .0625. (b) .0039.

3.72. **(a)** It is possible to solve for d but it is not possible to solve for c.
    (b) Since it is not possible to solve for c, no probabilities can be obtained.
    (c) $(2/3)c + 1$.

3.73. $E(X^2) = \sigma^2 + \mu^2$.

3.74. **(a)** $(1 - p)/(1 + 3p)$.
    (b) From inspection of the answer in part (a), it is obvious that the answer cannot exceed 1.

3.77. Two sets of z-scores might be computed, with each pair obtained so as to give $P(67$ inches $\leq$ height ≤ 68 inches) for men and for women, with the choice determined from the larger of the two probabilities.

3.79. .3935.

3.80. 8.47.

3.81. .632. This differs considerably from .50 because the exponential distribution is a right-skewed distribution, so most of the probability is at the lower end.

3.85. .009.

3.87. 50.

3.92. **(a)** The probability cannot be determined without a distribution being specified.
(b) No, there are no Bernoulli trials involved.
(c) The Poisson probability is .3935.

3.93. $\bar{x} = 72$ and $s = 20$.

3.95. $z_0 = 1.438$ and $\alpha = .472$.

3.99. $E(X) = 29/18$.

3.100. 100.

CHAPTER 4

4.2. Mean squared error (MSE) is variance plus squared bias. $Var(\bar{X}) = \sigma^2/4 = 4/4 = 1$. $Var(X^*) = Var(X_1 + 2X_2 + 2X_3 + X_4)/10 = \sigma^2/10 = 0.4$. $E(X^*) = 0.6\mu$ so $Bias(X^*) = -0.4\mu = -4$. Thus, $MSE(X^*) = 0.4 + 16 = 16.4$, which is much greater than $MSE(\bar{X}) = Var(\bar{X}) = 1$. Assuming μ positive, $MSE(\bar{X}) < MSE(X^*)$ when $\mu > 1.37$.

4.5. **(a)** $Var(\widehat{\mu}_1) = 2.0$ and $Var(\widehat{\mu}_2) = 0.9$.
(b) The first estimator is unbiased because the weights sum to 1.0; the second estimator is biased because the weights do not sum to 1.0.

4.9. It is not possible to obtain the maximum likelihood estimator of the mean because the mean is not a parameter in the *pdf*.

4.11. The distribution of the sample mean will also exhibit right skewness, but will have a smaller variance.

4.13. 225.

4.15. The mean, which is 2/3, is known because it is not a function any unknown parameters. Therefore, we cannot speak of a method of moments estimator for something that does not have to be estimated.

4.16. The sample mean is indeed a consistent estimator of the population mean since $\lim_{n\to\infty}(\bar{x} - \mu) = 0$.

4.18. For an arbitrary estimator θ, unbiased means that $E(\widehat{\theta}) = \theta$. This is a desirable property because it means that in repeated samples the average value of a point estimator is theoretically equal to the parameter that it estimates.

4.19. An estimator is a random variable; it has a mean, variance, and distribution. An estimate is a number in the case of a point estimate and an interval in the case of an interval estimate. Estimates do not have the properties that estimators do because estimates are simply numbers.

4.20. The square root of a linear operator is not involved. The same linear operator is being used; the square root is being taken of the statistic. In general, if $E(\widehat{\theta}) = \theta$, then $E\sqrt{\widehat{\theta}} \neq \sqrt{\theta}$.

4.26. **(a)** No, such a determination cannot be made from the information that is given.
(b) No, the assumption of normality doesn't tell us anything about how variable the data are.
(c) A rough estimate would be 5.

4.27. The variance cannot be determined without knowing the total number of people who entered the store since the variance is a function of that number.

4.28. **(a)** It is biased because $E(\widehat{\lambda}) \neq \lambda$.
(b) It is approximately unbiased when n is large.

4.31. Let $a = \bar{x} + \epsilon$. Then

$$\sum_{i=1}^{n} (x_i - a)^2 = \sum_{i=1}^{n} [x_i - (\bar{x} + \epsilon)]^2$$
$$= \sum_{i=1}^{n} [(x_i - \bar{x}) - \epsilon]^2$$
$$= \sum_{i=1}^{n} (x_i - \bar{x})^2 + n\epsilon^2 > \sum_{i=1}^{n} (x_i - \bar{x})^2$$

for any real number ϵ. [The middle term vanishes because $-2\epsilon \sum_{i=1}^{n} (x_i - \bar{x}) = 0$ since $\sum_{i=1}^{n} (x_i - \bar{x}) = 0$.] Yes, $\bar{x}$ is the least squares estimator of μ.

4.36. 12,500.

4.37. The new mean is $45,320. The new standard deviation is $2,080.

4.38. It costs money to take a sample, and the larger the sample the greater the cost. The inceased precision afforded by large samples must be weighed against the increased cost.

4.39. There is no maximum likelihood estimator for this *pdf* because there is no parameter to estimate in the *pdf*.

4.40. The maximum likelihood estimator is

$$\widehat{\alpha} = 1 - \frac{n}{\displaystyle\sum_{i=1}^{n} \log(x_i)}$$

4.43. Let the variance for the original data, in inches, be given by c. Then the variance in centimeters as a function of c is $(2.54)^2 c = 6.45c$.

CHAPTER 5

5.2. $z = (\overline{X} - \mu_0)/(\sigma/\sqrt{n}) < -z_{\alpha/2}$ since the null hypothesis was rejected and the test statistic was negative. Solving the inequality for μ_0, we then obtain $\mu_0 > \overline{X} + z_{\alpha/2}\sigma/\sqrt{n}$, with the latter being the upper limit of the confidence interval.

5.3. Since the null hypothesis is not rejected, $z = (\overline{X} - \mu_0)/(\sigma/\sqrt{n})$ is between $-z_{\alpha/2}$ and $z_{\alpha/2}$. Since this is the starting point for producing the form of the confidence interval, it follows that μ_0 must lie inside the interval.

5.5. .9456.

5.7. We wish to test $H_0: \mu = 60$ versus $H_a: \mu > 60$. Since $\overline{X} = 57.33$ is less than 60, the null hypothesis would not be rejected.

5.8. Hypothesis tests are constructed to test hypotheses for which there is at least a reasonable question of doubt. The sample information either provides evidence that the null hypothesis should be rejected, or it does not provide sufficient evidence. A null hypothesis can never be "proved." For example, it would be nonsensical to say that we accept $H_0: \mu = 50$ if $\overline{x} = 48$ for a sample of size 100.

5.11. Yes. No.

5.13. "less than 25."

5.15. You have to assume that the driving conditions during those 10 days are typical of the driving conditions that you will face in the future. The interval is (22.17, 24.63).

5.18. $H_0 : \sigma^2 = \sigma_0^2$ versus $H_a : \sigma^2 < \sigma_0^2$. The test could be performed using $\chi^2 (= (n - 1) S^2/\sigma_0^2)$ and it would be desirable to first test for approximate normality, since the test is based on the assumption of normality. The letter could be tested by constructing a normal probability plot.

5.19. The computed lower limit is negative because the percentage of unacceptable flatness measurements in the sample was small. This type of problem can be avoided by using a better method of constructing the interval, such as the one given in Eqs. (5.4a) and (5.4b).

5.21. The population mean μ is either between the two numbers or it isn't. In order to make a probability statement, a random variable must be involved, whereas the two endpoints of the intervals are numbers and μ is also a number.

5.23. 0.28.

5.27. 1.03 a.

5.30. Equality, as in a null hypothesis, can never be proved with a hypothesis test. Furthermore, equality almost certainly won't exist anyway if the percentage is stated to at least two decimal places.

5.31. "Statistically significant" means that use of the appropriate sample statistic(s) leads to rejection of the null hypothesis. Comparing the relative mass of Saturn with the relative mass of Earth is unrelated to hypothesis testing.

5.33. (a) The distribution should be less skewed for the larger sample size.
   (b) The relationship between the averages is not affected by the sample sizes. This is because $E(\overline{X})$ is a direct function of $E(X)$, with the latter being the mean of the population distribution, which is independent of any sample size.

5.35. In practice, not very many random variables can have negative values. A normal distribution can still be used as a model, however, because a normal distribution whose variance is small compared to the mean would have only a very tiny fraction of its area below zero, so the fact that the range of possible values for any normal distribution is from minus infinity to plus infinity is not necessarily relevant.

5.37. Information about variability of individual observations relative to prescribed limits are not contained in the confidence interval or hypothesis test for a mean. A tolerance interval is one approach that could be used, and a confidence interval on the proportion of observations above a specified value (such as 23.5 ounces in this case) could also be useful.

5.39. 2.08.

5.41. Because of the large sample size, it would be reasonable to construct a confidence interval using z, as only for some extreme distributions would a sample size of 100 be inadequate.

5.42. 0.4.

5.44. (a) 51.0.
   (b) It will not be exactly a 95% confidence interval because normality was not assumed; it would be such an interval if normality did exist.
   (c) No, because that probability is, trivially, either 0 or 1.

5.48. 100.

5.49. The explanation is vague and misleading. If we construct a large number of confidence intervals, approximately 95% of them will contain the parameter (or a function of multiple parameters) for which the interval was constructed.

5.51. Zero.

5.54. The experimenter should check to see if there is any prior information that would suggest an approximate value of σ. In the absence of such information, there is nothing that can be done since the distribution is not known, even approximately.

5.55. Especially when $\widehat{p}$ is small, the limits should be constructed using Eq. (5.4a) and (5.4b), which will result in the limits not being equidistant from $\widehat{p}$. The latter is not a shortcoming, however, and it is better to have limits that are not equidistant from $\widehat{p}$ than to not have a lower limit.

5.57. **(a)** If σ is unknown, there are two problems: a value of t is needed, which depends on the sample size, which can't be solved for without specifying a value for t; and a value for s must be used, but we don't have a value for s unless we have taken a sample.

(b) A prior estimate of σ would have to be used; otherwise the expected width could not be determined, with use of a prior estimate determining the expected width as $2\, t_{\alpha/2,15}\, \sigma^*/4$, with σ^* denoting the prior estimate.

5.58. It sounds as though the "confidence intervals" may actually be tolerance intervals since confidence intervals are constructed only for parameters but what the intervals are being constructed for are apparently not parameters.

5.60. **(a)** It would be a confidence interval on the average age of registered Republicans in Cobb County.

(b) It would probably be unwise to rely on the data as there would certainly be "measurement error."

5.61. $\overline{X} = 90.234$. The decision would be to not reject the null hypothesis.

5.63. The test statistic is

$$\chi^2 = \frac{(n-1)S^2}{\sigma_0^2} = \frac{99(.0035)^2}{(.0030)^2} = 134.75$$

This is a one-sided test and $P(\chi_{n-1}^2 > 134.75 \mid \sigma^2 = \sigma_0^2) = .0098$, so there is evidence that the standard deviation has increased and that the machine should be replaced.

5.65. 1.314.

5.67. Positive, 0.69.

5.69. The difference will be greater for the 95% confidence interval and the percentage difference will also be greater for the 95% interval. It is best to use the t-distribution whenever possible, especially for 90% intervals.

5.72. 107.

5.73. Zero.

5.74. $(\sigma/\sqrt{n})(1 + z_{\alpha/2})$.

5.78. The likely cause is that a normality-based approach was used but the random variable has a skewed distribution. An "exact" confidence interval could be used to try to correct the problem.

5.81. A p-value cannot be the probability that the null hypothesis is correct because the null hypothesis is a statement that is either correct or not correct (and we know that null hypotheses are virtually always incorrect), whereas a p-value is a nontrivial probability of some event occuring. The "probability," loosely speaking, of the null hypothesis being correct is either zero or one, and is almost certainly zero.

CHAPTER 6

6.1. The denominator of t^* is $\sqrt{s_1^2/n_1 + s_2^2/n_2}$, with this becoming $\sqrt{(s_1^2 + s_2^2)/n}$ when the sample sizes are equal. The denominator of the independent-sample t-test with equal sample sizes can be written

$$\sqrt{\frac{s_1^2 + s_2^2}{2}\left(\frac{1}{n} + \frac{1}{n}\right)}$$

which is obviously equal to the denominator of t^*.

6.2. Since $\widehat{p}_1 = X_1/n_1$ and $\widehat{p}_2 = X_2/n_2$, it follows that $X_1 + X_2 = n_1\widehat{p}_1 + n_2\widehat{p}_2$. Since this is the numerator of $\widehat{p}$ and $n_1 + n_2$ is the denominator, it follows that $\widehat{p}$ is a weighted average of the two sample proportions. The two sample proportions are combined to obtain a single estimate because the population proportions are equal under the null hypothesis.

6.3. It is not possible to make that test using one of the methods given in the chapter, based on the information that is given. Percentage increases are almost certainly not going to be even approximately normally distributed, and the sample sizes are not large enough, especially for the union mines, to be able to proceed with distributional information. Of course, information on variability would also have to be provided.

6.5. $\widehat{\sigma}^2 = s_{\text{pooled}}^2 = \dfrac{(n_1 - 1)s_1^2 + (n_2 - 1)s_2^2}{n_1 + n_2 - 2} = \dfrac{13(6) + 14(7)}{13 + 14} = 6.52$

6.7. The plotted points must slope upward because the cumulative probability by definition must increase as the variable increases, and similarly, z-scores increase as the variable increase.

6.9. (a) -15.
    (b) No, this estimate is not based on any distributional assumption.

6.11. The samples are dependent because the same people are being measured in each of the two samples.

6.13. 63.

6.19. No, such a confidence interval could not be constructed for various reasons; one of which is the fact that the data are population data, not sample data, for the year 2000.

6.20. It would be unwise to use one of the t-tests for these data since the sample sizes are so small. Instead, a nonparametric test (Chapter 15) might be used.

6.22. (a) Technically, there is no way to test for a "significant increase" without defining what that is.

(b) The data are paired so a paired t-test can be performed, provided that the differences are not strongly nonnormal. In fact, they are not, so

$$t = \frac{\bar{d} - 0}{s_{\bar{d}}} = \frac{1.889 - 0}{2.571/\sqrt{9}} = 2.20$$

and the p-value for the one-sided test is .029. Thus, there is some support for the alternative hypothesis, which is that respiration rate is increased.

6.23. Doing so would bias the results. The probabilities for Type I and Type II errors hold before a sample is obtained.

6.25. We will use an F-test because of the approximate normality. Thus,

$$F = \frac{1.85^2}{2.02^2} = 0.84$$

The p-value, which is $P(F < 0.84)$, is .1649. Thus, there is no evidence that the first mixture of gases is superior to the second one.

6.27. The lower one-sided limit will be greater than 23.4. This is because the t-value is smaller since .05 is in the lower tail of the distribution for the one-sided interval, whereas .025 is the lower-tail area for the two-sided interval. Thus, the t-value that is used for the one-sided interval is not as extreme as the one used for the two-sided interval, so the deduction from $\bar{x}_1 - \bar{x}_2$ that is of the form $t\, s\sqrt{1/n_1 + 1/n_2}$ is smaller for the one-sided interval.

6.29. The numerators are the same because $\bar{d}$ in the paired test equals $\bar{x}_1 - \bar{x}_2$, which is the numerator in the independent sample test. The denominators differ because with the independent test $Var(\bar{x}_1 - \bar{x}_2) = Var(\bar{x}_1) + Var(\bar{x}_2)$, whereas with the paired test if is $Var(\bar{d})$.

6.31. It isn't necessary to assume equal variances when t^* is used. In fact, it is used when the variances are believed to be unequal.

6.32. In order to perform the hypothesis test, it is necessary to substitute a value for $\mu_1 - \mu_2$. If we hypothesize that they are unequal, we do not have a value to substitute for the difference in the test statistic.

6.34. A t-test would not necessarily support the rule because the students may differ in ways other than campus versus off-campus. For example, are the courses of study comparable? Do many of those who commute do so from areas where secondary education is below par? Are many who live on campus from schools and areas outside the state that are very strong academically? These and other questions would have to be addressed.

6.35. Although the form of the confidence interval differs slightly from the form of the hypothesis test, we would nevertheless expect the confidence interval to include zero since the value of the test statistic is much less than $z_{\alpha/2} = 2.576$.

6.40. (a) The F-test for equal variances is highly sensitive to nonnormality. In the absence of information about the two distributions, it would be safer to use a test that is not based on the assumption of normality.
(b) It would be preferable to use a test like Levene's test that is not based on a normality assumption, but the raw data would have to be available in order to perform the test.

6.42. It should be about the same. When the samples are independent, the variance of the difference is equal to the sum of the variances, with the latter used in the independent sample test.

CHAPTER 7

7.1. None of the intervals in the chapter are of that form because that is the form of the confidence interval for a parameter and confidence intervals were not covered in this chapter.

7.2. (0.854, 3.946). Since this is also the general form of a tolerance interval in the case of known parameters, we could say that this is a tolerance interval with a confidence coefficient of 100 that covers (exactly) 95% of the population.

7.5. Yes, the prediction interval will always be wider because there is an additional element of uncertainty since the interval is for a random variable rather than a fixed parameter. More explicitly, $s\sqrt{1 + 1/n}$ is larger than $s/\sqrt{n}$ for any value of n.

7.6. The tolerance interval would be wider.

7.8. (b) confidence interval. As explained in Section 7.4.2.1, even extremely large samples won't help in the presence of considerable nonnormality.

7.10. A prediction interval is the type of interval that should be used, but it should not be of the form given in Eq. (7.1) because that is not based on any type of a regression relationship between GPA and factors that correlate with it. A prediction interval for this purpose should be constructed using appropriate regression data.

7.11. A 99% tolerance interval is an interval designed to cover 99% of the population values. This cannot be done with certainty, however, nor would we expect the interval to cover exactly 99% of the population, so a degree of cofidence is attached to the interval, with the degree of confidence indicating the probability that the interval contains, in this case, *at least* 99% of the population values. A 95% tolerance interval with 99% confidence is one for which the user is 99% confident that the constructed interval contains at least 95% of the population.

7.13. No, not using methodology presented in the chapter. Since a prediction interval is sensitive to nonnormality, regardless of the sample size, nonnormality must be considered, and here we have data involving percentages.

7.15. **(a)** $(0.67, 0.97)$.
  (b) There is no degree of confidence attached to the interval in part (a) because μ and σ are assumed known.

7.17. A new interval should be constructed using recent data.

7.20. No, as stated in Section 7.3, we must have $n \geq 93$ in order to accomplish this.

7.21. No, this cannot be done explicity because a t-variate must be used, which depends on the sample size. The sample size could be solved for explicity only if an assumed value of σ were used.

CHAPTER 8

8.1. A straight-line relationship is suggested only for the first and third datasets, with the relationship being weak for the first dataset and an obvious outlier present in the third dataset.

8.3. The number 2.38 is the estimate of the change in Y per unit change in X, for the range of X-values in the sample data used to obtain the regression equation.

8.8. The errors are assumed to be independent—of each other and of any predictor variable in or not in the model.

8.9. The least squares estimates minimize the sum of the squares of the *residuals*. The errors would be known only if the parameters were known, in which case they would of course not be estimated.

8.12. Simple linear regression models are generally not used when X is zero or even close to zero. Knowing that $Y = 0$ when $X = 0$ is not sufficient motivation for using regression through the origin. The determining factor should be the quality of the fit for the range of X covered by the data, and this will generally dictate the use of an intercept model.

8.13. This is the percent of the variation in Y that is *not* explained by the regression model.

8.15. The number must be 3.88. There is not enough information given to determine the prediction equation.

8.17. The prediction equation is $\widehat{Y} = 6.29$; since $\widehat{\beta}_1 = 0$, $R^2 = 0$. The graph shows a quadratic relationship between X and Y, with no evidence of linearity being part of that relationship. This illustrates that two-dimensional data must first be graphed before a regression model is fit.

8.20. The method of least squares causes the residuals to sum to zero, regardless of the model that is fit, as long as the model contains an intercept. This is simply a calculus-related result that is model independent.

8.23. -1.

8.24. 3.74.

8.25. The word "least" refers to the fact that a minimization is taking place, and what is being minimized is the sum of the squared residuals.

8.27. **(a)** 3.25.
   (b) Probably about 2.2–3.0.
   (c) GPA: 3.3–4.0; SAT-M: 620–730; SAT-V: 530–620. These of course are just estimates. We would have to have the data in order to determine the space of values for which the prediction equation could be used.

8.28. When a probability plot of any type is constructed, the plotted points must be realizations of random variable that have the same variance. The ordinary (raw) residuals have different variances, which is why a normal probability plot of ordinary residuals should not be constructed.

8.30. If we fit a bad model, a good data point may show up as an outlier, solely because of the difference between the true model and the model that was fit.

8.31. No, the value of $\widehat{\beta}_1$ cannot be determined from the value of $\widehat{\beta}_0$ and the other information that is given.

8.33. **(a)** Yes, $r_{xy} = .996$. **(b)** (3.21, 6.99).

8.37. **(a)** The value of the t-statistic is 257.92.
   (b) This value is far in excess of any value for t that might be considered the smallest value in order for the regression equation to be useful.

8.39. Preferably, a line should be fit to the data for X from 20 to 30. If the entire range is to be used, different lines should be fit from 10 to 20 and from 20 to 30. It is important to recognize, however, that lines or planes should not be fit in parts of the data space where there is no clear relationship.

8.43. The value of r_{xy} could certainly be reported, but it would be more meaningful to give the value of R^2. The value of r_{xy} cannot be determined from the prediction equation.

8.44. The line through (23.8, 10.2).

8.45. The consequence is simply that there won't be a very good fit since there is no way to accommodate pure error.

8.47. The value of R^2 will be the same because R^2 in the one-predictor case is simply the square of the correlation coefficient, and $r_{XY}^2 = r_{YX}^2$.

8.49. $r_{xy} = -0.87$.

8.51. Figure 8.4 suggests that a quadratic term should be used in the model, whereas Figure 8.3 suggests the use of a reciprocal term.

8.56. 0.56. The intercept would not change.

8.57. Neither variable would be a natural choice as the dependent variable. The high correlation can be explained by the fact that both variables are strongly related to overall intelligence.

8.58. Anything close to zero kernels would certainly not result in enough popcorn to satisfy the family, so whatever happens with close to zero kernels is irrelevant.

8.61. As one might expect, this won't help. Although $\sqrt{Var(\widehat{\beta_1})}$ is reduced by 12 in the conversion to inches, $\widehat{\beta_1}$ is reduced by the same amount. Thus, the parameter when the unit of measurement is inches is estimated with the same precision as the parameter when the unit of measurement is feet.

8.63. 1.0.

CHAPTER 9

9.4. There is no point in unnecessarily inflating $Var(\widehat{Y})$. If an experimenter continued adding variables, a point of diminishing returns would soon be reached, with small increases in R^2. Such small increases would be more than offset by the increases in $Var(\widehat{Y})$.

9.5. (a) The standard error for the coefficient of height is found by solving the equation $0.33893/SE = 2.61$, which leads to 0.1302 for the standard error (SE). The R^2 value is $7684.2/8106.1 = 94.8\%$, the mean square value is $421.9/28 = 15.1$, and the F-statistic is $3842.1/15.1 = 254.97$.

9.10. There will be 511 subsets because one of the 512 subsets is the empty set.

9.12. When the correlation between the predictors is this high, the predictor space will be a relatively narrow ellipse. The smaller the ellipse, the harder it will be to stay within the ellipse.

9.15. If the assumptions are approximately met, this will be lost if Y is transformed. This is why model improvement should usually center on predictor transformations.

9.18. It isn't necessary because a t-statistic is computed with the other variable(s) in the model, so conditioning is implied, but not explicitly stated.

9.19. (a) The regression equation is $\widehat{Y} = 4.59 + 0.190X_2$.
 (b) The regression equation with both predictors is $\widehat{Y} = 2.69 + 0.809X_1 - 0.425X_2$. The sign changes because X_1 is a much stronger predictor than X_2, and the predictors are correlated.

9.20. Scatter plots against individual predictors will be useful when the predictors have at most small pairwise correlations.

9.22. This is because $\widehat{\beta_i}$ is in the numerator of the statistic and the denominator is positive.

9.24. $Y = \beta_0 + X^{\beta_1} + \epsilon$.

9.28. $C_p = SSE_p/\widehat{\sigma}_{full}^2 - (n - 2p)$ with $\widehat{\sigma}_{full}^2 = SSE_{full}/(n - a)$, with a denoting the number of available predictors plus one. When all of the available predictors are in the model, $a = p$ and $SSE_p = SSE_{full}$. Then $C_p = (n - p) - (n - 2p) = p$.

9.29. These statements can be made only if the predictors are uncorrelated.

9.31. 2.379.

9.32. Using MINITAB, the 99% prediction interval is found to be $(2172.1, 3378.9)$.

9.34. Yes, this can be determined since an orthogonal experimental design was used and the sequential sum of squares is given in the output. Thus, $R^2 = (84{,}666 + 911925)/1180670 = 84.4\%$.

9.36. The values of t-statistics should not be employed to determine which predictors to use in the model when regression is applied to observational data, since correlations between the predictors can render such statistics useless.

9.38. The regression equation should be used only for values of the two predictors that fall within the predictor space. The fact that a fitted value could be negative for a small tree is irrelevant if there weren't any small trees in the dataset. In fact, the smallest tree had a diameter of 8.3 feet, so there weren't any small trees in the dataset.

9.40. It is possible, even likely, that the errors will be correlated, so this should be checked using the residuals.

9.41. The change from feet to inches will cause the regression coefficient for that term to be 1/12 of its magnitude when the term was in feet. The other regression coefficient will remain unchanged. This can be explained as follows. A linear transformation of a predictor does not change the relationship of that predictor to the response variable, nor does it change the relationship of it to the other predictor. Since these relationships remain unchanged, the only adjustment must be in the magnitude of the regression coefficient of the predictor for which there is a scale change.

9.42. Two out of 28 is not excessive under the assumption of a normal distribution, as the expected number is $28(.0456) = 1.28$.

9.44. In a given application, all or almost all of the parameter estimators will be biased because it is extremely unlikely that the true model will have been identified.

9.46. This doesn't work because the sum of columns constructed in this manner is the same as the column of ones that is used to represent the constant term, which prevents a unique solution to the normal equations.

9.47. Machine type is an indicator variable and the values that are used should be 0 and 1.

9.51. The slope of the line is $\widehat{\beta}_3$ and the intercept is zero.

CHAPTER 10

10.1. The first step would simply be to construct a scatter plot of the data. Since no subject matter knowledge is available, if the scatter plot exhibits nonlinearity, try to match the configuration of points with one of the curves in Ratkowsky (1990).

CHAPTER 11

11.1. 80.5.

11.3. The first step should be to look at the distribution of the 200 numbers and, in particular, determine if the distribution appears to be close to a normal distribution.

11.5. There will not be a 3-sigma lower control limit for subgroups of size 5. The user would be much better off using probability limits, which will ensure a lower control limit.

11.9. Individual observations obviously don't contain any information about variability.

11.11. All attributes control charts that are used should indeed have a LCL. Without it, it can be difficult to determine when quality improvement has occurred when a Shewhart chart is used.

11.12. The LCL on a p-chart won't exist whenever $\widehat{p} < 9/(9+n)$. Yes. This problem can and should be corrected by using the regression-based limits for these charts.

11.15. $C_{pk} = 1$.

11.17. A p-chart should be used, preferably with regression-based limits.

11.18. The p-chart limits are .267 and .408.

11.19. 55.2.

11.20. .9973.

11.25. .0027.

11.28. CUSUM and EWMA procedures are generally used for Phase II. They could be used in Phase I, but they would have to be reset to zero after each signal.

11.31. .0083.

11.33. 140.

11.35. 0.22% are nonconforming.

11.37. 62%.

11.39. Process capability and process control are essentially unrelated. A process can be in control but still not be a capable process, relative to the specification limits, because specification limits are not involved in the determination of control limits.

11.41. The two assumptions for an X-chart are that the observations are independent and (approximately) normally distributed.

11.43. (27.25, 44.95). The assumptions of independence and (approximate) normality should first be checked.

11.45. A CUSUM procedure uses all of the available data since the sums have been set or reset. This naturally includes data both before and after a parameter change. Since data after the change are thus used in the CUSUM computation, there is stronger evidence to suggest that a change has occurred than with a Shewhart chart, which would use out-of-control data only at a single point.

11.46. 3.718.

11.49. The objective should be to use the most appropriate estimator for each stage. In Phase I it is desirable to use estimators that are not unduly influenced by bad data points, whereas this is not a concern in Phase II. For the latter the objective should be to use unbiased estimators that have the smallest variance among all unbiased estimators that might be considered.

11.51. More than 6 units above its UCL.

11.53. This is because any additional criterion can only reduce the in-control ARL since a signal can then be received from a second source.

11.55. A signal is received on the third subgroup as $S_H = 5.01$. The fact that the $\overline{X}$ values are tending toward what would be the midline on the $\overline{X}$-chart suggests that the likely out-of-control condition is being rectified.

11.58. **(a)** There is actually not a LCL because zero cannot be a LCL for any type of range chart since ranges cannot be negative. There is no control limit unless plotted points can fall outside it.

   (b) Yes, technically all of the information about both location and variability are contained in the individual observations since that is all there is, and in this instance the two charts have very similar configurations of points.

   (c) This occurs because of the greater-than-usual change between consecutive points. The I-chart does not suggest that this is a problem, however, so I would not declare to Eaton that they have a variability problem. We should also remember that at least one false signal can easily occur when many points are plotted.

11.60. Many industrial personnel consider such conditions to be "normal." It would be better, however, to try to identify the cause of these intermittent conditions, difficult though that may be to do.

CHAPTER 12

12.2. **(a)** The smallest possible value of F is zero, which would occur if the average response value at each of the two levels is the same.

12.3. The conditional effects should be computed.

12.5. Residual plots can be used in various ways, including checking the equal variance assumption for replicated designs. In general, residual plots should be used in essentially the same way with Analysis of Variance models as are used with regression models.

12.6. 24.

12.8. The defining contrast is $I = ABC$. If the treatment combination bc is impossible, the other fraction should be used, assuming that it does not contain an impossible treatment combination.

12.11. **(a)** $AB = D = ABCEF = CDEF$.

   (b) A better choice would be $I = ABCD = CDEF$. Such a design would not confound main effects with two-factor interactions, which is unnecessary.

12.13. This means to set factors A and B at their high levels and factors C and D at their low levels.

12.14. 2.5.

12.15. All seven effects are estimable. They are A, B, C, AB, AC, BC, and ABC.

12.16. Analysis of Variance for Y

Source	DF	SS	MS	F	P
A	2	384.6	192.3	8.40	0.005
Error	12	274.8	22.9		
Total	14	658.9			

12.18. **(a)** What is gained with the second design is a reduction in the number of design points from 64 to 16, reducing the cost considerably. What is lost is the ability to estimate all two-factor interactions.

(b) This would not be a good design to use because $(ABCD)(CDEF) = ABEF$, so AB would be confounded with EF.

12.20. 4.77.

12.21. Each of the hypothesis tests would test the null hypothesis that the mean response at a given level is equal to the average of the mean responses, averaged over all five levels.

12.22. The decision lines in Analysis of Means should not be confused with limits on control charts. They apply only after an experiment has been performed and all of the effect estimates have been computed.

12.26. This will not suffice as the model coefficients of terms corresponding to factors that comprise large interactions can be misleadingly and inappropriately small. If the conditional effects are not small, the model will not be able to depict their effect.

12.29. A randomized block design should be used when an experimenter believes that the experimental units are not homogeneous. If such a design were used and the block totals were very close, this would suggest that blocking probably wasn't necessary.

12.31. A large three-factor interaction means that the two-factor interaction effects are unreliable, which in turn means that the main effect estimates are unreliable—a domino effect.

12.33. Using multiple readings as if they were replicates will likely underestimate the standard deviation of the error term in the model and cause factors to be erroneously declared significant.

12.34. The choice of factor levels to use can often be difficult. The statement refers to the fact that such decisions can be made much more easily after some experimental information is available.

12.37. 4.0.

12.40. When both factors are assumed random, they are tested against the AB interaction, whereas when the factors are fixed they are tested against the error term. A large interaction can cause the A and B main effects to appear small, and the problem could be compounded if those effects were improperly tested against the AB interaction.

12.42. Since the experiment was unreplicated, a normal probability plot analysis should be performed. That analysis shows the B and C effects to be significant, in addition to their interaction. The effect estimates are -132.5, -73.5, and 47.5, respectively. Since the interaction effect estimate is more than half of the C effect estimate and more than one-third of the B estimate, conditional main effect estimates should be reported.

12.44. The treatment combinations a, b, c, d, and abc will be in one fraction.

12.46. The coefficients for each of the two factors are orthogonal in a 2^2 design.

12.48. These are not true correlations because the treatment combinations are selected; they do not occur at random.

12.50. 42.34.

12.52. The hard-to-change factor could be fixed at one level while the treatment combinations for the other three factors were run in random order, and then this repeated for the second level of the hard-to-change factor.

12.54. There is no need to use stepwise regression for an orthogonal design as the p-values from the Analysis of Variance table can be relied on. Furthermore, the model that would be selected using stepwise regression would almost certainly agree with the model selected from the size of the p-values when all of the estimable effects are in the model.

12.56. The conditional effects should be computed, which should always be the case when interactions are large.

12.57. It would have to be a resolution III design. Such a desgin would generally be used as a screening design.

12.58. Using $I = ACDE$ would confound two-factor interactions with each other. This is unnecessary as using $I = ABCDE$ would result in two-factor interactions being confounded with three-factor interactions, a better arrangement.

12.60. The following order would be preferred: A, B, AB, C, AC, BC, ABC. This order would allow main effects to be easily compared to the magnitude of interactions that contain those factors.

12.61. The symbol (1) is used to represent the low level of each factor that is used in an experiment.

12.62. Effects are confounded (aliased) when fractional factorial designs are used. Additional experimental runs are sometimes made to "dealias" effects that are believed to be significant so that they can be estimated separately.

12.65. This could be accomplished with the treatment combinations 10, 10, 5, and 15, the first two numbers being at the low level of A.

12.66. No, there is no finite upper bound. This can be seen by constructing an interaction plot in the form of an "X" and then recognizing that the points that form the letter could have any possible values.

12.69. Hard-to-change factors and impossible or impractical factor-level combinations often occur and can obviate the random choice of a fraction. Instead, it may be necessary to select a fraction in which all of the treatment combinations are feasible.

12.70. **(a)** 24.
    (b) It would not be necessary to use a normal probability plot as significance tests could be performed.
    (c) 22.5.

12.72. Technically, there is no such thing as a 1×2 design, since a factor that has one level is thus fixed at that level in the experiment and is not a part of the experiment. Therefore, this would actually be a one-factor experiment with two levels, which would be of no value without replication. With replication, it would be an experiment that would be analyzed with a pooled t-test if the assumptions for that test were met.

12.74. **(a)** Only three observations for each of three levels would rarely be used and certainly it would be desirable to have more observations. In particular, evidence of nonhomogeneity of variance could be a false signal that resulted from simply not having enough data.
    (b) 10.9248.

12.78. Since the difference between the conditional effects is equal to the interaction effect estimate, the conditional effects will be almost the same when the interaction is very close to zero.

12.81. The predictors are correlated in that dataset, and for correlated predictors there is no such thing as an "effect estimate."

12.82. The experimenter must be able to assume that there are no real 3-factor interactions since main effects are confounded with 3-factor interactions.

CHAPTER 13

13.1. No, negative variance component estimates occur when F-statistics are small, such as when the F-statistic is less than one in one-factor ANOVA. F-statistics can be just large enough to result in positive variance component estimates, but be well short of the appropriate critical F-values.

13.2. (a) No.
(b) None of the effect estimates will be affected.

13.3. 2.38.

CHAPTER 14

14.1. $R(t) = 1 - F(t)$
$f(t) = (\beta/\alpha)(t/\alpha)^{\beta-1} \exp[-(t/\alpha)^\beta]$ so
$F(t) = \int_0^t (\beta/\alpha)(t/\alpha)^{\beta-1} \exp[-(t/\alpha)^\beta]dt$
Let $u = (t/\alpha)^\beta$. Then

$$R(t) = 1 - F(t) = 1 - \{1 - \exp[-(t/\alpha)^\beta]\} = \exp[-(t/\alpha)^\beta]$$

Thus, $R(t) = \exp[-(t/\alpha)^\beta]$, as given.

14.4. .984.

14.5. .328.

14.7. 62 (rounded to the nearest integer).

14.9. $R_s(t) = e^{-75t^2}$.

14.11. The hazard function cannot be obtained because an expression for $F(t)$ cannot be obtained for the normal distribution.

14.12. $h(t) = 0.096t^{-0.65}$. The failure rate is a strictly declining function of t because the exponent of t is negative. Thus, there is no bathtub curve, and furthermore this is unrealistic because products eventually give out.

14.15. $h(t) = \dfrac{f(t)}{R(t)} = \dfrac{(1/b)\exp[(t-a)/b]\exp[-\exp[(t-a)/b]]}{\exp[-\exp[(t-a)/b]]} = \dfrac{1}{b}\exp\left[\dfrac{(t-a)}{b}\right]$

which is clearly not a constant.

14.17. It is not counterintuitive because in a series system all of the components must function in order for the system to function. Since $R_s(t) = \prod_i R_i(t)$ and each $R_i(t) \leq 1$, the reliability of the system cannot be greater than the lowest component reliability, with the two being equal only if all of the other $R_i(t) = 1$.

14.18. For a parallel system, the system reliability is

$$R_s(t) = 1 - [F_1(t)][F_2(t)] \cdots [F_k(t)]$$

The highest component reliability is $R(t) = 1 - \min_i (F_i(t))$. Since all of the $F_i(t)$ are less than 1, it follows that $R_s(t) \geq R(t)$, with equality occuring only if all the other $F_i(t) = 1$ at a given value of t.

14.21. $R_s(t) = (1 - \{1 - \exp[-(t/\alpha)^\beta]\}^3 [\exp[-(t/\alpha)^\beta])$

14.25. The *MTTF* $= 1/\lambda = 1/(2.4 \times 10^{-5}) = 41{,}666.7$.

14.33. The *MTBF* is the average of the 10 failure times, which is 13,101.

CHAPTER 15

15.2. 29.5.

15.3. **(a)** It is true that this is *not* a sample of count data, so a conclusion may be drawn simply from the table. No inference is necessary (or appropriate) beacuse this is not a sample.
   (b) Such a position is rather shaky since such votes of impeachment are quite rare. Furthermore, this is not a sample from an *existing* population.

15.6. $\chi^2 = 2.931$ with a *p*-value of .087. Thus, we cannot reject the hypothesis of independence.

15.7. **(a)** Yes, a contingency table approach could be used as we may view the set of data as having come from a population and then been classified into each of the four cells.
   (b) We obtain $\chi^2 = 281.7$ with a *p*-value of .000. Thus, we conclude that the percentage of nonconforming units before the corrective action differs from the percentage after the corrective action.

15.9. This may be stated in one of various ways, including stating that the proportion of counts for the first column that is in the first row is the same proportion as for the second column.

15.12.

16	14
24	21

15.13. 21.6.

15.15. The expected frequencies in each row and column must add to the sum of the observed frequencies for each row and column. This means that the differences between observed and expected must add to zero for each row and column. Therefore, if the difference between the observed and expected in the first cell (i.e., first row and first column) is, say, k, then the other difference in the first column must be $-k$, as must the other difference in the first row. It then follows that the difference in the second row and column must be k. Thus, all of the differences are the same except for the sign.

15.16. The initial value is zero. No, the χ^2 value cannot be determined from ratios alone. The actual counts must be available.

15.17. **(a)** A contingency table analysis should not be performed because the data are for a population.
(b) $\chi^2 = 9.32$ and the p-value is .01. Thus, there is evidence that the views of men and women differ.

15.18. This is the minimum number of observations, which if all placed in the first two rows and first three columns, are needed to compute the remaining counts, from the row and column totals.

15.19. Zero. 32.

15.22. The χ^2 statistic will be zero. The value of the multiplier is irrelevant; any multiplier will produce a value of zero.

15.25. We can see from inspection that the ratio of the numbers in the first row to the corresponding numbers in the second row is almost constant. This will result in a small chi-square value and the null hypothesis will not be rejected.

CHAPTER 16

16.1. If a sign test were used, there would be the tacit assumption of a symmetric distribution, so that the probability of an observation being below the mean is the same as the probability of an observation falling above the mean. Such a test could not be implemented without some knowledge of the population distribution.

16.5. The Friedman test is the nonparametric counterpart to two-way Analysis of Variance and is thus used when two factors are involved in an experiment, whereas the Kruskal–Wallis test is the counterpart to one-way Analysis of Variance and is used when there is only a single factor.

16.8. **(a)** 5.

16.10. The Wilcoxon test is more sensitive to outliers since an outlier would be assigned an extreme rank, whereas an outlier has no more influence than any other value that lies on the same side of the median.

16.11. No, that was a three-factor dataset and no methods given in the chapter can be applied to such data.

16.14. The probability of at least 70 points plotting above the midline when the process is in control at the midline and the distribution is symmetric is .00004. Therefore, it is unlikely that the process is in control, under the assumption that the distribution is symmetric. Of course, there is no way to test that assumption with the information that is given.

16.15. .494.

16.17. The probability of at least 10 data points plotting above the assumed mean is .09 if the assumed mean is equal to the actual mean and the distribution is symmetric. This does not provide strong evidence against the assumed mean.

16.18. The Mann–Whitney test was used since nonnormality was suspected. Since evidence against the null hypothesis is of borderline strength, additional testing would be highly desirable. In particular, normality could be tested for the data from each plant, rather than assumed not to exist. If there was no evidence of more than slight-to-moderate nonnormality, a pooled t-test could be used. If nonnormality did appear to be severe, the Mann–Whitney test might be applied to additional data, if possible, in an effort to resolve the indecision.

CHAPTER 17

17.1. The Mann–Whitney test should be used.

17.2. The program appears to have been successful as a paired t-test results in a value for the t-statistic of 3.34 and a corresponding p-value of .009.

17.3. The graph suggests that a linear relationship exists for only a relatively narrow range of X. Clearly, a linear regression line should not be fit to the entire data set. One option would be to fit a line to only the subset of data that exhibits a linear relationship; another option would be to use a *loess* smooth.

17.4. These are excellent reasons why it would be a good idea to do both a parametric analysis and nonparametric analysis and use the result from the latter if the results differ. It's true that normality cannot be checked with only five observations, and testing the equality of variances when the sample variances are computed from only five observations is also of questionable value.

17.6. Parameters generally have to be estimated; the .0027 probability holds only if the parameter values are known.

17.9. The process may not be capable of meeting the target value. In general, neither the midline nor the control limits should be obtained from target values.

17.10. Histograms constructed using the $\sqrt{n}$ rule will have the greater variability.

17.11. Conditional effects will have to be used (estimated), for at least some of the factors. How the data are partitioned for the conditional effects will depend on the relative magnitude of the two interactions if conditional effects for C must be computed, and will also depend on for which factors the conditional effects must be computed.

17.13. All five available regressors should almost certainly not be used because of the intercorrelations. Since regression coefficients are essentially uninterpretable because of the intercorrelations, confidence intervals based on these estimates are also uninterpretable.

17.15. No, nonparametric methods are also based on certain assumptions, such as the observations having come from a symmetric distribution. Since assumptions cannot be dispensed with entirely when nonparametric methods are used, a user might as well test the assumptions for the corresponding parametric test, and use the latter, which is more powerful, when the parametric assumptions appear to be at least approximately met.

APPENDIX

Statistical Tables

TABLE A **Random Numbers**[a]

1559	9068	9290	8303	8508	8954	1051	6677	6415	0342
5550	6245	7313	0117	7652	5069	6354	7668	1096	5780
4735	6214	8037	1385	1882	0828	2957	0530	9210	0177
5333	1313	3063	1134	8676	6241	9960	5304	1582	6198
8495	2956	1121	8484	2920	7934	0670	5263	0968	0069
1947	3353	1197	7363	9003	9313	3434	4261	0066	2714
4785	6325	1868	5020	9100	0823	7379	7391	1250	5501
9972	9163	5833	0100	5758	3696	6496	6297	5653	7782
0472	4629	2007	4464	3312	8728	1193	2497	4219	5339
4727	6994	1175	5622	2341	8562	5192	1471	7206	2027
3658	3226	5981	9025	1080	1437	6721	7331	0792	5383
6906	9758	0244	0259	4609	1269	5957	7556	1975	7898
3793	6916	0132	8873	8987	4975	4814	2098	6683	0901
3376	5966	1614	4025	0721	1537	6695	6090	8083	5450
6126	0224	7169	3596	1593	5097	7286	2686	1796	1150
0466	7566	1320	8777	8470	5448	9575	4669	1402	3905
9908	9832	8185	8835	0384	3699	1272	1181	8627	1968
7594	3636	1224	6808	1184	3404	6752	4391	2016	6167
5715	9301	5847	3524	0077	6674	8061	5438	6508	9673
7932	4739	4567	6797	4540	8488	3639	9777	1621	7244
6311	2025	5250	6099	6718	7539	9681	3204	9637	1091
0476	1624	3470	1600	0675	3261	7749	4195	2660	2150
5317	3903	6098	9438	3482	5505	5167	9993	8191	8488
7474	8876	1918	9828	2061	6664	0391	9170	2776	4025
7460	6800	1987	2758	0737	6880	1500	5763	2061	9373
1002	1494	9972	3877	6104	4006	0477	0669	8557	0513
5449	6891	9047	6297	1075	7762	8091	7153	8881	3367
9453	0809	7151	9982	0411	1120	6129	5090	2053	7570

Modern Engineering Statistics   By Thomas P. Ryan
Copyright © 2007 John Wiley & Sons, Inc.

TABLE A   (*Continued*)

0471	2725	7588	6573	0546	0110	6132	1224	3124	6563
5469	2668	1996	2249	3857	6637	8010	1701	3141	6147
2782	9603	1877	4159	9809	2570	4544	0544	2660	6737
3129	7217	5020	3788	0853	9465	2186	3945	1696	2286
7092	9885	3714	8557	7804	9524	6228	7774	6674	2775
9566	0501	8352	1062	0634	2401	0379	1697	7153	6208
5863	7000	1714	9276	7218	6922	1032	4838	1954	1680
5881	9151	2321	3147	6755	2510	5759	6947	7102	0097
6416	9939	9569	0439	1705	4860	9881	7071	9596	8758
9568	3012	6316	9065	0710	2158	1639	9149	4848	8634
0452	9538	5730	1893	1186	9245	6558	9562	8534	9321
8762	5920	8989	4777	2169	7073	7082	9495	1594	8600
0194	0270	7601	0342	3897	4133	7650	9228	5558	3597
3306	5478	2797	1605	4996	0023	9780	9429	3937	7573
7198	3079	2171	6972	0928	6599	9328	0597	5948	5753
8350	4846	1309	0612	4584	4988	4642	4430	9481	9048
7449	4279	4224	1018	2496	2091	9750	6086	1955	9860
6126	5399	0852	5491	6557	4946	9918	1541	7894	1843
1851	7940	9908	3860	1536	8011	4314	7269	7047	0382
7698	4218	2726	5130	3132	1722	8592	9662	4795	7718
0810	0118	4979	0458	1059	5739	7919	4557	0245	4861
6647	7149	1409	6809	3313	0082	9024	7477	7320	5822
3867	7111	5549	9439	3427	9793	3071	6651	4267	8099
1172	7278	7527	2492	6211	9457	5120	4903	1023	5745
6701	1668	5067	0413	7961	7825	9261	8572	0634	1140
8244	0620	8736	2649	1429	6253	4181	8120	6500	8127
8009	4031	7884	2215	2382	1931	1252	8088	2490	9122
1947	8315	9755	7187	4074	4743	6669	6060	2319	0635
9562	4821	8050	0106	2782	4665	9436	4973	4879	8900
0729	9026	9631	8096	8906	5713	3212	8854	3435	4206
6904	2569	3251	0079	8838	8738	8503	6333	0952	1641

[a]Table A was produced using MINITAB, which is a registered trademark of Minitab, Inc., Quality Plaza, 1829 Pine Hall Rd., State College, PA 16801, (814) 238–3280.

TABLE B Normal Distribution[a] $[P(0 \leq Z \leq z)$ where $Z \sim N(0, 1)]$

z	0.00	0.01	0.02	0.03	0.04	0.05	0.06	0.07	0.08	0.09
0.0	0.00000	0.00399	0.00798	0.01197	0.01595	0.01994	0.02392	0.02790	0.03188	0.03586
0.1	0.03983	0.04380	0.04776	0.05172	0.05567	0.05962	0.06356	0.06749	0.07142	0.07535
0.2	0.07926	0.08317	0.08706	0.09095	0.09483	0.09871	0.10257	0.10642	0.11026	0.11409
0.3	0.11791	0.12172	0.12552	0.12930	0.13307	0.13683	0.14058	0.14431	0.14803	0.15173
0.4	0.15542	0.15910	0.16276	0.16640	0.17003	0.17364	0.17724	0.18082	0.18439	0.18793
0.5	0.19146	0.19497	0.19847	0.20194	0.20540	0.20884	0.21226	0.21566	0.21904	0.22240
0.6	0.22575	0.22907	0.23237	0.23565	0.23891	0.24215	0.24537	0.24857	0.25175	0.25490
0.7	0.25804	0.26115	0.26424	0.26730	0.27035	0.27337	0.27637	0.27935	0.28230	0.28524
0.8	0.28814	0.29103	0.29389	0.29673	0.29955	0.30234	0.30511	0.30785	0.31057	0.31327
0.9	0.31594	0.31859	0.32121	0.32381	0.32639	0.32894	0.33147	0.33398	0.33646	0.33891
1.0	0.34134	0.34375	0.34614	0.34849	0.35083	0.35314	0.35543	0.35769	0.35993	0.36214
1.1	0.36433	0.36650	0.36864	0.37076	0.37286	0.37493	0.37698	0.37900	0.38100	0.38298
1.2	0.38493	0.38686	0.38877	0.39065	0.39251	0.39435	0.39617	0.39796	0.39973	0.40147
1.3	0.40320	0.40490	0.40658	0.40824	0.40988	0.41149	0.41308	0.41466	0.41621	0.41774
1.4	0.41924	0.42073	0.42220	0.42364	0.42507	0.42647	0.42785	0.42922	0.43056	0.43189
1.5	0.43319	0.43448	0.43574	0.43699	0.43822	0.43943	0.44062	0.44179	0.44295	0.44408
1.6	0.44520	0.44630	0.44738	0.44845	0.44950	0.45053	0.45154	0.45254	0.45352	0.45449
1.7	0.45543	0.45637	0.45728	0.45818	0.45907	0.45994	0.46080	0.46164	0.46246	0.46327
1.8	0.46407	0.46485	0.46562	0.46638	0.46712	0.46784	0.46856	0.46926	0.46995	0.47062
1.9	0.47128	0.47193	0.47257	0.47320	0.47381	0.47441	0.47500	0.47558	0.47615	0.47670
2.0	0.47725	0.47778	0.47831	0.47882	0.47932	0.47982	0.48030	0.48077	0.48124	0.48169
2.1	0.48214	0.48257	0.48300	0.48341	0.48382	0.48422	0.48461	0.48500	0.48537	0.48574

(*Continued*)

TABLE B *(Continued)*

z	0.00	0.01	0.02	0.03	0.04	0.05	0.06	0.07	0.08	0.09
2.2	0.48610	0.48645	0.48679	0.48713	0.48745	0.48778	0.48809	0.48840	0.48870	0.48899
2.3	0.48928	0.48956	0.48983	0.49010	0.49036	0.49061	0.49086	0.49111	0.49134	0.49158
2.4	0.49180	0.49202	0.49224	0.49245	0.49266	0.49286	0.49305	0.49324	0.49343	0.49361
2.5	0.49379	0.49396	0.49413	0.49430	0.49446	0.49461	0.49477	0.49492	0.49506	0.49520
2.6	0.49534	0.49547	0.49560	0.49573	0.49585	0.49598	0.49609	0.49621	0.49632	0.49643
2.7	0.49653	0.49664	0.49674	0.49683	0.49693	0.49702	0.49711	0.49720	0.49728	0.49736
2.8	0.49744	0.49752	0.49760	0.49767	0.49774	0.49781	0.49788	0.49795	0.49801	0.49807
2.9	0.49813	0.49819	0.49825	0.49831	0.49836	0.49841	0.49846	0.49851	0.49856	0.49861
3.0	0.49865	0.49869	0.49874	0.49878	0.49882	0.49886	0.49889	0.49893	0.49896	0.49900
3.1	0.49903	0.49906	0.49910	0.49913	0.49916	0.49918	0.49921	0.49924	0.49926	0.49929
3.2	0.49931	0.49934	0.49936	0.49938	0.49940	0.49942	0.49944	0.49946	0.49948	0.49950
3.3	0.49952	0.49953	0.49955	0.49957	0.49958	0.49960	0.49961	0.49962	0.49964	0.49965
3.4	0.49966	0.49968	0.49969	0.49970	0.49971	0.49972	0.49973	0.49974	0.49975	0.49976
3.5	0.49977	0.49978	0.49978	0.49979	0.49980	0.49981	0.49981	0.49982	0.49983	0.49983
3.6	0.49984	0.49985	0.49985	0.49986	0.49986	0.49987	0.49987	0.49988	0.49988	0.49989
3.7	0.49989	0.49990	0.49990	0.49990	0.49991	0.49991	0.49992	0.49992	0.49992	0.49992
3.8	0.49993	0.49993	0.49993	0.49994	0.49994	0.49994	0.49994	0.49995	0.49995	0.49995
3.9	0.49995	0.49995	0.49996	0.49996	0.49996	0.49996	0.49996	0.49996	0.49997	0.49997

[a]These values were generated using MINITAB.

.000288

.5

TABLE C   *t* Distribution[a]

d.f. (ν)/α	0.40	0.25	0.10	0.05	0.025	0.01	0.005	0.0025	0.001	0.0005
1	0.325	1.000	3.078	6.314	12.706	31.820	63.655	127.315	318.275	636.438
2	0.289	0.816	1.886	2.920	4.303	6.965	9.925	14.089	22.327	31.596
3	0.277	0.765	1.638	2.353	3.182	4.541	5.841	7.453	10.214	12.923
4	0.271	0.741	1.533	2.132	2.776	3.747	4.604	5.597	7.173	8.610
5	0.267	0.727	1.476	2.015	2.571	3.365	4.032	4.773	5.893	6.869
6	0.265	0.718	1.440	1.943	2.447	3.143	3.707	4.317	5.208	5.959
7	0.263	0.711	1.415	1.895	2.365	2.998	3.499	4.029	4.785	5.408
8	0.262	0.706	1.397	1.860	2.306	2.896	3.355	3.833	4.501	5.041
9	0.261	0.703	1.383	1.833	2.262	2.821	3.250	3.690	4.297	4.781
10	0.260	0.700	1.372	1.812	2.228	2.764	3.169	3.581	4.144	4.587
11	0.260	0.697	1.363	1.796	2.201	2.718	3.106	3.497	4.025	4.437
12	0.259	0.695	1.356	1.782	2.179	2.681	3.055	3.428	3.930	4.318
13	0.259	0.694	1.350	1.771	2.160	2.650	3.012	3.372	3.852	4.221
14	0.258	0.692	1.345	1.761	2.145	2.624	2.977	3.326	3.787	4.140
15	0.258	0.691	1.341	1.753	2.131	2.602	2.947	3.286	3.733	4.073
16	0.258	0.690	1.337	1.746	2.120	2.583	2.921	3.252	3.686	4.015
17	0.257	0.689	1.333	1.740	2.110	2.567	2.898	3.222	3.646	3.965
18	0.257	0.688	1.330	1.734	2.101	2.552	2.878	3.197	3.610	3.922
19	0.257	0.688	1.328	1.729	2.093	2.539	2.861	3.174	3.579	3.883
20	0.257	0.687	1.325	1.725	2.086	2.528	2.845	3.153	3.552	3.849
21	0.257	0.686	1.323	1.721	2.080	2.518	2.831	3.135	3.527	3.819
22	0.256	0.686	1.321	1.717	2.074	2.508	2.819	3.119	3.505	3.792
23	0.256	0.685	1.319	1.714	2.069	2.500	2.807	3.104	3.485	3.768
24	0.256	0.685	1.318	1.711	2.064	2.492	2.797	3.091	3.467	3.745
25	0.256	0.684	1.316	1.708	2.060	2.485	2.787	3.078	3.450	3.725
26	0.256	0.684	1.315	1.706	2.056	2.479	2.779	3.067	3.435	3.707
27	0.256	0.684	1.314	1.703	2.052	2.473	2.771	3.057	3.421	3.690
28	0.256	0.683	1.313	1.701	2.048	2.467	2.763	3.047	3.408	3.674
29	0.256	0.683	1.311	1.699	2.045	2.462	2.756	3.038	3.396	3.659
30	0.256	0.683	1.310	1.697	2.042	2.457	2.750	3.030	3.385	3.646
40	0.255	0.681	1.303	1.684	2.021	2.423	2.704	2.971	3.307	3.551
60	0.254	0.679	1.296	1.671	2.000	2.390	2.660	2.915	3.232	3.460
100	0.254	0.677	1.290	1.660	1.984	2.364	2.626	2.871	3.174	3.391
Infinity	0.253	0.674	1.282	1.645	1.960	2.326	2.576	2.807	3.090	3.290

[a]These values were generated using MINITAB.

TABLE D **F-Distribution**[a,b]

$a.\ F_{v_1,v_2,.05}$

v_1	1	2	3	4	5	6	7	8	9	10	11	12	13	14	15
v_2															
1	161.44	199.50	215.69	224.57	230.16	233.98	236.78	238.89	240.55	241.89	242.97	243.91	244.67	245.35	245.97
2	18.51	19.00	19.16	19.25	19.30	19.33	19.35	19.37	19.39	19.40	19.40	19.41	19.42	19.42	19.43
3	10.13	9.55	9.28	9.12	9.01	8.94	8.89	8.85	8.81	8.79	8.76	8.74	8.73	8.71	8.70
4	7.71	6.94	6.59	6.39	6.26	6.16	6.09	6.04	6.00	5.96	5.94	5.91	5.89	5.87	5.86
5	6.61	5.79	5.41	5.19	5.05	4.95	4.88	4.82	4.77	4.74	4.70	4.68	4.66	4.64	4.62
6	5.99	5.14	4.76	4.53	4.39	4.28	4.21	4.15	4.10	4.06	4.03	4.00	3.98	3.96	3.94
7	5.59	4.74	4.35	4.12	3.97	3.87	3.79	3.73	3.68	3.64	3.60	3.57	3.55	3.53	3.51
8	5.32	4.46	4.07	3.84	3.69	3.58	3.50	3.44	3.39	3.35	3.31	3.28	3.26	3.24	3.22
9	5.12	4.26	3.86	3.63	3.48	3.37	3.29	3.23	3.18	3.14	3.10	3.07	3.05	3.03	3.01
10	4.96	4.10	3.71	3.48	3.33	3.22	3.14	3.07	3.02	2.98	2.94	2.91	2.89	2.86	2.85
11	4.84	3.98	3.59	3.36	3.20	3.09	3.01	2.95	2.90	2.85	2.82	2.79	2.76	2.74	2.72
12	4.75	3.89	3.49	3.26	3.11	3.00	2.91	2.85	2.80	2.75	2.72	2.69	2.66	2.64	2.62
13	4.67	3.81	3.41	3.18	3.03	2.92	2.83	2.77	2.71	2.67	2.63	2.60	2.58	2.55	2.53
14	4.60	3.74	3.34	3.11	2.96	2.85	2.76	2.70	2.65	2.60	2.57	2.53	2.51	2.48	2.46
15	4.54	3.68	3.29	3.06	2.90	2.79	2.71	2.64	2.59	2.54	2.51	2.48	2.45	2.42	2.40
16	4.49	3.63	3.24	3.01	2.85	2.74	2.66	2.59	2.54	2.49	2.46	2.42	2.40	2.37	2.35
17	4.45	3.59	3.20	2.96	2.81	2.70	2.61	2.55	2.49	2.45	2.41	2.38	2.35	2.33	2.31
18	4.41	3.55	3.16	2.93	2.77	2.66	2.58	2.51	2.46	2.41	2.37	2.34	2.31	2.29	2.27
19	4.38	3.52	3.13	2.90	2.74	2.63	2.54	2.48	2.42	2.38	2.34	2.31	2.28	2.26	2.23
20	4.35	3.49	3.10	2.87	2.71	2.60	2.51	2.45	2.39	2.35	2.31	2.28	2.25	2.22	2.20

v_2															
21	4.32	3.47	3.07	2.84	2.68	2.57	2.49	2.42	2.37	2.32	2.28	2.25	2.22	2.20	2.18
22	4.30	3.44	3.05	2.82	2.66	2.55	2.46	2.40	2.34	2.30	2.26	2.23	2.20	2.17	2.15
23	4.28	3.42	3.03	2.80	2.64	2.53	2.44	2.37	2.32	2.27	2.24	2.20	2.18	2.15	2.13
24	4.26	3.40	3.01	2.78	2.62	2.51	2.42	2.36	2.30	2.25	2.22	2.18	2.15	2.13	2.11
25	4.24	3.39	2.99	2.76	2.60	2.49	2.40	2.34	2.28	2.24	2.20	2.16	2.14	2.11	2.09
26	4.23	3.37	2.98	2.74	2.59	2.47	2.39	2.32	2.27	2.22	2.18	2.15	2.12	2.09	2.07
27	4.21	3.35	2.96	2.73	2.57	2.46	2.37	2.31	2.25	2.20	2.17	2.13	2.10	2.08	2.06
28	4.20	3.34	2.95	2.71	2.56	2.45	2.36	2.29	2.24	2.19	2.15	2.12	2.09	2.06	2.04
29	4.18	3.33	2.93	2.70	2.55	2.43	2.35	2.28	2.22	2.18	2.14	2.10	2.08	2.05	2.03
30	4.17	3.32	2.92	2.69	2.53	2.42	2.33	2.27	2.21	2.16	2.13	2.09	2.06	2.04	2.01
40	4.08	3.23	2.84	2.61	2.45	2.34	2.25	2.18	2.12	2.08	2.04	2.00	1.97	1.95	1.92

$b.\ F_{v_1, v_2, 0.01}$

v_2															
1	4052.45	4999.42	5402.96	5624.03	5763.93	5858.82	5928.73	5981.06	6021.73	6055.29	6083.22	6106.00	6125.37	6142.48	6157.06
2	98.51	99.00	99.17	99.25	99.30	99.33	99.35	99.38	99.39	99.40	99.41	99.41	99.42	99.42	99.43
3	34.12	30.82	29.46	28.71	28.24	27.91	27.67	27.49	27.35	27.23	27.13	27.05	26.98	26.92	26.87
4	21.20	18.00	16.69	15.98	15.52	15.21	14.98	14.80	14.66	14.55	14.45	14.37	14.31	14.25	14.20
5	16.26	13.27	12.06	11.39	10.97	10.67	10.46	10.29	10.16	10.05	9.96	9.89	9.82	9.77	9.72
6	13.74	10.92	9.78	9.15	8.75	8.47	8.26	8.10	7.98	7.87	7.79	7.72	7.66	7.60	7.56
7	12.25	9.55	8.45	7.85	7.46	7.19	6.99	6.84	6.72	6.62	6.54	6.47	6.41	6.36	6.31
8	11.26	8.65	7.59	7.01	6.63	6.37	6.18	6.03	5.91	5.81	5.73	5.67	5.61	5.56	5.52
9	10.56	8.02	6.99	6.42	6.06	5.80	5.61	5.47	5.35	5.26	5.18	5.11	5.05	5.01	4.96
10	10.04	7.56	6.55	5.99	5.64	5.39	5.20	5.06	4.94	4.85	4.77	4.71	4.65	4.60	4.56
11	9.65	7.21	6.22	5.67	5.32	5.07	4.89	4.74	4.63	4.54	4.46	4.40	4.34	4.29	4.25
12	9.33	6.93	5.95	5.41	5.06	4.82	4.64	4.50	4.39	4.30	4.22	4.16	4.10	4.05	4.01

(*Continued*)

TABLE D *(Continued)*

ν_1	1	2	3	4	5	6	7	8	9	10	11	12	13	14	15
13	9.07	6.70	5.74	5.21	4.86	4.62	4.44	4.30	4.19	4.10	4.02	3.96	3.91	3.86	3.82
14	8.86	6.51	5.56	5.04	4.69	4.46	4.28	4.14	4.03	3.94	3.86	3.80	3.75	3.70	3.66
15	8.68	6.36	5.42	4.89	4.56	4.32	4.14	4.00	3.89	3.80	3.73	3.67	3.61	3.56	3.52
16	8.53	6.23	5.29	4.77	4.44	4.20	4.03	3.89	3.78	3.69	3.62	3.55	3.50	3.45	3.41
17	8.40	6.11	5.18	4.67	4.34	4.10	3.93	3.79	3.68	3.59	3.52	3.46	3.40	3.35	3.31
18	8.29	6.01	5.09	4.58	4.25	4.01	3.84	3.71	3.60	3.51	3.43	3.37	3.32	3.27	3.23
19	8.18	5.93	5.01	4.50	4.17	3.94	3.77	3.63	3.52	3.43	3.36	3.30	3.24	3.19	3.15
20	8.10	5.85	4.94	4.43	4.10	3.87	3.70	3.56	3.46	3.37	3.29	3.23	3.18	3.13	3.09
21	8.02	5.78	4.87	4.37	4.04	3.81	3.64	3.51	3.40	3.31	3.24	3.17	3.12	3.07	3.03
22	7.95	5.72	4.82	4.31	3.99	3.76	3.59	3.45	3.35	3.26	3.18	3.12	3.07	3.02	2.98
23	7.88	5.66	4.76	4.26	3.94	3.71	3.54	3.41	3.30	3.21	3.14	3.07	3.02	2.97	2.93
24	7.82	5.61	4.72	4.22	3.90	3.67	3.50	3.36	3.26	3.17	3.09	3.03	2.98	2.93	2.89
25	7.77	5.57	4.68	4.18	3.85	3.63	3.46	3.32	3.22	3.13	3.06	2.99	2.94	2.89	2.85
26	7.72	5.53	4.64	4.14	3.82	3.59	3.42	3.29	3.18	3.09	3.02	2.96	2.90	2.86	2.81
27	7.68	5.49	4.60	4.11	3.78	3.56	3.39	3.26	3.15	3.06	2.99	2.93	2.87	2.82	2.78
28	7.64	5.45	4.57	4.07	3.75	3.53	3.36	3.23	3.12	3.03	2.96	2.90	2.84	2.79	2.75
29	7.60	5.42	4.54	4.04	3.73	3.50	3.33	3.20	3.09	3.00	2.93	2.87	2.81	2.77	2.73
30	7.56	5.39	4.51	4.02	3.70	3.47	3.30	3.17	3.07	2.98	2.91	2.84	2.79	2.74	2.70
40	7.31	5.18	4.31	3.83	3.51	3.29	3.12	2.99	2.89	2.80	2.73	2.66	2.61	2.56	2.52

[a]These values were generated using MINITAB.
[b]ν_2 = degrees of freedom for the denominator; ν_1 = degrees of freedom for the numerator.

TABLE E   Factors for Calculating Two-Sided 99% Statistical Intervals for a Normal Population to Contain at Least 100p% of the Population*

n	p		
	.90	.95	.99
5	6.65	7.87	10.22
10	3.62	4.29	5.61
15	2.97	3.53	4.62
20	2.68	3.18	4.17
25	2.51	2.98	3.91
30	2.39	2.85	3.74
40	2.25	2.68	3.52
60	2.11	2.51	3.30
∞	1.64	1.96	2.58

*These numbers are adapted from Table A.10b of *Statistical Intervals: A Guide for Practitioners* by G. J. Hahn and W. Q. Meeker and rounded to two decimal places. Reproduced with permission of John Wiley and Sons, Inc.

TABLE F Control Chart Constants

n	For Estimating Sigma		For $\overline{X}$ Chart		For $\overline{X}$ Chart (Standard Given)	For R Chart		For R Chart (Standard Given)		For s chart (Standard Given)			
	c_4	d_2	A_2	A_3	A	D_3	D_4	D_1	D_2	B_3	B_4	B_5	B_6
2	0.7979	1.128	1.880	2.659	2.121	0	3.267		3.686	0	3.267	0	2.606
3	0.8862	1.693	1.023	1.954	1.732	0	2.575		4.358	0	2.568	0	2.276
4	0.9213	2.059	0.729	1.628	1.500	0	2.282		4.698	0	2.266	0	2.088
5	0.9400	2.326	0.577	1.427	1.342	0	2.115		4.918	0	2.089	0	1.964
6	0.9515	2.534	0.483	1.287	1.225	0	2.004		5.078	0.030	1.970	0.029	1.874
7	0.9594	2.704	0.419	1.182	1.134	0.076	1.924	0.205	5.203	0.118	1.882	0.113	1.806
8	0.9650	2.847	0.373	1.099	1.061	0.136	1.864	0.387	5.307	0.185	1.815	0.179	1.751
9	0.9693	2.970	0.337	1.032	1.000	0.184	1.816	0.546	5.394	0.239	1.761	0.232	1.707
10	0.9727	3.078	0.308	0.975	0.949	0.223	1.777	0.687	5.469	0.284	1.716	0.276	1.669
15	0.9823	3.472	0.223	0.789	0.775	0.348	1.652	1.207	5.737	0.428	1.572	0.421	1.544
20	0.9869	3.735	0.180	0.680	0.671	0.414	1.586	1.548	5.922	0.510	1.490	0.504	1.470
25	0.9896	3.931	0.153	0.606	0.600	0.459	1.541	1.804	6.058	0.565	1.435	0.559	1.420

Author Index

Subject Index

Modern Engineering Statistics   By Thomas P. Ryan
Copyright © 2007 John Wiley & Sons, Inc.

579